5 Steps to a 5
AP Calculus AB/BC

Other books in McGraw-Hill's *5 Steps to a 5* Series include:

AP Biology
AP Chemistry
AP Computer Science
AP English Language
AP English Literature
AP Macroeconomics/Microeconomics
AP Physics B and C
AP Psychology
AP Spanish Language
AP Statistics
AP U.S. Government and Politics
AP U.S. History
AP World History
11 Practice Tests for the AP Exams
Writing the AP English Essay

McGRAW-HILL

5 Steps to a 5
AP Calculus AB/BC

SECOND EDITION

William Ma

BC Review and BC Practice Exams by Carolyn Wheater

New York   Chicago   San Francisco   Lisbon   London   Madrid
Mexico City   Milan   New Delhi   San Juan   Seoul
Singapore   Sydney   Toronto

The **McGraw·Hill** Companies

Copyright © 2007, 2002 by The McGraw-Hill Companies, Inc. All rights reserved.
Printed in the United States of America. Except as permitted under the United States
Copyright Act of 1976, no part of this publication may be reproduced or distributed in
any form or by any means, or stored in a data base or retrieval system, without the prior
written permission of the publisher.

1 2 3 4 5 6 7 8 9 0   QPD/QPD   0 1 0 9 8 7 6

ISBN 13: 978-0-07-147629-4
ISBN 10: 0-07-147629-6

The series editor was Grace Freedson, and the project editor was Don Reis.

Printed and bound by Quebecor/Dubuque.

This book was printed on acid-free paper.

McGraw-Hill books are available at special quantity discounts to use as premiums and
sales promotions, or for use in corporate training programs. For more information, please
write to the Director of Special Sales, McGraw-Hill Professional, Two Penn Plaza,
New York, NY 10121-2298. Or contact your local bookstore.

AP, Advanced Placement Program, and *College Board* are registered trademarks of the
College Entrance Examination Board, which was not involved in the production of, and
does not endorse, this product.

Dedication

To

My wife Mary

My daughters Janet and Karen

About the Authors

William Ma has taught calculus for many years. He received his B.A. and M.A. from Columbia University. He has been an adjunct lecturer at Baruch College, Fordham University, and Columbia University. He has also written, for the Internet, an interactive and multi-media review course for the New York State's Math A Regents. He is currently the chairman of the Math Department at the Herricks School District on Long Island, New York.

Carolyn Wheater, who added the Calculus BC topics and prepared the BC Practice Exams, teaches Middle School and Upper School Mathematics at The Nightingale-Bamford School in New York City. Educated at Marymount Manhattan College and the University of Massachusetts, Amherst, she has taught math and computer technology for thirty years to students from preschool through college.

Contents

Chapter 2   Differentiation / 60

Chapter 3   Graphs of Functions and Derivatives / 86

Chapter 4   Applications of Derivatives / 128

Chapter 5   More Applications of Derivatives / 150

Chapter 9   More Applications of Definite Integrals / 271

Chapter 10   Series / 304

PART IV   PRACTICE MAKES PERFECT

APPENDIXES

Preface

Congratulations! You are an AP Calculus student. Not too shabby! As you know, AP Calculus is one of the most challenging subjects in high school. You are studying mathematical ideas that helped change the world. Not that long ago, calculus was taught at the graduate level. Today, smart young people like yourself study calculus in high school. Most colleges will give you credit if you score a 3 or more on the AP Calculus Exam.

So how do you do well on the AP Exam? How do you get a 5? Well, you've already taken the first step. You're reading this book. The next thing you need to do is to make sure that you understand the materials and do the practice problems. In recent years, the AP Calculus exams have gone through many changes. For example, today the questions no longer stress long and tedious algebraic manipulations. Instead, you are expected to be able to solve a broad range of problems including problems presented to you in the form of a graph, a chart, or a word problem. For many of the questions, you are also expected to use your calculator to find the solutions.

After having taught AP Calculus for many years and having spoken to students and other calculus teachers, we understand some of the difficulties that students might encounter with the AP Calculus exams. For example, some students have complained about not being able to visualize what the question was asking and other students said that even when the solution was given, they could not follow the steps. Under these circumstances, who wouldn't be frustrated? In this book, we have addressed these issues. Whenever possible, problems are accompanied by diagrams and solutions are presented in a step-by-step manner. The graphing calculator is used extensively whenever it is permitted. The book also begins with a chapter on limits and continuity. These topics are normally taught in a pre-calculus course. If you're familiar with these concepts, you might skip this chapter and begin with Chapter 2.

So how do you get a 5 on the AP Calculus exam?

> Step 1: Pick one of the study plans from the book.
> Step 2: Study the chapters and do the practice problems as scheduled.
> Step 3: Take the Diagnostic Test and redo some of the practice problems.
> Step 4: Take the Practice Exams.
> Step 5: Get a good night's sleep the day before the exam.

As an old martial artist once said, "First you must understand. Then you must practice." Have fun and good luck!

Acknowledgments

I could not have written this book without the help of the following people:

My high school calculus teacher, *Michael Cantor*, who taught me calculus.

Professor *Leslie Beebe* who taught me how to write.

David Pickman who fixed my computer and taught me Equation Editor.

Jennifer Tobin, who tirelessly edited many parts of the manuscript and with whom I look forward to co-author a math book in the future.

Robert Teseo and his calculus students who field-tested many of the problems.

Allison Litvack, *Rich Peck*, and *Liz Spiegel* who proofread sections of the BC Practice Tests. And a special thanks to *Trisha Ho*, who edited Chapters 9 and 10.

Mark Reynolds who proofread part of the manuscript.

Maxine Lifshitz who offered many helpful comments and suggestions.

Grace Freedson, the series editor, *Don Reis*, the project manager, and *Aparna Shankar*, the project manager, for all their assistance.

Sam Lee and *Derek Ma* who were on 24-hour call for technical support.

My older daughter *Janet* for not killing me for missing one of her concerts.

My younger daughter *Karen* who helped me with many of the computer graphics.

My wife *Mary* who gave me many ideas for the book and who often has more confidence in me than I have in myself.

5 Steps to a 5
AP Calculus AB/BC

PART I

HOW TO USE THIS BOOK

HOW IS THIS BOOK ORGANIZED?

Part I contains an introduction to the Five-Step Program and three study plans for preparing for the AP Calculus AB exam.

Part II contains a diagnostic test. The diagnostic test covers all the major topics on the AB and BC exams.

Part III (Comprehensive Review) contains 10 chapters. Chapter 10 contains material covered only on the BC exam. At the end of each chapter you will find a set of practice problems and, beginning with Chapter 2, a set of cumulative review problems. These problems have been created to allow you to practice your skills. They have also been designed to avoid unnecessary duplications. At the end of each chapter, a Rapid Review gives you some of the highlights of the chapter.

Part IV contains two full-length AB practice tests and two BC practice tests, as well as the answers, explanations, and worksheets to compute your scores.

Note: The exercises in this book are done with the TI-89 graphing calculator.

INTRODUCTION TO THE FIVE-STEP PROGRAM

The Five-Step Program is designed to provide you with the skills and strategies vital to the exam and the practice that can help lead you to that perfect 5. Each of the five steps will provide you with the opportunity to get closer and closer to the "Holy Grail" 5.

Step One leads you through a brief process to help determine which type of exam preparation you want to commit yourself to.

1. Month-by-month: September through May.
2. The calendar year: January through May.
3. Basic training: Six weeks prior to the exam.

Step Two helps develop the knowledge you need to succeed on the exam.

1. A comprehensive review of the exam.
2. One "Diagnostic Test" which you can go through step-by-step and question-by-question to build your confidence level.
3. A summary of formulas related to the AP Calculus exam.
4. A list of interesting and related websites and a bibliography.

Step Three develops the skills necessary to take the exam and do well.

1. Practice multiple-choice questions.
2. Practice free-response questions.

Step Four helps you develop strategies for taking the exam.

1. Learning about the test itself.
2. Learning to read multiple-choice questions.
3. Learning how to answer multiple-choice questions, including whether or not to guess.
4. Learning how to plan and write the free-response questions.

Step Five will help you develop your confidence in using the skills demanded on the AP Calculus exam.

1. The opportunity to take a diagnostic exam.
2. Time management techniques/skills.
3. Two practice exams that test how well-honed your skills are.

THREE APPROACHES TO PREPARING FOR THE AP CALCULUS EXAM

Overview of the Three Plans

No one knows your study habits, likes and dislikes better than you. So, you are the only one who can decide which approach you want and/or need to adopt to prepare for the Advanced Placement Calculus exam. Look at the brief profiles below. These may help you to place yourself in a particular prep mode.

You are a full-year prep student (Approach A) if:

1. You are the kind of person who likes to plan for everything far in advance ... and I mean far ...;
2. You arrive at the airport 2 hours before your flight because, "you never know when these planes might leave early ...";
3. You like detailed planning and everything in its place;
4. You feel you must be thoroughly prepared;
5. You hate surprises.

You are a one-semester prep student (Approach B) if:

1. You get to the airport 1 hour before your flight is scheduled to leave;
2. You are willing to plan ahead to feel comfortable in stressful situations, but are okay with skipping some details;
3. You feel more comfortable when you know what to expect, but a surprise or two is cool;
4. You're always on time for appointments.

You are a 6-week prep student (Approach C) if:

1. You get to the airport just as your plane is announcing its final boarding;
2. You work best under pressure and tight deadlines;
3. You feel very confident with the skills and background you've learned in your AP Calculus class;
4. You decided late in the year to take the exam;
5. You like surprises;
6. You feel okay if you arrive 10–15 minutes late for an appointment.

CALENDAR FOR EACH PLAN

A Calendar for Approach A: A Year-Long Preparation for the AP Calculus Exam

Although its primary purpose is to prepare you for the AP Calculus Exam you will take in May, this book can enrich your study of calculus, your analytical skills and your problem solving techniques.

SEPTEMBER–OCTOBER (Check off the activities as you complete them.)

———— Determine into which student mode you would place yourself.

———— Carefully read Parts I and II.

———— Get on the web and take a look at the AP website(s).

———— Skim the Comprehensive Review section. (These areas will be part of your year-long preparation.)

———— Buy a few highlighters.

———— Flip through the entire book. Break the book in. Write in it. Toss it around a little bit… Highlight it.

———— Get a clear picture of what your own school's AP Calculus curriculum is.

———— Begin to use the book as a resource to supplement the classroom learning.

———— Read and study Chapter 1—Limits and Continuity.

———— Read and study Chapter 2—Differentiation.

———— Read and study Chapter 3—Graphs of Functions and Derivatives.

NOVEMBER (The first 10 weeks have elapsed.)

———— Read and study Chapter 4—Applications of Derivatives.

———— Read and study Chapter 5—More Applications of Derivatives.

DECEMBER

———— Read and study Chapter 6—Integration.

———— Review Chapters 1–3.

JANUARY (20 weeks have now elapsed.)

———— Read and study Chapter 7—Definite Integrals.

———— Review Chapters 4–6.

FEBRUARY

———— Read and study Chapter 8—Areas and Volumes.

———— Read and study Chapter 9—More Applications of Definite Integrals.

———— Take the Diagnostic Test.

———— Evaluate your strengths and weaknesses.

———— Study appropriate chapters to correct weaknesses.

MARCH (30 weeks have now elapsed.)

———— Read and study Chapter 10—Series (if you plan to take the BC exam)

———— Review Chapters 7–9.

APRIL

———— Take Practice Exam 1 in first week of April.

———— Evaluate your strengths and weaknesses.

———— Study appropriate chapters to correct weaknesses.

———— Review Chapters 1–10.

MAY—First Two Weeks (THIS IS IT!)

———— Take Practice Exam 2.

———— Score yourself.

———— Study appropriate chapters to correct weaknesses.

———— Get a good night's sleep the night before the exam. Fall asleep knowing you are well prepared.

GOOD LUCK ON THE TEST!

A Calendar for Approach B:
A Semester-Long Preparation for the AP Calculus Exam

Working under the assumption that you've completed one semester of calculus studies, the following calendar will use those skills you've been practicing to prepare you for the May exam.

JANUARY

_____ Carefully read Parts I and II.

_____ Read and study Chapter 1—Limits and Continuity.

_____ Read and study Chapter 2—Differentiation.

_____ Read and study Chapter 3—Graphs of Functions and Derivatives.

_____ Read and Study Chapter 4—Applications of Derivatives.

FEBRUARY

_____ Read and study Chapter 5—More Applications of Derivatives.

_____ Read and study Chapter 6—Integration.

_____ Read and study Chapter 7—Definite Integrals.

_____ Take the Diagnostic Test.

_____ Evaluate your strengths and weaknesses.

_____ Study appropriate chapters to correct weaknesses.

_____ Review Chapters 1–4.

MARCH (10 weeks to go.)

_____ Read and study Chapter 8—Areas and Volumes.

_____ Read and study Chapter 9—More Applications of Definite Integrals.

_____ Read and study Chapter 10—Series (if you plan to take the BC exam)

_____ Review Chapters 5–7.

APRIL

_____ Take Practice Exam 1 in first week of April.

_____ Evaluate your strengths and weaknesses.

_____ Study appropriate chapters to correct weaknesses.

_____ Review Chapters 1–10.

MAY—First Two Weeks (THIS IS IT!)

_____ Take Practice Exam 2.

_____ Score yourself.

_____ Study appropriate chapters to correct weaknesses.

_____ Get a good night's sleep the night before the exam. Fall asleep knowing you are well prepared.

GOOD LUCK ON THE TEST!

A Calendar for Approach C:
A Six-Week Preparation for the AP Calculus Exam

At this point, we are going to assume that you have been building your calculus knowledge base for more than six months. You will, therefore, use this book primarily as a specific guide to the AP Calculus Exam.

Given the time constraints, now it is not the time to try to expand your AP Calculus curriculum. Rather, it is the time to limit and refine what you already do know.

APRIL 1st–15th

_____ Skim Parts I and II.
_____ Skim Chapters 1–5.
_____ Carefully go over the "Rapid Review" sections of Chapters 1–5.
_____ Take the Diagnostic Test.
_____ Evaluate your strengths and weaknesses.
_____ Study appropriate chapters to correct weaknesses.

APRIL 16th–May 1st

_____ Skim Chapters 6–10.
_____ Carefully go over the "Rapid Review" sections of Chapters 6–10.

_____ Complete Practice Exam 1.
_____ Score yourself and analyze your errors.
_____ Study appropriate chapters to correct weaknesses.

MAY—First Two Weeks (THIS IS IT!)

_____ Complete Practice Exam 2.
_____ Score yourself and analyze your errors.
_____ Study appropriate chapters to correct weaknesses.
_____ Get a good night's sleep. Fall asleep knowing you are well prepared.

GOOD LUCK ON THE TEST!

Summary of the Three Study Plans

Month	Approach A: September Plan	Approach B: January Plan	Approach C: 6-Week Plan
September–October	Chapters 1, 2, & 3		
November	Chapters 4 & 5		
December	Chapter 6 Review Chapters 1–3		
January	Chapter 7 Review Chapters 4–6	Chapters 1, 2, 3, & 4	
February	Chapters 8 & 9 Diagnostic Test	Chapters 5, 6, & 7 Diagnostic Test Review Chapters 1–4	
March	Chapter 10* Review Chapters 7–9	Chapters 8, 9, & 10* Review Chapters 5–7	
April	Practice Exam 1 Review Chapters 1–10*	Practice Exam 1 Review Chapters 1–10*	Diagnostic Test Review Chapters 1–5 Practice Exam 1 Review Chapters 6–10*
May	Practice Exam 2	Practice Exam 2	Practice Exam 2

*Chapter 10 is only for the BC exam.

GRAPHICS USED IN THE BOOK

To emphasize particular skills, strategies, and practice, we use five sets of icons throughout this book. You will see these icons in the margins of Parts I, II, and III.

The first icon is an hourglass. We've chosen this to indicate the passage of time during the school year. This hourglass icon will be in the margin next to an item which may be of interest to one of the three types of students who are using this book.

For the student who plans to prepare for the AP Calculus exam during the entire school year, beginning in September through May, we use an hourglass which is full on the top.

For the student who decides to begin preparing for the exam in January of the calendar year, we use an hourglass which is half full on the top and half full on the bottom.

For the student who chooses to prepare during the final 6 weeks before the exam, we use an hourglass which is empty on the top and full on the bottom.

The second icon is a clock that indicates a timed practice activity or a time management strategy. It indicates on the face of the dial how much time to allow for a given exercise. The full dial will remind you that this is a strategy that can help you learn to manage your time on the test.

The third icon is an exclamation point that points to a very important idea, concept, or strategy point you should not pass over.

The fourth icon, a sun, indicates a tip that you might find useful.

The fifth icon identifies material covered only on the BC exam.

PART II

WHAT YOU NEED TO KNOW ABOUT THE AP CALCULUS EXAMS

BACKGROUND ON THE AP EXAMS

What Is Covered in the AP Calculus Exams?

The AP Calculus exams cover the following topics:

- Functions, Limits and Graphs of Functions, Continuity
- Definition and Computation of Derivatives, Second Derivatives, Relationship between the Graphs of Functions and their Derivatives, Applications of Derivatives
- Finding Antiderivatives, Definite Integrals, Applications of Integrals, Fundamental Theorem of Calculus, Numerical Approximations of Definite Integrals, and Separable Differential Equations.

The BC exam covers all of the topics on the AB exam as well as parametric, polar and vector functions, Euler's method, L'Hôpital's Rule, antiderivatives by parts and by partial fractions, improper integrals, logistic differential equations, and series.

Students are expected to be able to solve problems that are expressed graphically, numerically, analytically, and verbally. For a more detailed description of the topics covered in the AP Calculus exam, visit the College Board website at: *www.collegeboard.org/ap/calculus.*

What Is the Format of the AP Calculus Exam?

The AP Calculus exam has 2 sections:

Section I contains 45 multiple-choice questions with 105 minutes.

Section II contains 6 free-response questions with 90 minutes.

The time allotted for both sections is 3 hours and 15 minutes. Below is a summary of the different parts of each section.

Section I *Multiple-Choice*	Part A	28 questions	No Calculator	55 Minutes
	Part B	17 questions	Calculator	50 Minutes
Section II *Free-Response*	Part A	3 questions	Calculator	45 Minutes
	Part B	3 questions	No Calculator	45 Minutes

During the time allotted for Part B of Section II, students may continue to work on questions from Part A of Section II. However, they may not use a calculator at that time. Please note that you are not expected to be able to answer all the questions in order to receive a grade of 5. If you wish to see the specific instructions for each part of the test, visit the College Board website at: *www.collegeboard.org/ap/ calculus.*

What Are the Advanced Placement Exam Grades?

Advanced Placement Exam grades are given on a 5-point scale with 5 being the highest grade. The grades are described below:

5	Extremely Well Qualified
4	Well Qualified
3	Qualified
2	Possibly Qualified
1	No Recommendation

How Is the AP Calculus Exam Grade Calculated?

- The exam has a total raw score of 108 points: 54 points for the multiple-choice questions in Section I and 54 points for the free-response questions for Section II.
- Each correct answer in Section I is worth 1.2 points, an incorrect answer is worth $(¼)(-1.2)$ points, and no points for unanswered questions. For example, suppose your result in Section I is as follows:

Correct	Incorrect	Unanswered
36	4	5

Your score for Section I would be:

$$36 \times 1.2 - 4 \times (¼)(1.2) = 43.2 - 1.2 = 42. \text{ Not a bad score!}$$

- Each complete and correct solution for Section II is worth 9 points.
- The total raw score for both Section I and II is converted to a 5-point scale. The cut-off points for each grade (1–5) vary from year to year. Visit the College Board website at: *www.collegeboard.com/ap* for more information. Below is a rough estimate of the conversion scale:

Total Raw Score	Approximate AP Grade
75–108	5
60–74	4
45–59	3
31–44	2
0–30	1

Remember, these are approximate cut-off points.

Which Graphing Calculators Are Allowed for the Exam?

The following calculators are allowed:

Texas Instruments	Hewlett-Packard	Casio	Sharp
TI-82	HP-28 series	FX-9700 series	EL-9200 series
TI-83/TI-83 Plus	HP-38G	FX-9750 series	EL-9300 series
TI-85	HP-39G	CFX-9800 series	EL-9600 series
TI-86	HP-40G	CFX-9850 series	
TI-89	HP-48 series	CFX-9950 series	
	HP-49 series	CFX-9970 series	
		Algebra FX 2.0 series	

For a more complete list, visit the College Board website at: *www.collegeboard.com/ap*. If you wish to use a graphing calculator that is not on the approved list, your teacher must obtain written permission from the ETS before April 1st of the testing year.

Calculators and Other Devices Not Allowed for the AP Calculus Exam

- TI-92, HP-95 and devices with QWERTY keyboards
- Non-graphing scientific calculators
- Laptop computers
- Pocket organizers, electronic writing pads or pen-input devices

Other Restrictions on Calculators

- You may bring up to 2 (but no more than 2) approved graphing calculators to the exam.
- You may not share calculators with another student.
- You may store programs in your calculator.
- You are not required to clear the memories in your calculator for the exam.
- You may not use the memories of your calculator to store secured questions and take them out of the testing room.

How Much Work Do I Need to Show When I Use a Graphing Calculator in Section II, Free-Response Questions?

- When using a graphing calculator in solving a problem, you are required to write the setup that leads to the answer. For example, if you are finding the volume of a solid, you must write the definite integral and then use the calculator to compute the numerical value, e.g., Volume $= \pi \int_0^3 (5x)^2 dx = 225\pi$. Simply indicating the answer without writing the integral would only get you one point for the answer but no other credit for the work.
- You may *not* use calculator syntax to substitute for calculus notations. For example, you may *not* write "Volume $= \int (\pi)(5*x)^\wedge 2, \ x, \ 0, \ 3 = 225\pi$" instead of "Volume $= \pi \int_0^3 (5x)^2 dx = 225\pi$."
- You are permitted to use the following 4 built-in capabilities of your calculator to obtain an answer: plotting the graph of a function, finding the zeros of a function, finding the numerical derivative of a function, and evaluating a definite integral. All other capabilities of your calculator can only be used to *check* your answer. For example, you may *not* use the built-in Inflection function of your calculator to find points of inflection. You must use calculus using derivatives and showing change of concavity.

What Do I Need to Bring to the Exam?

- Several Number 2 pencils.
- A good eraser and a pencil sharpener.

- Two black or blue pens.
- One or two approved graphing calculators with fresh batteries. (Be careful when you change batteries so that you don't lose your programs.)
- A watch.
- An admissions card or a photo I.D. card if your school or the test site requires it.
- Your Social Security number.
- Your school code number if the test site is not at your school.
- A simple snack *if the test site permits it*. (Don't try anything you haven't eaten before. You might have an allergic reaction.)
- A light jacket if you know that the test site has strong air conditioning.
- Do *not* bring Wite Out or scrap paper.

TIPS FOR TAKING THE EXAM

General Tips

- Write legibly.
- Label all diagrams.
- Organize your solution so that the reader can follow your line of reasoning.
- Use complete sentences whenever possible. Always indicate what the final answer is.

More Tips

- Do easy questions first.
- Write out formulas and indicate all major steps.
- Guess if you can eliminate some of the choices in a multiple-choice question.
- Leave a multiple-choice question blank if you have no clue what the answer is.
- Be careful to bubble in the right grid, especially if you skip a question.
- Move on. Don't linger on a problem too long.
- Go with your first instinct if you are unsure.

Still More Tips

- Indicate units of measure.
- Simplify numeric or algebraic expressions only if the question asks you to do so.
- Carry all decimal places and round only at the end.
- Round to 3 decimal places unless the question indicates otherwise.
- Watch out for different units of measure, e.g., the radius, r, is 2 feet, find $\dfrac{dr}{dt}$ in inches per second.
- Use calculus notations and not calculator syntax, e.g., write $\int x^2 dx$ and not $\int (x^\wedge 2, x)$.
- Use only the four specified capabilities of your calculator to get your answer: plotting graph, finding zeros, calculating numerical derivatives, and evaluating definite integrals. All other built-in capabilities can only be used to *check* your solution.
- Answer all parts of a question from Section II even if you think your answer to an earlier part of the question might not be correct.

Enough Already... Just 3 More Tips

- Be familiar with the instructions for the different parts of the exam before the day of the exam. Visit the College Board website at: *www.collegeboard.com/ap* for more information.
- Get a good night sleep the night before.
- Have a light breakfast before the exam.

GETTING STARTED!

Taking the test that follows should help you assess your strengths and weaknesses as you begin preparing for the AP Calculus exams. Unlike the questions on the actual AP Calculus exam, these questions are arranged by topic. The questions here are similar to questions that might appear on Part I of either the AB or the BC exam, although on the actual exams, all the Part I questions would be multiple choice. Part II, or free-response questions, often combine several topics in one question. At the end of this test, you will find a group of questions that cover topics unique to the BC exam. These include all the material from Chapter 10 as well as selected topics from other chapters.

DIAGNOSTIC TEST

From Chapter 1

1. A function f is continuous on $[-2, 0]$ and some of the values of f are shown below.

x	-2	-1	0
f	4	b	4

If $f(x) = 2$ has no solution on $[-2, 0]$, then b could be

(A) 3

(B) 2

(C) 1

(D) 0

(E) -2

2. Evaluate $\lim\limits_{x \to -\infty} \dfrac{\sqrt{x^2 - 4}}{2x}$

3. If $h(x) = \begin{cases} \sqrt{x} & \text{if } x > 4 \\ x^2 - 12 & \text{if } x \leq 4 \end{cases}$ find $\lim\limits_{x \to 4} h(x)$.

4. If $f(x) = |2xe^x|$, what is the value of $\lim\limits_{x \to 0^+} f'(x)$?

From Chapter 2

5. If $f(x) = -2\csc(5x)$, find $f'\left(\dfrac{\pi}{6}\right)$.

6. Given the equation $y = (x + 1)(x - 3^2)$, what is the instantaneous rate of change of y at $x = -1$?

7. What is $\lim\limits_{\Delta x \to 0} \dfrac{\tan\left(\dfrac{\pi}{4} + \Delta x\right) - \tan\left(\dfrac{\pi}{4}\right)}{\Delta x}$?

From Chapter 3

8. The graph of f is shown in Figure D-1. Draw a possible graph of f' on (a, b).

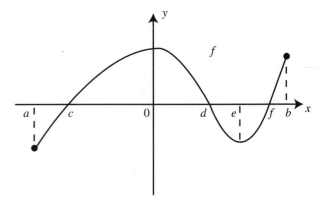

Figure D-1

9. The graph of the function g is shown in Figure D-2. Which of the following is true for g on (a, b)?

I. g is monotonic on (a, b).

II. g' is continuous on (a, b).

III. $g'' > 0$ on (a, b).

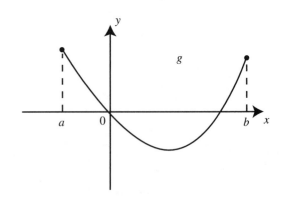

Figure D-2

10. The graph of f is shown in Figure D-3 and f is twice differentiable, which of the following statements is true?

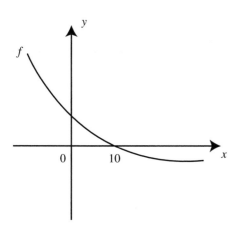

Figure D-3

(A) $f(10) < f'(10) < f''(10)$

(B) $f''(10) < f'(10) < f(10)$

(C) $f'(10) < f(10) < f''(10)$

(D) $f'(10) < f''(10) < f(10)$

(E) $f''(10) < f(10) < f'(10)$

11. The graph of f', the derivative of f is shown in Figure D-4. At which value(s) of x is the graph of f concave up?

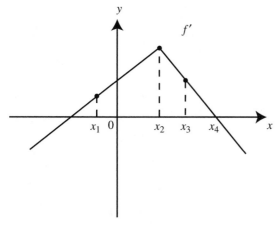

Figure D-4

12. How many points of inflection does the graph of $y = \sin(x^2)$ have on the interval $[-\pi, \pi]$?

13. If $g(x) = \int_a^x f(t)\,dt$ and the graph of f is shown in Figure D-5, which of the graphs in Figure D-6 on the next page is a possible graph of g?

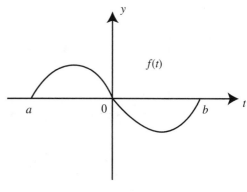

Figure D-5

14. The graphs of f', g', p', and q' are shown in Figure D-7. Which of the functions f, g, p, or q have a point of inflection on (a, b)?

From Chapter 4

15. When the area of a square is increasing four times as fast as the diagonals, what is the length of a side of the square?

16. If $g(x) = |x^2 - 4x - 12|$, which of the following statements about g is/are true?

   I. g has a relative maximum at $x = 2$.

   II. g is differentiable at $x = 6$.

   III. g has a point of inflection at $x = -2$.

From Chapter 5

17. Given the equation $y = \sqrt{x - 1}$, what is an equation of the normal line to the graph at $x = 5$?

18. What is the slope of the tangent to the curve $y = \cos(xy)$ at $x = 0$?

19. The velocity function of a moving particle on the x-axis is given as $v(t) = t^2 - t$. For what values of t is the particle's speed decreasing?

20. The velocity function of a moving particle is $v(t) = \dfrac{t^3}{3} - 2t^2 + 5$ for $0 \le t \le 6$. What is the maximum acceleration of the particle on the interval $0 \le t \le 6$?

21. Write an equation of the normal line to the graph of $y = x^3$ for $x \ge 0$ at the point where $f'(x) = 12$ and $y = f(x)$.

22. At what value(s) of x do the graphs of $f(x) = \dfrac{\ln x}{x}$ and $y = -x^2$ have perpendicular tangent lines?

23. Given a differentiable function f with $f\left(\dfrac{\pi}{2}\right) = 3$ and $f'\left(\dfrac{\pi}{2}\right) = -1$. Using a tangent

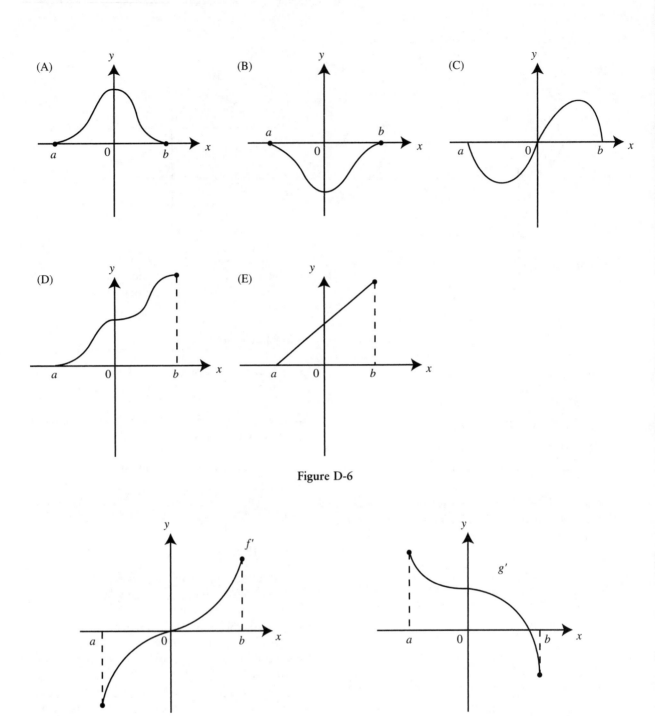

Figure D-6

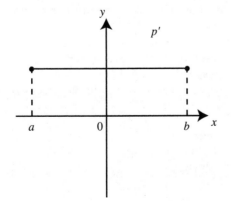

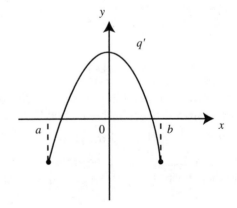

Figure D-7

line to the graph at $x = \dfrac{\pi}{2}$, find an approximate value of $f\left(\dfrac{\pi}{2} + \dfrac{\pi}{180}\right)$.

From Chapter 6

24. Evaluate $\displaystyle\int \dfrac{1 - x^2}{x^2}\, dx$

25. If $f(x)$ is an antiderivative of $\dfrac{e^x}{e^x + 1}$ and $f(0) = \ln(2)$, find $f(\ln 2)$.

26. Find the volume of the solid generated by revolving about the x-axis on the region bounded by the graph of $y = \sin 2x$ for $0 \le x \le \pi$ and the line $y = \dfrac{1}{2}$.

From Chapter 7

27. Evaluate $\displaystyle\int_1^4 \dfrac{1}{\sqrt{x}}\, dx$

28. If $\displaystyle\int_{-1}^k (2x - 3)\, dx = 6$, find k.

29. If $h(x) = \displaystyle\int_{\pi/2}^x \sqrt{\sin t}\, dt$, find $h'(\pi)$.

30. If $f'(x) = g(x)$ and g is a continuous function for all real values of x, then $\displaystyle\int_0^2 g(3x)\, dx$ is

(A) $\dfrac{1}{3} f(6) - \dfrac{1}{3} f(0)$

(B) $f(2) - f(0)$

(C) $f(6) - f(0)$

(D) $\dfrac{1}{3} f(0) - \dfrac{1}{3} f(6)$

(E) $3f(6) - 3f(0)$

31. Evaluate $\displaystyle\int_\pi^x \sin(2t)\, dt$

32. If a function f is continuous for all values of x, which of the following statements is/are always true?

I. $\displaystyle\int_a^c f(x)dx = \int_a^b f(x)dx + \int_b^c f(x)dx$

II. $\displaystyle\int_a^b f(x)dx = \int_a^c f(x)dx - \int_c^b f(x)dx$

III. $\displaystyle\int_b^c f(x)dx = \int_b^a f(x)dx - \int_c^a f(x)dx$

33. If $g(x) = \displaystyle\int_{\pi/2}^x 2 \sin t\, dt$ on $\left[\dfrac{\pi}{2}, \dfrac{5\pi}{2}\right]$, find the value(s) of x where g has a local minimum.

From Chapter 8

34. The graph of the velocity function of a moving particle is shown in Figure D-8. What is the total distance traveled by the particle during $0 \le t \le 6$?

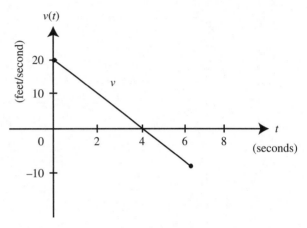

Figure D-8

35. The graph of f consists of four line segments, for $-1 \le x \le 5$ as shown in Figure D-9. What is the value of $\displaystyle\int_{-1}^5 f(x)\, dx$?

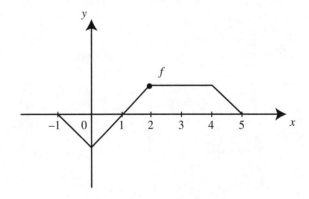

Figure D-9

36. Find the area of the region enclosed by the graph of $y = x^2 - x$ and the x-axis.

37. If $\displaystyle\int_{-k}^k f(x)\, dx = 0$ for all real values of k, then which of the graphs in Figure D-10 could be the graph of f?

(A)
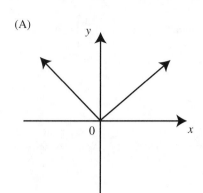

(B)
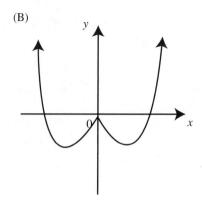

(C)
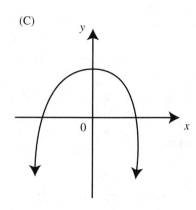

(D)
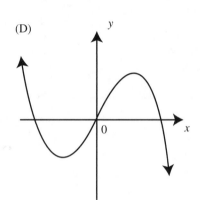

(E)
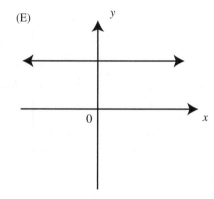

Figure D-10

38. The area under the curve $y = \sqrt{x}$ from $x = 1$ to $x = k$ is 8. Find the value of k.

39. For $0 \le x \le 3\pi$, find the area of the region bounded by the graphs of $y = \sin x$ and $y = \cos x$.

40. Let f be a continuous function on $[0, 6]$ that has selected values as shown below:

x	0	1	2	3	4	5	6
$f(x)$	1	2	5	10	17	26	37

Using three midpoint rectangles of equal lengths, find an approximate value of

$$\int_0^6 f(x)\,dx?$$

From Chapter 9

41. What is the average value of the function $y = e^{-4x}$ on $[-\ln 2, \ln 2]$?

42. If $\dfrac{dy}{dx} = 2 \sin x$ and at $x = \pi, y = 2$, find a solution to the differential equation.

43. Water is leaking from a tank at the rate of $f(t) = 10 \ln(t+1)$ *gallons per hour* for $0 \le t \le 10$, where t is measured in hours. How many gallons of water have leaked from the tank at exactly after *5 hours*?

44. Carbon-14 has a half-life of *5730 years*. If y is the amount of Carbon-14 present and y decays according to the equation $\dfrac{dy}{dt} = ky$, where k is a constant and t is measured in years, find the value of k.

45. What is the volume of the solid whose base is the region enclosed by the graphs of $y = x^2$ and $y = x + 2$ and whose cross sections perpendicular to the x-axis are squares?

 Note: The questions below focus on topics covered only on the BC exam.

From Chapter 2

46. Evaluate $\displaystyle\lim_{x \to \pi} \dfrac{e^x - e^\pi}{x^e - \pi^e}$

From Chapter 3

47. Find the rectangular equation of the curve defined by $x = 1 + e^{-t}$ and $y = 1 + e^t$.

From Chapter 5

48. An object moves in the plane on a path given by $x = 4t^2$ and $y = \sqrt{t}$. Find the acceleration vector when $t = 4$.
49. Find the equation of the tangent line to the curve defined by $x = 2t + 3$, $y = t^2 + 2t$ at $t = 1$.

From Chapter 6

50. Evaluate $\int_2^5 \dfrac{1}{x^2 + 2x - 3} \, dx$

51. Evaluate $\int x^2 \cos x \, dx$

From Chapter 7

52. Evaluate $\int_0^\infty e^{-x} dx$

From Chapter 8

53. Find the total area bounded by the curves $r = 2 \cos \theta$ and $r = 2 \sin \theta$.

54. Determine the length of curve defined by $x = 3t - t^3$ and $y = 3t^2$ from $t = 0$ to $t = 2$.

From Chapter 9

55. The growth of a colony of bacteria in a controlled environment is modeled by $\dfrac{dP}{dt} = .35P \left(1 - \dfrac{P}{4000} \right)$. If the initial population is 100, find the population when $t = 5$.

56. If $\dfrac{dy}{dx} = \dfrac{-y}{x^2}$ and $y = 3$ when $x = 2$, approximate y when $x = 3$.

From Chapter 10

57. Determine whether the series $\sum \dfrac{3}{(n+1)^4}$ converges or diverges.

58. For what values of x does the series $x - \dfrac{x^2}{2} + \dfrac{x^3}{3} - \dfrac{x^4}{4} + \ldots$ converge absolutely?

59. Find the Taylor Series expansion of $f(x) = \dfrac{1}{x}$ about the point $x = 2$.

60. Find the Maclaurin series for e^{-x^2}.

ANSWERS TO DIAGNOSTIC TEST

1. A

2. $-1/2$

3. Does not exist.

4. 2

5. $-20\sqrt{3}$

6. 16

7. 2

8. See Figure DS-3.

9. II & III

10. C

11. $x < x_2$

12. 8

13. A

14. q

15. $2\sqrt{2}$

16. I

17. $y = -4x + 22$

18. 0

19. $\left(\dfrac{1}{2}, 1\right)$

20. 12

21. $y = \dfrac{-1}{12}x + \dfrac{49}{6}$

22. 1.370

23. 2.983

24. $\dfrac{-1}{x} - x + c$

25. $\ln 3$

26. 1.503

27. 2

28. $\{-2, 5\}$

29. 0

30. A

31. $\dfrac{-1}{2}\cos(2x) + \dfrac{1}{2}$

32. I & III

33. 2π

34. 50 feet

35. 2

36. $\dfrac{1}{6}$

37. D

38. $13^{2/3}$

39. 5.657

40. 76

41. $\dfrac{255}{128 \ln 2}$

42. $y = -2\cos x$

43. 57.506

44. $\dfrac{-\ln 2}{5730}$

45. $\dfrac{81}{10}$

BC Topics Unique to the BC Exam

46. $\dfrac{e^{\pi - 1}}{\pi^{e-1}}$

47. $y = \dfrac{x}{x - 1}$

48. $\left\langle 8, -\dfrac{1}{32}\right\rangle$

49. $y = 2x - 7$

50. $\dfrac{1}{4}\ln\left(\dfrac{5}{2}\right)$

51. $x^2 \sin x + 2x\cos x - 2\sin x + C$

52. 1

53. $\dfrac{\pi}{2} - 1$

54. 14

55. 514.325

56. 2.415

57. Converges

58. $-1 < x < -1$

59. $\sum \dfrac{(-1)^n}{2^{n+1}}(x - 2)^n$

60. $\sum \dfrac{(-1)^n (x^{2n})}{n!}$

SOLUTIONS TO DIAGNOSTIC TEST

1. See Figure DS-1.
   If $b = 2$, then $x = -1$ would be a solution for $f(x) = 2$.
   If $b = 1, 0$ or -2, $f(x) = 2$ would have two solutions.
   Thus, $b = 3$, choice (A)

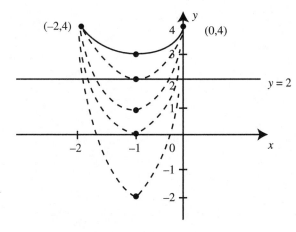

Figure DS-1

2. $\displaystyle \lim_{x \to -\infty} \frac{\sqrt{x^2 - 4}}{2x} = \lim_{x \to -\infty} \frac{\sqrt{x^2 - 4}/-\sqrt{x^2}}{2x/-\sqrt{x^2}}$

   (Note: as $x \to -\infty, x = -\sqrt{x^2}$)

   $\displaystyle = \lim_{x \to -\infty} \frac{-\sqrt{x^2 - 4/x^2}}{2}$

   $\displaystyle = \lim_{x \to -\infty} \frac{-\sqrt{1 - (4/x^2)}}{2}$

   $\displaystyle = -\frac{\sqrt{1}}{2} = -\frac{1}{2}$

3. $h(x) = \begin{cases} \sqrt{x} & \text{if } x > 4 \\ x^2 - 12 & \text{if } x \le 4 \end{cases}$

   $\displaystyle \lim_{x \to 4^+} h(x) = \lim_{x \to 4^+} \sqrt{x} = \sqrt{4} = 2$

   $\displaystyle \lim_{x \to 4^-} h(x) = \lim_{x \to 4^-} (x^2 - 12) = (4^2 - 12) = 4$

   Since $\displaystyle \lim_{x \to 4^+} h(x) \ne \lim_{x \to 4^-} h(x)$, thus $\displaystyle \lim_{x \to 4} h(x)$ does not exist.

4. $f(x) = |2xe^x| = \begin{cases} 2xe^x & \text{if } x \ge 0 \\ -2xe^x & \text{if } x < 0 \end{cases}$

   If $x \ge 0, f'(x) = 2e^x + e^x(2x) = 2e^x + 2xe^x$

   $\displaystyle \lim_{x \to 0^+} f'(x) = \lim_{x \to 0^+} (2e^x + 2xe^x) = 2e^0 + 0 = 2$

5. $f(x) = -2\csc(5x)$

   $f'(x) = -2(-\csc 5x)[\cot(5x)](5)$

   $= 10\csc(5x)\cot(5x)$

   $f'(x) = 10\csc\left(\frac{5\pi}{6}\right)\cot\left(\frac{5\pi}{6}\right)$

   $= 10(2)(-\sqrt{3}) = -20\sqrt{3}$

6. $y = (x + 1)(x - 3)^2;$

   $\dfrac{dy}{dx} = (1)(x - 3)^2 + 2(x - 3)(x + 1)$

   $= (x - 3)^2 + 2(x - 3)(x + 1)$

   $\left.\dfrac{dy}{dx}\right|_{x=-1} = (-1 - 3)^2 + 2(-1 - 3)(-1 + 1)$

   $= (-4)^2 + 0 = 16$

7. $f'(x_1) = \displaystyle\lim_{\Delta x \to 0} \frac{f(x_1 + \Delta x) - f(x_1)}{\Delta x}$

   Thus $\displaystyle\lim_{\Delta x \to 0} \frac{\tan\left(\frac{\pi}{4} + \Delta x\right) - \tan\left(\frac{\pi}{4}\right)}{\Delta x}$

   $= \dfrac{d}{dx}(\tan x)$ at $x = \dfrac{\pi}{4}$

   $= \sec^2\left(\dfrac{\pi}{4}\right) = (\sqrt{2})^2 = 2$

8. See Figure DS-2.

9. I. Since the graph of g is decreasing and then increasing, it is not monotonic.

   II. Since the graph of g is a smooth curve, g' is continuous,

   III. Since the graph of g is concave upward, $g'' > 0$.
   Thus only statements II and III are true.

10. The graph indicates that $(1) f(10) = 0$, $(2) f'(10) < 0$, since f is decreasing; and $(3) f''(10) > 0$, since f is concave upward. Thus $f'(10) < f(10) < f''(10)$, choice (C).

11. See Figure DS-3.
    The graph of f is concave upward for $x < x_2$.

12. See Figure DS-4.
    Enter $y_1 = \sin(x^2)$ Using the Inflection function of your calculator, you obtain four points of inflection on $[0, \pi]$. The points of

Based on the graph of f:

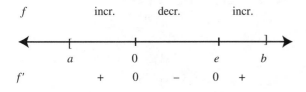

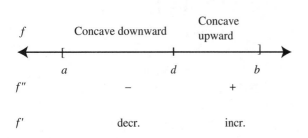

f' decr.   incr.

A possible graph of f'

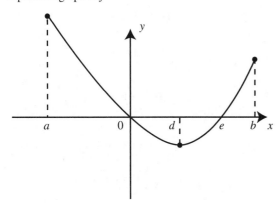

Figure DS-2

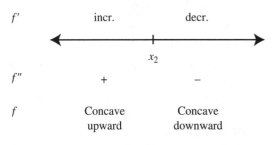

Figure DS-3

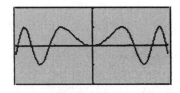

Figure DS-4

inflection occur at $x = 0.81, 1.81, 2.52,$ and 3.07. Since $y_1 = \sin(x^2)$ is an even function, there is a total of eight points of inflection on $[-\pi, \pi]$. An alternate solution is to enter

$$y_2 = \frac{d^2}{dx^2}(y_1(x), x, 2).$$ The graph of y_2 crosses

the x-axis eight times, thus eight zeros on $[-\pi, \pi]$.

13. Since $g(x) = \displaystyle\int_a^x f(t)\,dt,\ g'(x) = f(x)$

See Figure DS-5.
The only graph that satisfies the behavior of g is choice (A)

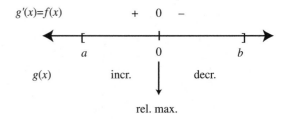

Figure DS-5

14. See Figure DS-6.
A change of concavity occurs at $x = 0$ for q. Thus, q has a point of inflection at $x = 0$. None of the other functions has a point of inflection.

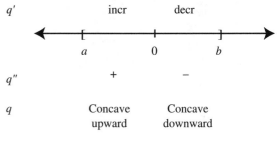

Figure DS-6

15. Let z be the diagonal of a square. Area of a

square $A = \dfrac{z^2}{2}$

$$\frac{dA}{dt} = \frac{2z}{2}\frac{dz}{dt} = z\frac{dz}{dt}$$

Since $\dfrac{dA}{dt} = 4\dfrac{dz}{dt};\ 4\dfrac{dz}{dt} = z\dfrac{dz}{dt} \Rightarrow z = 4$

Let s be a side of the square. Since the diagonal $z = 4, s^2 + s^2 = z^2$ or $2s^2 = 16$. Thus, $s^2 = 8$ or $s = 2\sqrt{2}$.

16. See Figure DS-7.
The graph of g indicates that a relative maximum occurs at $x = 2$; g is not

differentiable at $x = 6$, since there is a *cusp* at $x = 6$ and g does not have a point of inflection at $x = -2$, since there is no tangent line at $x = -2$. Thus, only statement I is true.

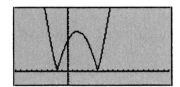

Figure DS-7

17. $y = \sqrt{x - 1} = (x - 1)^{1/2}$; $\dfrac{dy}{dx} = \dfrac{1}{2}(x - 1)^{-1/2}$

$$= \dfrac{1}{2(x - 1)^{1/2}}$$

$\dfrac{dy}{dx}\bigg|_{x=5} = \dfrac{1}{2(5 - 1)^{1/2}} = \dfrac{1}{2(4)^{1/2}} = \dfrac{1}{4}$

At $x = 5$, $y = \sqrt{x - 1} = \sqrt{5 - 1} = 2$; $(5, 2)$

Slope of normal line = negative reciprocal of

$\left(\dfrac{1}{4}\right) = -4$.

Equation of normal line:
$y - 2 = -4(x - 5) \Rightarrow y = -4(x - 5) + 2$ or
$\quad y = -4x + 22$.

18. $y = \cos(xy)$; $\dfrac{dy}{dx} = [-\sin(xy)]\left(1y + x\dfrac{dy}{dx}\right)$

$\dfrac{dy}{dx} = -y\sin(xy) - x\sin(xy)\dfrac{dy}{dx}$

$\dfrac{dy}{dx} + x\sin(xy)\dfrac{dy}{dx} = -y\sin(xy)$

$\dfrac{dy}{dx}[1 + x\sin(xy)] = -y\sin(xy)$

$\dfrac{dy}{dx} = \dfrac{-y\sin(xy)}{1 + x\sin(xy)}$

At $x = 0$, $y = \cos(xy) = \cos(0) = 1$; $(0, 1)$

$\dfrac{dy}{dx}\bigg|_{x=0, y=1} = \dfrac{-(1)\sin(0)}{1 + 0\sin(0)} = \dfrac{0}{1} = 0$

Thus the slope of the tangent at $x = 0$ is 0.

19. See Figure DS-8.

$v(t) = t^2 - t$

$\quad$ Set $v(t) = 0 \Rightarrow t(t - 1) = 0$

$\Rightarrow t = 0$ or $t = 1$

$a(t) = v'(t) = 2t - 1$

$\quad$ Set $a(t) = 0 \Rightarrow 2t - 1 = 0$ or $t = \dfrac{1}{2}$

Since $v(t) < 0$ and $a(t) > 0$ on $\left(\dfrac{1}{2}, 1\right)$, the speed of the particle is decreasing on $\left(\dfrac{1}{2}, 1\right)$.

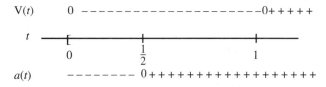

Figure DS-8

20. $v(t) = \dfrac{t^3}{3} - 2t^2 + 5$

$a(t) = v'(t) = t^2 - 4t$
See Figure DS-9.
The graph indicates that the maximum acceleration occurs at the endpoint $t = 6$.
$a(t) = t^2 - 4t$ and $a(6) = 6^2 - 4(6) = 12$

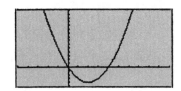

Figure DS-9

21. $\quad y = x^3, x \geq 0$; $\dfrac{dy}{dx} = 3x^2$

$\quad f'(x) = 12 \Rightarrow \dfrac{dy}{dx} = 3x^2 = 12$

$\quad\quad \Rightarrow x^2 = 4 \Rightarrow x = 2$

Slope of normal = negative reciprocal of slope of tangent $= -\dfrac{1}{12}$.

At $x = 2, y = x^3 = 2^3 = 8$; $(2, 8)$

$\quad\quad y - 8 = -\dfrac{1}{12}(x - 2)$

Equation of normal line: $\Rightarrow y = -\dfrac{1}{12}(x - 2) + 8$

$\quad\quad\quad$ or $y = -\dfrac{1}{12}x + \dfrac{49}{6}$

22. $f(x) = \dfrac{\ln x}{x}; \; f'x = \dfrac{(1/x)(x) - (1)\ln x}{x^2}$

$= \dfrac{1}{x^2} - \dfrac{\ln x}{x^2}$

$y = -x^2; \dfrac{dy}{dx} = -2x$

Perpendicular tangents $\Rightarrow (f'(x)) \left(\dfrac{dy}{dx} \right) = -1$

$\Rightarrow \left(\left(\dfrac{1}{x^2} \right) - \dfrac{\ln x}{x^2} \right)(-2x) = -1$

Using the Solve function on your calculator, you obtain $x \approx 1.37015 \approx 1.370$.

23. $f\left(\dfrac{\pi}{2} \right) = 3 \Rightarrow \left(\dfrac{\pi}{2}, 3 \right)$ is on the graph.

$f'\left(\dfrac{\pi}{2} \right) = -1 \Rightarrow$ slope of the tangent at $x = \dfrac{\pi}{2}$

is -1.

Equation of tangent line: $y - 3 = -1 \left(x - \dfrac{\pi}{2} \right)$

or $y = -x + \dfrac{\pi}{2} + 3$.

Thus $f\left(\dfrac{\pi}{2} + \dfrac{\pi}{180} \right) \approx -\left(\dfrac{\pi}{2} + \dfrac{\pi}{180} \right) + \dfrac{\pi}{2} + 3$

$\approx 3 - \dfrac{\pi}{180} \approx 2.98255$

≈ 2.983

24. $\displaystyle\int \dfrac{1 - x^2}{x^2} dx = \int \left(\dfrac{1}{x^2} - \dfrac{x^2}{x^2} \right) dx = \int \left(\dfrac{1}{x^2} - 1 \right) dx$

$= \displaystyle\int (x^{-2} - 1) dx = \dfrac{x^{-1}}{-1} - x + c$

$= -\dfrac{1}{x} - x + c$

! You can check the answer by differentiating your result.

25. Let $u = e^x + 1; du = e^x dx$

$f(x) = \displaystyle\int \dfrac{e^x}{e^x + 1} dx = \int \dfrac{1}{u} du$

$= \ln |u| + c = \ln |e^x + 1| + c$

$f(0) = \ln |e^0 + 1| + c = \ln(2) = c$ and

$f(0) = \ln 2 \Rightarrow \ln(2) + c = \ln 2 \Rightarrow c = 0$

Thus $f(x) = \ln(e^x + 1)$ and $f(\ln 2)$

$= \ln(e^{\ln 2} + 1) = \ln(2 + 1) = \ln 3$

26. See Figure DS-10.
To find the points of intersection, set

$\sin 2x = \dfrac{1}{2} \Rightarrow 2x = \sin^{-1} \left(\dfrac{1}{2} \right)$

$\Rightarrow 2x = \dfrac{\pi}{6}$ or $2x = \dfrac{5\pi}{6} \Rightarrow x = \dfrac{\pi}{12}$ or $x = \dfrac{5\pi}{12}$

Volume of solid $= \pi \displaystyle\int_{\pi/12}^{5\pi/12} \left[(\sin 2x)^2 - \left(\dfrac{1}{2} \right)^2 \right] dx$

Using your calculator, you obtain:
Volume of solid $\approx (0.478306)\pi \approx 1.50264 \approx 1.503$

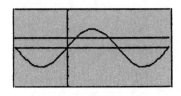

Figure DS-10

27. $\displaystyle\int_1^4 \dfrac{1}{\sqrt{x}} dx = \int_1^4 x^{-1/2} dx = \dfrac{x^{1/2}}{1/2} \Big]_1^4 = 2x^{1/2} \Big]_1^4$

$= 2(4)^{1/2} - 2(1)^{1/2} = 4 - 2 = 2$

28. $\displaystyle\int_{-1}^k (2x - 3) dx = x^2 - 3x \Big]_{-1}^k$

$= (k^2 - 3k) - ((-1)^2 - 3(-1))$

$= k^2 - 3k - (1 + 3)$

$= k^2 - 3k - 4$

Set $k^2 - 3k - 4 = 6 \Rightarrow k^2 - 3k - 10 = 0$

$\Rightarrow (k - 5)(k + 2) = 0 \Rightarrow k = 5$ or $k = -2$

! You can check your answer by evaluating

$\displaystyle\int_{-1}^{-2} (2x - 3) dx$ and $\displaystyle\int_{-1}^6 (2x - 3) dx$

29. $h(x) = \displaystyle\int_{\pi/2}^{\pi} \sqrt{\sin t} \; dt \Rightarrow h'(x) = \sqrt{\sin x}$

$h'(\pi) = \sqrt{\sin \pi} = \sqrt{0} = 0$

30. Let $u = 3x; du = 3dx$ or $\dfrac{du}{3} = dx$

$\displaystyle\int g(3x) dx = \int g(u) \dfrac{du}{3} = \dfrac{1}{3} \int g(u) du$

$= \dfrac{1}{3} f(u) + c = \dfrac{1}{3} f(3x) + c$

$$\int_0^2 g(3x)dx = \frac{1}{3}\left[f(3x)\right]_0^2 = \frac{1}{3}f(6) - \frac{1}{3}f(0)$$

Thus, the correct choice is (A).

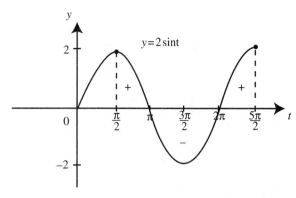

Figure DS-11

31. $\int_\pi^x \sin(2t)dt = \left[\dfrac{-\cos(2t)}{2}\right]_\pi^x$

$= \dfrac{-\cos(2x)}{2} - \left(-\dfrac{\cos(2\pi)}{2}\right)$

$= -\dfrac{1}{2}\cos(2x) + \dfrac{1}{2}$

32. I. $\int_a^c f(x)dx = \int_a^b f(x)dx + \int_b^c f(x)dx$

The statement is true, since the upper and lower limits of the integrals are in sequence, i.e. $a \to c = a \to b \to c$.

II. $\int_a^b f(x)dx = \int_a^c f(x)dx - \int_c^b f(x)dx$

$= \int_a^c f(x)dx + \int_b^c f(x)dx$

The statement is not always true.

III. $\int_b^c f(x)dx = \int_b^a f(x)dx - \int_c^a f(x)dx$

$= \int_b^a f(x)dx + \int_a^c f(x)dx$

The statement is true.
Thus only statements I and III are true.

33. Since $g(x) = \int_{\pi/2}^x 2\sin t\, dt$, then $g'(x) = 2\sin x$.
Set $g'(x) = 0 \Rightarrow 2\sin x = 0 \Rightarrow x = \pi$ or 2π
$g''(x) = 2\cos x$ and $g''(\pi) = 2\cos\pi = -2$ and $g''(2\pi) = 1$.
Thus g has a local minimum at $x=2\pi$. You can also approach the problem geometrically by looking at the area under the curve. See Figure DS-11.

34. Total distance $= \int_0^4 v(t) + \left|\int_4^5 v(t)dt\right|$

$= \dfrac{1}{2}(4)(20) + \left|\dfrac{1}{2}(2)(-10)\right|$

$= 40 + 10 = 50$ feet

35. $\int_{-1}^5 f(x)dx = \int_{-1}^1 f(x)dx + \int_1^5 f(x)dx$

$= -\dfrac{1}{2}(2)(1) + \dfrac{1}{2}(2 + 4)(1)$

$= -1 + 3 = 2$

36. To find points of intersection, set
$y = x^2 - x = 0$
$\Rightarrow x(x - 1) = 0 \Rightarrow x = 0$ or $x = 1$.
See Figure DS-12.

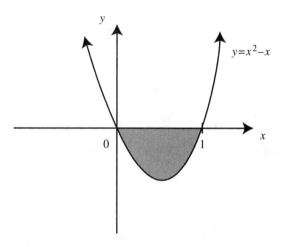

Figure DS-12

$\text{Area} = \left|\int_0^1 \left(x^2 - x\right) dx\right| = \left|\dfrac{x^3}{3} - \dfrac{x^2}{2}\right]_0^1\right|$

$= \left|\left(\dfrac{1}{3} - \dfrac{1}{2}\right) - 0\right| = \left|-\dfrac{1}{6}\right|$

$= \dfrac{1}{6}$

37. $\int_{-k}^{k} f(x)\,dx = 0 \Rightarrow f(x)$ is odd function, i.e., $f(x) = -f(-x)$. Thus the graph in choice (D) is the only odd function.

38. Area $= \int_{1}^{k} \sqrt{x}\,dx = \int_{1}^{k} x^{1/2}\,dx = \left[\dfrac{x^{3/2}}{3/2}\right]_{1}^{k}$

$= \left[\dfrac{2}{3}x^{3/2}\right]_{1}^{k} = \dfrac{2}{3}k^{3/2} - \dfrac{2}{3}(1)^{3/2}$

$= \dfrac{2}{3}k^{3/2} - \dfrac{2}{3} = \dfrac{2}{3}\left(k^{3/2} - 1\right)$

Since A $= 8$, set $\dfrac{2}{3}\left(k^{3/2} - 1\right) = 8 \Rightarrow k^{3/2} - 1$
$= 12 \Rightarrow k^{3/2} = 13$ or $k = 13^{2/3}$

39. See Figure DS-13.

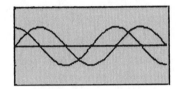

Figure DS-13

Using the Intersection function of the calculator, you obtain the intersection points at $x = 0.785398$, 3.92699, and 7.06858.

Area $= \displaystyle\int_{0.785398}^{3.92699} (\sin x - \cos x)\,dx$

$+ \displaystyle\int_{3.92699}^{7.06858} (\cos x - \sin x)\,dx$

$= 2.82843 + 2.82843 \approx 5.65685$

You can also find the area by:

Area $= \displaystyle\int_{.785398}^{7.06858} |\sin x - \cos x|\,dx$

$\approx 5.65685 \approx 5.657$

40. Length of a rectangle $= \dfrac{6 - 0}{3} = 2$.
Midpoints are $x = 1, 3$, and 5 and $f(1) = 2$,
$f(3) = 10$ and $f(5) = 26$

$\displaystyle\int_{0}^{6} f(x)\,dx \approx 2(2 + 10 + 26) \approx 2(38) = 76$

41. Average value $= \dfrac{1}{\ln 2 - (-\ln 2)} \displaystyle\int_{-\ln 2}^{\ln 2} e^{-4x}\,dx$.

Let $u = -4x$; $du = -4dx$, or $\dfrac{-du}{4} = dx$

$\displaystyle\int e^{-4x}\,dx = \int e^{u}\left(\dfrac{-du}{4}\right) = \dfrac{-1}{4}e^{u} + c$

$= \dfrac{-1}{4}e^{-4x} + c$

Average value $= \dfrac{1}{2\ln 2}\left[\dfrac{e^{-4x}}{-4}\right]_{-\ln 2}^{\ln 2}$

$= \dfrac{1}{2\ln 2}\left[\left(\dfrac{e^{-4\ln 2}}{-4}\right) - \left(\dfrac{e^{-4(-\ln 2)}}{-4}\right)\right]$

$= \dfrac{1}{2\ln 2}\left[\dfrac{\left(e^{\ln 2}\right)^{-4}}{-4} + \dfrac{\left(e^{\ln 2}\right)^{4}}{4}\right]$

$= \dfrac{1}{2\ln 2}\left[\dfrac{2^{-4}}{-4} + \dfrac{2^{4}}{4}\right] = \dfrac{1}{2\ln 2}\left(\dfrac{1}{-64} + 4\right)$

$= \dfrac{1}{2\ln 2}\left(\dfrac{255}{64}\right) = \dfrac{255}{128\ln 2}$

42. $\dfrac{dy}{dx} = 2\sin x \Rightarrow dy = 2\sin x\,dx$

$\displaystyle\int dy = \int 2\sin x\,dx \Rightarrow y = -2\cos x + c$

At $x = \pi, y = 2 \Rightarrow 2 = -2\cos \pi + c$

$\Rightarrow 2 = (-2)(-1) + c$

$\Rightarrow 2 = 2 + c = 0$

Thus $y = -2\cos x$

43. Amount of Water Leaked $= \displaystyle\int_{0}^{5} 10\ln(t + 1)\,dt$
Using your calculator, you obtain $10(6\ln 6 - 5)$
which is approximately *57.506 gallons*.

44. $\dfrac{dy}{dx} = ky \Rightarrow y = y_0 e^{kt}$

Half-life $= 5730 \Rightarrow y = \dfrac{1}{2}y_0$ when $t = 5730$

Thus, $\dfrac{1}{2}y_0 = y_0 e^{k(5730)} \Rightarrow \dfrac{1}{2} = e^{5730k}$

$\ln\left(\dfrac{1}{2}\right) = \ln\left(e^{5730k}\right) \Rightarrow \ln\left(\dfrac{1}{2}\right) = 5730k$

$$\ln 1 - \ln 2 = 5730k \Rightarrow -\ln 2 = 5730k$$

$$k = \frac{-\ln 2}{5730}$$

45. See Figure DS-14.

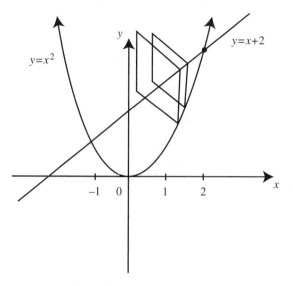

Figure DS-14

To find points of intersection, set $x^2 = x + 2$
$\Rightarrow x^2 - x - 2 = 0 \Rightarrow x = 2$ or $x = -1$
Area of cross section $= ((x + 2) - x^2)^2$
Volume of solid, $V = \int_{-1}^{2} \left(x + 2 - x^2\right)^2 dx$
Using your calculator, you obtain: $V = \dfrac{81}{10}$

 Material Unique to the BC Exam

From Chapter 2

46. By L'Hôpital's Rule, $\lim\limits_{x \to \pi} \dfrac{e^x - e^\pi}{x^e - \pi^e} = \lim\limits_{x \to \pi} \dfrac{e^x}{ex^{e-1}}$

$= \lim\limits_{x \to \pi} \dfrac{e^{x-1}}{x^{e-1}} = \dfrac{e^{\pi-1}}{\pi^{e-1}}$

From Chapter 3

47. Solve $x = 1 + e^{-t}$ for t. $x - 1 = e^{-t} \Rightarrow$
$-\ln(x - 1) = t$. Substitute in $y = 1 + e^t$.

$$y = 1 + e^{-\ln(x-1)} \Rightarrow y = 1 + \frac{1}{x - 1}$$

$$\Rightarrow y = \frac{x}{x - 1}$$

From Chapter 5

48. Position is given by $x = 4t^2$ and $y = \sqrt{t}$,
so velocity is $\dfrac{dx}{dt} = 8t$ and $\dfrac{dy}{dt} = \dfrac{1}{2\sqrt{t}}$. The
acceleration will be $\dfrac{d^2x}{dt^2} = 8$, $\dfrac{d^2y}{dt^2} = -\dfrac{1}{4}t^{-3/2}$

Evaluate $-\dfrac{1}{4}t^{-3/2}\Big]_{t=4} = -\dfrac{1}{4}\left(\dfrac{1}{8}\right) = -\dfrac{1}{32}$ to get

the acceleration vector $\left\langle 8, -\dfrac{1}{32}\right\rangle$.

49. The slope of the tangent line is $\dfrac{dy}{dx}\Big|_{t=1}$ and

$\dfrac{dx}{dt} = 2$, $\dfrac{dy}{dt} = 2t + 2$, so $\dfrac{dy}{dx}\Big|_{t=1} =$
$\dfrac{2t + 2}{2}\Big|_{t=1} =$
$t + 1|_{t=1} = 2$. At $t = 1, x = 5, y = 3, \Rightarrow (5, 3)$,
So the equation of the tangent line is
$y - 3 = 2(x - 5) \Rightarrow y = 2(x - 5) + 3 \Rightarrow y = 2x - 7$.

From Chapter 6

50. $\displaystyle\int_2^5 \dfrac{1}{x^2 + 2x - 3}dx = \int_2^5 \dfrac{1}{(x + 3)(x - 1)}dx$
Use a partial fraction decomposition with
$\dfrac{A}{(x + 3)} + \dfrac{B}{(x - 1)} = \dfrac{1}{(x + 3)(x - 1)}$, which gives
$A = \dfrac{-1}{4}$ and $B = \dfrac{1}{4}$. Then the integral becomes

$= \displaystyle\int_2^5 \dfrac{-1/4}{(x + 3)}dx + \int_2^5 \dfrac{1/4}{(x - 1)}dx$

$= \dfrac{-1}{4}\displaystyle\int_2^5 \dfrac{1}{(x + 3)}dx + \dfrac{1}{4}\int_2^5 \dfrac{1}{(x - 1)}dx$

$= \dfrac{-1}{4}\ln|x + 3| + \dfrac{1}{4}\ln|x - 1|\Big]_2^5$

$= \dfrac{1}{4}\left(\ln\left|\dfrac{x - 1}{x + 3}\right|\right)\Big]_2^5$

$= \dfrac{1}{4}\left[\left(\ln\left|\dfrac{4}{8}\right|\right) - \left(\ln\left|\dfrac{1}{5}\right|\right)\right]$

$= \dfrac{1}{4}\ln\left(\dfrac{5}{2}\right)$

51. Integrate $\int x^2 \cos x \, dx$ by parts with $u = x^2$,
$du = 2x \, dx$, $dv = \cos x \, dx$, and $v = \sin x$.
The integral becomes
$= x^2 \sin x - \int \sin x(2x) \, dx$
$= x^2 \sin x - 2 \int x \sin x \, dx$.

Use parts again for the remaining integral, letting $u = x$, $du = dx$, $dv = \sin x\, dx$, and $v = -\cos x$. The integral

$$= x^2 \sin x - 2\left[-x\cos x - \int (-\cos x)\, dx\right]$$

simplifies to $= x^2 \sin x + 2x\cos x - 2\int \cos x\, dx$

and the final integration gives you
$$= x^2 \sin x + 2x\cos x - 2\sin x + C.$$

From Chapter 7

52. $\displaystyle\int_0^\infty e^{-x}dx = \lim_{k\to\infty}\int_0^k e^{-x}dx = \lim_{k\to\infty}\left[-e^{-x}\right]_0^k$

$= \lim_{k\to\infty}\left[-e^{-k} + e^0\right] = \lim_{k\to\infty}\left[-e^{-k} + 1\right] = 1.$

From Chapter 8

53. The intersection of the circles $r = 2\cos\theta$ and $r = 2\sin\theta$ can be found by adding the area swept out by $r = 2\sin\theta$ for $0 \le \theta \le \dfrac{\pi}{4}$ and the area swept by $r = 2\cos\theta$ for $\dfrac{\pi}{4} \le \theta \le \dfrac{\pi}{2}$.

$A = \dfrac{1}{2}\int_0^{\pi/4} 4\sin^2\theta\, d\theta + \dfrac{1}{2}\int_{\pi/4}^{\pi/2} 4\cos^2\theta\, d\theta$

$= 2\int_0^{\pi/4} \sin^2\theta\, d\theta + 2\int_{\pi/4}^{\pi/2}\cos^2\theta\, d\theta$

$= \int_0^{\pi/4} (1 - \cos 2\theta)\, d\theta + \int_{\pi/4}^{\pi/2}(1 + \cos 2\theta)\, d\theta$

$= \left(\theta - \dfrac{1}{2}\sin 2\theta\right)\Big|_0^{\pi/4} + \left(\theta + \dfrac{1}{2}\sin 2\theta\right)\Big|_{\pi/4}^{\pi/2}$

$= \left[\dfrac{\pi}{4} - \dfrac{1}{2} - 0\right] + \left[\left(\dfrac{\pi}{2} - 0\right) - \left(\dfrac{\pi}{4} + \dfrac{1}{2}\right)\right]$

$= \dfrac{\pi}{2} - 1$

54. Differentiate $x = 3t - t^3 \Rightarrow \dfrac{dx}{dt} = 3 - 3t^2$ and $y = 3t^2 \Rightarrow \dfrac{dy}{dt} = 6t$. The length of the curve from $t = 0$ to $t = 2$ is

$L = \displaystyle\int_0^2 \sqrt{\left(3 - 3t^2\right)^2 + (6t)^2}\, dt$

$= \displaystyle\int_0^2 \sqrt{9 - 18t^2 + 9t^4 + 36t^2}\, dt$

$= \displaystyle\int_0^2 \sqrt{9 + 18t^2 + 9t^4}\, dt$

$= \displaystyle\int_0^2 3\sqrt{\left(1 + t^2\right)^2}\, dt = 3\int_0^2\left(1 + t^2\right) dt$

$= 3\left(t + \dfrac{1}{3}t^3\right)\Big|_0^2 = 3t + t^3\Big|_0^2$

$= (6 + 8) - (0) = 14.$

From Chapter 9

55. Separate and simplify $\dfrac{dP}{dt} = .35P\left(1 - \dfrac{P}{4000}\right)$.

$\dfrac{1}{P\left(1 - \dfrac{P}{4000}\right)}dP = .35dt$

$\dfrac{4000}{P(4000 - P)}dP = .35dt.$

Integrate with a partial fraction decomposition.

$\displaystyle\int \dfrac{4000}{P(4000 - P)}dP = \int .35dt$

$\displaystyle\int \dfrac{dP}{P} + \int \dfrac{dP}{4000 - P} = \int .35dt$

$\ln|P| - \ln|4000 - P| = .35t + C_1$

$\ln\left|\dfrac{P}{4000 - P}\right| = .35t + C_1$

$\dfrac{P}{4000 - P} = C_2 e^{.35t}$

Population at $t = 0$ is 100, so

$\dfrac{100}{4000 - 100} = \dfrac{100}{3900} = \dfrac{1}{39} = C_2.$

The population model is

$\Rightarrow 39P = e^{.35t}(4000 - P)$

$\Rightarrow 39P + e^{.35t}P = 4000e^{.35t}$

$\dfrac{P}{4000 - P} = \dfrac{e^{.35t}}{39} \Rightarrow P\left(39 + e^{.35t}\right) = 4000e^{.35t}$

$\Rightarrow P = \dfrac{4000e^{.35t}}{39 + e^{.35t}}$

$\Rightarrow P = \dfrac{4000}{39e^{-.35t} + 1}$

When $t = 5$, $P = \dfrac{4000}{39e^{-.35(5)} + 1}$

$\Rightarrow P = \dfrac{4000}{39e^{-1.75} + 1} \approx 514.325.$

56. If $\dfrac{dy}{dx} = \dfrac{-y}{x^2}$ and $y = 3$ and $x = 2$, approximate y when $x = 3$. Use Euler's Method with an increment of 0.5.

$y(2) = 3$ and $\left(\dfrac{dy}{dx}\right)_{x=2,\,y=3} = \dfrac{-3}{4}$ so

$$y(2.5) = y(2) + 0.5\left(\dfrac{dy}{dx}\right)_{x=2,\,y=3}$$

$$= 3 + 0.5(-0.75) = 2.625$$

$$\left(\dfrac{dy}{dx}\right)_{x=2.5,\,y=2.625} = \dfrac{-2.625}{(2.5)^2}$$

$$= \dfrac{-21}{8(6.25)} = \dfrac{-21}{50} = -0.42$$

$$y(3) = y(2.5) + 0.5\left(\dfrac{dy}{dx}\right)_{x=2.5,\,y=2.625}$$

$$= 2.625 + 0.5(-0.42)$$

$$= 2.625 - 0.21 = 2.415$$

From Chapter 10

57. The series $\sum \dfrac{3}{(n+1)^4}$ is a series with positive terms, which can be compared to the series $\sum \dfrac{3}{n^4}$. $\sum \dfrac{3}{n^4} = 3\sum \dfrac{1}{n^4}$ and $\sum \dfrac{1}{n^4}$ is a p-series with $p = 4$, and therefore convergent. $\sum \dfrac{3}{(n+1)^4}$ is term by term smaller than $\sum \dfrac{3}{n^4}$ and so $\sum \dfrac{3}{(n+1)^4}$ converges.

58. The series $x - \dfrac{x^2}{2} + \dfrac{x^3}{3} - \dfrac{x^4}{4} + \cdots$ is an alternating series with general term $\dfrac{(-1)^{n-1}x^n}{n}$. Using the ratio test for absolute convergence, we have $\lim\limits_{n\to\infty}\left|\dfrac{x^{n+1}}{n+1} \cdot \dfrac{n}{x^n}\right| = |x|$

$\lim\limits_{n\to\infty} = \left(\dfrac{n}{n+1}\right) = |x|$. The series will converge absolutely when $|x| < 1 \Rightarrow -1 < x < 1$. We do not consider the end points since the question asks for absolute convergence.

59. Investigate the first few derivatives of

$f(x) = \dfrac{1}{x}$, $f'(x) = \dfrac{-1}{x^2}$, $f''(x) = \dfrac{2}{x^3}$, $f'''(x) = \dfrac{-6}{x^4}$,

$f^{(4)}(x) = \dfrac{24}{x^5}$ and, in general, $f^{(n)}(x) = \dfrac{(-1)^n n!}{x^{n+1}}$.

Evaluate the derivatives at $x = 2$. $f(2) = \dfrac{1}{2}$,

$f'(2) = \dfrac{-1}{4}$, $f''(2) = \dfrac{2}{8}$, $f'''(2) = \dfrac{-6}{16}$, $f^{(4)}(x) = \dfrac{24}{32}$

and, in general, $f^{(n)}(2) = \dfrac{(-1)^n n!}{2^{n+1}}$.

The Taylor Series is $f(x) = \dfrac{1}{x} = \sum \dfrac{\dfrac{(-1)^n n!}{2^{n+1}}}{n!}(x-2)^n$

$= \sum \dfrac{(-1)^n}{2^{n+1}}(x-2)^n$

$= \dfrac{1/2}{0!}(x-2)^0 + \dfrac{-1/4}{1!}(x-2)^1 + \dfrac{2/8}{2!}(x-2)^2$

$+ \dfrac{-6/16}{3!}(x-2)^3 + \dfrac{24/32}{4!}(x-2)^4 + \cdots$

$= \dfrac{1}{2} - \dfrac{1}{4}(x-2) + \dfrac{1}{8}(x-2)^2 - \dfrac{1}{16}(x-2)^3$

$+ \dfrac{1}{32}(x-2)^4 - \cdots$

60. Begin with the Maclaurin series for e^x. If $f(x) = e^x$, then $f'(x) = e^x$, $f''(x) = e^x$, and $f''(x) = e^x$. Thus $e^x = 1 + x + \dfrac{x^2}{2!} + \dfrac{x^3}{3!} + \cdots$. Replacing x by $-x^2$, we have

$e^{-x^2} = 1 - x^2 + \dfrac{x^4}{2!} - \dfrac{x^6}{3!} + \dfrac{x^8}{4!} - \cdots$.

Thus $e^x = \sum \dfrac{(-1)^n (x^{2n})}{n!}$.

PART III

COMPREHENSIVE REVIEW

Chapter 1

Limits and Continuity

1.1   THE LIMIT OF A FUNCTION

Main Concepts: Definition and Properties of Limits, Evaluating Limits, One-sided Limits, Squeeze Theorem

Definition and Properties of Limits

Definition of Limit

Let f be a function defined on an open interval containing a, except possibly at a itself. Then $\lim\limits_{x \to a} f(x) = L$ (read as the limit of $f(x)$ as x approaches a is L) if for any $\varepsilon > 0$, there exists a $\delta > 0$ such that $|f(x) - L| < \varepsilon$ whenever $|x - a| < \delta$.

Properties of Limits

Given $\lim\limits_{x \to a} f(x) = L$ and $\lim\limits_{x \to a} g(x) = M$ and L, M, a, c, and n are real numbers, then

1. $\lim\limits_{x \to a} c = c$

2. $\lim\limits_{x \to a}[cf(x)] = c \lim\limits_{x \to a} f(x) = cL$

3. $\lim\limits_{x \to a}[f(x) \pm g(x)] = \lim\limits_{x \to a} f(x) \pm \lim\limits_{x \to a} g(x) = L + M$

4. $\lim\limits_{x \to a}[f(x) \cdot g(x)] = \lim\limits_{x \to a} f(x) \cdot \lim\limits_{x \to a} g(x) = L \cdot M$

5. $\lim\limits_{x \to a} \dfrac{f(x)}{g(x)} = \dfrac{\lim\limits_{x \to a} f(x)}{\lim\limits_{x \to a} g(x)} = \dfrac{L}{M}, M \neq 0$

6. $\lim\limits_{x \to a} [f(x)]^n = \left(\lim\limits_{x \to a} f(x)\right)^n = L^n$

Evaluating Limits

If f is a continuous function on an open interval containing the number a, then $\lim_{x \to a} f(x) = f(a)$.

Common techniques in evaluating limits are:

1. Substituting directly
2. Factoring and simplifying
3. Multiplying the numerator and denominator of a rational function by the conjugate of either the numerator or denominator
4. Using a graph or a table of values of the given function.

Example 1

Find the limit: $\lim_{x \to 5} \sqrt{3x + 1}$

Substituting directly: $\lim_{x \to 5} \sqrt{3x + 1} = \sqrt{3(5) + 1} = 4.$

Example 2

Find the limit: $\lim_{x \to \pi} 3x \sin x$

Using the product rule, $\lim_{x \to \pi} 3x \sin x = \left(\lim_{x \to \pi} 3x \right) \left(\lim_{x \to \pi} \sin x \right) = (3\pi)(\sin \pi) = (3\pi)(0) = 0.$

Example 3

Find the limit: $\lim_{t \to 2} \dfrac{t^2 - 3t + 2}{t - 2}$

Factoring and simplifying: $\lim_{t \to 2} \dfrac{t^2 - 3t + 2}{t - 2} = \lim_{t \to 2} \dfrac{(t - 1)(t - 2)}{(t - 2)}$

$$= \lim_{t \to 2}(t - 1) = (2 - 1) = 1$$

(Note that had you substituted $t = 2$ directly in the original expression, you would have obtained a zero in both the numerator and denominator.)

Example 4

Find the limit: $\lim_{x \to b} \dfrac{x^5 - b^5}{x^{10} - b^{10}}$

Factoring and simplifying: $\lim_{x \to b} \dfrac{x^5 - b^5}{x^{10} - b^{10}} = \lim_{x \to b} \dfrac{x^5 - b^5}{(x^5 - b^5)(x^5 + b^5)}$

$$= \lim_{x \to b} \dfrac{1}{x^5 + b^5} = \dfrac{1}{b^5 + b^5} = \dfrac{1}{2b^5}$$

Example 5

Find the limit: $\lim_{t \to 0} \dfrac{\sqrt{t + 2} - \sqrt{2}}{t}$

Multiplying both the numerator and the denominator by the conjugate of the numerator,

$\left(\sqrt{t + 2} + \sqrt{2} \right)$, yields $\lim_{t \to 0} \dfrac{\sqrt{t + 2} - \sqrt{2}}{t} \left(\dfrac{\sqrt{t + 2} + \sqrt{2}}{\sqrt{t + 2} + \sqrt{2}} \right)$

$$= \lim_{t \to 0} \frac{t + 2 - 2}{t \left(\sqrt{t + 2} + \sqrt{2} \right)}$$

$$= \lim_{t \to 0} \frac{t}{\left(\sqrt{t + 2} + \sqrt{2} \right)} = \lim_{t \to 0} \frac{1}{\left(\sqrt{t + 2} + \sqrt{2} \right)} = \frac{1}{\sqrt{0 + 2} + \sqrt{2}} = \frac{1}{2\sqrt{2}}$$

$$= \frac{1}{2\sqrt{2}} \left(\frac{\sqrt{2}}{\sqrt{2}} \right) = \frac{\sqrt{2}}{4}.$$ (Note that substituting 0 directly into the original expression would have produced a 0 in both the numerator and denominator.)

Example 6

Find the limit: $\lim\limits_{x \to 0} \dfrac{3 \sin 2x}{2x}$

Enter $y1 = \dfrac{3 \sin 2x}{2x}$ in the calculator. You see that the graph of $f(x)$ approaches 3 as x approaches 0. Thus, the $\lim\limits_{x \to 0} \dfrac{3 \sin 2x}{2x} = 3$. (Note that had you substituted $x = 0$ directly in the original expression, you would have obtained a zero in both the numerator and denominator.) (See Figure 1.1-1.)

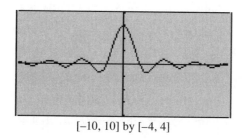

[−10, 10] by [−4, 4]

Figure 1.1-1

Example 7

Find the limit: $\lim\limits_{x \to 3} \dfrac{1}{x - 3}$

Enter $y1 = \dfrac{1}{x - 3}$ into your calculator. You notice that as x approaches 3 from the right, the graph of $f(x)$ goes higher and higher, and that as x approaches 3 from the left, the graph of $f(x)$ goes lower and lower. Therefore, $\lim\limits_{x \to 3} \dfrac{1}{x - 3}$ is undefined. (See Figure 1.1-2.)

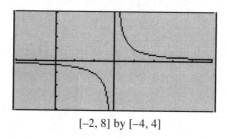

[−2, 8] by [−4, 4]

Figure 1.1-2

> • Always indicate what the final answer is e.g., "The maximum value of f is 5."
> Use complete sentences whenever possible.

One-Sided Limits

Let f be a function and a is a real number. Then the right-hand limit: $\lim\limits_{x \to a^+} f(x)$ represents the limit of f as x approaches a from the right, and left-hand limit: $\lim\limits_{x \to a^-} f(x)$ represents the limit of f as x approaches a from the left.

Existence of a Limit

Let f be a function and let a and L be real numbers. Then the two-sided limit $\lim\limits_{x \to a} f(x) = L$ if and only if the one-sided limits exist and $\lim\limits_{x \to a^+} f(x) = \lim\limits_{x \to a^-} f(x) = L$.

Example 1

Given $f(x) = \dfrac{x^2 - 2x - 3}{x - 3}$, find the limits (a) $\lim\limits_{x \to 3^+} f(x)$, (b) $\lim\limits_{x \to 3^-} f(x)$, and (c) $\lim\limits_{x \to 3} f(x)$. Substituting $x = 3$ into $f(x)$ leads to a 0 in both the numerator and denominator. Factor $f(x)$ as $\dfrac{(x-3)(x+1)}{(x-3)}$ which is equivalent to $(x + 1)$ where $x \neq 3$. Thus, (a) $\lim\limits_{x \to 3^+} f(x) = \lim\limits_{x \to 3^+} (x+1) = 4$, (b) $\lim\limits_{x \to 3^-} f(x) = \lim\limits_{x \to 3^-} (x+1) = 4$, and (c) since the one-sided limits exist and are equal, $\lim\limits_{x \to 3^+} f(x) = \lim\limits_{x \to 3^-} f(x) = 4$, therefore the two-sided limit $\lim\limits_{x \to 3} f(x)$ exists and $\lim\limits_{x \to 3} f(x) = 4$. (Note that $f(x)$ is undefined at $x = 3$, but the function gets arbitrarily close to 4 as x approaches 3. Therefore the limit exists.) (See Figure 1.1-3.)

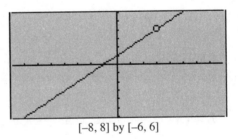

[–8, 8] by [–6, 6]

Figure 1.1-3

Example 2

Given $f(x)$ as illustrated in the accompanying diagram (Figure 1.1-4), find the limits (a) $\lim\limits_{x \to 0^+} f(x)$, (b) $\lim\limits_{x \to 0^-} f(x)$, and (c) $\lim\limits_{x \to 0} f(x)$.

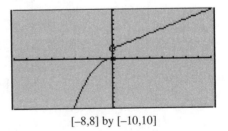

[–8,8] by [–10,10]

Figure 1.1-4

(a) As x approaches 0 from the left, $f(x)$ gets arbitrarily close to 0. Thus, $\lim\limits_{x \to 0^-} f(x) = 0$.

(b) As x approaches 0 from the right, $f(x)$ gets arbitrarily close to 2. Therefore, $\lim\limits_{x \to 0^+} f(x) = 2$. Note that $f(0) \neq 2$.

(c) Since $\lim\limits_{x \to 0^+} f(x) \neq \lim\limits_{x \to 0^-} f(x)$, $\lim\limits_{x \to 0} f(x)$ does not exist.

Example 3

Given the greatest integer function $f(x) = [x]$, find the limits (a) $\lim\limits_{x \to 1^+} f(x)$, (b) $\lim\limits_{x \to 1^-} f(x)$, and (c) $\lim\limits_{x \to 1} f(x)$.

(a) Enter $y1 = \text{int}(x)$ in your calculator. You see that as x approaches 1 from the right, the function stays at 1. Thus, $\lim\limits_{x \to 1^+} [x] = 1$. Note that $f(1)$ is also equal 1.

(b) As x approaches 1 from the left, the function stays at 0. Therefore, $\lim\limits_{x \to 1^-} [x] = 0$. Notice that $\lim\limits_{x \to 1^-} [x] \neq f(1)$.

(c) Since $\lim\limits_{x \to 1^-} [x] \neq \lim\limits_{x \to 1^+} [x]$, therefore, $\lim\limits_{x \to 1} [x]$ does not exist. (See Figure 1.1-5.)

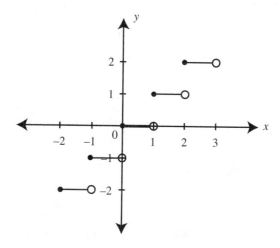

Figure 1.1-5

Example 4

Given $f(x) = \dfrac{|x|}{x}, x \neq 0$, find the limits (a) $\lim\limits_{x \to 0^+} f(x)$, (b) $\lim\limits_{x \to 0^-} f(x)$, and (c) $\lim\limits_{x \to 0} f(x)$.

(a) From inspecting the graph, $\lim\limits_{x \to 0^+} = \dfrac{|x|}{x} = 1$, (b) $\lim\limits_{x \to 0^-} = \dfrac{|x|}{x} = -1$ and (c) since $\lim\limits_{x \to 0^+} \dfrac{|x|}{x} \neq \lim\limits_{x \to 0^-} \dfrac{|x|}{x}$, therefore, $\lim\limits_{x \to 0} = \dfrac{|x|}{x}$ does not exist. (See Figure 1.1-6.)

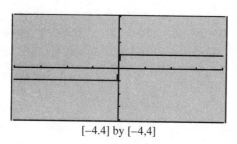

[−4.4] by [−4,4]

Figure 1.1-6

Example 5

If $f(x) = \begin{cases} e^{2x} & \text{for } -4 \le x < 0 \\ xe^x & \text{for } 0 \le x \le 4 \end{cases}$, find $\lim\limits_{x \to 0} f(x)$.

$\lim\limits_{x \to 0^+} f(x) = \lim\limits_{x \to 0^+} xe^x = 0$ and $\lim\limits_{x \to 0^-} f(x) = \lim\limits_{x \to 0^-} e^{2x} = 1$.

Thus $\lim\limits_{x \to 0} f(x)$ does not exist.

- Remember $\ln(e) = 1$ and $e^{\ln 3} = 3$ since $y = \ln x$ and $y = e^x$ are inverse functions.

Squeeze Theorem

If f, g, and h are functions defined on some open interval containing a such that $g(x) \le f(x) \le h(x)$ for all x in the interval except possibly at a itself, and $\lim\limits_{x \to a} g(x) = \lim\limits_{x \to a} h(x) = L$, then $\lim\limits_{x \to a} f(x) = L$.

Theorems on Limits

(1) $\lim\limits_{x \to 0} \dfrac{\sin x}{x} = 1$ and (2) $\lim\limits_{x \to 0} \dfrac{\cos x - 1}{x} = 0$

Example 1

Find the limit if it exists: $\lim\limits_{x \to 0} \dfrac{\sin 3x}{x}$

Substituting 0 into the expression would lead to $\dfrac{0}{0}$. Rewrite $\dfrac{\sin 3x}{x}$ as $\dfrac{3}{3} \cdot \dfrac{\sin 3x}{x}$ and thus, $\lim\limits_{x \to 0} \dfrac{\sin 3x}{x} = \lim\limits_{x \to 0} \dfrac{3 \sin 3x}{3x} = 3 \lim\limits_{x \to 0} \dfrac{\sin 3x}{3x}$. As x approaches 0, so does $3x$. Therefore, $3 \lim\limits_{x \to 0} \dfrac{\sin 3x}{3x} = 3 \lim\limits_{3x \to 0} \dfrac{\sin 3x}{3x} = 3(1) = 3$. (Note that $\lim\limits_{3x \to 0} \dfrac{\sin 3x}{3x}$ is equivalent to $\lim\limits_{x \to 0} \dfrac{\sin x}{x}$ by replacing $3x$ by x.) Verify your result with a calculator. (See Figure 1.1-7.)]

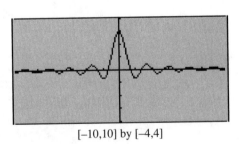

[–10,10] by [–4,4]

Figure 1.1-7

Example 2

Find the limit if it exists: $\lim\limits_{h \to 0} \dfrac{\sin 3h}{\sin 2h}$.

Rewrite $\dfrac{\sin 3h}{\sin 2h}$ as $\dfrac{3\left(\dfrac{\sin 3h}{3h}\right)}{2\left(\dfrac{\sin 2h}{2h}\right)}$. As h approaches 0, so do $3h$ and $2h$. Therefore,

$\lim\limits_{h \to 0} \dfrac{\sin 3h}{\sin 2h} = \dfrac{3 \lim\limits_{3h \to 0} \dfrac{\sin 3h}{3h}}{2 \lim\limits_{2h \to 0} \dfrac{\sin 2h}{2h}} = \dfrac{3(1)}{2(1)} = \dfrac{3}{2}$. (Note that substituting $h = 0$ into the original expression would have produced 0/0). Verify your result with a calculator. (See Figure 1.1-8.)

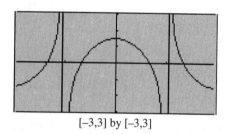

[–3,3] by [–3,3]

Figure 1.1-8

Example 3

Find the limit if it exists: $\lim\limits_{y \to 0} \dfrac{y^2}{1 - \cos y}$.

Substituting 0 in the expression would lead to 0/0. Multiplying both the numerator and denominator by the conjugate $(1 + \cos y)$ produces $\lim\limits_{y \to 0} \dfrac{y^2}{1 - \cos y} \cdot \dfrac{(1 + \cos y)}{(1 + \cos y)} =$

$\lim\limits_{y \to 0} \dfrac{y^2(1 + \cos y)}{1 - \cos^2 y} = \lim\limits_{y \to 0} \dfrac{y^2(1 + \cos y)}{\sin^2 y} = \lim\limits_{y \to 0} \dfrac{y^2}{\sin^2 y} \cdot \lim\limits_{y \to 0} (1 + \cos^2 y) = \lim\limits_{y \to 0} \left(\dfrac{y}{\sin y}\right)^2 \cdot$

$\lim\limits_{y \to 0} (1 + \cos^2 y) = \left(\lim\limits_{y \to 0} \dfrac{y}{\sin y}\right)^2 \cdot \lim\limits_{y \to 0} (1 + \cos^2 y) = (1)^2(1 + 1) = 2$. (Note that

$\lim\limits_{y \to 0} \dfrac{y}{\sin y} = \lim\limits_{y \to 0} \dfrac{1}{\dfrac{\sin y}{y}} = \dfrac{\lim\limits_{y \to 0} (1)}{\lim\limits_{y \to 0} \dfrac{\sin y}{y}} = \dfrac{1}{1} = 1$). Verify your result with a calculator.

(See Figure 1.1-9.)

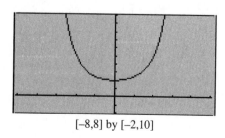

[–8,8] by [–2,10]

Figure 1.1-9

Example 4

Find the limit if it exists: $\lim\limits_{x\to 0} \dfrac{3x}{\cos x}$.

Using the quotient rule for limits, you have $\lim\limits_{x\to 0} \dfrac{3x}{\cos x} = \dfrac{\lim\limits_{x\to 0}(3x)}{\lim\limits_{x\to 0}(\cos x)} = \dfrac{0}{1} = 0$. Verify

your result with a calculator. (See Figure 1.1-10.)

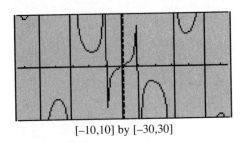

[−10,10] by [−30,30]

Figure 1.1-10

1.2 LIMITS INVOLVING INFINITIES

Main Concepts: *Infinite Limits (as* x → a*), Limits at Infinity (as* x → ∞*), Horizontal and Vertical Asymptotes*

Infinite Limits (as $x \to a$)

If f is a function defined at every number in some open interval containing a, except possibly at a itself, then

(1) $\lim\limits_{x\to a} f(x) = \infty$ means that $f(x)$ increases without bound as x approaches a.

(2) $\lim\limits_{x\to a} f(x) = -\infty$ means that $f(x)$ decreases without bound as x approaches a.

Limit Theorems

(1) If n is a positive integer, then

    (a) $\lim\limits_{x\to 0^+} \dfrac{1}{x^n} = \infty$

    (b) $\lim\limits_{x\to 0^-} \dfrac{1}{x^n} = \begin{cases} \infty & \text{if } n \text{ is even} \\ -\infty & \text{if } n \text{ is odd} \end{cases}$

(2) If the $\lim\limits_{x\to a} f(x) = c,\ c > 0$, and $\lim\limits_{x\to a} g(x) = 0$, then

$$\lim\limits_{x\to a} \frac{f(x)}{g(x)} = \begin{cases} \infty & \text{if } g(x) \text{ approaches } 0 \text{ through positive values} \\ -\infty & \text{if } g(x) \text{ approaches } 0 \text{ through negative values} \end{cases}$$

(3) If the $\lim\limits_{x \to a} f(x) = c$, $c < 0$, and $\lim\limits_{x \to a} g(x) = 0$, then

$$\lim\limits_{x \to a} \frac{f(x)}{g(x)} = \begin{cases} -\infty & \text{if } g(x) \text{ approaches 0 through positive values} \\ \infty & \text{if } g(x) \text{ approaches 0 through negative values} \end{cases}$$

(Note that limit theorems 2 and 3 hold true for $x \to a^+$ and $x \to a^-$.)

Example 1

Evaluate the limit: (a) $\lim\limits_{x \to 2^+} \dfrac{3x - 1}{x - 2}$ and (b) $\lim\limits_{x \to 2^-} \dfrac{3x - 1}{x - 2}$.

The limit of the numerator is 5 and the limit of the denominator is 0 through positive values. Thus, $\lim\limits_{x \to 2^+} \dfrac{3x - 1}{x - 2} = \infty$. (b) The limit of the numerator is 5 and the limit of the denominator is 0 through negative values. Therefore, $\lim\limits_{x \to 2^-} \dfrac{3x - 1}{x - 2} = -\infty$. Verify your result with a calculator. (See Figure 1.2-1.)

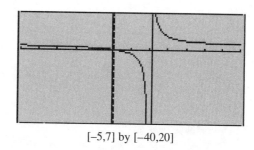

[−5,7] by [−40,20]

Figure 1.2-1

Example 2

Find: $\lim\limits_{x \to 3^-} \dfrac{x^2}{x^2 - 9}$

Factor the denominator obtaining $\lim\limits_{x \to 3^-} \dfrac{x^2}{x^2 - 9} = \lim\limits_{x \to 3^-} \dfrac{x^2}{(x - 3)(x + 3)}$. The limit of the numerator is 9 and the limit of the denominator is $(0)(6) = 0$ through negative values. Therefore, $\lim\limits_{x \to 3^-} \dfrac{x^2}{x^2 - 9} = -\infty$. Verify your result with a calculator. (See Figure 1.2-2.)

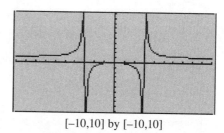

[−10,10] by [−10,10]

Figure 1.2-2

Example 3

Find: $\lim\limits_{x \to 5^-} \dfrac{\sqrt{25 - x^2}}{x - 5}$

Substituting 5 into the expression leads to 0/0. Factor the numerator $\sqrt{25 - x^2}$ into $\sqrt{(5 - x)(5 + x)}$. As $x \to 5^-, (x - 5) < 0$. Rewrite $(x - 5)$ as $-(5 - x)$

as $x \to 5^-$, $(5 - x) > 0$ and thus, you may express $(5 - x)$ as $\sqrt{(5 - x)^2} = \sqrt{(5 - x)(5 - x)}$. Therefore, $(x - 5) = -(5 - x) = -\sqrt{(5 - x)(5 - x)}$. Substituting these equivalent expressions into the original problem, you have $\lim\limits_{x \to 5^-} \dfrac{\sqrt{25 - x^2}}{x - 5} =$

$\lim\limits_{x \to 5^-} \dfrac{\sqrt{(5 - x)(5 + x)}}{\sqrt{(5 - x)(5 - x)}} = -\lim\limits_{x \to 5^-} \dfrac{\sqrt{(5 - x)(5 + x)}}{(5 - x)(5 - x)} = -\lim\limits_{x \to 5^-} \sqrt{\dfrac{(5 + x)}{(5 - x)}}$. The limit of the numerator is 10 and the limit of the denominator is 0 through positive values. Thus,

the $\lim\limits_{x \to 5^-} \dfrac{\sqrt{25 - x^2}}{x - 5} = -\infty$.

Example 4

Find: $\lim\limits_{x \to 2^-} \dfrac{[x] - x}{2 - x}$, where $[x]$ is the greatest integer value of x.

As $x \to 2^-$, $[x] = 1$. The limit of the numerator is $(1 - 2) = -1$. As $x \to 2^-$, $(2 - x) = 0$ through positive values. Thus, $\lim\limits_{x \to 2^-} \dfrac{[x] - x}{2 - x} = -\infty$.

- Do easy questions first. The easy ones are worth the same number of points as the hard ones.

Limits at Infinity (as $x \to \pm\infty$)

If f is a function defined at every number in some interval (a, ∞), then $\lim\limits_{x \to \infty} f(x) = L$ means that L is the limit of $f(x)$ as x increases without bound.

If f is a function defined at every number in some interval $(-\infty, a)$, then $\lim\limits_{x \to -\infty} f(x) = L$ means that L is the limit of $f(x)$ as x decreases without bound.

Limit Theorem

If n is a positive integer, then

(a) $\lim\limits_{x \to \infty} \dfrac{1}{x^n} = \infty$

(b) $\lim\limits_{x \to -\infty} \dfrac{1}{x^n} = 0$

Example 1

Evaluate the limit: $\lim\limits_{x \to \infty} \dfrac{6x - 13}{2x + 5}$

Divide every term in the numerator and denominator by the highest power of x, and in this case, it is x and obtain:

$$\lim\limits_{x \to \infty} \frac{6x - 13}{2x + 5} = \lim\limits_{x \to \infty} \frac{6 - \dfrac{13}{x}}{2 + \dfrac{5}{x}} = \frac{\lim\limits_{x \to \infty}(6) - \lim\limits_{x \to \infty}\dfrac{13}{x}}{\lim\limits_{x \to \infty}(2) + \lim\limits_{x \to \infty}\left(\dfrac{5}{x}\right)} = \frac{\lim\limits_{x \to \infty}(6) - 13\lim\limits_{x \to \infty}\left(\dfrac{1}{x}\right)}{\lim\limits_{x \to \infty}(2) + 5\lim\limits_{x \to \infty}\left(\dfrac{1}{x}\right)}$$

$$= \frac{6 - 13(0)}{2 + 5(0)} = 3.$$

Verify your result with a calculator. (See Figure 1.2-3.)

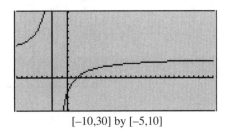

[−10,30] by [−5,10]

Figure 1.2-3

Example 2

Evaluate the limit: $\displaystyle\lim_{x \to -\infty} \frac{3x - 10}{4x^3 + 5}$

Divide every term in the numerator and denominator by the highest power of x and in this case, it is x^3. Thus, $\displaystyle\lim_{x \to -\infty} \frac{3x - 10}{4x^3 + 5} = \lim_{x \to -\infty} \frac{\dfrac{3}{x^2} - \dfrac{10}{x^3}}{4 + \dfrac{5}{x^3}} = \frac{0 - 0}{4 + 0} = 0.$

Verify your result with a calculator. (See Figure 1.2-4.)

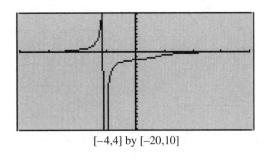

[−4,4] by [−20,10]

Figure 1.2-4

Example 3

Evaluate the limit: $\displaystyle\lim_{x \to \infty} \frac{1 - x^2}{10x + 7}$

Divide every term in the numerator and denominator by the highest power of x and in this case, it is x^2. Therefore, $\displaystyle\lim_{x \to \infty} \frac{1 - x^2}{10x + 7} = \lim_{x \to \infty} \frac{\dfrac{1}{x^2} - 1}{\dfrac{10}{x} + \dfrac{7}{x^2}} = \frac{\displaystyle\lim_{x \to \infty}\left(\dfrac{1}{x^2}\right) - \lim_{x \to \infty}(1)}{\displaystyle\lim_{x \to \infty}\left(\dfrac{10}{x}\right) + \lim_{x \to \infty}\dfrac{7}{x^2}}.$

The limit of the numerator is -1 and the limit of the denominator is 0. Thus, $\displaystyle\lim_{x \to \infty} \frac{1 - x^2}{10x + 7} = -\infty.$ Verify your result with a calculator. (See Figure 1.2-5.)

Example 4

Evaluate the limit: $\displaystyle\lim_{x \to -\infty} \frac{2x + 1}{\sqrt{x^2 + 3}}$

As $x \to -\infty$, $x < 0$ and thus, $x = -\sqrt{x^2}$. Divide the numerator and denominator by x (not x^2 since the denominator has a square root). Thus, you have $\displaystyle\lim_{x \to -\infty} \frac{2x + 1}{\sqrt{x^2 + 3}}$

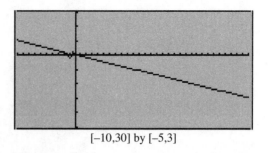

[−10,30] by [−5,3]

Figure 1.2-5

$$= \lim_{x \to -\infty} \frac{\dfrac{2x+1}{x}}{\dfrac{\sqrt{x^2+3}}{x}}. \text{ Replacing the } x \text{ below } \sqrt{x^2+3} \text{ by } (-\sqrt{x^2}), \text{ you have } \lim_{x \to -\infty} \frac{2x+1}{\sqrt{x^2+3}}$$

$$= \lim_{x \to -\infty} \frac{\dfrac{2x+1}{x}}{\dfrac{\sqrt{x^2+3}}{-\sqrt{x^2}}} \lim_{x \to -\infty} \frac{2 + \dfrac{1}{x}}{-\sqrt{1 + \dfrac{3}{x^2}}} = \frac{\lim\limits_{x \to -\infty}(2) - \lim\limits_{x \to -\infty}\dfrac{1}{x}}{-\sqrt{\lim\limits_{x \to -\infty}(1) + \lim\limits_{x \to -\infty}\left(\dfrac{3}{x^2}\right)}} = \frac{2}{-1} = -2.$$

Verify your result with a calculator. (See Figure 1.2-6.)

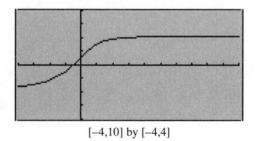

[−4,10] by [−4,4]

Figure 1.2-6

- Remember that $\ln\left(\dfrac{1}{x}\right) = \ln(1) - \ln x = -\ln x$ and $y = e^{-x} = \dfrac{1}{e^x}$.

Vertical and Horizontal Asymptotes

A line $y = b$ is called a horizontal asymptote for the graph of a function f if either $\lim\limits_{x \to \infty} f(x) = b$ or $\lim\limits_{x \to -\infty} f(x) = b$.

A line $x = a$ is called a vertical asymptote for the graph of a function f if either $\lim\limits_{x \to a^+} f(x) = +\infty$ or $\lim\limits_{x \to a^-} f(x) = +\infty$.

Example 1

Find the horizontal and vertical asymptotes of the function $f(x) = \dfrac{3x + 5}{x - 2}$.

To find the horizontal asymptotes, examine the $\lim\limits_{x \to \infty} f(x)$ and the $\lim\limits_{x \to -\infty} f(x)$.

The $\lim\limits_{x \to \infty} f(x) = \lim\limits_{x \to \infty} \dfrac{3x+5}{x-2} = \lim\limits_{x \to \infty} \dfrac{3+\dfrac{5}{x}}{1-\dfrac{2}{x}} = \dfrac{3}{1} = 3$, and the $\lim\limits_{x \to -\infty} f(x) =$

$\lim\limits_{x \to -\infty} \dfrac{3x+5}{x-2} = \lim\limits_{x \to -\infty} \dfrac{3+\dfrac{5}{x}}{1-\dfrac{2}{x}} = \dfrac{3}{1} = 3.$

Thus, $y = 3$ is a horizontal asymptote.
To find the vertical asymptotes, look for x values such that the denominator $(x-2)$ would be 0, in this case, $x = 2$. Then examine:

(a) $\lim\limits_{x \to 2^+} f(x) = \lim\limits_{x \to 2^+} \dfrac{3x+5}{x-2} = \dfrac{\lim\limits_{x \to 2^+} (3x+5)}{\lim\limits_{x \to 2^+} (x-2)}$, the limit of the numerator is 11

and the limit of the denominator is 0 through positive values, and thus, $\lim\limits_{x \to 2^+} \dfrac{3x+5}{x-2} = \infty.$

(b) $\lim\limits_{x \to 2^-} f(x) = \lim\limits_{x \to 2^-} \dfrac{3x+5}{x-2} = \dfrac{\lim\limits_{x \to 2^-} (3x+5)}{\lim\limits_{x \to 2^-} (x-2)}$, the limit of the numerator is 11

and the limit of the denominator is 0 through negative values, and thus, $\lim\limits_{x \to 2^-} \dfrac{3x+5}{x-2} = -\infty.$

Therefore, $x = 2$ is a vertical asymptote.

Example 2

Using your calculator, find the horizontal and vertical asymptotes of the function $f(x) = \dfrac{x}{x^2 - 4}.$

Enter $y1 = \dfrac{x}{x^2 - 4}$. The graphs shows that as $x \to \pm\infty$, the function approaches 0, thus $\lim\limits_{x \to \infty} f(x) = \lim\limits_{x \to -\infty} f(x) = 0$. Therefore, a horizontal asymptote is $y = 0$ (or the x-axis).

For vertical asymptotes, you notice that $\lim\limits_{x \to 2^+} f(x) = \infty$, $\lim\limits_{x \to 2^-} f(x) = -\infty$, and $\lim\limits_{x \to -2^+} f(x) = \infty$, $\lim\limits_{x \to -2^-} f(x) = -\infty$. Thus, the vertical asymptotes are $x = -2$ and $x = 2$. (See Figure 1.2-7.)

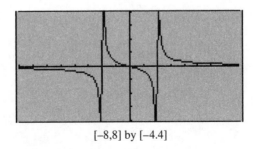

[–8,8] by [–4.4]

Figure 1.2-7

Example 3

Using your calculator, find the horizontal and vertical asymptotes of the function $f(x) = \dfrac{x^3 + 5}{x}.$

Enter $y1 = \dfrac{x^3 + 5}{x}$. The graph of $f(x)$ shows that as x increases in the first quadrant, $f(x)$ goes higher and higher without bound. As x moves to the left in the 2nd quadrant, $f(x)$ again goes higher and higher without bound. Thus, you may conclude that $\lim\limits_{x \to \infty} f(x) = \infty$ and $\lim\limits_{x \to -\infty} f(x) = \infty$ and thus, $f(x)$ has no horizontal asymptote. For vertical asymptotes, you notice that $\lim\limits_{x \to 0^+} f(x) = \infty$, and $\lim\limits_{x \to 0^-} f(x) = -\infty$, Therefore, the line $x = 0$ (or the y-axis) is a vertical asymptote. (See Figure 1.2-8.)

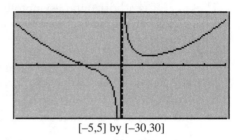

[−5,5] by [−30,30]

Figure 1.2-8

Relationship between the limits of rational functions as $x \to \infty$ and horizontal asymptotes:

Given $f(x) = \dfrac{p(x)}{q(x)}$, then

(1) If the degree of $p(x)$ is same as the degree of $q(x)$, then $\lim\limits_{x \to \infty} f(x) = \lim\limits_{x \to -\infty} f(x) = \dfrac{a}{b}$ where a is the coefficient of the highest power of x in $p(x)$ and b is the coefficient of the highest power of x in $q(x)$. The line $y = \dfrac{a}{b}$ is a horizontal asymptote. See example 1, above.

(2) If the degree of $p(x)$ is smaller than the degree of $q(x)$, then $\lim\limits_{x \to \infty} f(x) = \lim\limits_{x \to -\infty} f(x) = 0$. The line $y = 0$ (or x-axis) is a horizontal asymptote. See example 2, above.

(3) If the degree of $p(x)$ is greater than the degree of $q(x)$, then $\lim\limits_{x \to \infty} f(x) = \pm\infty$ and $\lim\limits_{x \to -\infty} f(x) = \pm\infty$. Thus, $f(x)$ has no horizontal asymptote. See example 3, above.

Example 4

Using your calculator, find the horizontal asymptotes of the function $f(x) = \dfrac{2 \sin x}{x}$.

Enter $y1 = \dfrac{2 \sin x}{x}$. The graph shows that $f(x)$ oscillates back and forth about the x-axis. As $x \to \pm\infty$, the graph gets closer and closer to the x-axis which implies that $f(x)$ approaches 0. Thus, the line $y = 0$ (or the x-axis) is a horizontal asymptote. (See Figure 1.2-9.)

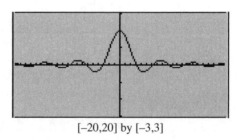

[−20,20] by [−3,3]

Figure 1.2-9

> • When entering a rational function into a calculator, use parentheses for both the numerator and denominator, e.g., $(x - 2) + (x + 3)$.

1.3 CONTINUITY OF A FUNCTION

Continuity of a Function at a Number: A function f is said to be continuous at a number a if the following three conditions are satisfied:

1. $f(a)$ exists
2. $\lim_{x \to a} f(x)$ exists
3. $\lim_{x \to a} f(x) = f(a)$

The function f is said to be discontinuous at a if one or more of these three conditions are not satisfied and a is called the point of discontinuity.

Continuity of a Function over an Interval: A function is continuous over an interval if it is continuous at every point in the interval.

Theorems on Continuity

1. If the functions f and g are continuous at a, then the functions $f + g, f - g, f \cdot g$ and f/g, $g(a) \neq 0$, are also continuous at a.
2. A polynomial function is continuous everywhere.
3. A rational function is continuous everywhere, except at points where the denominator is zero.
4. *Intermediate-Value Theorem:* If a function f is continuous on a closed interval $[a, b]$ and k is a number with $f(a) \leq k \leq f(b)$, then there exists a number c in $[a, b]$ such that $f(c) = k$.

Example 1

Find the points of discontinuity of the function $f(x) = \dfrac{x + 5}{x^2 - x - 2}$.

Since $f(x)$ is a rational function, it is continuous everywhere, except at points where the denominator is 0. Factor the denominator and set it equal to 0: $(x - 2)(x + 1) = 0$. Thus $x = 2$ or $x = -1$. The function $f(x)$ is undefined at $x = -1$ and at $x = 2$. Therefore, $f(x)$ is discontinuous at these points. Verify your result with a calculator. (See Figure 1.3-1.)

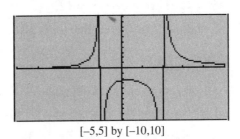

[−5,5] by [−10,10]

Figure 1.3-1

Example 2

Determine the intervals on which the given function is continuous:

$$f(x) = \begin{cases} \dfrac{x^2 + 3x - 10}{x - 2}, & x \neq 2 \\ 10, & x = 2 \end{cases}$$

Check the three conditions of continuity at $x = 2$:

Condition 1: $f(2) = 10$.

Condition 2: $\displaystyle\lim_{x \to 2} \frac{x^2 + 3x - 10}{x - 2} = \lim_{x \to 2} \frac{(x + 5)(x - 2)}{x - 2} = \lim_{x \to 2} (x + 5) = 7$.

Condition 3: $f(2) \neq \displaystyle\lim_{x \to 2} f(x)$. Thus, $f(x)$ is discontinuous at $x = 2$.

The function is continuous on $(-\infty, 2)$ and $(2, \infty)$. Verify your result with a calculator. (See Figure 1.3-2.)

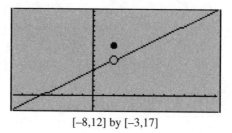

[−8,12] by [−3,17]

Figure 1.3-2

- Remember that $\dfrac{d}{dx}\left(\dfrac{1}{x}\right) = -\dfrac{1}{x^2}$ and $\displaystyle\int \dfrac{1}{x}\,dx = \ln|x| + c$.

Example 3

For what value of k is the function $f(x) = \begin{cases} x^2 - 2x, & x \leq 6 \\ 2x + k, & x > 6 \end{cases}$ continuous at $x = 6$?

For $f(x)$ to be continuous at $x = 6$, it must satisfy the three conditions of continuity.

Condition 1: $f(6) = 6^2 - 2(6) = 24$.

Condition 2: $\displaystyle\lim_{x \to 6^-} (x^2 - 2x) = 24$; thus $\displaystyle\lim_{x \to 6^-} (2x + k)$ must also be 24 in order for the $\displaystyle\lim_{x \to 6} f(x)$ to equal 24. Thus, $\displaystyle\lim_{x \to 6^-} (2x + k) = 24$ which implies $2(6) + k = 24$ and $k = 12$. Therefore, if $k = 12$,

Condition (3): $f(6) = \displaystyle\lim_{x \to 6} f(x)$ is also satisfied.

Example 4

Given $f(x)$ as shown in Figure 1.3-3, (a) find $f(3)$ and $\displaystyle\lim_{x \to 3} f(x)$, and (b) determine if $f(x)$ is continuous at $x = 3$? Explain why.

The graph of $f(x)$ shows that $f(3) = 5$ and the $\displaystyle\lim_{x \to 3} f(x) = 1$. Since $f(3) \neq \displaystyle\lim_{x \to 3} f(x)$, $f(x)$ is discontinuous at $x = 3$.

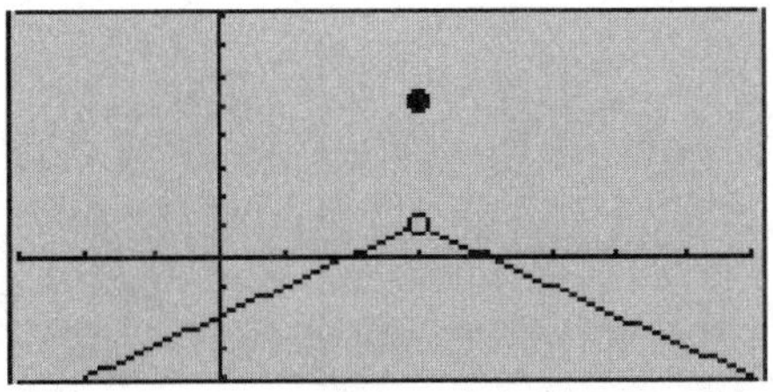

[–3,8] by [–4,8]

Figure 1.3-3

Example 5

If $g(x) = x^2 - 2x - 15$, using the Intermediate Value Theorem show that $g(x)$ has a root in the interval $[1, 7]$.

Begin by finding $g(1)$ and $g(7)$, and $g(1) = -16$ and $g(7) = 20$. If $g(x)$ has a root, then $g(x)$ crosses the x-axis, i.e., $g(x) = 0$. Since $-16 \leq 0 \leq 20$, by the Intermediate Value Theorem, there exists at least one number c in $[1, 7]$ such that $g(c) = 0$. The number c is a root of $g(x)$.

Example 6

A function f is continuous on $[0, 5]$ and some of the values of f are shown below.

x	0	3	5
f	–4	b	–4

If $f(x) = -2$ has no solution on $[0, 5]$ then b could be

(A) 3   (B) 1   (C) 0   (D) –2   (E) –5

If $b = -2$, then $x = 3$ would be a solution for $f(x) = -2$.

If $b = 0, 1$, or 3, $f(x) = -2$ would have two solutions for $f(x) = -2$.

Thus, $b = -5$, choice (E). (See Figure 1.3-4.)

1.4 RAPID REVIEW

1. Find $f(2)$ and $\lim\limits_{x \to 2} f(x)$ and determine if f is continuous at $x = 2$. (See Figure 1.4-1 on page 54.)

   Answer: $f(2) = 2$, $\lim\limits_{x \to 2} f(x) = 4$, and f is discontinuous at $x = 2$.

2. Evaluate $\lim\limits_{x \to a} \dfrac{x^2 - a^2}{x - a}$.

   Answer: $\lim\limits_{x \to a} \dfrac{(x + a)(x - a)}{x - a} = 2a$.

3. Evaluate $\lim\limits_{x \to \infty} \dfrac{1 - 3x^2}{x^2 + 100x + 99}$.

   Answer: The limit is -3, since the polynomials in the numerator and denominator have the same degree.

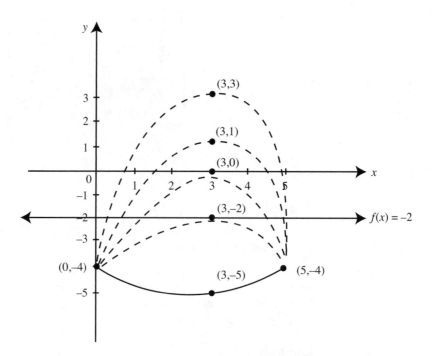

Figure 1.3-4

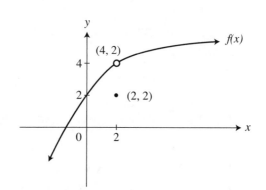

Figure 1.4-1

4. Determine if $f(x) = \begin{cases} x + 6 & \text{for } x < 3 \\ x^2 & \text{for } x \geq 3 \end{cases}$ is continuous at $x = 3$.

   Answer: The function f is continuous, since $f(3) = 9$, $\lim\limits_{x \to 3^+} f(x) = \lim\limits_{x \to 3^-} f(x) = 9$ and $f(3) = \lim\limits_{x \to 3} f(x)$.

5. If $f(x) = \begin{cases} e^x & \text{for } x \neq 0 \\ 5 & \text{for } x = 0 \end{cases}$, find $\lim\limits_{x \to 0} f(x)$.

   Answer: $\lim\limits_{x \to 0} f(x) = 1$, since $\lim\limits_{x \to 0^+} f(x) = \lim\limits_{x \to 0^-} f(x) = 1$.

6. Evaluate $\lim\limits_{x \to 0} \dfrac{\sin 6x}{\sin 2x}$.

   Answer: The limit is $\dfrac{6}{2} = 3$, since $\lim\limits_{x \to 0} \dfrac{\sin x}{x} = 1$.

7. Evaluate $\lim\limits_{x \to 5^-} \dfrac{x^2}{x^2 - 25}$.

Answer: The limit is $-\infty$, since $(x^2 - 25)$ approaches 0 through negative values.

8. Find the vertical and horizontal asymptotes of $f(x) = \dfrac{1}{x^2 - 25}$.

Answer: The vertical asymptotes are $x = \pm 5$, and the horizontal asymptote is $y = 0$, since $\lim\limits_{x \to \pm\infty} f(x) = 0$.

 # 1.5 PRACTICE PROBLEMS

Part A—The use of a calculator is not allowed.

Find the limits of the following:

1. $\lim\limits_{x \to 0} (x - 5) \cos x$

2. If $b \neq 0$, evaluate $\lim\limits_{x \to b} \dfrac{x^3 - b^3}{x^6 - b^6}$

3. $\lim\limits_{x \to 0} \dfrac{2 - \sqrt{4 - x}}{x}$

4. $\lim\limits_{x \to \infty} \dfrac{5 - 6x}{2x + 11}$

5. $\lim\limits_{x \to -\infty} \dfrac{x^2 + 2x - 3}{x^3 + 2x^2}$

6. $\lim\limits_{x \to \infty} \dfrac{3x^2}{5x + 8}$

7. $\lim\limits_{x \to -\infty} \dfrac{3x}{\sqrt{x^2 - 4}}$

8. If $f(x) = \begin{cases} e^x & \text{for } 0 \leq x < 1 \\ x^2 e^x & \text{for } 1 \leq x \leq 5 \end{cases}$, find $\lim\limits_{x \to 1} f(x)$.

9. $\lim\limits_{x \to \infty} \dfrac{e^x}{1 - x^3}$

10. $\lim\limits_{x \to 0} \dfrac{\sin 3x}{\sin 4x}$

11. $\lim\limits_{x \to 3^+} \dfrac{\sqrt{t^2 - 9}}{t - 3}$

12. The graph of a function f is shown in Figure 1.5-1.
Which of the following statements is/are true?

    I. $\lim\limits_{x \to 4^-} f(x) = 5$

    II. $\lim\limits_{x \to 4} f(x) = 2$

    III. $x = 4$ is not in the domain of f

Part B—Calculators are allowed.

13. Find the horizontal and vertical asymptotes of the graph of the function $f(x) = \dfrac{1}{x^2 + x - 2}$.

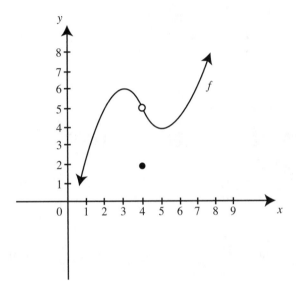

Figure 1.5-1

14. Find the limit: $\lim\limits_{x \to 5^+} \dfrac{5 + [x]}{5 - x}$ when $[x]$ is the greatest integer of x.

15. Find the points of discontinuity of the function
$$f(x) = \dfrac{x + 1}{x^2 + 4x - 12}.$$

16. For what value of k is the function
$$g(x) = \begin{cases} x^2 + 5, & x \leq 3 \\ 2x - k, & x > 3 \end{cases} \text{ continuous at } x = 3?$$

17. Determine if
$$f(x) = \begin{cases} \dfrac{x^2 + 5x - 14}{x - 2}, & \text{if } x \neq 2 \\ 12, & \text{if } x = 2 \end{cases}$$
is continuous at $x = 2$. Explain why or why not.

18. Given $f(x)$ as shown in Figure 1.5-2, find

    (a) $f(3)$

    (b) $\lim\limits_{x \to 3^+} f(x)$

    (c) $\lim\limits_{x \to 3^-} f(x)$

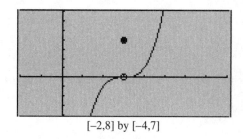

[−2,8] by [−4,7]

Figure 1.5-2

(d) $\lim\limits_{x\to 3} f(x)$

(e) Is $f(x)$ continuous at $x = 3$? Explain why or why not.

19. A function f is continuous on $[-2, 2]$ and some of the values of f are shown below:

x	−2	0	2
$f(x)$	3	b	4

If f has only one root, r, on the closed interval $[-2, 2]$, and $r \neq 0$, then a possible value of b is

(A) −3   (B) −2   (C) −1   (D) 0   (E) 1

20. Evaluate $\lim\limits_{x\to 0} \dfrac{1 - \cos x}{\sin^2 x}$.

 1.6   CUMULATIVE REVIEW PROBLEMS

21. Write an equation of the line passing through the point $(2, -4)$ and perpendicular to the line $3x - 2y = 6$.

22. The graph of a function f is shown in Figure 1.6-1. Which of the following statements is/are true?

   I. $\lim\limits_{x\to 4^-} f(x) = 3$.

   II. $x = 4$ is not in the domain of f.

   III. $\lim\limits_{x\to 4} f(x)$ does not exist.

23. Evaluate $\lim\limits_{x\to 0} \dfrac{|3x - 4|}{x - 2}$

24. Find $\lim\limits_{x\to 0} \dfrac{\tan x}{x}$.

25. Find the horizontal and vertical asymptotes of $f(x) = \dfrac{x}{\sqrt{x^2 + 4}}$.

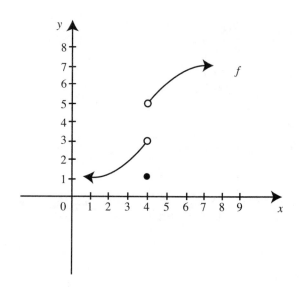

Figure 1.6-1

1.7   SOLUTIONS TO PRACTICE PROBLEMS

Part A—No calculators are permitted.

1. Using the product rule,
$$\lim_{x\to 0}(x - 5)(\cos x) = \left[\lim_{x\to 0}(x - 5)\right]\left[\lim_{x\to 0}(\cos x)\right]$$
$$= (0 - 5)(\cos 0) = (-5)(1) = -5.$$
(Note that $\cos 0 = 1$.)

2. Rewrite $\lim\limits_{x\to b} \dfrac{x^3 - b^3}{x^6 - b^6}$ as

$$\lim_{x\to b} \frac{x^3 - b^3}{(x^3 - b^3)(x^3 + b^3)} = \lim_{x\to b} \frac{1}{x^3 + b^3}.$$

Substitute $x = b$ and obtain $\dfrac{1}{b^3 + b^3} = \dfrac{1}{2b^3}$.

3. Substituting $x = 0$ into the expression $\dfrac{2 - \sqrt{4 - x}}{x}$ leads to 0/0 which is an indeterminate form. Thus, multiply both the

numerator and denominator by the conjugate $\left(2 + \sqrt{4 - x}\right)$ and obtain

$$\lim_{x \to 0} \frac{2 - \sqrt{4 - x}}{x} \left(\frac{2 + \sqrt{4 - x}}{2 + \sqrt{4 - x}}\right)$$

$$= \lim_{x \to 0} \frac{4 - (4 - x)}{x\left(2 + \sqrt{4 - x}\right)}$$

$$= \lim_{x \to 0} \frac{x}{x\left(2 + \sqrt{4 - x}\right)} = \lim_{x \to 0} \frac{1}{\left(2 + \sqrt{4 - x}\right)}$$

$$= \frac{1}{\left(2 + \sqrt{4 - (0)}\right)} = \frac{1}{4}.$$

4. Since the degree of the polynomial in the numerator is the same as the degree of the polynomial in the denominator,
$$\lim_{x \to \infty} \frac{5 - 6x}{2x + 11} = -\frac{6}{2} = -3.$$

5. Since the degree of the polynomial in the numerator is 2 and the degree of the polynomial in the denominator is 3, $\lim\limits_{x \to -\infty} \dfrac{x^2 + 2x - 3}{x^3 + 2x^2} = 0.$

6. The degree of the monomial in the numerator is 2 and the degree of the binomial in the denominator is 1. Thus, $\lim\limits_{x \to \infty} \dfrac{3x^2}{5x + 8} = \infty.$

7. Divide every term in both the numerator and denominator by the highest power of x and in this case, it is x. Thus, you have $\lim\limits_{x \to -\infty} \dfrac{\frac{3x}{x}}{\sqrt{x^2 - 4}}$.

As $x \to \infty, x = -\sqrt{x^2}$. Since the denominator involves a radical, rewrite the expression as

$$\lim_{x \to -\infty} \frac{\frac{3x}{x}}{\frac{\sqrt{x^2 - 4}}{-\sqrt{x^2}}} = \lim_{x \to -\infty} \frac{3}{-\sqrt{1 - \frac{4}{x^2}}}$$

$$= \frac{3}{-\sqrt{1 - 0}} = -3$$

8. $\lim\limits_{x \to 1^+} f(x) = \lim\limits_{x \to 1^+} \left(x^2 e^x\right) = e$ and $\lim\limits_{x \to 1^-} f(x) = \lim\limits_{x \to 1^-} \left(e^x\right) = e.$ Thus, $\lim\limits_{x \to 1} f(x) = e.$

9. $\lim\limits_{x \to \infty} e^x = \infty$ and $\lim\limits_{x \to \infty} \left(1 - x^3\right) = \infty.$ However, as $x \to \infty$, the rate of increase of e^x is much greater than the rate of decrease of $(1 - x^3)$. Thus, $\lim\limits_{x \to \infty} \dfrac{e^x}{1 - x^3} = -\infty.$

10. Divide both numerator and denominator by x and obtain $\lim\limits_{x \to 0} \dfrac{\frac{\sin 3x}{x}}{\frac{\sin 4x}{x}}$. Now rewrite the limit as

$$\lim_{x \to 0} \frac{3 \frac{\sin 3x}{3x}}{4 \frac{\sin 4x}{4x}} = \frac{3}{4} \lim_{x \to 0} \frac{\frac{\sin 3x}{3x}}{\frac{\sin 4x}{4x}}.$$ As x approaches 0, so do $3x$ and $4x$. Thus, you have

$$\frac{3}{4} \frac{\lim\limits_{3x \to 0} \frac{\sin 3x}{3x}}{\lim\limits_{4x \to 0} \frac{\sin 4x}{4x}} = \frac{3(1)}{4(1)} = \frac{3}{4}.$$

11. As $t \to 3^+$, $(t - 3) > 0$ and thus $(t - 3) = \sqrt{(t - 3)^2}$. Rewrite the limit as

$$\lim_{t \to 3^+} \frac{\sqrt{(t - 3)(t + 3)}}{\sqrt{(t - 3)^2}} = \lim_{t \to 3^+} \frac{\sqrt{(t + 3)}}{\sqrt{(t - 3)}}.$$ The limit

of the numerator is $\sqrt{6}$ and the denominator is approaching 0 through positive values. Thus,
$$\lim_{t \to 3^+} \frac{\sqrt{t^2 - 9}}{t - 3} = \infty.$$

12. The graph of f indicates that:

   I. $\lim\limits_{x \to 4^-} f(x) = 5$ is true.

   II. $\lim\limits_{x \to 4} f(x) = 2$ is false. (The $\lim\limits_{x \to 4} f(x) = 5.$)

   III. "$x = 4$ is not in the domain of f" is false since $f(4) = 2$.

Part B—Calculators are permitted.

13. Examining the graph in your calculator, you notice that the function approaches the x-axis as $x \to \infty$ or as $x \to -\infty$. Thus, the line $y = 0$ (the x-axis) is a horizontal asymptote. As x approaches 1 from either side, the function increases or decreases without bound. Similarly, as x approaches -2 from either side, the function increases or decreases without bound. Therefore, $x = 1$ and $x = -2$ are vertical asymptotes. (See Figure 1.7-1.)

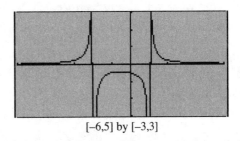

[−6,5] by [−3,3]

Figure 1.7-1

14. As $x \to 5^+$, the limit of the numerator $(5 + [5])$ is 10 and as $x \to 5^+$, the denominator approaches 0 through negative values. Thus, the
$$\lim_{x \to 5^+} \frac{5 + [x]}{5 - x} = -\infty.$$

15. Since $f(x)$ is a rational function, it is continuous everywhere except at values where the denominator is 0. Factoring and setting the denominator equal to 0, you have $(x + 6)$ $(x - 2) = 0$. Thus, the points of discontinuity are at $x = -6$ and $x = 2$. Verify your result with a calculator. (See Figure 1.7-2.)

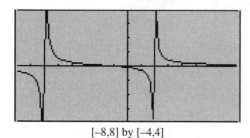

[–8,8] by [–4,4]

Figure 1.7-2

16. In order for $g(x)$ to be continuous at $x = 3$, it must satisfy the three conditions of continuity: (1) $g(3) = 3^2 + 5 = 14$, (2) $\lim_{x \to 3^+} (x^2 + 5) = 14$ and $\lim_{x \to 3^-} (2x - k) = 6 - k$, and the two one-sided limits must be equal in order for $\lim_{x \to 3} g(x)$ to exist. Therefore, $6 - k = 14$ and $k = -8$. Now, $g(3) = \lim_{x \to 3} g(x)$ and condition 3 is satisfied.

17. Checking with the three conditions of continuity:
Condition 1 $f(2) = 12$,
Condition 2 $\lim_{x \to 2} \dfrac{x^2 + 5x - 14}{x - 2} =$
$\lim_{x \to 2} \dfrac{(x + 7)(x - 2)}{x - 2} = \lim_{x \to 2} (x + 7) = 9$, and
Condition 3 $f(2) \neq \lim_{x \to 2} (x + 7)$. Therefore, $f(x)$ is discontinuous at $x = 2$.

18. The graph indicates that (a) $f(3) = 4$, (b) $\lim_{x \to 3^+} f(x) = 0$, (c) $\lim_{x \to 3^-} f(x) = 0$, (d) $\lim_{x \to 3} f(x) = 0$, and (e) therefore, $f(x)$ is not continuous at $x = 3$ since $f(3) \neq \lim_{x \to 3} f(x)$.

19. (See Figure 1.7-3.) If $b = 0$, then $r = 0$, but r cannot be 0. If $b = -3, -2,$ or -1 f would have more than one root. Thus $b = 1$. Choice (e).

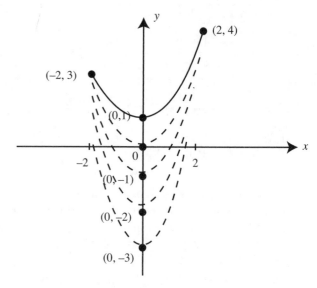

Figure 1.7-3

20. Substituting $x = 0$ would lead to 0/0. Substitute $(1 - \cos^2 x)$ in place of $\sin^2 x$ and obtain
$$\lim_{x \to 0} \frac{1 - \cos x}{\sin^2 x} = \lim_{x \to 0} \frac{1 - \cos x}{(1 - \cos^2 x)}$$
$$= \lim_{x \to 0} \frac{1 - \cos x}{(1 - \cos x)(1 + \cos x)} = \lim_{x \to 0} \frac{1}{(1 + \cos x)}$$
$$= \frac{1}{1 + 1} = \frac{1}{2}.$$

Verify your result with a calculator. (See Figure 1.7-4)

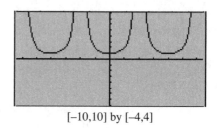

[–10,10] by [–4,4]

Figure 1.7-4

1.8 SOLUTIONS TO CUMULATIVE REVIEW PROBLEMS

21. Rewrite $3x - 2y = 6$ in $y = mx + b$ form which is $y = \dfrac{3}{2}x - 3$. The slope of this line whose equation is $y = \dfrac{3}{2}x - 3$ is $m = \dfrac{3}{2}$. Thus, the slope of a line perpendicular to this line is

$m = -\frac{2}{3}$. Since the perpendicular line passes through the point $(2, -4)$, therefore, an equation of the perpendicular line is

$y - (-4) = -\dfrac{2}{3}(x - 2)$ which is equivalent to

$y + 4 = -\dfrac{2}{3}(x - 2)$.

22. The graph indicates that $\lim\limits_{x \to 4^-} f(x) = 3$, $f(4) = 1$, and $\lim\limits_{x \to 4} f(x)$ does not exist. Therefore, only statements I and III are true.

23. Substituting $x = 0$ into $\dfrac{|3x - 4|}{x - 2}$, you obtain $\dfrac{4}{-2} = -2$.

24. Rewrite $\lim\limits_{x \to 0} \dfrac{\tan x}{x}$ as $\lim\limits_{x \to 0} \dfrac{\sin x / \cos x}{x}$ which is equivalent to $\lim\limits_{x \to 0} \dfrac{\sin x}{x \cos x}$ which is equal to $\lim\limits_{x \to 0} \dfrac{\sin x}{x} \cdot \lim\limits_{x \to 0} \dfrac{1}{\cos x} = (1)(1) = 1$.

25. To find horizontal asymptotes, examine the $\lim\limits_{x \to \infty} f(x)$ and the $\lim\limits_{x \to -\infty} f(x)$. The

$\lim\limits_{x \to \infty} f(x) = \lim\limits_{x \to \infty} \dfrac{x}{\sqrt{x^2 + 4}}$. Dividing by the highest power of x (and in this case, it's x), you obtain

$\lim\limits_{x \to \infty} \dfrac{x/x}{\sqrt{x^2 + 4}/x}$. As $x \to \infty$, $\sqrt{x^2}$ thus, you have

$\lim\limits_{x \to \infty} \dfrac{x/x}{\sqrt{x^2 + 4}/\sqrt{x^2}} = \lim\limits_{x \to \infty} \dfrac{1}{\sqrt{\dfrac{x^2 + 4}{x^2}}}$

$= \lim\limits_{x \to \infty} \dfrac{1}{\sqrt{1 + \dfrac{4}{x^2}}} = 1$. Thus, the line $y = 1$ is a horizontal asymptote. The

$\lim\limits_{x \to -\infty} f(x) = \lim\limits_{x \to -\infty} \dfrac{x}{\sqrt{x^2 + 4}}$. As $x \to \infty$,

$x = -\sqrt{x^2}$. Thus, $\lim\limits_{x \to -\infty} \dfrac{x}{\sqrt{x^2 + 4}}$

$= \lim\limits_{x \to -\infty} \dfrac{x/x}{\sqrt{x^2 + 4}/-\sqrt{x^2}} = \lim\limits_{x \to -\infty} \dfrac{1}{-\sqrt{1 + \dfrac{4}{x^2}}} = -1$.

Therefore, the line $y = -1$ is a horizontal asymptote. As for vertical asymptotes, $f(x)$ is continuous and defined for all real numbers. Thus, there is no vertical asymptote.

Chapter 2

Differentiation

2.1 DERIVATIVES OF ALGEBRAIC FUNCTIONS

Main Concepts: *Definition of the Derivative of a Function; Power Rule; The Sum, Difference, Product, and Quotient Rules; The Chain Rule*

Definition of the Derivative of a Function

The derivative of a function f, written as f', is defined as

$$f'(x) = \lim_{h \to 0} \frac{f(x+h) - f(x)}{h},$$

if this limit exists. (Note that $f'(x)$ is read as f prime of x.)
Other symbols of the derivative of a function are:

$$D_x f, \frac{d}{dx} f(x), \text{ and if } y = f(x), y', \frac{dy}{dx}, \text{ and } D_x y.$$

Let m_{tangent} be the slope of the tangent to a curve $y = f(x)$ at a point on the curve. Then

$$m_{\text{tangent}} = f'(x) = \lim_{h \to 0} \frac{f(x+h) - f(x)}{h}$$

$$m_{\text{tangent}}(\text{at } x = a) = f'(a) = \lim_{h \to 0} \frac{f(a+h) - f(a)}{h} \text{ or } \lim_{x \to a} \frac{f(x) - f(a)}{x - a}. \text{ (See Figure 2.1-1.)}$$

Given a function f, if $f'(x)$ exists at $x = a$, then the function f is said to be differentiable at $x = a$. If a function f is differentiable at $x = a$, then f is continuous at $x = a$. (Note that the converse of the statement is not necessarily true, i.e., if a function f is continuous at $x = a$, then f may or may not be differentiable at $x = a$.) Here are several examples of functions that are not differentiable at a given number $x = a$. (See Figures 2.1-2–2.1-5 on page 61.)

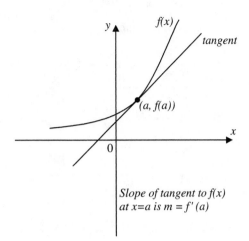

*Slope of tangent to f(x)
at x=a is m = f'(a)*

Figure 2.1-1

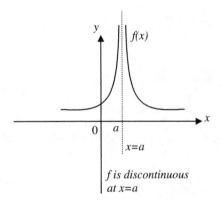

*f is discontinuous
at x=a*

Figure 2.1-2

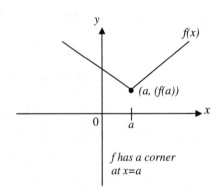

*f has a corner
at x=a*

Figure 2.1-3

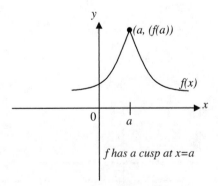

f has a cusp at x=a

Figure 2.1-4

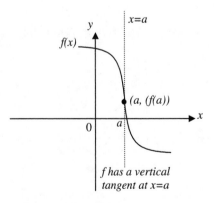

*f has a vertical
tangent at x=a*

Figure 2.1-5

Example 1

If $f(x) = x^2 - 2x - 3$, find (a) $f'(x)$ using the definition of derivative, (b) $f'(0)$, (c) $f'(1)$, and (d) $f'(3)$.

(a) Using the definition of derivative, $f'(x) = \lim\limits_{h \to 0} \dfrac{f(x+h) - f(x)}{h}$

$$= \lim\limits_{h \to 0} \frac{\left[(x+h)^2 - 2(x+h) - 3\right] - \left[x^2 - 2x - 3\right]}{h}$$

$$= \lim\limits_{h \to 0} \frac{\left[x^2 + 2xh + h^2 - 2x - 2h - 3\right] - \left[x^2 - 2x - 3\right]}{h}$$

$$= \lim\limits_{h \to 0} \frac{2xh + h^2 - 2h}{h}$$

$$= \lim\limits_{h \to 0} \frac{h(2x + h - 2)}{h}$$

$$= \lim\limits_{h \to 0}(2x + h - 2) = 2x - 2.$$

(b) $f'(0) = 2(0) - 2 = -2$, (c) $f'(1) = 2(1) - 2 = 0$ and (d) $f'(3) = 2(3) - 2 = 4$.

Example 2

Evaluate $\lim\limits_{h \to 0} \dfrac{\cos(\pi + h) - \cos(\pi)}{h}$

The expression $\lim\limits_{h \to 0} \dfrac{\cos(\pi + h) - \cos(\pi)}{h}$ is equivalent to the derivative of the function $f(x) = \cos x$ at $x = \pi$, i.e., $f'(\pi)$. The derivative of $f(x) = \cos x$ at $x = \pi$ is equivalent to the slope of the tangent to curve of $\cos x$ at $x = \pi$. The tangent is parallel to the x-axis. Thus, the slope is 0 or $\lim\limits_{h \to 0} \dfrac{\cos(\pi + h) - \cos(\pi)}{h} = 0$.

Or, using an algebraic method, note that $\cos(a+b) = \cos(a)\cos(b) - \sin(a)\sin(b)$.

Then rewrite $\lim\limits_{h \to 0} \dfrac{\cos(\pi+h) - \cos(\pi)}{h} = \lim\limits_{h \to 0} \dfrac{\cos(\pi)\cos(h) - \sin(\pi)\sin(h) - \cos(\pi)}{h}$

$$= \lim\limits_{h \to 0} \frac{-\cos(h) - (-1)}{h} = \lim\limits_{h \to 0} \frac{-\cos(h) + 1}{h} = \lim\limits_{h \to 0} \frac{-\left[\cos(h) - 1\right]}{h} = -\lim\limits_{h \to 0} \frac{\left[\cos(h) - 1\right]}{h} = 0.$$

(See Figure 2.1-6.)

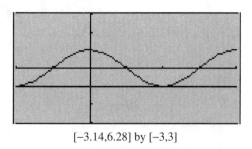

[−3.14,6.28] by [−3,3]

Figure 2.1-6

Example 3

If the function $f(x) = x^{2/3} + 1$, find all points where f is not differentiable.

    The function $f(x)$ is continuous for all real numbers and the graph of $f(x)$ forms a "cusp" at the point $(0, 1)$. Thus, $f(x)$ is not differentiable at $x = 0$. See Figure 2.1-7.

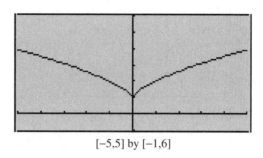

[−5,5] by [−1,6]

Figure 2.1-7

Example 4

Using a calculator, find the derivative of $f(x) = x^2 + 4x$ at $x = 3$.

    There are several ways to find $f'(3)$, using a calculator. One way is to use the *n*Deriv function of the calculator. From the main (Home) screen, select F3-Calc and then select *n*Deriv. Enter *n*Deriv $(x^2 + 4x,\ x)|x = 3$. The result is 10.

• Always write out all formulas in your solutions.

Power Rule

If $f(x) = c$ where c is a constant, then $f'(x) = 0$.
If $f(x) = x^n$ where n is a real number, then $f'(x) = nx^{n-1}$.
If $f(x) = cx^n$ where c is a constant and n is a real number, then $f'(x) = cnx^{n-1}$.

Summary of Derivatives of Algebraic Functions:

$$\frac{d}{dx}(c) = 0, \quad \frac{d}{dx}(x^n) = nx^{n-1}, \quad \text{and} \quad \frac{d}{dx}(cx^n) = cnx^{n-1}$$

Example 1

If $f(x) = 2x^3$, find (a) $f'(x)$, (b) $f'(1)$ and (c) $f'(0)$
Note that (a) $f'(x) = 6x^2$, (b) $f'(1) = 6(1)^2 = 6$, and (c) $f'(0) = 0$.

Example 2

If $y = \dfrac{1}{x^2}$, find (a) $\dfrac{dy}{dx}$ and (b) $\left.\dfrac{dy}{dx}\right|_{x=0}$ $\left(\text{which represents } \dfrac{dy}{dx} \text{ at } x = 0.\right)$

Note that (a) $y = \dfrac{1}{x^2} = x^{-2}$ and thus, $\dfrac{dy}{dx} = -2x^{-3} = \dfrac{-2}{x^3}$ and (b) $\left.\dfrac{dy}{dx}\right|_{x=0}$ does not exist because the expression $\dfrac{-2}{0}$ is undefined.

Example 3

Here are several examples of algebraic functions and their derivatives:

Function	Written in cx^n form	Derivative	Derivative with Positive Exponents
$3x$	$3x^1$	$3x^0 = 3$	3
$-5x^7$	$-5x^7$	$-35x^6$	$-35x^6$
$8\sqrt{x}$	$8x^{\frac{1}{2}}$	$4x^{-\frac{1}{2}}$	$\dfrac{4}{x^{\frac{1}{2}}}$ or $\dfrac{4}{\sqrt{x}}$
$\dfrac{1}{x^2}$	x^{-2}	$-2x^{-3}$	$\dfrac{-2}{x^3}$
$\dfrac{-2}{\sqrt{x}}$	$\dfrac{-2}{x^{\frac{1}{2}}} = -2x^{-\frac{1}{2}}$	$x^{-\frac{3}{2}}$	$\dfrac{1}{x^{\frac{3}{2}}}$ or $\dfrac{1}{\sqrt{x^3}}$
4	$4x^0$	0	0
π^2	$(\pi^2)x^0$	0	0

Example 4

Using a calculator, find $f'(x)$ and $f'(3)$ if $f(x) = \dfrac{1}{\sqrt{x}}$.

There are several ways of finding $f'(x)$ and $f'(9)$ using a calculator. One way to use the d(differentiate) function. Go to Home screen. Select F3-Calc and then select d(differentiate). Enter $d(1/\sqrt{(x)},\ x)$. The result is $f'(x) = \dfrac{-1}{2x^{\frac{3}{2}}}$. To find $f'(3)$, enter $d(1/\sqrt{(x)},\ x)\big| x = 3$. The result is $f'(3) = \dfrac{-1}{54}$.

The Sum, Difference, Product, and Quotient Rules

If u and v are two differentiable functions, then

$$\frac{d}{dx}(u \pm v) = \frac{du}{dx} \pm \frac{dv}{dx} \qquad \text{Sum \& Difference Rules}$$

$$\frac{d}{dx}(uv) = v\frac{du}{dx} + u\frac{dv}{dx} \qquad \text{Product Rule}$$

$$\frac{d}{dx}\left(\frac{u}{v}\right) = \frac{v\dfrac{du}{dx} - u\dfrac{dv}{dx}}{v^2}, v \neq 0 \quad \text{Quotient Rule}$$

Summary of Sum, Difference, Product, and Quotient Rules:

$$(u \pm v)' = u' \pm v' \quad (uv)' = u'v + v'u \quad \&\ \left(\frac{u}{v}\right)' = \frac{u'v - v'u}{v^2}$$

Example 1

Find $f'(x)$ if $f(x) = x^3 - 10x + 5$.

Using the sum and difference rules, you can differentiate each term and obtain $f'(x) = 3x^2 - 10$. Or using your calculator, select the d(differentiate) function and enter $d(x^3 - 10x + 5, x)$ and obtain $3x^2 - 10$.

Example 2

If $y = (3x - 5)(x^4 + 8x - 1)$, find $\dfrac{dy}{dx}$.

Using the product rule $\dfrac{d}{dx}(uv) = v\dfrac{du}{dx} + u\dfrac{dv}{dx}$, let $u = (3x - 5)$ and $v = (x^4 + 8x - 1)$.

Then $\dfrac{dy}{dx} = (3)(x^4 + 8x - 1) + (4x^3 + 8)(3x - 5) = (3x^4 + 24x - 3) + (12x^4 - 20x^3 + 24x - 40) = 15x^4 - 20x^3 + 48x - 43$. Or you can use your calculator and enter $d((3x - 5)(x^4 + 8x - 1), x)$ and obtain the same result.

Example 3

If $f(x) = \dfrac{2x - 1}{x + 5}$, find $f'(x)$.

Using the quotient rule $\left(\dfrac{u}{v}\right)' = \dfrac{u'v - v'u}{v^2}$, let $u = 2x - 1$ and $v = x + 5$. Then $f'(x) = \dfrac{(2)(x + 5) - (1)(2x - 1)}{(x + 5)^2} = \dfrac{2x + 10 - 2x + 1}{(x + 5)^2} = \dfrac{11}{(x + 5)^2}, x \neq -5$. Or you can use your calculator and enter $d((2x - 1)/(x + 5), x)$ and obtain the same result.

Example 4

Using your calculator, find an equation of the tangent to the curve $f(x) = x^2 - 3x + 2$ at $x = 5$.

Find the slope of the tangent to the curve at $x = 5$ by entering $d(x^2 - 3x + 2, x)|x = 5$. The result is 7. Compute $f(5) = 12$. Thus, the point $(5, 12)$ is on the curve of $f(x)$. An equation of the line whose slope $m = 7$ and passing through the point $(5, 12)$ is $y - 12 = 7(x - 5)$.

- Remember that $\dfrac{d}{dx}\ln x = \dfrac{1}{x}$ and $\int \ln x \; dx = x \ln x - x + c$. The integral formula is not usually tested in the AB exam.

The Chain Rule

If $y = f(u)$ and $u = g(x)$ are differentiable functions of u and x respectively, then $\dfrac{d}{dx}[f(g(x))] = f'(g(x)) \cdot g'(x)$ or $\dfrac{dy}{dx} = \dfrac{dy}{du} \cdot \dfrac{du}{dx}$.

Example 1

If $y = (3x - 5)^{10}$, find $\dfrac{dy}{dx}$.

Using the chain rule, let $u = 3x - 5$ and thus, $y = u^{10}$. Then, $\dfrac{dy}{du} = 10u^9$ and $\dfrac{du}{dx} = 3$.

Since $\dfrac{dy}{dx} = \dfrac{dy}{du} \cdot \dfrac{du}{dx}, \dfrac{dy}{dx} = (10u^9)(3) = 10(3x - 5)^9(3) = 30(3x - 5)^9$. Or you can use your calculator and enter $d((3x - 5)^{10}, x)$ and obtain the same result.

Example 2

If $f(x) = 5x\sqrt{25 - x^2}$, find $f'(x)$.

Rewrite $f(x) = 5x\sqrt{25 - x^2}$ as $f(x) = 5x(25 - x^2)^{\frac{1}{2}}$. Using the product rule, $f'(x) = (25 - x^2)^{\frac{1}{2}}\dfrac{d}{dx}(5x) + (5x)\dfrac{d}{dx}(25 - x^2)^{\frac{1}{2}} = 5(25 - x^2)^{\frac{1}{2}} + (5x)\dfrac{d}{dx}(25 - x^2)^{\frac{1}{2}}$.

To find $\dfrac{d}{dx}(25 - x^2)^{\frac{1}{2}}$, use the chain rule and let $u = 25 - x^2$.

Thus, $\dfrac{d}{dx}(25 - x^2)^{\frac{1}{2}} = \dfrac{1}{2}(25 - x^2)^{-1/2}(-2x) = \dfrac{-x}{(25 - x^2)^{\frac{1}{2}}}$. Substituting this quantity back into $f'(x)$, you have

$$f'(x) = 5(25 - x^2)^{\frac{1}{2}} + (5x)\left(\dfrac{-x}{(25 - x^2)^{\frac{1}{2}}}\right) = \dfrac{5(25 - x^2) - 5x^2}{(25 - x^2)^{\frac{1}{2}}} = \dfrac{125 - 10x^2}{(25 - x^2)^{\frac{1}{2}}}.$$

Or you can use your calculator and enter $d(5x\sqrt{25 - x^2}, x)$ and obtain the same result.

Example 3

If $y = \left(\dfrac{2x - 1}{x^2}\right)^3$, find $\dfrac{dy}{dx}$.

Using the chain rule, let $u = \left(\dfrac{2x - 1}{x^2}\right)$. Then $\dfrac{dy}{dx} = 3\left(\dfrac{2x - 1}{x^2}\right)^2\dfrac{d}{dx}\left(\dfrac{2x - 1}{x^2}\right)$.

To find $\dfrac{d}{dx}\left(\dfrac{2x - 1}{x^2}\right)$, use the quotient rule.

Thus, $\dfrac{d}{dx}\left(\dfrac{2x - 1}{x^2}\right) = \dfrac{(2)(x^2) - (2x)(2x - 1)}{(x^2)^2} = \dfrac{-2x^2 + 2x}{x^4}$. Substituting this quantity back into $\dfrac{dy}{dx} = 3\left(\dfrac{2x - 1}{x^2}\right)^2\dfrac{d}{dx}\left(\dfrac{2x - 1}{x^2}\right) = 3\left(\dfrac{2x - 1}{x^2}\right)^2\dfrac{-2x^2 + 2x}{x^4}$

$$= \dfrac{-6(x - 1)(2x - 1)^2}{x^7}.$$

An alternate solution is to use the product rule and rewrite $y = \left(\dfrac{2x - 1}{x^2}\right)^3$ as $y = \dfrac{(2x - 1)^3}{(x^2)^3} = \dfrac{(2x - 1)^3}{x^6}$ and use the quotient rule. Another approach is to express $y = (2x - 1)^3(x^{-6})$ and use the product rule. Of course, you can always use your calculator if you are permitted to do so.

2.2 DERIVATIVES OF TRIGONOMETRIC, INVERSE TRIGONOMETRIC, EXPONENTIAL, AND LOGARITHMIC FUNCTIONS

Main Concepts: *Derivatives of Trigonometric Functions, Derivatives of Inverse Trigonometric Functions, Derivatives of Exponential and Logarithmic Functions*

Derivatives of Trigonometric Functions

Summary of Derivatives of Trigonometric Functions:

$$\frac{d}{dx}(\sin x) = \cos x \qquad \frac{d}{dx}(\cos x) = -\sin x$$

$$\frac{d}{dx}(\tan x) = \sec^2 x \qquad \frac{d}{dx}(\cot x) = -\csc^2 x$$

$$\frac{d}{dx}(\sec x) = \sec x \tan x \qquad \frac{d}{dx}(\csc x) = -\csc x \cot x.$$

Note that the derivatives of *cosine*, *cotangent*, and *cosecant* all have a negative sign.

Example 1

If $y = 6x^2 + 3 \sec x$, find $\dfrac{dy}{dx}$.

$\dfrac{dy}{dx} = 12x + 3 \sec x \tan x.$

Example 2

Find $f'(x)$ if $f(x) = \cot(4x - 6)$.
Using the chain rule, let $u = 4x - 6$. Then $f'(x) = [-\csc^2(4x-6)][4] = -4 \csc^2(4x-6)$.
Or using your calculator, enter $d(1/\tan(4x - 6), x)$ and obtain $\dfrac{-4}{\sin^2(4x - 6)}$ which is an equivalent form.

Example 3

Find $f'(x)$ if $f(x) = 8 \sin(x^2)$.
Using the chain rule, let $u = x^2$. Then $f'(x) = [8 \cos(x^2)][2x] = 16x \cos(x^2)$.

Example 4

If $y = \sin x \cos(2x)$, find $\dfrac{dy}{dx}$.

Using the product rule, let $u = \sin x$ and $v = \cos(2x)$.

Then $\dfrac{dy}{dx} = \cos x \cos(2x) + [-\sin(2x)](2)(\sin x) = \cos x \cos(2x) - 2 \sin x \sin(2x).$

Example 5

If $y = \sin[\cos(2x)]$, find $\dfrac{dy}{dx}$.

Using the chain rule, let $u = \cos(2x)$. Then

$$\frac{dy}{dx} = \frac{dy}{du} \cdot \frac{du}{dx} = \cos[\cos(2x)]\frac{d}{dx}[\cos(2x)].$$

To evaluate $\dfrac{d}{dx}[\cos(2x)]$, use the chain rule again by making another u-substitution, this time for $2x$. Thus, $\dfrac{d}{dx}[\cos(2x)] = [-\sin(2x)]2 = -2\sin(2x)$. Therefore, $\dfrac{dy}{dx}\cos[\cos(2x)](-2\sin(2x)) = -2\sin(2x)\cos[\cos(2x)].$

Example 6

Find $f'(x)$ if $f(x) = 5x \csc x$.
Using the product rule, let $u = 5x$ and $v = \csc x$. Then $f'(x) = 5\csc x + (-\csc x \cot x)\times (5x) = 5\csc x - 5x(\csc x)(\cot x)$.

Example 7

If $y = \sqrt{\sin x}$, find $\dfrac{dy}{dx}$.

Rewrite $y = \sqrt{\sin x}$ as $y = (\sin x)^{1/2}$. Using the chain rule, let $u = \sin x$. Thus,
$\dfrac{dy}{dx} = \dfrac{1}{2}(\sin x)^{-\frac{1}{2}}(\cos x) = \dfrac{\cos x}{2(\sin x)^{\frac{1}{2}}} = \dfrac{\cos x}{2\sqrt{\sin x}}.$

Example 8

If $y = \dfrac{\tan x}{1 + \tan x}$, find $\dfrac{dy}{dx}$.

Using the quotient rule, let $u = \tan x$ and $v = (1 + \tan x)$. Then,

$$\frac{dy}{dx} = \frac{(\sec^2 x)(1 + \tan x) - (\sec^2 x)(\tan x)}{(1 + \tan x)^2}$$

$$= \frac{\sec^2 x + (\sec^2 x)(\tan x) - (\sec^2 x)(\tan x)}{(1 + \tan x)^2}$$

$$= \frac{\sec^2 x}{(1 + \tan x)^2}, \text{ which is equivalent to } \frac{\dfrac{1}{(\cos x)^2}}{1 + \left(\dfrac{\sin x}{\cos x}\right)^2}$$

$$= \frac{\dfrac{1}{(\cos x)^2}}{\left(\dfrac{\cos x + \sin x}{\cos x}\right)^2} = \frac{1}{(\cos x + \sin x)^2}.$$

Note: For all of the above exercises, you can find the derivatives by using a calculator provided that you are permitted to do so.

Derivatives of Inverse Trigonometric Functions

Summary of Derivatives of Inverse Trigonometric Functions:

Let u be a differentiable function of x, then

$$\frac{d}{dx}\sin^{-1}u = \frac{1}{\sqrt{1-u^2}}\frac{du}{dx}, \ |u| < 1 \qquad \frac{d}{dx}\cos^{-1}u = \frac{-1}{\sqrt{1-u^2}}\frac{du}{dx}, \ |u| < 1$$

$$\frac{d}{dx}\tan^{-1}u = \frac{1}{\sqrt{1+u^2}}\frac{du}{dx} \qquad \frac{d}{dx}\cot^{-1}u = \frac{-1}{\sqrt{1-u^2}}\frac{du}{dx}$$

$$\frac{d}{dx}\sec^{-1}u = \frac{1}{|u|\sqrt{u^2-1}}\frac{du}{dx}, \ |u| > 1 \quad \frac{d}{dx}\csc^{-1}u = \frac{-1}{|u|\sqrt{u^2-1}}\frac{du}{dx}, \ |u| > 1.$$

Note that the derivatives of $\cos^{-1}x$, $\cot^{-1}x$, and $\csc^{-1}x$ all have a "−1" in their numerators.

Example 1

If $y = 5\sin^{-1}(3x)$, find $\frac{dy}{dx}$.

Let $u = 3x$. Then $\frac{dy}{dx} = (5)\frac{1}{\sqrt{1-(3x)^2}}\frac{du}{dx} = \frac{5}{\sqrt{1-(3x)^2}}(3) = \frac{15}{\sqrt{1-9x^2}}$.

Or using a calculator, enter $d[5\sin^{-1}(3x),\ x]$ and obtain the same result.

Example 2

Find $f'(x)$ if $f(x) = \tan^{-1}\sqrt{x}$.

Let $u = \sqrt{x}$. Then $f'(x) = \frac{1}{1+(\sqrt{x})^2}\frac{du}{dx} = \frac{1}{1+x}\left(\frac{1}{2}x^{-\frac{1}{2}}\right) = \frac{1}{1+x}\left(\frac{1}{2\sqrt{x}}\right)$

$$= \frac{1}{2\sqrt{x}(1+x)}.$$

Example 3

If $y = \sec^{-1}(3x^2)$, find $\frac{dy}{dx}$.

Let $u = 3x^2$. Then $\frac{dy}{dx} = \frac{1}{|3x^2|\sqrt{(3x^2)^2-1}}\frac{du}{dx} = \frac{1}{3x^2\sqrt{9x^4-1}}(6x) = \frac{2}{x\sqrt{9x^4-1}}$.

Example 4

If $y = \cos^{-1}\left(\frac{1}{x}\right)$, find $\frac{dy}{dx}$.

Let $u = \left(\frac{1}{x}\right)$. Then $\frac{dy}{dx} = \frac{-1}{\sqrt{1-\left(\frac{1}{x}\right)^2}}\frac{du}{dx}$.

Rewrite $u = \left(\dfrac{1}{x}\right)$ as $u = x^{-1}$. Then $\dfrac{du}{dx} = -1x^{-2} = \dfrac{-1}{x^2}$.

Therefore, $\dfrac{dy}{dx} = \dfrac{-1}{\sqrt{1-\left(\dfrac{1}{x}\right)^2}}\dfrac{du}{dx} = \dfrac{-1}{\sqrt{1-\left(\dfrac{1}{x}\right)^2}}\dfrac{-1}{x^2} = \dfrac{1}{\sqrt{\dfrac{x^2-1}{x^2}}(x^2)} = \dfrac{1}{\dfrac{\sqrt{x^2-1}(x^2)}{|x|}}$

$$= \dfrac{1}{|x|\sqrt{x^2-1}}.$$

Note: For all of the above exercises, you can find the derivatives by using a calculator provided that you are permitted to do so.

Derivatives of Exponential and Logarithmic Functions

Summary of Derivatives of Exponential and Logarithmic Functions:

Let u be a differentiable function of x, then

$$\frac{d}{dx}(e^u) = e^u\frac{du}{dx} \qquad \frac{d}{dx}(a^u) = a^u\ln a\frac{du}{dx}, \; a > 0 \; \& \; a \neq 1$$

$$\frac{d}{dx}(\ln u) = \frac{1}{u}\frac{du}{dx}, \; u > 0 \qquad \frac{d}{dx}(\log u) = \frac{1}{u\ln a}\frac{du}{dx}, \; a > 0 \; \& \; a \neq 1.$$

For the following examples, find $\dfrac{dy}{dx}$ and verify your result with a calculator.

Example 1

$y = e^{3x} + 5xe^3 + e^3$

$\dfrac{dy}{dx} = (e^{3x})(3) + 5e^3 + 0 = 3e^{3x} + 5e^3$ (Note that e^3 is a constant.)

Example 2

$y = xe^x - x^2e^x$

Using the product rule for both terms, you have

$$\frac{dy}{dx} = (1)e^x + (e^x)x - \left[(2x)e^x + (e^x)x^2\right] = e^x + xe^x - 2xe^x - x^2e^x = e^x - xe^x - x^2e^x$$

$$= -x^2e^x - xe^x + e^x = e^x(-x^2 - x + 1).$$

Example 3

$y = 3^{\sin x}$

Let $u = \sin x$. Then, $\dfrac{dy}{dx} = (3^{\sin x})(\ln 3)\dfrac{du}{dx} = (3^{\sin x})(\ln 3)\cos x = (\ln 3)(3^{\sin x})\cos x.$

Example 4

$y = e^{(x^3)}$

Let $u = x^3$. Then, $\dfrac{dy}{dx} = \left[e^{(x^3)}\right]\dfrac{du}{dx} = \left[e^{(x^3)}\right]3x^2 = 3x^2e^{(x^3)}.$

Example 5

$y = (\ln x)^5$

Let $u = \ln x$. Then, $\dfrac{dy}{dx} = 5(\ln x)^4 \dfrac{du}{dx} = 5(\ln x)^4 \left(\dfrac{1}{x}\right) = \dfrac{5(\ln x)^4}{x}$.

Example 6

$y = \ln(x^2 + 2x - 3) + \ln 5$

Let $u = x^2 + 2x - 3$. Then, $\dfrac{dy}{dx} = \dfrac{1}{x^2 + 2x - 3} \dfrac{du}{dx} + 0 = \dfrac{1}{x^2 + 2x - 3}(2x + 2)$

$= \dfrac{2x + 2}{x^2 + 2x - 3}$.

(Note that $\ln 5$ is a constant. Thus the derivative of $\ln 5$ is 0.)

Example 7

$y = 2x \ln x + x$

Using the product rule for the first term,

you have $\dfrac{dy}{dx} = (2) \ln x + \left(\dfrac{1}{x}\right)(2x) + 1 = 2 \ln x + 2 + 1 = 2 \ln x + 3$.

Example 8

$y = \ln(\ln x)$

Let $u = \ln x$. Then $\dfrac{dy}{dx} = \dfrac{1}{\ln x} \dfrac{du}{dx} = \dfrac{1}{\ln x} \left(\dfrac{1}{x}\right) = \dfrac{1}{x \ln x}$.

Example 9

$y = \log_5(2x + 1)$

Let $u = 2x + 1$. Then $\dfrac{dy}{dx} = \dfrac{1}{(2x + 1) \ln 5} \dfrac{du}{dx} = \dfrac{1}{(2x + 1) \ln 5} \cdot (2) = \dfrac{2}{(2x + 1) \ln 5}$.

Example 10

Write an equation of the line tangent to the curve of $y = e^x$ at $x = 1$.

The slope of the tangent to the curve $y = e^x$ at $x = 1$ is equivalent to the value of the derivative of $y = e^x$ evaluated at $x = 1$. Using your calculator, enter $d(e^\wedge(x), x)|x = 1$ and obtain e. Thus, $m = e$, the slope of the tangent to the curve at $x = 1$. At $x = 1$, $y = e^1 = e$, and thus the point on the curve is $(1, e)$. Therefore, the equation of the tangent is $y - e = e(x - 1)$ or $y = ex$. (See Figure 2.2-1.)

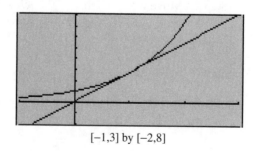

[−1,3] by [−2,8]

Figure 2.2-1

• Guess, if you can eliminate some of the choices in a multiple-choice question.

2.3 IMPLICIT DIFFERENTIATION

!

Procedure for Implicit Differentiation:

Given an equation containing the variables x and y for which you cannot easily solve for y in terms of x, you can find $\dfrac{dy}{dx}$ by doing the following:

Steps 1: Differentiate each term of the equation with respect to x.

2: Move all terms containing $\dfrac{dy}{dx}$ to the left side of the equation and all other terms to the right side.

3: Factor out $\dfrac{dy}{dx}$ on the left side of the equation.

4: Solve for $\dfrac{dy}{dx}$.

Example 1

Find $\dfrac{dy}{dx}$ if $y^2 - 7y + x^2 - 4x = 10$.

Step 1: Differentiate each term of the equation with respect to x. (Note that y is treated as a function of x.) $2y\dfrac{dy}{dx} - 7\dfrac{dy}{dx} + 2x - 4 = 0$

Step 2: Move all terms containing $\dfrac{dy}{dx}$ to the left side of the equation and all other terms to the right: $2y\dfrac{dy}{dx} - 7\dfrac{dy}{dx} = -2x + 4$

Step 3: Factor out $\dfrac{dy}{dx}$: $\dfrac{dy}{dx}(2y - 7) = -2x + 4$

Step 4: Solve for $\dfrac{dy}{dx}$: $\dfrac{dy}{dx} = \dfrac{-2x + 4}{(2y - 7)}$

Example 2

Given $x^3 + y^3 = 6xy$, find $\dfrac{dy}{dx}$.

Step 1: Differentiate each term with respect to x: $3x^2 + 3y^2\dfrac{dy}{dx} = (6)y + \left(\dfrac{dy}{dx}\right)(6x)$

Step 2: Move all $\dfrac{dy}{dx}$ terms to the left side: $3y^2\dfrac{dy}{dx} - 6x\dfrac{dy}{dx} = 6y - 3x^2$

Step 3: Factor out $\dfrac{dy}{dx}$: $\dfrac{dy}{dx}(3y^2 - 6x) = 6y - 3x^2$

Step 4: Solve for $\dfrac{dy}{dx}$: $\dfrac{dy}{dx} = \dfrac{6y - 3x^2}{3y^2 - 6x} = \dfrac{2y - x^2}{y^2 - 2x}$

Example 3

Find $\dfrac{dy}{dx}$ if $(x + y)^2 - (x - y)^2 = x^5 + y^5$

Step 1: Differentiate each term with respect to x:

$$2(x + y)\left(1 + \dfrac{dy}{dx}\right) - 2(x - y)\left(1 - \dfrac{dy}{dx}\right) = 5x^4 + 5y^4\dfrac{dy}{dx}$$

Distributing $2(x + y)$ and $-2(x - y)$, you have

$$2(x + y) + 2(x + y)\frac{dy}{dx} - 2(x - y) + 2(x - y)\frac{dy}{dx} = 5x^4 + 5y^4\frac{dy}{dx}$$

Step 2: Move all $\frac{dy}{dx}$ terms to the left side:

$$2(x + y)\frac{dy}{dx} + 2(x - y)\frac{dy}{dx} - 5y^4\frac{dy}{dx} = 5x^4 - 2(x + y) + 2(x - y)$$

Step 3: Factor out $\frac{dy}{dx}$:

$$\frac{dy}{dx}[2(x + y) + 2(x - y) - 5y^4] = 5x^4 - 2x - 2y + 2x - 2y$$

$$\frac{dy}{dx}[2x + 2y + 2x - 2y - 5y^4] = 5x^4 - 4y$$

$$\frac{dy}{dx}[4x - 5y^4] = 5x^4 - 4y$$

Step 4: Solve for $\frac{dy}{dx}$: $\frac{dy}{dx} = \frac{5x^4 - 4y}{4y - 5y^4}$

Example 4

Write an equation of the tangent to the curve $x^2 + y^2 + 19 = 2x + 12y$ at $(4, 3)$
The slope of the tangent to the curve at $(4, 3)$ is equivalent to the derivative $\frac{dy}{dx}$ at $(4, 3)$.

Using implicit differentiation, you have

$$2x + 2y\frac{dy}{dx} = 2 + 12\frac{dy}{dx}$$

$$2y\frac{dy}{dx} - 12\frac{dy}{dx} = 2 - 2x$$

$$\frac{dy}{dx}(2y - 12) = 2 - 2x$$

$$\frac{dy}{dx} = \frac{2 - 2x}{2y - 12} = \frac{1 - x}{y - 6} \text{ and } \frac{dy}{dx}\Big|_{(4,3)} = \frac{1 - 4}{3 - 6} = 1$$

Thus, the equation of the tangent is $y - 3 = (1)(x - 4)$ or $y - 3 = x - 4$.

Example 5

Find $\frac{dy}{dx}$, if $\sin(x + y) = 2x$

$$\left[\cos(x + y)\left(1 + \frac{dy}{dx}\right)\right] = 2$$

$$1 + \frac{dy}{dx} = \frac{2}{\cos(x + y)}$$

$$\frac{dy}{dx} = \frac{2}{\cos(x + y)} - 1$$

2.4 APPROXIMATING A DERIVATIVE

Given a continuous and differentiable function, you can find the approximate value of a derivative at a given point numerically. Here are two examples.

Example 1

The graph of a function f on $[0, 5]$ is shown in Figure 2.4-1. Find the approximate value of $f'(3)$. (See Figure 2.4-1.)

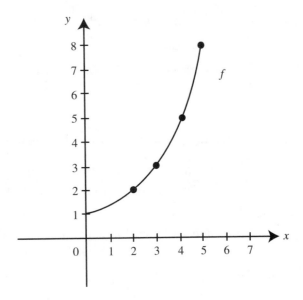

Figure 2.4-1

Since $f'(3)$ is equivalent to the slope of the tangent to $f(x)$ at $x = 3$, there are several ways you can find its approximate value.

Method 1: Using the slope of the line segment joining the points at $x = 3$ and $x = 4$.

$$f(3) = 3 \text{ and } f(4) = 5$$

$$m = \frac{f(4) - f(3)}{4 - 3} = \frac{5 - 3}{4 - 3} = 2$$

Method 2: Using the slope of the line segment joining the points at $x = 2$ and $x = 3$.

$$f(2) = 2 \text{ and } f(3) = 3$$

$$m = \frac{f(3) - f(2)}{3 - 2} = \frac{3 - 2}{3 - 2} = 1$$

Method 3: Using the slope of the line segment joining the points at $x = 2$ and $x = 4$.

$$f(2) = 2 \text{ and } f(4) = 5$$

$$m = \frac{f(4) - f(2)}{4 - 2} = \frac{5 - 2}{4 - 2} = \frac{3}{2}$$

Note that $\frac{3}{2}$ is the average of the results from methods 1 and 2.

Thus, $f'(3) \approx 1, 2$ or $\frac{3}{2}$ depending on which line segment you use.

Example 2

Let f be a continuous and differentiable function. Selected values of f are shown below. Find the approximate value of f' at $x = 1$.

x	-2	-1	0	1	2	3
f	1	0	1	1.59	2.08	2.52

You can use the difference quotient $\dfrac{f(a + h) - f(a)}{h}$ to approximate $f'(a)$.

Let $h = 1$; $f'(1) \approx \dfrac{f(2) - f(1)}{2 - 1} \approx \dfrac{2.08 - 1.59}{1} \approx 0.49$

Let $h = 2$; $f'(1) \approx \dfrac{f(3) - f(1)}{3 - 1} \approx \dfrac{2.52 - 1.59}{2} \approx 0.465$

Or, you can use the symmetric difference quotient $\dfrac{f(a + h) - f(a - h)}{2h}$ to approximate $f'(a)$.

Let $h = 1$; $f'(1) \approx \dfrac{f(2) - f(0)}{2 - 0} \approx \dfrac{2.08 - 1}{2} \approx 0.54$

Let $h = 2$; $f'(1) \approx \dfrac{f(3) - f(-1)}{3 - (-1)} \approx \dfrac{2.52 - 0}{4} \approx 0.63$

Thus, $f'(3) \approx 0.49, 0.465, 0.54,$ or 0.63 depending on your method.

Note that f is decreasing on $(-2, -1)$ and increasing on $(-1, 3)$. Using the symmetric difference quotient with $h = 3$ would not be accurate. (See Figure 2.4-2.)

- Remember that the $\lim\limits_{x \to 0} \dfrac{\sin 6x}{\sin 2x} = \dfrac{6}{2} = 3$ because the $\lim\limits_{x \to 0} \dfrac{\sin x}{x} = 1$.

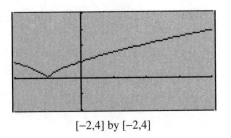

[−2,4] by [−2,4]

Figure 2.4-2

2.5 DERIVATIVES OF INVERSE FUNCTIONS

Let f be a one-to-one differentiable function with inverse function f^{-1}. If $f'(f^{-1}(a)) \neq 0$, then the inverse function f^{-1} is differentiable at a and $(f^{-1})'(a) = \dfrac{1}{f'(f^{-1}(a))}$. (See Figure 2.5-1.)

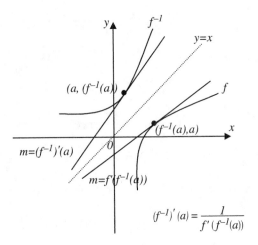

Figure 2.5-1

If $y = f^{-1}(x)$ so that $x = f(y)$, then $\dfrac{dy}{dx} = \dfrac{1}{dx/dy}$ with $\dfrac{dy}{dx} \neq 0$.

Example 1

If $f(x) = x^3 + 2x - 10$, find $(f^{-1})'(x)$

Step 1: Check if $(f^{-1})'(x)$ exists. $f'(x) = 3x^2 + 2$ and $f'(x) > 0$ for all real values of x. Thus $f(x)$ is strictly increasing which implies that $f(x)$ is $1 - 1$. Therefore, $(f^{-1})'(x)$ exists.

Step 2: Let $y = f(x)$ and thus $y = x^3 + 2x - 10$.

Step 3: Interchange x and y to obtain the inverse function $x = y^3 + 2y - 10$.

Step 4: Differentiate with respect to y: $\dfrac{dx}{dy} = 3y^2 + 2$

Step 5: Apply formula $\dfrac{dy}{dx} = \dfrac{1}{dx/dy}$.

$$\frac{dy}{dx} = \frac{1}{dx/dy} = \frac{1}{3y^2 + 2}. \text{ Thus, } (f^{-1})'(x) = \frac{1}{3y^2 + 2}$$

Example 2

Example 1 could have been done by using implicit differentiation.

Step 1: Let $y = f(x)$, and thus $y = x^3 + 2x - 10$.

Step 2: Interchange x and y to obtain the inverse function $x = y^3 + 2y - 10$.

Step 3: Differentiate each term implicitly with respect to x.

$$\frac{d}{dx}(x) = \frac{d}{dx}(y^3) + \frac{d}{dx}(2y) - \frac{d}{dx}(-10)$$

$$1 = 3y^2\frac{dy}{dx} + 2\frac{dy}{dx} - 0$$

Step 4: Solve for $\frac{dy}{dx}$.

$$1 = \frac{dy}{dx}(3y^2 + 2)$$

$$\frac{dy}{dx} = \frac{1}{3y^2 + 2}. \text{ Thus } (f^{-1})'(x) = \frac{1}{3y^2 + 2}$$

Example 3

If $f(x) = 2x^5 + x^3 + 1$, find (a) $f(1)$ and $f'(1)$ and (b) $(f^{-1})(4)$ and $(f^{-1})'(4)$.
Enter $y1 = 2x^5 + x^3 + 1$. Since $y1$ is strictly increasing, thus $f(x)$ has an inverse.

(a) $f(1) = 2(1)^5 + (1)^3 + 1 = 4$
$f'(x) = 10x^4 + 3x^2$
$f'(x) = 10(1)^4 + 3(1)^2 = 13$

(b) Since $f(1) = 4$ implies the point $(1, 4)$ is on the curve $f(x) = 2x^5 + x^3 + 1$, therefore the point $(4, 1)$ (which is the reflection of $(1, 4)$ on $y = x$) is on the curve $(f^{-1})(x)$. Thus $(f^{-1})(4) = 1$

$$(f^{-1})'(4) = \frac{1}{f'(1)} = \frac{1}{13}$$

Example 4

If $f(x) = 5x^3 + x + 8$, find $(f^{-1})'(8)$.
Enter $y1 = 5x^3 + x + 8$. Since $y1$ is strictly increasing near $x = 8$, $f(x)$ has an inverse near $x = 8$.
Note that $f(0) = 5(0)^3 + 0 + 8 = 8$ which implies the point $(0, 8)$ is on the curve of $f(x)$.
Thus, the point $(8, 0)$ is on the curve of $(f^{-1})(x)$.

$$f'(x) = 15x^2 + 1$$

$$f'(0) = 1$$

Therefore $(f^{-1})'(8) = \dfrac{1}{f'(0)} = \dfrac{1}{1} = 1$.

• Leave a multiple-choice question blank, if you have no clue. You don't have to answer every question to get a 5 on the AP Calculus AB exam.

2.6 HIGHER ORDER DERIVATIVES

If the derivative f' of a function f is differentiable, then the derivative of f' is the second derivative of f represented by f'' (reads as f double prime). You can continue to differentiate f as long as there is differentiability.

Some of the Symbols of Higher Order Derivatives:

$$f'(x), f''(x), f'''(x), f^{(4)}(x)$$

$$\frac{dy}{dx}, \frac{d^2y}{dx^2}, \frac{d^3y}{dx^3}, \frac{d^4y}{dx^4}$$

$$y', y'', y''', y^{(4)}$$

$$D_x(y), D_x^2(y), D_x^3(y), D_x^4(y)$$

Note that $\dfrac{d^2y}{dx^2} = \dfrac{d}{dx}\left(\dfrac{dy}{dx}\right)$ or $\dfrac{dy'}{dx}$.

Example 1

If $y = 5x^3 + 7x - 10$, find the first four derivatives.

$$\frac{dy}{dx} = 15x^2 + 7; \frac{d^2y}{dx^2} = 30x; \frac{d^3y}{dx^3} = 30; \frac{d^4y}{dx^4} = 0$$

Example 2

If $f(x) = \sqrt{x}$, find $f''(4)$.

Rewrite: $f(x) = \sqrt{x} = x^{1/2}$ and differentiate: $f'(x) = \dfrac{1}{2}x^{-1/2}$

Differentiate again:

$$f''(x) = -\frac{1}{4}x^{-3/2} = \frac{-1}{4x^{3/2}} = \frac{-1}{4\sqrt{x^3}} \text{ and } f''(4) = \frac{-1}{4\sqrt{4^3}} = -\frac{1}{32}$$

Example 3

If $y = x\cos x$, find y''.

Using the product rule, $y' = (1)(\cos x) + (x)(-\sin x) = \cos x - x\sin x$

$$y'' = -\sin x - [(1)(\sin x) + (x)(\cos x)] = -\sin x - \sin x - x\cos x$$

$$= -2\sin x - x\cos x$$

Or, you can use a calculator and enter $d(x^* \cos x, x, 2)$ and obtain the same result.

2.7 INDETERMINATE FORMS

Main Concepts: *L'Hôpital's Rule for indeterminate forms*

L'Hôpital's Rule

Suppose $f(x)$ and $g(x)$ are differentiable, and $g'(x) \neq 0$ near c, except possibly at c, and suppose $\lim_{x \to c} f(x) = 0$ and $\lim_{x \to c} g(x) = 0$ so that $\lim_{x \to c} \dfrac{f(x)}{g(x)} = \lim_{x \to c} \dfrac{f(x)}{g(x)} \to \dfrac{0}{0}$ is an indeterminate form. In this case $\lim_{x \to c} \dfrac{f(x)}{g(x)} = \lim_{x \to c} \dfrac{f'(x)}{g'(x)}$. If $\lim_{x \to c} f(x) = \pm\infty$ and $\lim_{x \to c} g(x) = \pm\infty$, the same rule applies.

Example 1

Find $\lim_{x \to 0} \dfrac{1 - \cos x}{x^2}$, if it exists.

Since $\lim_{x \to 0}(1 - \cos x) = 0$ and $\lim_{x \to 0}(x^2) = 0$, this limit is an inderminate form. Taking the derivatives, $\dfrac{d}{dx}(1 - \cos x) = \sin x$ and $\dfrac{d}{dx}(x^2) = 2x$. By L'Hôpital's Rule,

$$\lim_{x \to 0} \frac{1 - \cos x}{x^2} = \lim_{x \to 0} \frac{\sin x}{2x} = \frac{1}{2} \lim_{x \to 0} \frac{\sin x}{x} = \frac{1}{2}$$

Example 2

Find $\lim_{x \to \infty} x^3 e^{-x^2}$, if it exists.

Rewriting $\lim_{x \to \infty} x^3 e^{-x^2}$ as $\lim_{x \to \infty} \left(\dfrac{x^3}{e^{x^2}} \right)$ shows that the limit is an indeterminate form, since $\lim_{x \to \infty} (x^3) = \infty$ and $\lim_{x \to \infty} \left(e^{x^2} \right) = \infty$. Differentiating and applying L'Hôpital's Rule means that $\lim_{x \to \infty} \left(\dfrac{x^3}{e^{x^2}} \right) = \lim_{x \to \infty} \left(\dfrac{3x^2}{2xe^{x^2}} \right) = \dfrac{3}{2} \lim_{x \to \infty} \left(\dfrac{x}{e^{x^2}} \right)$. Unfortunately, this new limit is also indeterminate. However, it is possible to apply L'Hôpital's Rule again, so $\dfrac{3}{2} \lim_{x \to \infty} \left(\dfrac{x}{e^{x^2}} \right)$ will be equal to $\dfrac{3}{2} \lim_{x \to \infty} \left(\dfrac{1}{2xe^{x^2}} \right)$. This expression approaches zero as x becomes large, so $\lim_{x \to \infty} x^3 e^{-x^3} = 0$.

2.8 RAPID REVIEW

1. If $y = e^{x^3}$, find $\dfrac{dy}{dx}$.

   Answer: Using the chain rule, $\dfrac{dy}{dx} = \left(e^{x^3} \right)(3x^2)$.

2. Evaluate $\lim_{h \to 0} \dfrac{\cos\left(\dfrac{\pi}{6} + h\right) - \cos\left(\dfrac{\pi}{6}\right)}{h}$.

   Answer: The limit is equivalent to $\dfrac{d}{dx}\cos x \Big|_{x = \frac{\pi}{6}} = -\sin\left(\dfrac{\pi}{6}\right) = -\dfrac{1}{2}$.

3. Find $f'(x)$ if $f(x) = \ln(3x)$

   Answer: $f'(x) = \dfrac{1}{3x}(3) = \dfrac{1}{x}$.

4. Find the approximate value of $f'(3)$. (See Figure 2.8-1.)

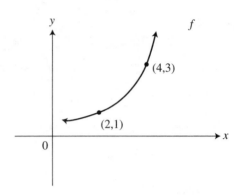

Figure 2.8-1

   Answer: Using the slope of the line segment joining $(2, 1)$ and $(4, 3)$,
   $$f'(3) = \frac{3-1}{4-2} = 1.$$

5. Find $\dfrac{dy}{dx}$ if $xy = 5x^2$.

   Answer: Using implicit differentiation, $1y + x\dfrac{dy}{dx} = 10x$. Thus $\dfrac{dy}{dx} = \dfrac{10x - y}{x}$.
   Or simply solve for y leading to $y = 5x$ and thus $\dfrac{dy}{dx} = 5$.

6. If $y = \dfrac{5}{x^2}$, find $\dfrac{d^2y}{dx^2}$.

   Answer: Rewrite $y = 5x^{-2}$. Then $\dfrac{dy}{dx} = -10x^{-3}$ and $\dfrac{d^2y}{dx^2} = 30x^{-4} = \dfrac{30}{x^4}$.

7. Using a calculator, write an equation of the line tangent to the graph $f(x) = -2x^4$ at the point where $f'(x) = -1$.

   Answer: $f'(x) = -8x^3$. Using a calculator, enter Solve $(-8x^\wedge 3 = -1, x)$ and obtain $x = \dfrac{1}{2} \Rightarrow f'\left(\dfrac{1}{2}\right) = -1$. Using the calculator $f\left(\dfrac{1}{2}\right) = -\dfrac{1}{8}$. Thus tangent is $y + \dfrac{1}{8} = -1\left(x - \dfrac{1}{2}\right)$.

8. $\displaystyle \lim_{x \to 2} \dfrac{x^2 + x - 6}{x^2 - 4}$

   Answer: Since $\dfrac{x^2 + x - 6}{x^2 - 4} \to \dfrac{0}{0}$, consider $\displaystyle \lim_{x \to 2} \dfrac{2x + 1}{2x} = \dfrac{5}{4}$

9. $\displaystyle \lim_{x \to \infty} \dfrac{\ln x}{x}$

   Answer: Since $\dfrac{\ln x}{x} \to \dfrac{\infty}{\infty}$, consider $\displaystyle \lim_{x \to \infty} \dfrac{1/x}{1} = \lim_{x \to \infty} \dfrac{1}{x} = 0$

2.9 PRACTICE PROBLEMS

Part A—The use of a calculator is not allowed.

Find the derivative of each of the following functions.

1. $y = 6x^5 - x + 10$

2. $f(x) = \dfrac{1}{x} + \dfrac{1}{\sqrt[3]{x^2}}$

3. $y = \dfrac{5x^6 - 1}{x^2}$

4. $y = \dfrac{x^2}{5x^6 - 1}$

5. $f(x) = (3x - 2)^5(x^2 - 1)$

6. $y = \sqrt{\dfrac{2x + 1}{2x - 1}}$

7. $y = 10\cot(2x - 1)$

8. $y = 3x\sec(3x)$

9. $y = 10\cos\left\lfloor \sin(x^2 - 4) \right\rfloor$

10. $y = 8\cos^{-1}(2x)$

11. $y = 3e^5 + 4xe^x$

12. $y = \ln(x^2 + 3)$

Part B—Calculators are allowed.

13. Find $\dfrac{dy}{dx}$, if $x^2 + y^3 = 10 - 5xy$.

14. The graph of a function f on $[1, 5]$ is shown in Figure 2.9-1. Find the approximate value of $f'(4)$.

15. Let f be a continuous and differentiable function. Selected values of f are shown below. Find the approximate value of f' at $x = 2$.

x	-1	0	1	2	3
f	6	5	6	9	14

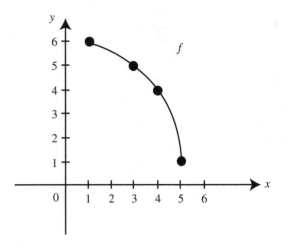

Figure 2.9-1

16. If $f(x) = x^5 + 3x - 8$, find $(f^{-1})'(-8)$.

17. Write an equation of the tangent to the curve $y = \ln x$ at $x = e$.

18. If $y = 2x\sin x$, find $\dfrac{d^2y}{dx^2}$ at $x = \dfrac{\pi}{2}$.

19. If the function $f(x) = (x - 1)^{2/3} + 2$, find all points where f is not differentiable.

20. Write an equation of the normal line to the curve $x\cos y = 1$ at $\left(2, \dfrac{\pi}{3}\right)$.

 21. $\displaystyle\lim_{x \to 3} \dfrac{x^2 - 3x}{x^2 - 9}$

22. $\displaystyle\lim_{x \to 0^+} \dfrac{\ln(x + 1)}{\sqrt{x}}$

23. $\displaystyle\lim_{x \to 0} \dfrac{e^x - 1}{\tan 2x}$

24. $\displaystyle\lim_{x \to 0} \dfrac{\cos(x) - 1}{\cos(2x) - 1}$

25. $\displaystyle\lim_{x \to 0} \dfrac{5x + 2\ln x}{x + 3\ln x}$

2.10   CUMULATIVE REVIEW PROBLEMS

("Calculator" indicates that calculators are permitted)

26. Find $\lim\limits_{h\to 0} \dfrac{\sin\left(\frac{\pi}{2}+h\right)-\sin\left(\frac{\pi}{2}\right)}{h}$

27. If $f(x)=\cos^2(\pi - x)$, find $f'(0)$.

28. Find $\lim\limits_{x\to\infty} \dfrac{x-25}{10+x-2x^2}$.

29. (Calculator) Let f be a continuous and differentiable function. Selected values of f are

shown below. Find the approximate value of f' at $x=2$.

x	0	1	2	3	4	5
f	3.9	4	4.8	6.5	8.9	11.8

30. (Calculator) If $f(x)=\begin{cases} \dfrac{x^2-9}{x-3}, & x\neq 3, \\ 3, & x=3 \end{cases}$

determine if $f(x)$ is continuous at $(x=3)$. Explain why or why not?

2.11   SOLUTIONS TO PRACTICE PROBLEMS

Part A—No calculators.

1. Applying the power rule, $\dfrac{dy}{dx}=30x^4-1$.

2. Rewrite $f(x)=\dfrac{1}{x}+\dfrac{1}{\sqrt[3]{x^2}}$ as $f(x)=x^{-1}+x^{-2/3}$.
   Differentiate,
   $f'(x)=-x^{-2}-\dfrac{2}{3}x^{-5/3}=-\dfrac{1}{x^2}-\dfrac{2}{3\sqrt[3]{x^5}}$.

3. Rewrite
   $y=\dfrac{5x^6-1}{x^2}$ as $y=\dfrac{5x^6}{x^2}-\dfrac{1}{x^2}=5x^4-x^{-2}$.
   Differentiate,
   $\dfrac{dy}{dx}=20x^3-(-2)x^{-3}=20x^3+\dfrac{2}{x^3}$.
   An alternate method is to differentiate
   $y=\dfrac{5x^6-1}{x^2}$ directly, using the quotient rule.

4. Applying the quotient rule,
   $\dfrac{dy}{dx}=\dfrac{(2x)(5x^6-1)-(30x^5)(x^2)}{(5x^6-1)^2}$
   $=\dfrac{10x^7-2x-30x^7}{(5x^6-1)^2}$
   $=\dfrac{-20x^7-2x}{(5x^6-1)^2}=\dfrac{-2x(10x^6+1)}{(5x^6-1)^2}$

5. Applying the product rule, $u=(3x-2)^5$ and $v=(x^2-1)$ and then the chain rule,

   $f'(x)=[5(3x-2)^4(3)][x^2-1]+[2x][(3x-2)^5]$
   $=15(x^2-1)(3x-2)^4+2x(3x-2)^5$
   $=(3x-2)^4[15(x^2-1)+2x(3x-2)]$
   $=(3x-2)^4[15x^2-15+6x^2-4x]$
   $=(3x-2)^4(21x^2-4x-15)$

6. Rewrite $y=\sqrt{\dfrac{2x+1}{2x-1}}$ as $y=\left(\dfrac{2x+1}{2x-1}\right)^{1/2}$.

   Applying first the chain rule and then the quotient rule,

   $\dfrac{dy}{dx}=\dfrac{1}{2}\left(\dfrac{2x+1}{2x-1}\right)^{-1/2}\left[\dfrac{(2)(2x-1)-(2)(2x+1)}{(2x-1)^2}\right]$

   $=\dfrac{1}{2}\dfrac{1}{\left(\dfrac{2x+1}{2x-1}\right)^{1/2}}\left[\dfrac{-4}{(2x-1)^2}\right]$

   $=\dfrac{1}{2}\dfrac{\dfrac{1}{(2x+1)^{1/2}}}{(2x-1)^{1/2}}\left[\dfrac{-4}{(2x-1)^2}\right]$

   $=\dfrac{-2}{(2x+1)^{1/2}(2x-1)^{3/2}}$

Note: $\left(\dfrac{2x+1}{2x-1}\right)^{1/2} = \dfrac{(2x+1)^{1/2}}{(2x-1)^{1/2}}$,

if $\dfrac{2x+1}{2x-1} > 0$ which implies $x < -\dfrac{1}{2}$

or $x > \dfrac{1}{2}$.

An alternate method of solution is to write

$y = \dfrac{\sqrt{2x+1}}{2x-1}$ and use the quotient rule. Another

method is to write $y = (2x+1)^{1/2}(2x-1)^{1/2}$ and use the product rule.

7. Let $u = 2x - 1$, $\dfrac{dy}{dx} = 10\left[-\csc^2(2x-1)\right](2)$

$= -20\csc^2(2x-1)$

8. Using the product rule,

$\dfrac{dy}{dx} = (3[\sec(3x)]) + [\sec(3x)\tan(3x)](3)[3x]$

$= 3\sec(3x) + 9x\sec(3x)\tan(3x)$

$= 3\sec(3x)[1 + 3x\tan(3x)]$

9. Using the chain rule, let $u = \sin(x^2 - 4)$.

$\dfrac{dy}{dx} = 10(-\sin[\sin(x^2-4)])[\cos(x^2-4)](2x)$

$= -20x\cos(x^2-4)\sin\left[\sin(x^2-4)\right]$

10. Using the chain rule, let $u = 2x$.

$\dfrac{dy}{dx} = 8\left(\dfrac{-1}{\sqrt{1-(2x)^2}}\right)(2) = \dfrac{-16}{\sqrt{1-4x^2}}$

11. Since $3e^5$ is a constant, thus its derivative is 0.

$\dfrac{dy}{dx} = 0 + (4)(e^x) + (e^x)(4x)$

$= 4e^x + 4xe^x = 4e^x(1 + x)$

12. Let $u = (x^2 + 3)$, $\dfrac{dy}{dx} = \left(\dfrac{1}{x^2+3}\right)(2x)$

$= \dfrac{2x}{x^2+3}$

Part B—Calculators are permitted.

13. Using implicit differentiation, differentiate each term with respect to x.

$2x + 3y^2\dfrac{dy}{dx} = 0 - \left[(5)(y) + \dfrac{dy}{dx}(5x)\right]$

$2x + 3y^2\dfrac{dy}{dx} = -5y - 5x\dfrac{dy}{dx}$

$3y^2\dfrac{dy}{dx} + 5x\dfrac{dy}{dx} = -5y - 2x$

$\dfrac{dy}{dx} = (3y^2 + 5x) = -5y - 2x$

$\dfrac{dy}{dx} = \dfrac{-5y - 2x}{3y^2 + 5x}$ or $\dfrac{dy}{dx} = \dfrac{-(2x + 5y)}{5x + 3y^2}$

14. Since $f'(4)$ is equivalent to the slope of the tangent to $f(x)$ at $x = 4$, there are several ways you can find its approximate value.

Method 1: Using the slope of the line segment joining the points at $x = 4$ and $x = 5$.

$f(5) = 1$ and $f(4) = 4$

$m = \dfrac{f(5) - f(4)}{5 - 4} = \dfrac{1 - 4}{1} = -3$

Method 2: Using the slope of the line segment joining the points at $x = 3$ and $x = 4$.

$f(3) = 5$ and $f(4) = 4$

$m = \dfrac{f(4) - f(3)}{4 - 3} = \dfrac{4 - 5}{4 - 3} = -1$

Method 3: Using the slope of the line segment joining the points at $x = 3$ and $x = 5$.

$f(3) = 5$ and $f(5) = 1$

$m = \dfrac{f(5) - f(3)}{5 - 3} = \dfrac{1 - 5}{5 - 3} = -2$

Note that -2 is the average of the results from methods 1 and 2. Thus $f'(4) \approx -3, -1$, or -2 depending on which line segment you use.

15. You can use the difference quotient

$\dfrac{f(a+h) - f(a)}{h}$ to approximate $f'(a)$.

Let $h = 1$; $f'(2) \approx \dfrac{f(3) - f(2)}{3 - 2} \approx \dfrac{14 - 9}{3 - 2} \approx 5$

Or, you can use the symmetric difference quotient $\dfrac{f(a + h) - f(a - h)}{2h}$ to approximate $f'(a)$.

Let $h = 1$; $f'(2) \approx \dfrac{f(3) - f(1)}{2 - 0} \approx \dfrac{14 - 6}{2} \approx 4$

Thus, $f'(2) \approx 4$ or 5 depending on your method.

16. Enter $y1 = x^5 + 3x - 8$. The graph of $y1$ is strictly increasing. Thus $f(x)$ has an inverse. Note that $f(0) = -8$. Thus the point $(0, -8)$ is on the graph of $f(x)$ which implies that the point $(-8, 0)$ is on the graph of $f^{-1}(x)$.

$f'(x) = 5x^4 + 3$ and $f'(0) = 3$

Since $(f^{-1})'(-8) = \dfrac{1}{f'(0)}$, thus $(f^{-1})'(-8) = \dfrac{1}{3}$.

17. $\dfrac{dy}{dx} = \dfrac{1}{x}$ and $\dfrac{dy}{dx}\Big|_{x=e} = \dfrac{1}{e}$

Thus the slope of the tangent to $y = \ln x$ at $x = e$ is $\dfrac{1}{e}$. At $x = e$, $y = \ln x = \ln e = 1$, which means the point $(e, 1)$ is on the curve of $y = \ln x$. Therefore, an equation of the tangent is $y - 1 = \dfrac{1}{e}(x - e)$ or $y = \dfrac{x}{e}$. [See Figure 2.11-1.]

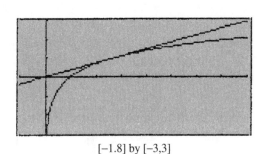

[-1.8] by [-3,3]

Figure 2.11-1

18. $\dfrac{dy}{dx} = (2)(\sin x) + (\cos x)(2x) = 2\sin x + 2x\cos x$

$\dfrac{d^2y}{dx^2} = 2\cos x + [(2)(\cos x) + (-\sin x)(2x)]$

$\quad = 2\cos x + 2\cos x - 2x\sin x$

$\quad = 4\cos x - 2x\sin x$

$\dfrac{d^2y}{dx^2}\Big|_{x=\pi/2} = 4\cos\left(\dfrac{\pi}{2}\right) - 2\left(\dfrac{\pi}{2}\right)\left(\sin\left(\dfrac{\pi}{2}\right)\right)$

$\quad = 0 - 2\left(\dfrac{\pi}{2}\right)(1) = -\pi$

Or, using a calculator, enter $d(2x - \sin(x), x, 2)$ $x = \dfrac{\pi}{2}$ and obtain $-\pi$.

19. Enter $y1 = (x - 1)^{2/3} + 2$ in your calculator. The graph of $y1$ forms a cusp at $x = 1$. Therefore, f is not differentiable at $x = 1$.

20. Differentiate with respect to x:

$(1)\cos y + \left[(-\sin y)\dfrac{dy}{dx}\right](x) = 0$

$\cos y - x\sin y\dfrac{dy}{dx} = 0$

$\dfrac{dy}{dx} = \dfrac{\cos y}{x\sin y}$

$\dfrac{dy}{dx}\Big|_{x=2,y=\pi/3} = \dfrac{\cos(\pi/3)}{(2)\sin(\pi/3)} = \dfrac{1/2}{2\left(\sqrt{3}/2\right)} = \dfrac{1}{2\sqrt{3}}$

Thus, the slope of the tangent to the curve at $(2, \pi/3)$ is $m = \dfrac{1}{2\sqrt{3}}$.

The slope of the normal line to the curve at $(2, \pi/3)$ is $m = -\dfrac{2\sqrt{3}}{1} = -2\sqrt{3}$.

Therefore an equation of the normal line is $y - \pi/3 = -2\sqrt{3}(x - 2)$.

21. $\displaystyle\lim_{x\to 3}\dfrac{x^2 - 3x}{x^2 - 9} = \lim_{x\to 3}\dfrac{2x - 3}{2x} = \dfrac{1}{2}$

22. $\displaystyle\lim_{x\to 0^+}\dfrac{\ln(x + 1)}{\sqrt{x}} = \lim_{x\to 0^+}\dfrac{1/(x + 1)}{1/(2\sqrt{x})}$

$\quad = \displaystyle\lim_{x\to 0^+}\dfrac{2\sqrt{x}}{x + 1} = 0$

23. $\displaystyle\lim_{x\to 0}\dfrac{e^x - 1}{\tan 2x} = \lim_{x\to 0}\dfrac{e^x}{2\sec^2 2x} = \dfrac{1}{2}$

24. $\displaystyle\lim_{x\to 0}\dfrac{\cos(x) - 1}{\cos(2x) - 1} = \lim_{x\to 0}\dfrac{-\sin x}{-2\sin(2x)}$

$\quad = \displaystyle\lim_{x\to 0}\dfrac{-\cos x}{-4\cos(2x)} = \dfrac{1}{4}$

25. $\displaystyle\lim_{x\to\infty}\dfrac{5x + 2\ln x}{x + 3\ln x} = \lim_{x\to\infty}\dfrac{5 + (2/x)}{5 + (3/x)} = 5$

2.12   SOLUTIONS TO CUMULATIVE REVIEW PROBLEMS

26. The expression $\lim\limits_{h \to 0} \dfrac{\sin(\pi/2 + h) - \sin(\pi/2)}{h}$ is
the derivative of $\sin x$ at $x = \pi/2$ which is the
slope of the tangent to $\sin x$ at $x = \pi/2$.
The tangent to $\sin x$ at $x = \pi/2$ is parallel
to the x-axis.
Therefore the slope is 0, i.e.,
$$\lim\limits_{h \to 0} \frac{\sin(\pi/2 + h) - \sin(\pi/2)}{h} = 0.$$
An alternate method is to expand $\sin(\pi/2 + h)$
as $\sin(\pi/2) \cos h + \cos(\pi/2) \sin h$.
Thus, $\lim\limits_{h \to 0} \dfrac{\sin(\pi/2 + h) - \sin(\pi/2)}{h}$

$= \lim\limits_{h \to 0} \dfrac{\sin(\pi/2) \cos h + \cos(\pi/2) \sin h - \sin(\pi/2)}{h}$

$= \lim\limits_{h \to 0} \dfrac{\sin(\pi/2)[\cos h - 1] + \cos(\pi/2) \sin h}{h}$

$= \lim\limits_{h \to 0} \sin\left(\dfrac{\pi}{2}\right)\left(\dfrac{\cos h - 1}{h}\right) - \lim\limits_{h \to 0} \cos\left(\dfrac{\pi}{2}\right)\left(\dfrac{\sin h}{h}\right)$

$= \sin\left(\dfrac{\pi}{2}\right) \lim\limits_{h \to 0} \left(\dfrac{\cos h - 1}{h}\right) - \cos\left(\dfrac{\pi}{2}\right) \lim\limits_{h \to 0} \left(\dfrac{\sin h}{h}\right)$

$= \left[\sin\left(\dfrac{\pi}{2}\right)\right] 0 + \cos\left(\dfrac{\pi}{2}\right)(1) = \cos\left(\dfrac{\pi}{2}\right) = 0$

27. Using the chain rule, let $u = (\pi - x)$.
Then, $f'(x) = 2 \cos(\pi - x)[-\sin(\pi - x)](-1)$

$= 2 \cos(\pi - x) \sin(\pi - x)$

$f'(0) = 2 \cos \pi \sin \pi = 0.$

28. Since the degree of the polynomial in the
denominator is greater than the degree of the
polynomial in the numerator, the limit is 0.

29. You can use the difference quotient
$\dfrac{f(a + h) - f(a)}{h}$ to approximate $f'(a)$.

Let $h = 1$; $f'(2) \approx \dfrac{f(3) - f(2)}{3 - 2}$

$\approx \dfrac{6.5 - 4.8}{1} \approx 1.7.$

Let $h = 2$; $f'(2) \approx \dfrac{f(4) - f(2)}{4 - 2}$

$\approx \dfrac{8.9 - 4.8}{2} \approx 2.05.$

Or, you can use the symmetric difference
quotient $\dfrac{f(a + h) - f(a - h)}{2h}$ to
approximate $f'(a)$.

Let $h = 1$; $f'(2) \approx \dfrac{f(3) - f(1)}{3 - 1}$

$\approx \dfrac{6.5 - 4}{2} \approx 1.25$

Let $h = 2$; $f'(2) \approx \dfrac{f(4) - f(0)}{4 - 0}$

$\approx \dfrac{8.9 - 3.9}{4} \approx 1.25$

Thus, $f'(2) = 1.7, 2.05$ or 1.25 depending on
your method.

30. (See Figure 2.12–1.) Checking the three
conditions of continuity:

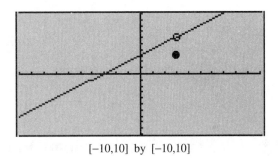

[−10,10] by [−10,10]

Figure 2.12-1

(1) $f(3) = 3$

(2) $\lim\limits_{x \to 3} \dfrac{x^2 - 9}{x - 3} = \lim\limits_{x \to 3} \left(\dfrac{(x + 3)(x - 3)}{(x - 3)}\right)$

$= \lim\limits_{x \to 3}(x + 3) = (3) + 3 = 6.$

(3) Since $f(3) \neq \lim\limits_{x \to 3} f(x)$, $f(x)$ is discontinuous
at $x = 3$.

Chapter 3

Graphs of Functions and Derivatives

3.1 ROLLE'S THEOREM, MEAN VALUE THEOREM, AND EXTREME VALUE THEOREM

Main Concepts: *Rolle's Theorem, Mean Value Theorem, Extreme Value Theorem*

- Set your calculator to Radians and change it to Degree if/when you need to. Don't forget to change it back to Radians after you've finished using it in Degrees.

Rolle's Theorem and Mean Value Theorem

Rolle's Theorem: If f is a function that satisfies the following three conditions:

1. f is continuous on a closed interval $[a, b]$
2. f is differentiable on the open interval (a, b)
3. $f(a) = f(b) = 0$

then there exists a number c in (a, b) such that $f'(c) = 0$. (See Figure 3.1-1.)

Note that if you change condition 3 from $f(a) = f(b) = 0$ to $f(a) = f(b)$, the conclusion of Rolle's Theorem is still valid.

Mean Value Theorem: If f is a function that satisfies the following conditions:

1. f is continuous on a closed interval $[a, b]$
2. f is differentiable on the open interval (a, b)

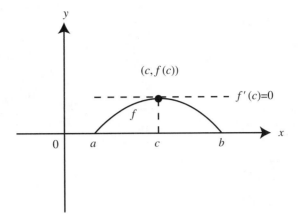

Figure 3.1-1

then there exists a number c in (a, b) such that $f'(c) = \dfrac{f(b) - f(a)}{b - a}$. (See Figure 3.1-2.)

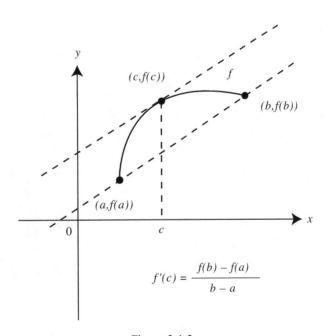

$$f'(c) = \frac{f(b) - f(a)}{b - a}$$

Figure 3.1-2

Example 1

If $f(x) = x^2 + 4x - 5$, show that the hypotheses of Rolle's Theorem are satisfied on the interval $[-4, 0]$ and find all values of c that satisfy the conclusion of the theorem. Check the three conditions in the hypothesis of Rolle's Theorem:

(1) $f(x) = x^2 + 4x - 5$ is continuous everywhere since it is polynomial.
(2) The derivative $f'(x) = 2x + 4$ is defined for all numbers and thus is differentiable on $(-4, 0)$.
(3) $f(0) = f(-4) = -5$. Therefore, there exists a c in $(-4, 0)$ such that $f'(c) = 0$. To find c, set $f'(x) = 0$. Thus $2x + 4 = 0 \Rightarrow x = -2$, i.e., $f'(-2) = 0$. (See Figure 3.1-3.)

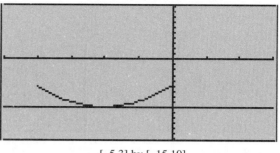

[−5,3] by [−15,10]

Figure 3.1-3

Example 2

Let $f(x) = \dfrac{x^3}{3} - \dfrac{x^2}{2} - 2x + 2$. Using Rolle's Theorem, show that there exists a number c in the domain of f such that $f'(c) = 0$. Find all values of c.

Note $f(x)$ is a polynomial and thus $f(x)$ is continuous and differentiable everywhere. Enter $y1 = \dfrac{x^3}{3} - \dfrac{x^2}{2} - 2x + 2$. The zeros of $y1$ are approximately -2.3, 0.9, and 2.9 i.e., $f(-2.3) = f(0.9) = f(2.9) = 0$. Therefore, there exists at least one c in the interval $(-2.3, 0.9)$ and at least one c in the interval $(0.9, 2.9)$ such that $f'(c) = 0$. Use d(differentiate) to find $f'(x)$: $f'(x) = x^2 - x - 2$. Set $f'(x) = 0 \Rightarrow x^2 - x - 2 = 0$ or $(x - 2)(x + 1) = 0$.

Thus $x = 2$ or $x = -1$, which implies $f'(2) = 0$ and $f'(-1) = 0$. Therefore the values of c are -1 and 2. (See Figure 3.1-4.)

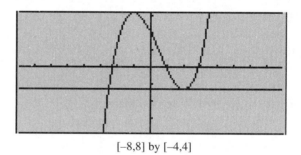

[−8,8] by [−4,4]

Figure 3.1-4

Example 3

The points $P(1, 1)$ and $Q(3, 27)$ are on the curve $f(x) = x^3$. Using the Mean Value Theorem, find c in the interval $(1, 3)$ such that $f'(c)$ is equal to the slope of the secant $\overline{PQ}$.

The slope of secant $\overline{PQ}$ is $m = \dfrac{27 - 1}{3 - 1} = 13$. Since $f(x)$ is defined for all real numbers, $f(x)$ is continuous on $[1, 3]$. Also $f'(x) = 3x^2$ is defined for all real numbers. Thus $f(x)$ is differentiable on $(1, 3)$. Therefore, there exists a number c in $(1, 3)$ such that $f'(c) = 13$. Set $f'(c) = 13 \Rightarrow 3(c)^2 = 13$ or $c^2 = \dfrac{13}{3}$ $c = \pm\sqrt{\dfrac{13}{3}}$. Since only $\sqrt{\dfrac{13}{3}}$ is in the interval $(1, 3)$, thus $c = \sqrt{\dfrac{13}{3}}$. (See Figure 3.1-5.)

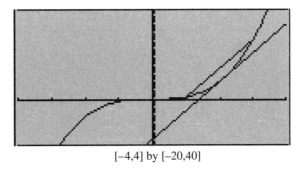

[−4,4] by [−20,40]

Figure 3.1-5

Example 4

Let f be the function $f(x) = (x - 1)^{2/3}$. Determine if the hypotheses of the Mean Value Theorem are satisfied on the interval $[0, 2]$ and if so, find all values of c that satisfy the conclusion of the theorem.

Enter $y1 = (x - 1)^{2/3}$. The graph $y1$ shows that there is a cusp at $x = 1$. Thus, $f(x)$ is not differentiable on $(0, 2)$ which implies there may or may not exist a c in $(0, 2)$ such that $f'(c) = \dfrac{f(2) - f(0)}{2 - 0}$. The derivative $f'(x) = \dfrac{2}{3}(x - 1)^{-1/3}$ and $\dfrac{f(2) - f(0)}{2 - 0} = \dfrac{1 - 1}{2} = 0$. Set $\dfrac{2}{3}(x - 1)^{1/3} = 0 \Rightarrow x = 1$. Note f is not differentiable $(a + x = 1)$. Therefore c does not exist. (See Figure 3.1-6.)

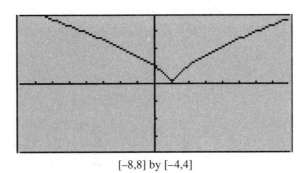

[−8,8] by [−4,4]

Figure 3.1-6

- The formula for finding the area of an equilateral triangle is $area = \dfrac{s^2\sqrt{3}}{4}$ where s is the length of a side. You might need this to find the volume of a solid whose cross sections are equilateral triangles.

Extreme Value Theorem

Extreme Value Theorem: If f is a continuous function on a closed interval $[a, b]$, then f has both a maximum and a minimum value on the interval.

Example 1

If $f(x) = x^3 + 3x^2 - 1$, find the maximum and minimum values of f on $[-2, 2]$. Since $f(x)$ is a polynomial, it is a continuous function everywhere. Enter $y1 = x^3 + 3x^2 - 1$.

The graph of $y1$ indicates that f has a minimum of -1 at $x = 0$ and a maximum value of 19 at $x = 2$. (See Figure 3.1-7.)

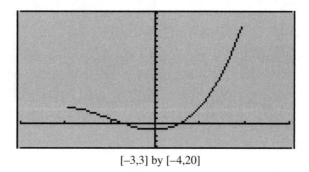

[−3,3] by [−4,20]

Figure 3.1-7

Example 2

If $f(x) = \dfrac{1}{x^2}$, find any maximum and minimum values of f on $[0, 3]$. Since $f(x)$ is a rational function, it is continuous everywhere except at values where the denominator is 0. In this case, at $x = 0$, $f(x)$ is undefined. Since $f(x)$ is not continuous on $[0, 3]$, the Extreme Value Theorem may not be applicable. Enter $y1 = \dfrac{1}{x^2}$. The graph of $y1$ shows that as $x \to 0^+$, $f(x)$ increases without bound (i.e., $f(x)$ goes to infinity). Thus f has no maximum value. The minimum value occurs at the endpoint $x = 3$ and the minimum value is $\dfrac{1}{9}$. (See Figure 3.1-8.)

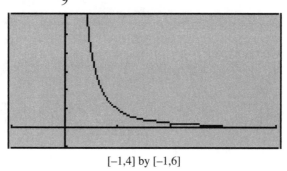

[−1,4] by [−1,6]

Figure 3.1-8

3.2 DETERMINING THE BEHAVIOR OF FUNCTIONS

Main Concepts: *Test for Increasing and Decreasing Functions, First Derivative Test and Second Derivative Test for Relative Extrema, Test for Concavity and Points of Inflection*

Test for Increasing and Decreasing Functions

Let f be a continuous function on the closed interval $[a, b]$ and differentiable on the open interval (a, b).

1. If $f'(x) > 0$ on (a, b), then f is increasing on $[a, b]$
2. If $f'(x) < 0$ on (a, b), then f is decreasing on $[a, b]$
3. If $f'(x) = 0$ on (a, b), then f is constant on $[a, b]$

Definition: Let f be a function defined at a number c. Then c is a critical number of f if either $f'(c) = 0$ or $f'(c)$ does not exist. (See Figure 3.2-1.)

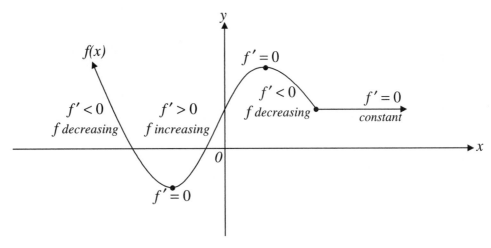

Figure 3.2-1

Example 1

Find the critical numbers of $f(x) = 4x^3 + 2x^2$.

To find the critical numbers of $f(x)$, you have to determine where $f'(x) = 0$ and where $f'(x)$ does not exist. Note $f'(x) = 12x^2 + 4x$, and $f'(x)$ is defined for all real numbers. Let $f'(x) = 0$ and thus $12x^2 + 4x = 0$ which implies $4x(3x + 1) = 0 \Rightarrow x = -1/3$ or $x = 0$. Therefore the critical numbers of f are 0 and $-1/3$. (See Figure 3.2-2.)

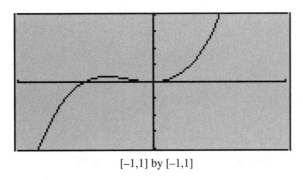

[–1,1] by [–1,1]

Figure 3.2-2

Example 2

Find the critical numbers of $f(x) = (x - 3)^{2/5}$.

$f'(x) = \dfrac{2}{5}(x - 3)^{-3/5} = \dfrac{2}{5(x - 3)^{3/5}}$. Note that $f'(x)$ is undefined at $x = 3$ and that $f'(x) \neq 0$. Therefore, 3 is the only critical number of f. (See Figure 3.2-3.)

Example 3

The graph of f' on (1, 6) is shown in Figure 3.2-4. Find the intervals on which f is increasing or decreasing.

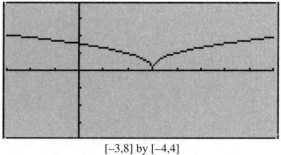

[−3,8] by [−4,4]

Figure 3.2-3

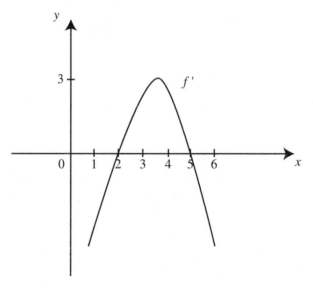

Figure 3.2-4

Solution: See Figure 3.2-5.

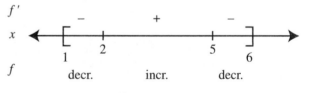

Figure 3.2-5

Thus, f is decreasing on [1, 2] and [5, 6] and increasing on [2, 5].

Example 4

Find the open intervals on which $f(x) = (x^2 - 9)^{2/3}$ is increasing or decreasing.

Step 1: Find the critical numbers of f.

$$f'(x) = \frac{2}{3}(x^2 - 9)^{-1/3}(2x) = \frac{4x}{3(x^2 - 9)^{1/3}}$$

Set $f'(x) = 0 \Rightarrow 4x = 0$ or $x = 0$.

Since $f'(x)$ is a rational function, $f'(x)$ is undefined at values where the denominator is 0. Thus, set $x^2 - 9 = 0 \Rightarrow x = 3$ or $x = -3$. Therefore the critical numbers are -3, 0, and 3.

Step 2: Determine intervals.

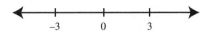

Intervals are $(-\infty, -3)$, $(-3, 0)$, $(0, 3)$ and $(3, \infty)$.

Step 3: Set up a table.

Interval	$(-\infty, -3)$	$(-3, 0)$	$(0, 3)$	$(3, \infty)$
Test Point	-5	-1	1	5
$f'(x)$	$-$	$+$	$-$	$+$
$f(x)$	decr	incr	decr	incr

Step 4: Write a conclusion. Therefore $f(x)$ is increasing on $[-3, 0]$ and $[3, \infty)$ and decreasing on $(-\infty, -3]$ and $[0, 3]$. (See Figure 3.2-6.)

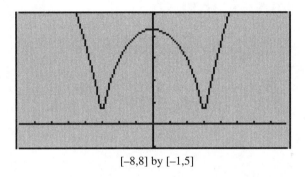

[−8,8] by [−1,5]

Figure 3.2-6

Example 5

The derivative of a function f is given as $f'(x) = \cos(x^2)$. Using a calculator, find the values of x on $\left[-\dfrac{\pi}{2}, \dfrac{\pi}{2}\right]$ such that f is increasing. (See Figure 3.2-7.)

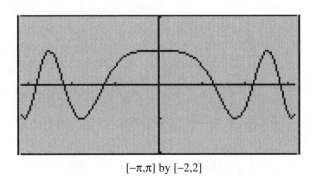

[−π,π] by [−2,2]

Figure 3.2-7

Using the zero function of the calculator, you obtain $x = 1.25331$ is a zero of f' on $\left[0, \dfrac{\pi}{2}\right]$. Since $f'(x) = \cos(x^2)$ is an even function, $x = -1.25331$ is also a zero on $\left[-\dfrac{\pi}{2}, 0\right]$. (See Figure 3.2-8.)

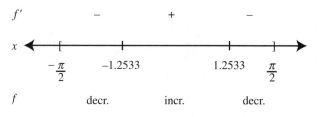

Figure 3.2-8

Thus f is increasing on $[-1.2533, 1.2533]$.

> • Bubble in the right grid. You have to be careful in filling in the bubbles especially when you skip a question.

First Derivative Test and Second Derivative Test for Relative Extrema

First Derivative Test for Relative Extrema:

Let f be a continuous function and c be a critical number of f. (Figure 3.2-9.)

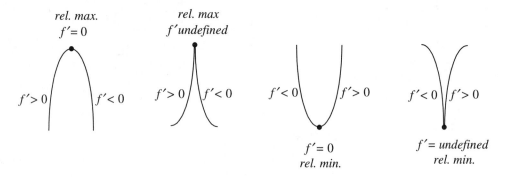

Figure 3.2-9

1. If $f'(x)$ changes from positive to negative at $x = c$ ($f' > 0$ for $x < c$ and $f' < 0$ for $x > c$), then f has a relative maximum at c.
2. If $f'(x)$ changes from negative to positive at $x = c$ ($f' < 0$ for $x < c$ and $f' > 0$ for $x > c$), then f has a relative minimum at c.

Second Derivative Test for Relative Extrema:

Let f be a continuous function at a number c.

1. If $f'(c) = 0$ and $f''(c) < 0$, then $f(c)$ is a relative maximum.
2. If $f'(c) = 0$ and $f''(c) > 0$, then $f(c)$ is a relative minimum.
3. If $f'(c) = 0$ and $f''(c) = 0$, then the test is inconclusive. Use the first Derivative Test.

Example 1

See Figure 3.2-10. The graph of f', the derivative of a function f is shown in Figure 3.2-10. Find the relative extrema of f.

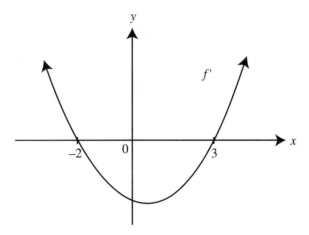

Figure 3.2-10

Solution: See Figure 3.2-11.

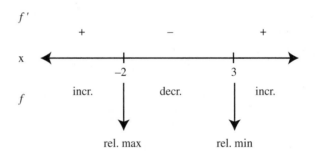

Figure 3.2-11

Thus f has a relative maximum at $x = -2$, and a relative minimum at $x = 3$.

Example 2

Find the relative extrema for the function $f(x) = \dfrac{x^3}{3} - x^2 - 3x$.

Step 1: Find $f'(x)$.

$$f'(x) = x^2 - 2x - 3$$

Step 2: Find all critical numbers of $f(x)$.
Note that $f'(x)$ is defined for all real numbers.
Set $f'(x) = 0$: $x^2 - 2x - 3 = 0 \Rightarrow (x - 3)(x + 1) = 0 \Rightarrow x = 3$ or $x = -1$.

Step 3: Find $f''(x)$: $f''(x) = 2x - 2$.

Step 4: Apply the Second Derivative Test.
$f''(3) = 2(3) - 2 = 4 \Rightarrow f(3)$ is a relative minimum.
$f''(-1) = 2(-1) - 2 = -4 \Rightarrow f(-1)$ is a relative maximum.

$$f(3) = \frac{3^3}{3} - (3)^2 - 3(3) = -9 \text{ and } f(-1) = \frac{5}{3}.$$

Therefore, -9 is a relative minimum value of f and $\dfrac{5}{3}$ is a relative maximum value. (See Figure 3.2-12.)

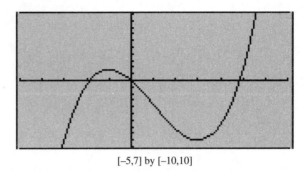

[−5,7] by [−10,10]

Figure 3.2-12

Example 3

Find the relative extrema for the function $f(x) = (x^2 - 1)^{2/3}$.

Using the First Derivative Test:

Step 1: Find $f'(x)$.

$$f'(x) = \frac{2}{3}(x^2 - 1)^{-1/3}(2x) = \frac{4x}{3(x^2 - 1)^{1/3}}.$$

Step 2: Find all critical numbers of f.
Set $f'(x) = 0$. Thus $4x = 0$ or $x = 0$.
Set $x^2 - 1 = 0$. Thus $f'(x)$ is undefined at $x = 1$ and $x = -1$. Therefore the critical numbers are -1, 0 and 1.

Step 3: Determine intervals.

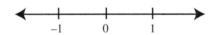

The intervals are $(-\infty, -1)$, $(-1, 0)$, $(0, 1)$ and $(1, \infty)$.

Step 4: Set up a table.

Interval	$(-\infty, -1)$	$x = -1$	$(-1, 0)$	$x = 0$	$(0, 1)$	$x = 1$	$(1, \infty)$
Test Point	-2		$-1/2$		$1/2$		2
$f'(x)$	$-$	undefined	$+$	0	$-$	undefined	$+$
$f(x)$	decr	rel min	incr	rel max	decr	rel min	incr

Step 5: Write a conclusion
Using the First Derivative Test, note that $f(x)$ has a relative maximum at $x = 0$ and relative minimum at $x = -1$ and $x = 1$.

Note that $f(-1) = 0$, $f(0) = 1$ and $f(1) = 0$. Therefore, 1 is a relative maximum value and 0 is a relative minimums value. (See Figure 3.2-13.)

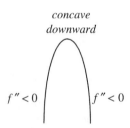

[–3,3] by [–2,5]

Figure 3.2-13

> • Don't forget the constant, C, when you write the antiderivative after evaluating an indefinite integral, e.g., $\int \cos x\,dx = \sin x + C$.

Test for Concavity and Points of Inflection

Test for Concavity:

Let f be a differentiable function.

1. If $f'' > 0$ on an interval I, then f is concave upward on I.
2. If $f'' < 0$ on an interval I, then f is concave downward on I.

See Figures 3.2-14 and 3.2-15.

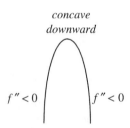

concave downward

$f'' < 0 \qquad f'' < 0$

Figure 3.2-14

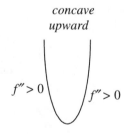

concave upward

$f'' > 0 \qquad f'' > 0$

Figure 3.2-15

A point P on a curve is a point of inflection if

1. the curve has a tangent line at P, and
2. the curve changes concavity at P (from concave upward to downward or from concave downward to upward).

See Figures 3.2-16–3.2-18.

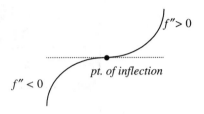

Figure 3.2-16

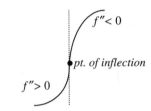

Figure 3.2-17

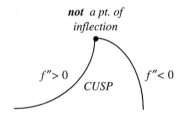

Figure 3.2-18

Note that if a point $(a, f(a))$ is a point of inflection, then $f''(c) = 0$ or $f''(c)$ does not exist. (The converse of the statement is not necessarily true.)

Note: There are some textbooks that define a point of inflection as a point where the concavity changes and do not require the existence of a tangent at the point of inflection. In that case, the point at the *cusp* in Figure 3.2-18 would be a point of inflection.

Example 1

See Figure 3.2-19. The graph of f', the derivative of a function f, is shown in Figure 3.2-19 on page 99. Find the points of inflection of f and determine where the function f is concave upward and where it is concave downward on $[-3, 5]$.

Solution: See Figure 3.2-20 on page 99.

Thus f is concave upward on $[-3, 0)$ and $(3, 5]$, and is concave downward on $(0, 3)$.

There are two points of inflection: one at $x = 0$ and the other at $x = 3$.

Example 2

Using a calculator, find the values of x at which the graph of $y = x^2 e^x$ changes concavity.

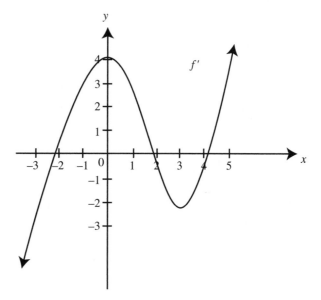

Figure 3.2-19

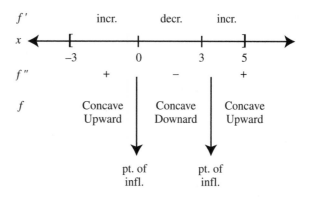

Figure 3.2-20

Enter $y_1 = x^\wedge 2 * e^\wedge x$ and $y_2 = d(y_1(x), x, 2)$. The graph of y_2, the second derivative of y, is shown in Figure 3.2-21. Using the Zero function, you obtain $x = -3.41421$ and $x = -0.585786$. (See Figures 3.2-21 and 3.2-22.)

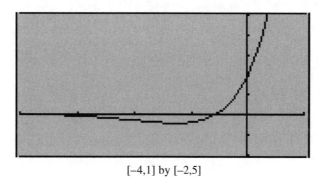

[−4,1] by [−2,5]

Figure 3.2-21

Thus, f changes concavity at $x = -3.41421$ and $x = -0.585786$.

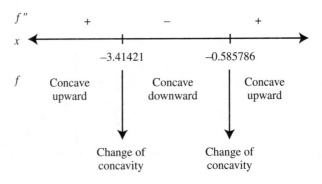

Figure 3.2-22

Example 3

Find the points of inflection of $f(x) = x^3 - 6x^2 + 12x - 8$ and determine the intervals where the function f is concave upward and where it is concave downward.

Step 1: Find $f'(x)$ and $f''(x)$.
$f'(x) = 3x^2 - 12x + 12$
$f''(x) = 6x - 12$

Step 2: Set $f''(x) = 0$
$6x - 12 = 0$
$x = 2$
Note that $f''(x)$ is defined for all real numbers.

Step 3: Determine intervals.

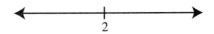

The intervals are $(-\infty, 2)$ and $(2, \infty)$.

Step 4: Set up a table.

Interval	$(-\infty, 2)$	$x = 2$	$(2, \infty)$
Test Point	0		5
$f''(x)$	−	0	+
$f(x)$	concave downward	point of inflection	concave upward

Since $f(x)$ has change of concavity at $x = 2$, the point $(2, f(2))$ is a point of inflection. $f(2) = (2)^3 - 6(2)^2 + 12(2) - 8 = 0$.

Step 5: Write a conclusion.
Thus $f(x)$ is concave downward on $(-\infty, 2)$, concave upward on $(2, \infty)$ and $f(x)$ has a point of inflection at $(2, 0)$. (See Figure 3.2-23.)

Example 4

Find the points of inflection of $f(x) = (x - 1)^{2/3}$ and determine the intervals where the function f is concave upward and where it is concave downward.

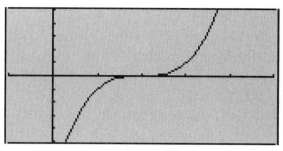

[−1,5] by [−5,5]

Figure 3.2-23

Step 1: Find $f'(x)$ and $f''(x)$.

$$f'(x) = \frac{2}{3}(x-1)^{-1/3} = \frac{2}{3(x-1)^{1/3}}$$

$$f''(x) = -\frac{2}{9}(x-1)^{-4/3} = \frac{-2}{9(x-1)^{4/3}}$$

Step 2: Find all values of x where $f''(x) = 0$ or $f''(x)$ is undefined.
Note that $f''(x) \neq 0$ and that $f''(1)$ is undefined.

Step 3: Determine intervals.

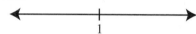

The intervals are $(-\infty, 1)$, and $(1, \infty)$.

Step 4: Set up a table.

Interval	$(-\infty, 1)$	$x = 1$	$(1, \infty)$
Test Point	0		2
$f''(x)$	−	undefined	−
$f(x)$	concave downward	no change of cancavity	concave downward

Note that since $f(x)$ has no change of concavity at $x = 1$, f does not have a point of inflection.

Step 5: Write a conclusion.
Therefore $f(x)$ is concave downward on $(-\infty, \infty)$ and has no point of inflection. (See Figure 3.2-24.)

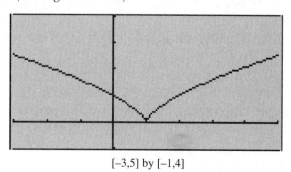

[−3,5] by [−1,4]

Figure 3.2-24

Example 5

The graph of f is shown in Figure 3.2-25 and f is twice differentiable. Which of the following statements is true:

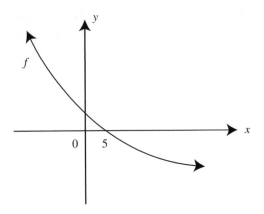

Figure 3.2-25

(A) $f(5) < f'(5) < f''(5)$

(B) $f''(5) < f'(5) < f(5)$

(C) $f'(5) < f(5) < f''(5)$

(D) $f'(5) < f''(5) < f(5)$

(E) $f''(5) < f(5) < f'(5)$

The graph indicates that (1) $f(5) = 0$, (2) $f'(5) < 0$, since f is decreasing; and (3) $f''(5) < 0$, since f is concave upward. Thus $f'(5) < f(5) < f''(5)$, choice (C).

- Move on. Don't linger on a problem too long. You can earn many more points from other problems.

3.3   SKETCHING THE GRAPHS OF FUNCTIONS

Main Concepts: *Graphing without Calculators, Graphing with Calculators*

Graphing without Calculators

General Procedure for Sketching the Graph of a Function

Steps:

1. Determine the domain and if possible the range of the function $f(x)$.
2. Determine if the function has any symmetry, i.e., if the function is even ($f(x) = f(-x)$); odd ($f(x) = -f(-x)$) or periodic ($f(x + p) = f(x)$).
3. Find $f'(x)$ and $f''(x)$.
4. Find all critical numbers ($f'(x) = 0$ or $f'(x)$ is undefined) and possible points of inflection ($f''(x) = 0$ or $f''(x)$ is undefined).

5. Using the numbers in Step 4, determine the intervals on which to analyze $f(x)$.
6. Set up a table using the intervals, to
   (a) determine where $f(x)$ is increasing or decreasing.
   (b) find relative and absolute extrema.
   (c) find points of inflection.
   (d) determine the concavity of $f(x)$ on each interval.
7. Find any horizontal, vertical, or slant asymptotes.
8. If necessary, find the x-intercepts, the y-intercepts, and a few selected points.
9. Sketch the graph.

Example 1

Sketch the graph of $f(x) = \dfrac{x^2 - 4}{x^2 - 25}$.

Step 1: Domain: all real numbers $x \neq \pm 5$.

Step 2: Symmetry: $f(x)$ is an even function $(f(x) = f(-x))$; symmetrical with respect to y-axis.

Step 3: $f'(x) = \dfrac{(2x)(x^2 - 25) - (2x)(x^2 - 4)}{(x^2 - 25)^2} = \dfrac{-42x}{(x^2 - 25)^2}$

$f''(x) = \dfrac{-42(x^2 - 25)^2 - 2(x^2 - 25)(2x)(-42x)}{(x^2 - 25)^4} = \dfrac{42(3x^2 + 25)}{(x^2 - 25)^3}$

Step 4: Critical numbers:
$f'(x) = 0 \Rightarrow -42x = 0$ or $x = 0$
$f'(x)$ is undefined at $x = \pm 5$ which are not in the domain.

Possible points of inflection:
$f''(x) \neq 0$ and $f''(x)$ is undefined at $x = \pm 5$ which are not in the domain.

Step 5: Determine intervals:

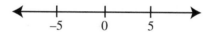

Intervals are $(-\infty, -5)$, $(5, 0)$, $(0, 5)$ & $(5, \infty)$

Step 6: Set up a table:

Intervals	$(-\infty, -5)$	$x = -5$	$(-5, 0)$	$x = 0$	$(0, 5)$	$x = 5$	$(5, \infty)$
$f(x)$		undefined		4/25		undefined	
$f'(x)$	+	undefined	+	0	−	undefined	−
$f''(x)$	+	undefined	−	−	−	undefined	+
conclusion	incr concave upward		incr concave downward	rel max	decr concave downward		decr concave upward

Step 7: Vertical asymptote: $x = 5$ and $x = -5$
Horizontal asymptote: $y = 1$

Step 8: y-intercept: $\left(0, \dfrac{4}{25}\right)$

x-intercept: $(-2, 0)$ and $(2, 0)$

See Figure 3.3-1.

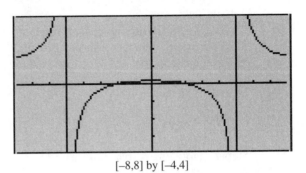

[−8,8] by [−4,4]

Figure 3.3-1

Graphing with Calculators

Example 1

Using a calculator, sketch the graph of $f(x) = -x^{5/3} + 3x^{2/3}$ indicating all relative extrema, points of inflection, horizontal and vertical asymptotes, intervals where $f(x)$ is increasing or decreasing, and intervals where $f(x)$ is concave upward or downward.

1. Domain: all real numbers; Range: all real numbers
2. No symmetry
3. Relative Minimum $(0, 0)$; Relative Maximum $(1.2, 2.03)$; Points of Inflection $(-0.6, 2.56)$
4. No asymptote.
5. $f(x)$ is decreasing on $(-\infty, 0]$, $[1.2, \infty)$; and increasing on $(0, 1.2)$
6. Evaluating the $f''(x)$ on either side of the point of inflection $(-0.6, 2.56)$

$$d\left(-x \wedge \left(\frac{5}{3}\right) + 3 * x \wedge \left(\frac{2}{3}\right), x, 2\right) \; x = -2 \rightarrow 0.19$$

$$d\left(-x \wedge \left(\frac{5}{3}\right) + 3 * x \wedge \left(\frac{2}{3}\right), x, 2\right) \; x = -1 \rightarrow -4.66$$

$\Rightarrow f(x)$ is concave upward on $(-\infty, -0.6)$ and concave downward on $(-0.6, \infty)$. (See Figure 3.3-2.)

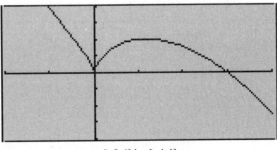

[−2,4] by [−4,4]

Figure 3.3-2

Example 2

Using a calculator, sketch the graph of $f(x) = e^{-x^2/2}$, indicating all relative minimum and maximum points, points of inflection, vertical and horizontal asymptotes, intervals on which $f(x)$ is increasing, decreasing, concave upward or concave downward.

1. Domain: all real numbers; Range $(0, 1]$
2. Symmetry: $f(x)$ is an even function, and thus is symmetrical with respect to the y-axis.
3. Relative maximum: $(0, 1)$
   No relative minimum
   Points of inflection: $(-1, 0.6)$ and $(1, 0.6)$
4. $y = 0$ is a horizontal asymptote; no vertical asymptote
5. $f(x)$ is increasing on $(-\infty, 0]$ and decreasing on $[0, \infty)$
6. $f(x)$ is concave upward on $(-\infty, -1)$ and $(1, \infty)$; and concave downward on $(-1, 1)$

See Figure 3.3-3.

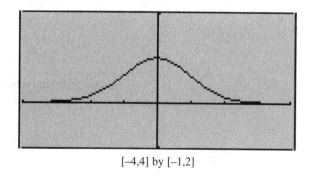

[−4,4] by [−1,2]

Figure 3.3-3

> • When evaluating a definite integral, you don't have to write a constant C, e.g., $\int_1^3 2x\,dx = x^2\big|_1^3 = 8$. Notice, no C.

3.4 GRAPHS OF DERIVATIVES

The functions f, f', and f'' are interrelated, and so are their graphs. Therefore, you can usually infer from the graph of one of the three functions (f, f', or f'') and obtain information about the other two. Here are some examples.

Example 1

The graph of a function f is shown in Figure 3.4-1. Which of the following is true for f on (a, b)?

I. $f' \geq 0$ on (a, b)
II. $f'' > 0$ on (a, b)

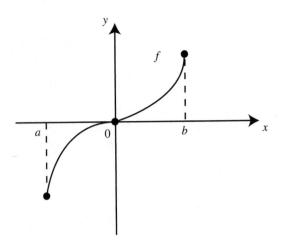

Figure 3.4-1

Solution:

I. Since f is strictly increasing, $f' \geq 0$ on (a, b) is true.

II. The graph is concave downward on $(a, 0)$ and upward on $(0, b)$. Thus $f'' > 0$ on $(0, b)$ only. Therefore only statement I is true.

Example 2

Given the graph of f' in Figure 3.4-2, find where the function f (a) has its relative maximum(s) or relative minimums, (b) is increasing or decreasing, (c) has its point(s) of inflection, (d) is concave upward or downward, and (e) if $f(-2) = f(2) = 1$ and $f(0) = -3$, draw a sketch of f.

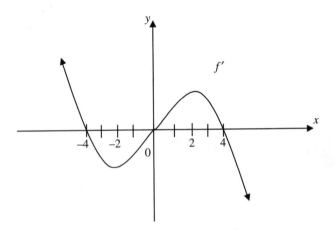

Figure 3.4-2

(a) Summarize the information of f' on a number line:

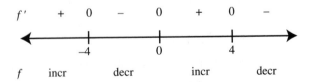

The function f has a relative maximum at $x = -4$ and at $x = 4$, and a relative minimum at $x = 0$.

(b) The function f is increasing on interval $(-\infty, -4]$ and $[0, 4]$, and f is decreasing on $[-4, 0]$ and $[4, \infty)$.

(c) Summarize the information of f'' on a number line:

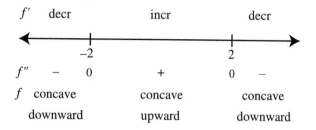

A change of concavity occurs at $x = -2$ and at $x = 2$ and f' exists at $x = -2$ and at $x = 2$, which implies that there is a tangent line to the graph of f at $x = -2$ and at $x = 2$. Therefore, f has a point of inflection at $x = -2$ and at $x = 2$.

(d) The graph of f is concave upward on the interval $(-2, 2)$ and concave downward on $(-\infty, -2)$ and $(2, \infty)$.

(e) A sketch of the graph of f is shown in Figure 3.4-3.

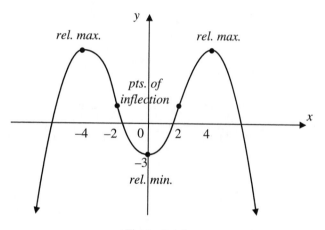

Figure 3.4-3

Example 3

Given the graph of f' in Figure 3.4-4, find where the function f (a) has a horizontal tangent, (b) has its relative extrema, (c) is increasing or decreasing, (d) has a point of inflection, and (e) is concave upward or downward.

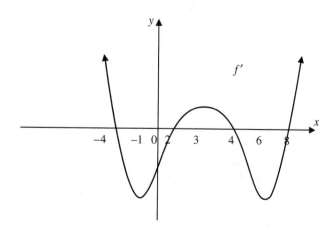

Figure 3.4-4

(a) $f'(x) = 0$ at $x = -4, 2, 4, 8$. Thus f has a horizontal tangent at these values.
(b) Summarize the information of f' on a number line:

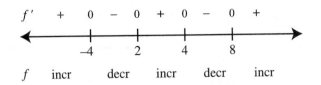

The first Derivative Test indicates that f has a relative maximums at $x = -4$ and 4; and f has relative minimums at $x = 2$ and 8.
(c) The function f is increasing on $(-8, -4]$, $[2, 4]$, and $[8, \infty)$ and is decreasing on $[-4, 2]$ and $[4, 8]$.
(d) Summarize the information of f'' on a number line:

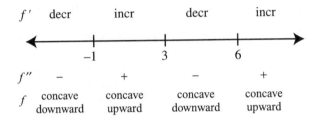

A change of concavity occurs at $x = -1, 3$, and 6. Since $f'(x)$ exists, f has a tangent at every point. Therefore, f has a point of inflection at $x = -1, 3$, and 6.
(e) The function f is concave upward on $(-1, 3)$ and $(6, \infty)$ and concave downward on $(-\infty, -1)$ and $(3, 6)$.

Example 4

A function f is continuous on the interval $[-4, 3]$ with $f(-4) = 6$ and $f(3) = 2$ and the following properties:

Intervals	$(-4, -2)$	$x = -2$	$(-2, 1)$	$x = 1$	$(1, 3)$
f'	$-$	0	$-$	undefined	$+$
f''	$+$	0	$-$	undefined	$-$

(a) Find the intervals on which f is increasing or decreasing.
(b) Find where f has its absolute extrema.
(c) Find where f has the points of inflection.
(d) Find the intervals on where f is concave upward or downward.
(e) Sketch a possible graph of f.

Solution:

(a) The graph of f is increasing on $[1, 3]$ and decreasing on $[-4, -2]$ and $[-2, 1]$.
(b) At $x = -4$, $f(x) = 6$. The function decreases until $x = 1$ and increases back to 2 at $x = 3$. Thus, f has its absolute maximum at $x = -4$ and its absolute minimum at $x = 1$.
(c) A change of concavity occurs at $x = -2$, and since $f'(-2) = 0$ which implies a tangent line exists at $x = -2$, thus f has a point of inflection at $x = -2$.

(d) The graph of f is concave upward on $(-4, -2)$ and concave downward on $(-2, 1)$ and $(1, 3)$.

(e) A possible sketch of f is shown in Figure 3.4-5.

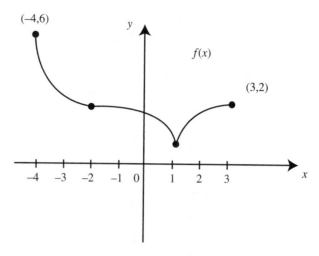

Figure 3.4-5

Example 5

If $f(x) = |\ln(x+1)|$, *find* $\lim\limits_{x \to 0^-} f'(x)$. (See Figure 3.4-6.)

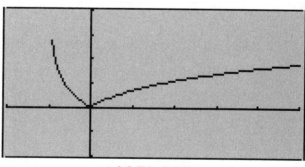

$[-2,5]$ by $[-2,4]$

Figure 3.4-6

The domain of f is $(-1, \infty)$.

$$f(0) = |\ln(0+1)| = |\ln(1)| = 0$$

$$f(x) = |\ln(x+1)| = \begin{cases} \ln(x+1) & \text{if } x \geq 0 \\ -\ln(x+1) & \text{if } x < 0 \end{cases}$$

Thus, $f'(x) = \begin{cases} \dfrac{1}{x+1} & \text{if } x \geq 0 \\ -\dfrac{1}{x+1} & \text{if } x < 0 \end{cases}$

Therefore, $\lim\limits_{x \to 0^-} f'(x) = \lim\limits_{x \to 0^-} \left(-\dfrac{1}{x+1} \right) = -1$.

BC 3.5 PARAMETRIC, POLAR, AND VECTOR REPRESENTATIONS

> Main Concepts: *Parametric Curves, Polar Equations, Types of Polar Graphs, Symmetry of Polar Graphs, Vectors, Vector Arithmetic*

Parametric Curves

Parametric curves are relations $(x(t), y(t))$ for which both x and y are defined as functions of a third variable, t, that is, $x = f(t)$ and $y = g(t)$.

Example 1

A particle is moving in the coordinate plane in such a way that $x(t) = 2t - 5$ and $y(t) = 4 \sin\left(\dfrac{\pi}{t+1}\right)$ for $0 \le t \le 5$. Sketch the path of the particle and indicate the direction of motion.

Step 1: Create a table of values.

t	0	1	2	3	4	5
$x(t)$	-5	-3	-1	1	3	5
$y(t)$	0	4	3.464	2.828	2.351	2

Step 2: Plot the points and sketch the path of a particle as a smooth curve. Place arrows to indicate the direction of motion.

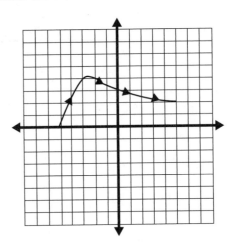

Example 2

A parametric curve is defined by $x = 2 + e^t$ and $y = e^{3t}$. Find the Cartesian equation of the curve.

Step 1: Solve $x = 2 + e^t$ for t. $x - 2 = e^t$ so $t = \ln(x - 2)$.

Step 2: Substitute $t = \ln(x - 2)$ into $y = e^{3t}$. $y = e^{3\ln(x-2)} = (x - 2)^3$.

Step 3: Note that $t = \ln(x-2)$ is defined only when $x > 2$. The equation of the curve is $y = (x-2)^3$ with domain $(2, \infty)$.

Polar Equations

The polar coordinate system locates points by a distance from the origin or pole, and an angle of rotation. Points are represented by a coordinate pair (r, θ). If conversions between polar and Cartesian representations are necessary, make the appropriate substitutions and simplify.

$$x = r\cos\theta \quad y = r\sin\theta \qquad r = \sqrt{x^2 + y^2} \quad \theta = \tan^{-1}\left(\frac{y}{x}\right)$$

Example 1

Convert $r = 4\sin\theta$ to Cartesian coordinates.

Step 1: Substitute in $r = 4\sin\theta$ to get $\sqrt{x^2 + y^2} = 4\sin\left(\tan^{-1}\frac{y}{x}\right)$

Step 2: Since $\sin\left(\tan^{-1}\frac{y}{x}\right) = \dfrac{y}{\sqrt{x^2 + y^2}}$, this becomes $\sqrt{x^2 + y^2} = 4\dfrac{y}{\sqrt{x^2 + y^2}}$.
Multiplying through by $\sqrt{x^2 + y^2}$, gives $x^2 + y^2 = 4y$.

Step 3: Complete the square on $x^2 + y^2 - 4y = 0$ to produce $x^2 + (y-2)^2 = 4$.

Example 2

Find the polar representation of $\dfrac{x^2}{4} + \dfrac{y^2}{9} = 1$.

Step 1: Substitute in $\dfrac{x^2}{4} + \dfrac{y^2}{9} = 1$ to produce $\dfrac{(r\cos\theta)^2}{4} + \dfrac{(r\sin\theta)^2}{9} = 1$

Step 2: Simplify and clear denominators to get $9r^2\cos^2\theta + 4r^2\sin^2\theta = 36$, then factor for $r^2(9\cos^2\theta + 4\sin^2\theta) = 36$.

Step 3: Divide to isolate $r^2 = \dfrac{36}{9\cos^2\theta + 4\sin^2\theta}$

Step 4: Apply the Pythagorean identity to the denominator $r^2 = \dfrac{36}{5\cos^2\theta + 4}$.

Types of Polar Graphs

Shape	Typical Equation	Notes
Line	$\theta = k$	
Circle	$r = a$ $r = 2a\cos\theta$ $r = 2a\sin\theta$	Radius of the circle $= a$

(continued)

Shape	Typical Equation	Notes
Rose	$r = a\sin(n\theta)$ $r = a\cos(n\theta)$	Length of petal $= a$ If n is odd, n petals. If n is even, $2n$ petals.
Cardiod	$r = a \pm a\sin\theta$ $r = a \pm a\cos\theta$	
Limaçon	$r = a \pm b\sin\theta$ $r = a \pm b\cos\theta$	If $\dfrac{a}{b} < 1$, limaçon has an inner loop.
Spirals	$r = a\theta$ $r = a\sqrt{\theta}$ $r = \dfrac{a}{\theta}$ $r = \dfrac{a}{\sqrt{\theta}}$ $r = ae^{b\theta}$	

Example 1

Classify each of the following equations according to the shape of its graph.
(a) $r = 5 + 7\cos\theta$, (b) $r = \dfrac{4}{\theta}$, (c) $r = 4 - 4\sin\theta$

The equation in (a) is a limaçon, and since $\dfrac{5}{7} < 1$, it will have an inner loop. The equation in (b) is a spiral. Equation (c) appears at first glance to be a limaçon; however, since the coefficients are equal, it is a cardiod.

Example 2

Sketch the graph of $r = 3\cos(2\theta)$. The equation $r = 3\cos(2\theta)$ is a polar rose with four petals each 3 units long. Since $3\cos(0) = 3$, the tip of a petal sits at 3 on the polar axis.

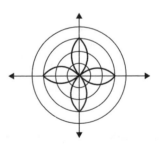

Symmetry of Polar Graphs

A polar curve of the form $r = f(\theta)$ will be symmetric about the polar, or horizontal, axis if $f(\theta) = f(-\theta)$, symmetric about the line $\theta = \dfrac{\pi}{2}$ if $f(\theta) = f(\pi - \theta)$, and symmetric about the pole if $f(\theta) = f(\theta + \pi)$.

Example 1

Determine the symmetry, if any, of the graph of $r = 2 + 4\cos\theta$

Step 1: Since $2 + 4\cos(-\theta) = 2 + 4\cos\theta$ the graph is symmetric about the polar axis.

Step 2: $2 + 4\cos(\pi - \theta) = 2 - 4\cos\theta$ so the graph is not symmetric about the line $\theta = \dfrac{\pi}{2}$.

Step 3: Since $2 + 4\cos(\theta + \pi) = 2 + 4[\cos\theta\cos\pi - \sin\theta\sin\pi] = 2 - 4\cos\theta$, the graph is not symmetric about the pole.

Example 2

Determine the symmetry, if any, of the graph $r = 3 - 3\sin\theta$.

Step 1: Since $3 - 3\sin(-\theta) = 3 + 3\sin\theta$ is not equal to $r = 3 - 3\sin\theta$, the graph is not symmetric about the polar axis.

Step 2: $3 - 3\sin(\pi - \theta) = 3 - 3\sin\theta$ so the graph is symmetric about the line $\theta = \pi/2$.

Step 3: Since $3 - 3\sin(\theta + \pi) = 3 - 3[\sin\theta\cos\pi + \sin\pi\cos\theta] = 3 + 3\sin\theta$, the graph is not symmetric about the pole.

Example 3

Determine the symmetry, if any, of the graph of $r = 5\cos(4\theta)$.

Step 1: Since $5\cos(4(-\theta)) = 5\cos 4\theta$, the graph is symmetric about the polar axis.

Step 2: $5\cos(4(\pi - \theta)) = 5\cos(4\pi - 4\theta)$ which, by identity, is equal to $5[\cos 4\pi \cos 4\theta + \sin 4\pi \sin\theta]$ or $5\cos 4\theta$ so the graph is symmetric about the line $\theta = \pi/2$.

Step 3: Since $5\cos 4(\theta + \pi) = 5\cos(4\theta + 4\pi) = 5\cos(4\theta)$ the graph is symmetric about the pole.

Vectors

A vector represented a displacement, both magnitude and direction. The length, r, of the vector is its magnitude, and the angle, θ, it makes with the x-axis gives its direction. The vector can be resolved into a horizontal and a vertical component. $x = \|r\|\cos\theta$ and $y = \|r\|\sin\theta$.

A unit vector is a vector of magnitude 1. If $i = \langle 1, 0 \rangle$ is the unit vector parallel to the positive x-axis, that is, a unit vector with direction angle $\theta = 0$, and $j = \langle 0, 1 \rangle$ is the unit vector parallel to the y-axis, with an angle $\theta = \dfrac{\pi}{2}$, then any vector in the plane can be represented as $xi + yj$ or simply as the ordered pair $\langle x, y \rangle$. The magnitude of the vector is $\|r\| = \sqrt{x^2 + y^2}$, and the direction can be found from $\tan\theta = \dfrac{y}{x}$. Since $\tan^{-1}\left(\dfrac{y}{x}\right)$ will return values in quadrant I or quadrant IV, if the terminal point of the vector falls in quadrant II or quadrant III the direction angle will be equal to $\tan^{-1}\left(\dfrac{y}{x}\right) + \pi$.

Example 1

Find the magnitude and direction of the vector represented by $\langle 6, -3 \rangle$.

Step 1: Calculate the magnitude $\|r\| = \sqrt{(6)^2 + (-3)^2} = \sqrt{45} = 3\sqrt{5}$

Step 2: The terminal point of the vector is in the fourth quadrant. Calculate $\theta = \tan^{-1}\left(\dfrac{-3}{6}\right) = \tan^{-1}\left(\dfrac{-1}{2}\right) \approx -.464$ radians. This angle falls in quadrant IV.

Example 2

Find the magnitude and direction of the vector represented by $\langle -5, -5 \rangle$.

Step 1: Calculate the magnitude $\|r\| = \sqrt{(-5)^2 + (-5)^2} = \sqrt{50} = 5\sqrt{2}$

Step 2: The terminal point of the vector is in the third quadrant. Calculate $\tan^{-1}\left(\dfrac{-5}{-5}\right) = \tan^{-1}(1) = \dfrac{\pi}{4}$ radians. The direction angle is $\theta = \dfrac{\pi}{4} + \pi = \dfrac{5\pi}{4}$.

Example 3

Find the magnitude and direction of the vector represented by $\langle -1, \sqrt{3} \rangle$.

Step 1: Calculate the magnitude $\|r\| = \sqrt{(-1)^2 + (\sqrt{3})^2} = \sqrt{4} = 2$

Step 2: The terminal point of the vector is in the second quadrant. Calculate $\tan^{-1}\left(\dfrac{\sqrt{3}}{-1}\right) = \tan^{-1}(-\sqrt{3}) = -\dfrac{\pi}{3}$ radians. The direction angle is $\theta = -\dfrac{\pi}{3} + \pi = \dfrac{2\pi}{3}$.

Example 4

Find the ordered pair representation of a vector of magnitude 12 and direction $\dfrac{-\pi}{4}$.

$x = 12\cos\left(\dfrac{-\pi}{4}\right) = 6\sqrt{2}$ and $y = 12\sin\left(\dfrac{-\pi}{4}\right) = -6\sqrt{2}$ so the vector is $\langle 6\sqrt{2}, -6\sqrt{2} \rangle$.

Vector Arithmetic

If c is a constant, $r_1 = \langle x_1, y_1 \rangle$ and $r_2 = \langle x_2, y_2 \rangle$, then:
  Addition: $r_1 + r_2 = \langle x_1 + x_2, y_1 + y_2 \rangle$
  Subtraction: $r_1 - r_2 = \langle x_1 - x_2, y_1 - y_2 \rangle$
  Scalar Multiplication: $cr_1 = \langle cx_1, cy_1 \rangle$
      Note: $\|cr_1\| = \|c\| \cdot \|r_1\|$
  Dot Product: The dot product of two vectors is $r_1 \cdot r_2 = \|r_1\| \cdot \|r_2\| \cdot \cos\theta$
  or $r_1 \cdot r_2 = x_1 x_2 + y_1 y_2$.

Parallel and Perpendicular Vectors

If $r_2 = cr_1$, then r_1 and r_2 are parallel.
If $r_1 \cdot r_2 = 0$, then r_1 and r_2 are perpendicular or orthogonal.
The angle between two vectors can be found by $\cos\theta = \dfrac{r_1 \cdot r_2}{\|r_1\| \cdot \|r_2\|}$.

Example 1

Given $r_1 = \langle 4, -7 \rangle$, $r_2 = \langle -3, -2 \rangle$ and $r_2 = \langle -1, 5 \rangle$, find $3r_1 - 5r_2 + 2r_3$.

$3r_1 - 5r_2 - 2r_3 = 3\langle 4, -7 \rangle - 5\langle -3, -2 \rangle + 2\langle -1, 5 \rangle = \langle 12, -21 \rangle - \langle -15, -10 \rangle + \langle -2, 10 \rangle = \langle 27, -11 \rangle - \langle -2, 10 \rangle = \langle 29, -21 \rangle$.

Example 2

Determine whether the vectors $r_1 = \langle 4, -7 \rangle$ and $r_2 = \langle -3, -2 \rangle$ are orthogonal. If the vectors are not orthogonal, approximate the angle between them.

Step 1. Find the dot product $r_1 \cdot r_2 = 4(-3) + (-7)(-2) = 2$. Since the dot product is not equal to zero, the vectors are not orthogonal.

Step 2. If θ is the angle between the vectors, then $\cos\theta = \dfrac{r_1 \cdot r_2}{\|r_1\| \cdot \|r_2\|}$. The dot product is 2, $\|r_1\| = \sqrt{65}$, and $\|r_2\| = \sqrt{13}$ so $\cos\theta = \dfrac{2}{\sqrt{65} \cdot \sqrt{13}} = \dfrac{2\sqrt{5}}{65} \approx 0.0688$ and $\theta \approx 1.5019$ radians.

3.6 RAPID REVIEW

1. If $f'(x) = x^2 - 4$, find the intervals where f is decreasing. (See Figure 3.6-1.)

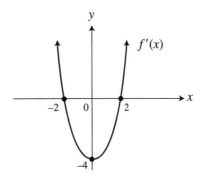

Figure 3.6-1

Answer: Since $f'(x) < 0$ if $-2 < x < 2$, f is decreasing on $(-2, 2)$.

2. If $f''(x) = 2x - 6$ and f' is continuous, find the values of x where f has a point of inflection. (See Figure 3.6-2.)

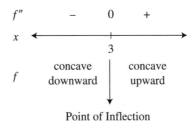

Figure 3.6-2

Answer: Thus, f has a point of inflection at $x = 3$.

3. See Figure 3.6-3. Find the values of x where f has change of concavity.

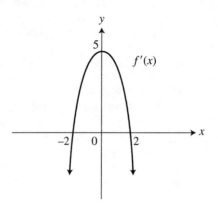

Figure 3.6-3

Answer: f has a change of concavity at $x = 0$. See Figure 3.6-4.

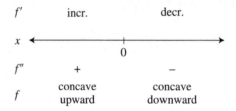

Figure 3.6-4

4. See Figure 3.6-5. Find the values of x where f has a relative minimum.

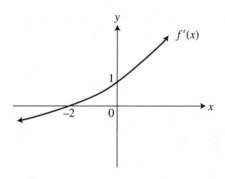

Figure 3.6-5

Answer: Thus, f has a relative minimum at $x = -2$. See Figure 3.6-6.

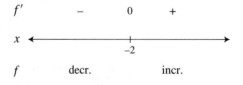

Figure 3.6-6

5. See Figure 3.6-7. Given f is twice differentiable, arrange $f(10)$, $f'(10)$, $f''(10)$ from smallest to largest.

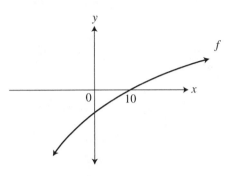

Figure 3.6-7

Answer: $f(10) = 0$, $f'(10) > 0$ since f is increasing and $f''(10) < 0$ since f is concave downward. Thus the order is $f''(10)$, $f(10)$, $f'(10)$.

6. See Figure 3.6-8. Find the values of x where f' is concave up.

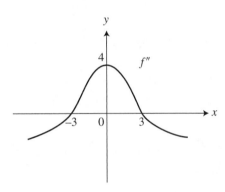

Figure 3.6-8

Answer: Thus, f' is concave upward on $(-\infty, 0)$. See Figure 3.6-9.

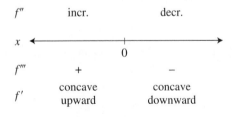

Figure 3.6-9

7. The path of an object is defined by $x = 2t$, $y = t + 1$. Find the location of the object when $t = 5$.

Answer: When $t = 5$, $x = 2(5) = 10$ and $y = 5 + 1 = 6$ so the location is $(10, 6)$.

8. Identify the shape of each equation: (a) $r = \dfrac{\theta}{5}$; (b) $r = 6\cos 3\theta$

Answers: (a) spiral, (b) rose

9. Find the magnitude of the vector $\langle 1, -\sqrt{3} \rangle$ and the angle it makes with the positive x-axis.

   Answers: Magnitude $= \sqrt{1^2 + (-\sqrt{3})^2} = \sqrt{1+3} = 2$. Angle $= \tan^{-1}\left(\dfrac{-\sqrt{3}}{1}\right) = \dfrac{-\pi}{3}$.

10. If $a = \langle 4, -2 \rangle$, $b = \langle -3, 1 \rangle$, and $c = \langle 0, 5 \rangle$, find $3a - 2b + c$.

    Answer: $3\langle 4, -2 \rangle - 2\langle -3, 1 \rangle + \langle 0, 5 \rangle = \langle 12, -6 \rangle + \langle 6, -2 \rangle + \langle 0, 5 \rangle = \langle 18, -8 \rangle + \langle 0, 5 \rangle = \langle 18, -3 \rangle$.

3.7 PRACTICE PROBLEMS

Part A—The use of a calculator is not allowed.

1. If $f(x) = x^3 - x^2 - 2x$, show that the hypotheses of Rolle's Theorem are satisfied on the interval $[-1, 2]$ and find all values of c that satisfy the conclusion of the theorem.

2. Let $f(x) = e^x$. Show that the hypotheses of the Mean Value Theorem are satisfied on $[0, 1]$ and find all values of c that satisfy the conclusion of the theorem.

3. Determine the intervals in which the graph of $f(x) = \dfrac{x^2 + 9}{x^2 - 25}$ is concave upward or downward.

4. Given $f(x) = x + \sin x \; 0 \le x \le 2\pi$, find all points of inflection of f.

5. Show that the absolute minimum of $f(x) = \sqrt{25 - x^2}$ on $[-5, 5]$ is 0 and the absolute maximum is 5.

6. Given the function f in Figure 3.7-1, identify the points where:

   (a) $f' < 0$ and $f'' > 0$,   (b) $f' < 0$ and $f'' < 0$,
   (c) $f' = 0$,   (d) f'' does not exist.

7. Given the graph of f'' in Figure 3.7-2, determine the values of x at which the function f has a point of inflection. (See Figure 3.7-2.)

8. If $f''(x) = x^2(x + 3)(x - 5)$, find the values of x at which the graph of f has a change of concavity.

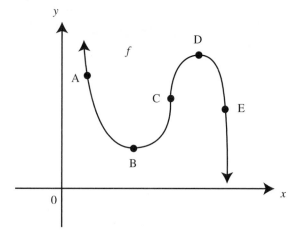

Figure 3.7-1

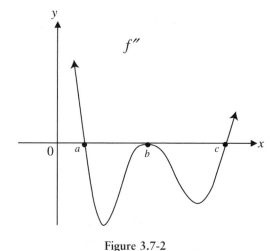

Figure 3.7-2

9. The graph of f' on $[-3, 3]$ is shown in Figure 3.7-3. Find the values of x on $[-3, 3]$ such that (a) f is increasing and (b) f is concave downward.

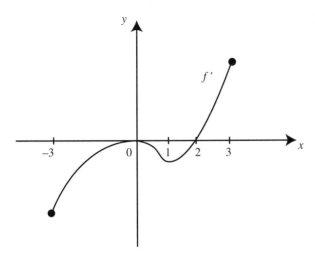

Figure 3.7-3

10. The graph of f is shown in Figure 3.7-4 and f is twice differentiable. Which of the following has the largest value:

(A) $f(-1)$
(B) $f'(-1)$
(C) $f''(-1)$
(D) $f(-1)$ and $f'(-1)$
(E) $f'(-1)$ and $f''(-1)$

Sketch the graphs of the following functions indicating any relative and absolute extrema, points of inflection, intervals on which the function is increasing, decreasing, concave upward or concave downward.

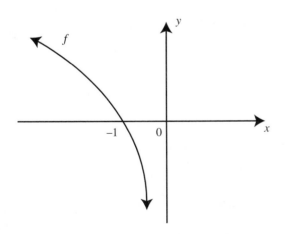

Figure 3.7-4

11. $f(x) = x^4 - x^2$

12. $f(x) = \dfrac{x+4}{x-4}$

Part B—Calculators are permitted.

13. Given the graph of f' in Figure 3.7-5, determine at which of the four values of x (x_1, x_2, x_3, x_4) f has:

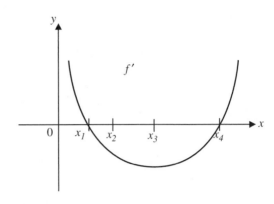

Figure 3.7-5

(a) the largest value,
(b) the smallest value,
(c) a point of inflection,
(d) and at which of the four values of x does f'' have the largest value.

14. Given the graph of f in Figure 3.7-6, determine at which values of x is

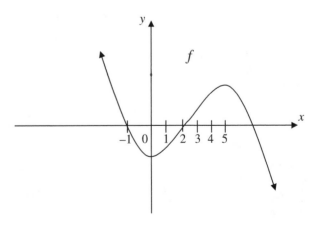

Figure 3.7-6

(a) $f'(x) = 0$
(b) $f''(x) = 0$
(c) f' a decreasing function.

15. A function f is continuous on the interval $[-2, 5]$ with $f(-2) = 10$ and $f(5) = 6$ and the

following properties:

Intervals	$(-2, 1)$	$x = 1$	$(1, 3)$	$x = 3$	$(3, 5)$
f'	+	0	−	undefined	+
f''	−	0	−	undefined	+

(a) Find the intervals on which f is increasing or decreasing.
(b) Find where f has its absolute extrema.
(c) Find where f has points of inflection.
(d) Find the intervals where f is concave upward or downward.
(e) Sketch a possible graph of f.

16. Given the graph of f' in Figure 3.7-7, find where the function f(a) has its relative extrema, (b) is increasing or decreasing, (c) has its point(s) of inflection, (d) is concave upward or downward, and (e) if $f(0) = 1$ and $f(6) = 5$, draw a sketch of f.

17. If $f(x) = |x^2 - 6x - 7|$, which of the following statements about f are true?

   I. f has a relative maximum at $x = 3$.
   II. f is differentiable at $x = 7$.
   III. f has a point of inflection at $x = -1$.

18. How many points of inflection does the graph of $y = \cos(x^2)$ have on the interval $[-\pi, \pi]$?

Sketch the graphs of the following functions indicating any relative extrema, points of inflection, asymptotes, and intervals where the function is increasing, decreasing, concave upward or concave downward.

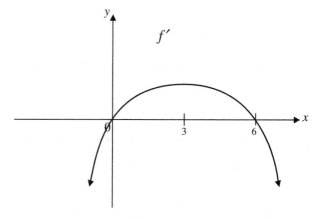

Figure 3.7-7

19. $f(x) = 3e^{-x^2/2}$

20. $f(x) = \cos x \sin^2 x \ [0, 2\pi]$

 21. Find the Cartesian equation of the curve defined by $x = \dfrac{t}{2}, \ y = t^2 - 4t + 1$.

22. Find the polar equation of the line with Cartesian equation $y = 3x - 5$.

23. Identify the type of graph defined by the equation $r = 2 - \sin\theta$ and determine its symmetry, if any.

24. Find the value of k so that the vectors $\langle 3, -2 \rangle$ and $\langle 1, k \rangle$ are orthogonal.

25. Determine whether the vectors $\langle 5, -3 \rangle$ and $\langle 5, 3 \rangle$ are orthogonal. If not, find the angle between the vectors.

 # 3.8 CUMULATIVE REVIEW PROBLEMS

"Calculator" indicates that calculators are permitted.

26. Find $\dfrac{dy}{dx}$ if $(x^2 + y^2)^2 = 10xy$

27. Evaluate $\lim\limits_{x \to 0} \dfrac{\sqrt{x+9} - 3}{x}$

28. Find $\dfrac{d^2y}{dx^2}$ if $y = \cos(2x) + 3x^2 - 1$

29. (Calculator) Determine the value of k such that the function
$$f(x) = \begin{cases} x^2 - 1, & x \le 1 \\ 2x + k, & x > 1 \end{cases}$$
is continuous for all real numbers.

30. A function f is continuous on the interval $[-1, 4]$ with $f(-1) = 0$ and $f(4) = 2$ and the following properties:

Intervals	$(-1, 0)$	$x = 0$	$(0, 2)$	$x = 2$	$(2, 4)$
f'	$+$	undefined	$+$	0	$-$
f''	$+$	undefined	$-$	0	$-$

(a) Find the intervals on which f is increasing or decreasing.

(b) Find where f has its absolute extrema.

(c) Find where f has points of inflection.

(d) Find intervals on which f is concave upward or downward.

(e) Sketch a possible graph of f.

 31. Evaluate $\lim\limits_{x \to \pi} \dfrac{2x}{\sin x}$

32. Evaluate $\lim\limits_{x \to \frac{1}{2}} \dfrac{3 - 6x}{4x^2 - 1}$

33. Find the polar equation of the ellipse $x^2 + 4y^2 = 4$.

3.9 SOLUTIONS TO PRACTICE PROBLEMS

Part A—No calculators.

1. Condition 1: Since $f(x)$ is a polynomial, it is continuous on $[-1, 2]$.

   Condition 2: Also, $f(x)$ is differentiable on $(-1, 2)$ because $f'(x) = 3x^2 - 2x - 2$ is defined for all numbers in $[-1, 2]$.

   Condition 3: $f(-1) = f(2) = 0$. Thus $f(x)$ satisfies the hypotheses of Rolle's Theorem which means there exists a c in $[-1, 2]$ such that $f'(c) = 0$. Set $f'(x) = 3x^2 - 2x - 2 = 0$. Solve $3x^2 - 2x - 2 = 0$, using the quadratic formula and obtain $x = \dfrac{1 \pm \sqrt{7}}{3}$. Thus $x \approx 1.215$ or -0.549 and both values are in the interval $(-1, 2)$. Therefore $c = \dfrac{1 \pm \sqrt{7}}{3}$.

2. Condition 1: $f(x) = e^x$ is continuous on $[0, 1]$.

   Condition 2: $f(x)$ is differentiable on $(0, 1)$ since $f'(x) = e^x$ is defined for all numbers in $[0, 1]$. Thus, there exists a number c in $[0, 1]$ such that $f'(c) = \dfrac{e^1 - e^0}{1 - 0} = (e - 1)$. Set $f'(x) = e^x = (e - 1)$. Thus $e^x = (e - 1)$. Take ln of both sides. $\ln(e^x) = \ln(e - 1) \Rightarrow x = \ln(e - 1)$. Thus $x \approx 0.541$ which is in the $(0, 1)$. Therefore $c = \ln(e - 1)$.

3. $f(x) = \dfrac{x^2 + 9}{x^2 - 25}$,

   $f'(x) = \dfrac{2x(x^2 - 25) - (2x)(x^2 + 9)}{(x^2 - 25)^2}$

   $= \dfrac{-68x}{(x^2 - 25)^2}$ and

   $f''(x) = \dfrac{-68(x^2 - 25)^2 - 2(x^2 - 25)(2x)(-68x)}{(x^2 - 25)^4}$

   $= \dfrac{68(3x^2 + 25)}{(x^2 - 25)^3}$

   Set $f'' > 0$. Since $(3x^2 + 25) > 0$, $\Rightarrow (x^2 - 25)^3 > 0 \Rightarrow x^2 - 25 > 0$, $x < -5$ or $x > 5$. Thus $f(x)$ is concave upward on $(-\infty, -5)$ and $(5, \infty)$ and concave downward on $(-5, 5)$.

4. Step 1: $f(x) = x + \sin x$, $f'(x) = 1 + \cos x$, $f'' = -\sin x$.

   Step 2: Set $f''(x) = 0 \Rightarrow -\sin x = 0$ or $x = 0, \pi, 2\pi$

   Step 3: Check intervals.

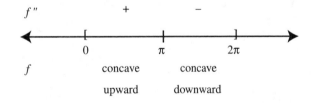

   Step 4: Check for tangent line: At $x = \pi$, $f'(x) = 1 + (-1) \Rightarrow 0$ there is a tangent line at $x = \pi$.

   Step 5: Thus (π, π) is a point of inflection.

5. Step 1: Rewrite $f(x)$ as $f(x) = (25 - x^2)^{1/2}$

   Step 2: $f'(x) = \dfrac{1}{2}(25 - x^2)^{-1/2}(-2x)$

   $= \dfrac{-x}{(25 - x^2)^{1/2}}$

Step 3: Find critical numbers. $f'(x) = 0$; at $x = 0$; and $f'(x)$ is undefined at $x = \pm 5$.

Step 4:

$$f''(x) = \frac{(-1)\sqrt{(25 - x^2)} - \dfrac{(-2x)(-x)}{2\sqrt{(25 - x^2)}}}{(25 - x^2)}$$

$$= \frac{-1}{(25 - x^2)^{1/2}} - \frac{x^2}{(25 - x^2)^{3/2}}$$

$f'(0) = 0$ and $f''(0) = \dfrac{1}{5}$ (and $f(0) = 5$) $\Rightarrow (0, 5)$ is a relative maximum. Since $f(x)$ is continuos on $[-5, 5]$, $f(x)$ has both a maximum and a minimum value on $[-5, 5]$ by the Extreme Value Theorem. And since the point $(0,5)$ is the only relative extremum, it is an absolute extramum. Thus $(0,5)$ is an absolute maximum point and 5 is the maximum value. Now we check the end points, $f(-5) = 0$ and $f(5) = 0$. Therefore $(-5, 0)$ and $(5, 0)$ are the lowest points for f on $[-5, 5]$. Thus 0 is the absolute minimum value.

6. (a) Point A $f' < 0 \Rightarrow$ decreasing and $f'' > 0 \Rightarrow$ concave upward.
   (b) Point E $f' < 0 \Rightarrow$ decreasing and $f'' < 0 \Rightarrow$ concave downward.
   (c) Points B and D $f' = 0 \Rightarrow$ horizontal tangent.
   (d) Point C f'' does not exist $\Rightarrow$ vertical tangent.

7. A change in concavity $\Rightarrow$ a point of inflection. At $x = a$, there is a change of concavity; f'' goes from positive to negative $\Rightarrow$ concavity changes from upward to downward. At $x = c$, there is a change of concavity; f'' goes from negative to positive $\Rightarrow$ concavity changes from downward to upward. Therefore f has two points of inflection, one at $x = a$ and the other at $x = c$.

8. Set $f''(x) = 0$. Thus $x^2(x + 3)(x - 5) = 0 \Rightarrow$ $x = 0$, $x = -3$ or $x = 5$. (See Figure 3.9-1.)

   Thus f has a change of concavity at $x = -3$ and at $x = 5$.

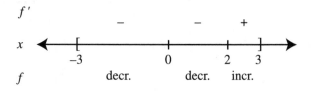

Figure 3.9-1

9. See Figure 3.9-2
   Thus f is increasing on $[2, 3]$ and concave downward on $(0, 1)$

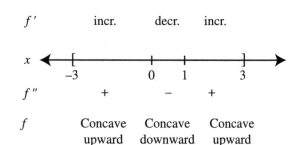

Figure 3.9-2

10. The correct answer is (A)
    $f(-1) = 0$; $f'(0) < 0$ since f is decreasing and $f''(-1) < 0$ since f is concave downward. Thus $f(-1)$ has the largest value.

11. Step 1: Domain: all real numbers.

    Step 2: Symmetry: Even function ($f(x) = f(-x)$); symmetrical with respect to the y-axis.

    Step 3: $f'(x) = 4x^3 - 2x$ and $f''(x) = 12x^2 - 2$.

    Step 4: Critical numbers:
    $f'(x)$ is defined for all real numbers. Set $f'(x) = 4x^3 - 2x = 0 \Rightarrow 2x(2x^2 - 1) = 0 \Rightarrow x = 0$ or $x = \pm\sqrt{1/2}$.
    Possible points of inflection:
    $f''(x)$ is defined for all real numbers. Set $f''(x) = 12x^2 - 2 = 0 \Rightarrow 2(6x^2 - 1) = 0 \Rightarrow x = \pm\sqrt{1/6}$.

Step 5: Determine intervals:

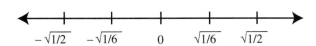

Intervals are: $\left(-\infty, -\sqrt{1/2}\right)$, $\left(-\sqrt{1/2}, -\sqrt{1/6}\right)$, $\left(-\sqrt{1/6}, 0\right)$, $\left(0, \sqrt{1/6}\right)$, $\left(\sqrt{1/6}, \sqrt{1/2}\right)$, $\left(\sqrt{1/2}, \infty\right)$. Since $f'(x)$ is symmetrical with respect to the y-axis, you only need to examine half of the intervals.

Step 6: Set up a table. (Table 3.9-1) The function has an absolute minimum value of $(-1/4)$ and no absolute maximum value.

Step 7: Sketch the graph. (See Figure 3.9-3.)

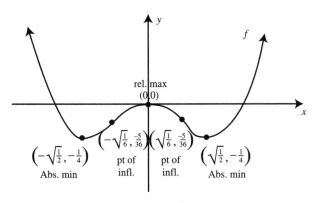

Figure 3.9-3

12. Step 1: Domain: all real numbers $x \neq 4$.

Step 2: Symmetry: none

Step 3: Find f' and f''.

$$f'(x) = \frac{(1)(x-4) - (1)(x+4)}{(x-4)^2}$$

$$= \frac{-8}{(x-4)^2}, \quad f''(x) = \frac{16}{(x-4)^3}$$

Step 4: Critical numbers: $f'(x) \neq 0$ and $f'(x)$ is undefined at $x = 4$.

Step 5: Determine intervals.

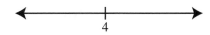

Intervals are $(-\infty, 4)$ and $(4, \infty)$.

Step 6: Set up table as below:

Interval	$(-\infty, 4)$	$(4, \infty)$
f'	–	–
f''	–	+
conclusion	Decr Concave downward	Incr Concave upward

Step 7: Horizontal asymptote: $\lim\limits_{x \to \pm\infty} \dfrac{x+4}{x-4} = 1$; Thus, $y = 1$ is a horizontal asymptote.
Vertical asymptote: $\lim\limits_{x \to 4^+} \dfrac{x+4}{x-4} = \infty$ and $\lim\limits_{x \to 4^-} \dfrac{x+4}{x-4} = -\infty$; Thus, $x = 4$ is a vertical asymptote.

Step 8: x-intercept: Set $f'(x) = 0 \Rightarrow x + 4 = 0$; $x = -4$
y-intercept: Set $x = 0 \Rightarrow f(x) = -1$.

Table 3.9-1

Intervals	$x = 0$	$(0, \sqrt{1/6})$	$x = \sqrt{1/6}$	$(\sqrt{1/6}, \sqrt{1/2})$	$x = \sqrt{1/2}$	$(\sqrt{1/2}, \infty)$
$f(x)$	0		$-5/36$		$-1/4$	
$f'(x)$	0	–	–	–	0	+
$f''(x)$	–	–	0	+	+	+
conclusion	rel max	decr concave downward	decr pt. of inflection	decr concave upward	rel min	incr concave upward

Step 9: Sketch the graph. (See Figure 3.9-4.)

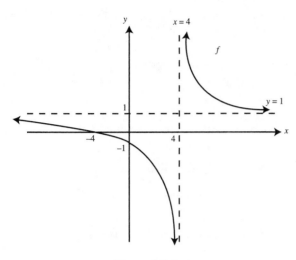

Figure 3.9-4

13. (a)

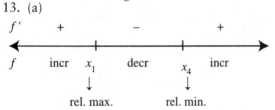

The function f has the largest value (of the four choices) at $x = x_1$. (See Figure 3.9-5.)

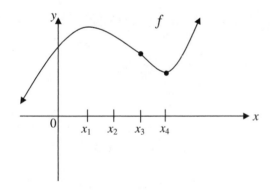

Figure 3.9-5

(b) And f has the smallest value at $x = x_4$.

(c)

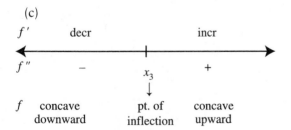

A change of concavity occurs at $x = x_3$, and $f'(x_3)$ exists which implies there is a tangent to f at $x = x_3$. Thus, at $x = x_3$, f has a point of inflection.

(d) The function f'' represents the slope of the tangent to f'. The slope of the tangent to f' is the largest at $x = x_4$.

14. (a) Since $f'(x)$ represents the slope of the tangent, $f'(x) = 0$ at $x = 0$, and $x = 5$.

(b) At $x = 2$, f has a point of inflection which implies that if $f''(x)$ exists, $f''(x) = 0$. Since $f'(x)$ is differentiable for all numbers in the domain, $f''(x)$ exists, and $f''(x) = 0$ at $x = 2$.

(c) Since the function f is concave downwards on $(2, \infty)$, $f'' < 0$ on $(2, \infty)$ which implies f' is decreasing on $(2, \infty)$.

15. (a) The function f is increasing on the intervals $(-2, 1)$ and $(3, 5)$ and decreasing on $(1, 3)$.

(b) The absolute maximum occurs at $x = 1$, since it is a relative maximum, $f(1) > f(-2)$ and $f(5) < f(-2)$. Similarly, the absolute minimum occurs at $x = 3$, since it is a relative minimum, and $f(3) < f(5) < f(-2)$.

(c) No point of inflection. (Note that at $x = 3$ f has a cusp.)
Note: Some textbooks define a point of inflection as a point where the concavity changes and do not require the existence of a tangent. In that case, at $x = 3$, f has a point of inflection.

(d) Concave upward on $(3, 5)$ and concave downward on $(-2, 3)$.

(e) A possible graph is shown in Figure 3.9-6.

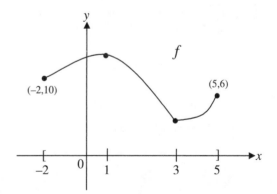

Figure 3.9-6

16. (a)

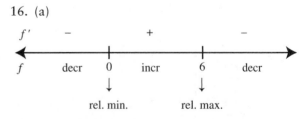

The function f has its relative minimum at $x = 0$ and its relative maximum at $x = 6$.

(b) The function f is increasing on $[0, 6]$ and decreasing on $(-\infty, 0]$ and $[6, \infty)$.

(c)

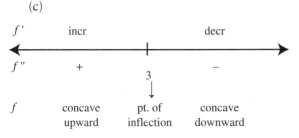

Since $f'(3)$ exists and a change of concavity occurs at $x = 3$, f has a point of inflection at $x = 3$.

(d) Concave upward on $(-\infty, 3)$ and downward on $(3, \infty)$.

(e) Sketch a graph. (See Figure 3.9-7.)

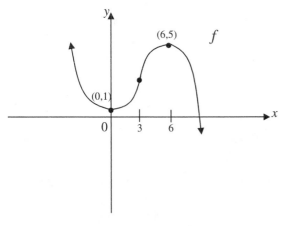

Figure 3.9-7

17. See Figure 3.9-8.

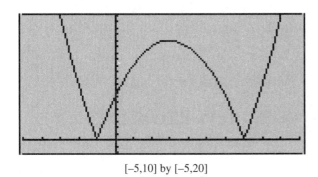

$[-5,10]$ by $[-5,20]$

Figure 3.9-8

The graph of f indicates that a relative maximum occurs at $x = 3$, f is not differentiable at $x = 7$, since there is a *cusp* at $x = 7$ and f does not have a point of inflection at $x = -1$, since there is no tangent line at $x = -1$. Thus, only statement I is true.

18. See Figure 3.9-9.

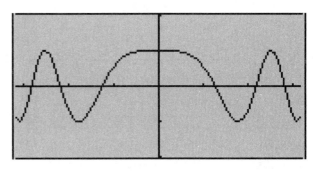

$[-\pi,\pi]$ by $[-2,2]$

Figure 3.9-9

Enter $y_1 = \cos(x^2)$

Using the Inflection function of your calculator, you obtain three points of inflection on $[0, \pi]$. The points of inflection occur at $x = 1.35521$, 2.1945 and 2.81373. Since $y_1 = \cos(x^2)$, is an even function; there is a total of 6 points of inflection on $[-\pi, \pi]$. An alternate solution is to enter $y_2 = \dfrac{d^2}{dx^2}(y_1(x), x, 2)$. The graph of y_2 indicates that there are 6 zero's on $[-\pi, \pi]$.

19. Enter $y_1 = 3 * e \wedge (-x \wedge 2/2)$. Note that the graph has a symmetry about the y-axis. Using the functions of the calculator, you will find:

(a) a relative maximum point at $(0, 3)$, which is also the absolute maximum point;

(b) points of inflection at $(-1, 1.819)$ and $(1, 1.819)$;

(c) $y = 0$ (the x-axis) a horizontal asymptote;

(d) y_1 increasing on $(-\infty, 0]$ and decreasing on $[0, \infty)$; and

(e) y_1 concave upward on $(-\infty, -1)$ and $(1, \infty)$ and concave downward on $(-1, 1)$. (See Figure 3.9-10.)

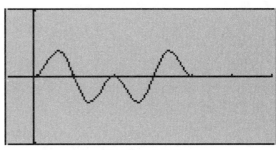

$[-4,4]$ by $[-1,4]$

Figure 3.9-10

20. (See Figure 3.9-11.) Enter $y_1 = \cos(x) *$ $(\sin(x)) \wedge 2$. A fundamental domain of y_1 is $[0, 2\pi]$. Using the functions of the calculator, you will find:

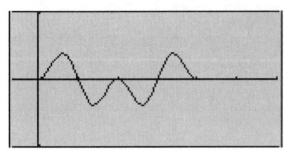

[−1,9.4] by [−1,1]

Figure 3.9-11

(a) relative maximum points at (0.955, 0.385), $(\pi, 0)$ and (5.328, 0.385) and relative minimum points at (2.186, −0.385) and (4.097, −0.385);

(b) points of inflection at (0.491, 0.196), $\left(\frac{\pi}{2}, 0\right)$, (2.651, −0.196), (3.632, −0.196), $\left(\frac{3\pi}{2}, 0\right)$ and (5.792, 0.196);

(c) no asymptote;

(d) function is increasing on intervals (0, 0.955), (2.186, π) and (4.097, 5.328) and decreasing on intervals (0.955, 2.186), (π, 4.097) and (5.328, 2π);

(e) function is concave upward on intervals (0, 0.491), $\left(\frac{\pi}{2}, 2.651\right)$, $\left(3.632, \frac{3\pi}{2}\right)$, and (5.792, 2$\pi$) and concave downward on the intervals $\left(0.491, \frac{\pi}{2}\right)$, (2.651, 3.632) and $\left(\frac{3\pi}{2}, 5.792\right)$.

21. Solve $x = \frac{t}{2}$ for $t = 2x$ and substitute into $y = t^2 - 4t + 1$. $y = (2x)^2 - 4(2x) + 1 = 4x^2 - 8x + 1$.

22. Since $x = r\cos\theta$ and $y = r\sin\theta$, $y = 3x - 5$ becomes $r\sin\theta = 3r\cos\theta - 5$. Solving for r produces $r(\sin\theta - 3\cos\theta) = -5$ and $r = \dfrac{-5}{\sin\theta - 3\cos\theta}$.

23. The equation $r = 2 - \sin\theta$ is of the form $r = a - b\sin\theta$ with $\frac{a}{b} > 1$, so the graph is a limaçon with no inner loop. Since $r(-\theta) = 2 + \sin\theta \neq r(\theta)$, the graph is not symmetric about the polar axis. However, $r(\pi - \theta) = 2 - \sin(\pi - \theta)$ is equal to $2\sin\theta = r(\theta)$, so the graph is symmetric about the line $x = \frac{\pi}{2}$. Finally, $r(\theta + \pi) = 2 - \sin(\theta + \pi)$ $= 2 + \sin\theta$ and so the graph is not symmetric about the pole.

24. The vectors $\langle 3, -2 \rangle$ and $\langle 1, k \rangle$ will be orthogonal if the dot product is equal to zero. $\langle 3, -2 \rangle \cdot \langle 1, k \rangle = 3.1 - 2k$ will be equal to zero when $2k = 3$ so $k = \frac{3}{2}$.

25. The dot product of $\langle 5, -3 \rangle$ and $\langle 5, 3 \rangle$ is $5 \cdot 5 + -3 \cdot 3 = 25 - 9 = 16$, so the vectors are not orthogonal. To find the angle between the vectors, begin by dividing the dot product by the product of the magnitudes of the two vectors. Both vectors have a magnitude of $\sqrt{34}$, so the quotient becomes $\frac{16}{34} = \frac{8}{17}$. The angle between the vectors is $\theta = \cos^{-1}\left(\frac{8}{17}\right) \approx 1.081$ radians.

3.10 SOLUTIONS TO CUMULATIVE REVIEW PROBLEMS

26. $(x^2 + y^2)^2 = 10xy$

$2\left(x^2 + y^2\right)\left(2x + 2y\frac{dy}{dx}\right)$

$= 10y + (10x)\frac{dy}{dx}$

$4x\left(x^2 + y^2\right) + 4y\left(x^2 + y^2\right)\frac{dy}{dx}$

$= 10y + (10x)\frac{dy}{dx}$

$4y\left(x^2 + y^2\right)\frac{dy}{dx} - (10x)\frac{dy}{dx}$
$\quad = 10y - 4x\left(x^2 + y^2\right)$

$\frac{dy}{dx}\left(4y\left(x^2 + y^2\right) - 10x\right)$
$\quad = 10y - 4x\left(x^2 + y^2\right)$

$\frac{dy}{dx} = \frac{10y - 4x\left(x^2 + y^2\right)}{4y\left(x^2 + y^2\right) - 10x} = \frac{5y - 2x\left(x^2 + y^2\right)}{2y\left(x^2 + y^2\right) - 5x}$

27. $\lim\limits_{x\to 0}\dfrac{\sqrt{x+9}-3}{x}=$

$\lim\limits_{x\to 0}\dfrac{(\sqrt{x+9}-3)}{x}\cdot\dfrac{(\sqrt{x+9}+3)}{(\sqrt{x+9}+3)}$

$=\lim\limits_{x\to 0}\dfrac{(x+9)-9}{x\left(\sqrt{x+9}+3\right)}$

$=\lim\limits_{x\to 0}\dfrac{x}{x\left(\sqrt{x+9}+3\right)}$

$=\lim\limits_{x\to 0}\dfrac{1}{\sqrt{x+9}+3}=\dfrac{1}{\sqrt{0+9}+3}$

$=\dfrac{1}{3+3}=\dfrac{1}{6}.$

28. $y=\cos(2x)+3x^2-1$

$\dfrac{dy}{dx}=[-\sin(2x)](2)+6x=-2\sin(2x)+6x$

$\dfrac{d^2y}{dx^2}=-2(\cos(2x))(2)+6=-4\cos(2x)+6.$

29. (Calculator) The function f is continuous everywhere for all values of k except possibly at $x=1$. Checking with the three conditions of continuity at $x=1$:

(1) $f(1)=(1)^2-1=0$

(2) $\lim\limits_{x\to 1^+}(2x+k)=2+k,\ \lim\limits_{x\to 1^-}(x^2-1)=0;$
thus $2+k=0\Rightarrow k=-2$. Since $\lim\limits_{x\to 1^+}f(x)=\lim\limits_{x\to 1^-}f(x)=0$, therefore $\lim\limits_{x\to 1}f(x)=0.$

(3) $f(1)=\lim\limits_{x\to 1}f(x)=0$. Thus, $k=-2$.

30. (a) Since $f'>0$ on $(-1,0)$ and $(0,2)$, the function f is increasing on the intervals $[-1,0]$ and $[0,2]$. Since $f'<0$ on $(2,4)$, f is decreasing on $[2,4]$.

(b) The absolute maximum occurs at $x=2$, since it is a relative maximum and it is the only relative extremum on $(-1,4)$. The absolute minimum occurs at $x=-1$, since $f(-1)<f(4)$ and the function has no relative minimum on $[-1,4]$.

(c) A change of concavity occurs at $x=0$. However, $f'(0)$ is undefined, which implies f may or may not have a tangent at $x=0$. Thus f may or may not have a point of inflection at $x=0$.

(d) Concave upward on $(-1,0)$ and concave downward on $(0,4)$.

(e) A possible graph is shown in Figure 3.10-1.

31. $\lim\limits_{x\to\pi}\dfrac{2x}{\sin x}\to\dfrac{2\pi}{\sin\pi}\to\dfrac{2\pi}{0}$. Note that L'Hôpital's Rule does not apply, since the form is not $\dfrac{0}{0}$.

$\lim\limits_{x\to\pi^-}\dfrac{2x}{\sin x}=\infty$ but $\lim\limits_{x\to\pi^+}\dfrac{2x}{\sin x}=-\infty$; therefore $\lim\limits_{x\to\pi}\dfrac{2x}{\sin x}$ does not exist.

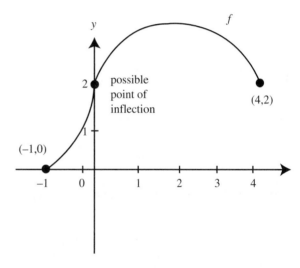

Figure 3.10-1

32. $\lim\limits_{x\to\frac12}\dfrac{3-6x}{4x^2-1}\to\dfrac{0}{0}$ but by L'Hôpital's Rule

$\lim\limits_{x\to\frac12}\dfrac{3-6x}{4x^2-1}=\lim\limits_{x\to\frac12}\dfrac{-6}{8x}=\dfrac{-6}{4}=\dfrac{-3}{2}$

33. To convert $x^2+4y^2=4$ to a polar representation, recall that $x=r\cos\theta$ and $y=r\sin\theta$. Then $(r\cos\theta)^2+4(r\sin\theta)^2=4$. Simplifying gives $r^2\cos^2\theta+4r^2\sin^2\theta=4$

$r=\dfrac{2}{\sqrt{\cos^2\theta+4\sin^2\theta}}=\dfrac{2}{\sqrt{1+3\sin^2\theta}}$

Chapter 4

Applications of Derivatives

4.1 RELATED RATE

Main Concepts: *General Procedure for Solving Related Rate Problems, Common Related Rate Problems, Inverted Cone (Water Tank) Problem, Shadow Problem, Angle of Elevation Problem*

General Procedure for Solving Related Rate Problems

1. Read the problem and if appropriate, draw a diagram.
2. Represent the given information and the unknowns by mathematical symbols.
3. Write an equation involving the rate of change to be determined. (If the equation contains more than one variable, it may be necessary to reduce the equation to one variable.)
4. Differentiate each term of the equation with respect to time.
5. Substitute all known values and known rates of change into the resulting equation.
6. Solve the resulting equation for the desired rate of change.
7. Write the answer and indicate the units of measure.

Common Related Rate Problems

Example 1

When the area of a square is increasing twice as fast as its diagonals, what is the length of a side of the square?

Let z represent the diagonal of the square. The area of a square is $A = \dfrac{z^2}{2}$.

$$\frac{dA}{dt} = 2z\frac{dz}{dt}\left(\frac{1}{2}\right) = z\frac{dz}{dt}.$$

Since $\dfrac{dA}{dt} = 2\dfrac{dz}{dt}, 2\dfrac{dz}{dt} = z\dfrac{dz}{dt} \Rightarrow z = 2$

Let s be a side of the square. Since the diagonal $z = 2$, then $s^2 + s^2 = z^2$
$$\Rightarrow 2s^2 = 4 \Rightarrow s^2 = \frac{4}{2} \Rightarrow s^2 = 2 \text{ or } s = \sqrt{2}.$$

Example 2

Find the surface area of a sphere at the instant when the rate of increase of the volume of the sphere is nine times the rate of increase of the radius.

Volume of a sphere: $V = \dfrac{4}{3}\pi r^3$; Surface area of a sphere: $S = 4\pi r^2$

$$V = \frac{4}{3}\pi r^3; \quad \frac{dV}{dt} = 4r^2 \frac{dr}{dt}$$

Since $\dfrac{dV}{dt} = 9\dfrac{dr}{dt}$, you have $9\dfrac{dr}{dt} = 4\pi r^2 \dfrac{dr}{dt}$ or $9 = 4\pi r^2$.

Since $S = 4\pi r^2$, the surface area is $S = 9$ square units.

Note: At $9 = 4\pi r^2$, you could solve for r and obtain $r^2 = \dfrac{9}{4\pi}$ or $r = \dfrac{3}{2}\dfrac{1}{\sqrt{\pi}}$. You could

then substitute $r = \dfrac{3}{2}\dfrac{1}{\sqrt{\pi}}$ into the formula for surface area $S = 4\pi r^2$ and obtain 9.

These steps are of course correct but not necessary.

Example 3

The height of a right circular cone is always three times the radius. Find the volume of the cone at the instant when the rate of increase of the volume is twelve times the rate of increase of the radius.

Let r, h be the radius and height of the cone respectively.

Since $h = 3r$, the volume of the cone $V = \dfrac{1}{3}\pi r^2 h = \dfrac{1}{3}\pi r^2 (3r) = \pi r^3$.

$$V = \pi r^3; \quad \frac{dV}{dt} = 3\pi r^2 \frac{dr}{dt}.$$

When $\dfrac{dV}{dt} = 12\dfrac{dr}{dt}$, $12\dfrac{dr}{dt} = 3\pi r^2\dfrac{dr}{dt} \Rightarrow 4 = \pi r^2 \Rightarrow r = \dfrac{2}{\sqrt{\pi}}$.

Thus $V = \pi r^3 = \pi\left(\dfrac{2}{\sqrt{\pi}}\right)^3 = \pi\left(\dfrac{8}{\pi\sqrt{\pi}}\right) = \dfrac{8}{\sqrt{\pi}}$.

> • Go with your first instinct if you are unsure. Usually that's the correct one.

Inverted Cone (Water Tank) Problem

A water tank is in the shape of an inverted cone. The height of the cone is 10 meters and the diameter of the base is 8 meters as shown in Figure 4.1-1. Water is being pumped into the tank at the rate of 2 m³/min. How fast is the water level rising when the water is 5 meters deep? (See Figure 4.1-1 on page 130.)

Solution:

Step 1: Define the variables. Let V be the volume of water in the tank; h be the height of the water level at t minutes; r be the radius of surface of the water at t minutes; and t be the time in minutes.

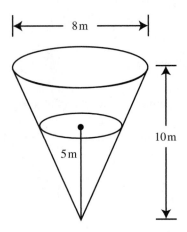

Figure 4.1-1

Step 2: Given: $\dfrac{dV}{dt} = 2$ m^3/min. Height $= 10$ m, diameter $= 8$ m

Find: $\dfrac{dh}{dt}$ at $h=5$.

Step 3: Set up an equation $V = \dfrac{1}{3}\pi r^2 h$.

Using similar triangles, you have $\dfrac{4}{10} = \dfrac{r}{h} \Rightarrow 4h = 10r$; or $r = \dfrac{2h}{5}$. See Figure 4.1-2.

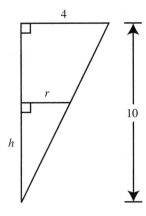

Figure 4.1-2

Thus, you can reduce the equation to one variable:

$$V = \frac{1}{3}\pi \left(\frac{2h}{5}\right)^2 h = \frac{4}{75}\pi h^3$$

Step 4: Differentiate both sides of the equation with respect to t.

$$\frac{dV}{dt} = \frac{4}{75}\pi(3)h^2\frac{dh}{dt} = \frac{4}{25}\pi h^2\frac{dh}{dt}$$

Step 5: Substitute known values.

$$2 = \frac{4}{25}\pi h^2\frac{dh}{dt}; \; \frac{dh}{dt} = \left(\frac{25}{2}\right)\frac{1}{\pi h^2} \text{ m/min}$$

$$\text{Evaluating } \frac{dh}{dt} \text{ at } h = 5; \left. \frac{dh}{dt} \right|_{h=5} = \left(\frac{25}{2} \right) \frac{1}{\pi(5)^2} \text{ m/min}$$

$$= \frac{1}{2\pi} \text{ m/min}.$$

Step 6: Thus, the water level is rising at $\frac{1}{2\pi}$ m/min when the water is 5 m high.

Shadow Problem

A light on the ground 100 feet from a building is shining at a 6-foot tall man walking away from the streetlight and toward the building at the rate of 4 ft/sec. How fast is his shadow on the building becoming shorter when he is 40 feet from the building? See Figure 4.1-3.

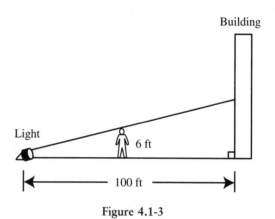

Figure 4.1-3

Solution:

Step 1: Let s be the height of the man's shadow; x be the distance between the man and the light; and t be the time in seconds.

Step 2: Given: $\frac{dx}{dt} = 4$ ft/sec; man is 6 ft tall; distance between light and building = 100 ft. Find $\frac{ds}{dt}$ at $x = 60$.

Step 3: See Figure 4.1-4. Write an equation using similar triangles, you have:

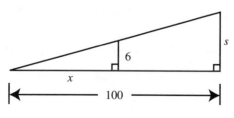

Figure 4.1-4

$$\frac{6}{s} = \frac{x}{100}; s = \frac{600}{x} = 600x^{-1}$$

Step 4: Differentiate both sides of the equation with respect to t.

$$\frac{ds}{dt} = (-1)(600)x^{-2}\frac{dx}{dt} = \frac{-600}{x^2}\frac{dx}{dt} = \frac{-600}{x^2}(4) = \frac{-2400}{x^2} \text{ ft/sec}$$

Step 5: Evaluate $\frac{ds}{dt}$ at $x = 60$.

Note: when the man is 40 ft from the building, x (distance from the light) is 60 ft.

$$\frac{ds}{dt}\bigg|_{x=60} = \frac{-2400}{(60)^2} \text{ ft/sec} = -\frac{2}{3} \text{ ft/sec}$$

Step 6: The height of the man's shadow on the building is changing at $-\frac{2}{3}$ ft/sec.

• Indicate units of measure, e.g., the velocity is 5 m/sec *or* the volume is 25 in³.

Angle of Elevation Problem

A camera on the ground 200 meters away from a hot air balloon records the balloon rising into the sky at a constant rate of 10 m/sec. How fast is the camera's angle of elevation changing when the balloon is 150 m in the air? See Figure 4.1-5.

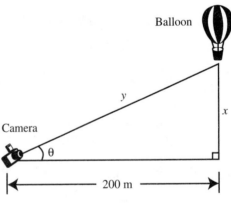

Balloon

Camera

y

x

θ

200 m

Figure 4.1-5

Step 1: Let x be the distance between the balloon and the ground; θ be the camera's angle of elevation; and t be the time in seconds.

Step 2: Given: $\frac{dx}{dt} = 10$ m/sec; distance between camera and the point on the ground where the balloon took off is 200 m, $\tan\theta = \frac{x}{200}$.

Step 3: Find $\frac{d\theta}{dt}$ at $x = 150$ m.

Step 4: Differentiate both sides with respect to t.

$$\sec^2\theta\frac{d\theta}{dt} = \frac{1}{200}\frac{dx}{dt}; \quad \frac{d\theta}{dt} = \frac{1}{200}\left(\frac{1}{\sec^2\theta}\right)(10) = \frac{1}{20\sec^2\theta}$$

Step 5: $\sec \theta = \dfrac{y}{200}$ and at $x = 150$.

Using the Pythagorean Theorem: $y^2 = x^2 + (200)^2$

$$y^2 = (150)^2 + (200)^2$$

$$y = \pm 250$$

Since $y > 0$, then $y = 250$. Thus, $\sec \theta = \dfrac{250}{200} = \dfrac{5}{4}$.

Evaluating $\left. \dfrac{d\theta}{dt} \right|_{x=150} = \dfrac{1}{20 \sec^2 \theta} = \dfrac{1}{20 \left(\dfrac{5}{4} \right)^2}$ radian/sec

$$= \dfrac{1}{20 \left(\dfrac{5}{4} \right)^2} = \dfrac{1}{20 \left(\dfrac{25}{16} \right)} = \dfrac{1}{\dfrac{125}{4}} = \dfrac{4}{125} \text{ radian/sec}$$

or .032 radian/sec

$= 1.833$ deg/sec

Step 6: The camera's angle of elevation changes at approximately 1.833 deg/sec when the balloon is 150 m in the air.

4.2 APPLIED MAXIMUM AND MINIMUM PROBLEMS

Main Concepts: *General Procedure for Solving Applied Maximum and Minimum Problems, Distance Problems, Area and Volume Problems, Business Problems*

General Procedure for Solving Applied Maximum and Minimum Problems

Steps:
1. Read the problem carefully and if appropriate, draw a diagram.
2. Determine what is given and what is to be found and represent these quantities by mathematical symbols.
3. Write an equation that is a function of the variable representing the quantity to be maximized or minimized.
4. If the equation involves other variables, reduce the equation to a single variable that represents the quantity to be maximized or minimized.
5. Determine the appropriate interval for the equation (i.e., the appropriate domain for the function) based on the information given in the problem.
6. Differentiate to obtain the first derivative and to find critical numbers.
7. Apply the First Derivative Test or the Second Derivative Test by finding the second derivative.
8. Check the function values at the end points of the interval.
9. Write the answer(s) to the problem and, if given, indicate the units of measure.

Distance Problems

Find the shortest distance between the point A (19, 0) and the parabola $y = x^2 - 2x + 1$.

Solution:

Step 1: Draw a diagram. See Figure 4.2-1.

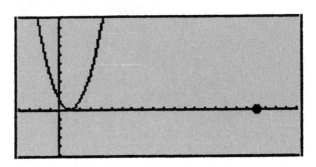

Figure 4.2-1

Step 2: Let $P(x, y)$ be the point on the parabola and let Z represent the distance between points $P(x, y)$ and $A(19, 0)$.

Step 3: Using the distance formula,

$$Z = \sqrt{(x - 19)^2 + (y - 0)^2} = \sqrt{(x - 19)^2 + (x^2 - 2x + 1 - 0)^2}$$

$$= \sqrt{(x - 19)^2 + \left((x - 1)^2\right)^2} = \sqrt{(x - 19)^2 + (x - 1)^4}$$

(Special case: In distance problems, the distance and the square of the distance have the same maximum and minimum points.) Thus, to simplify computations, let $L = Z^2 = (x - 19)^2 + (x - 1)^4$. The domain of L is $(-\infty, \infty)$.

Step 4: Differentiate $\dfrac{dL}{dx} = 2(x - 19)(1) + 4(x - 1)^3(1)$

$$= 2x - 38 + 4x^3 - 12x^2 + 12x - 4 = 4x^3 - 12x^2 + 14x - 42$$

$$= 2(2x^3 - 6x^2 + 7x - 21)$$

$\dfrac{dL}{dx}$ is defined for all real numbers.

Set $\dfrac{dL}{dx} = 0$; $2x^3 - 6x^2 + 7x - 21 = 0$. The factors of 21 are $\pm 1, \pm 3, \pm 7$ and ± 21.

Using Synthetic Division, $2x^3 - 6x^2 + 7x - 21 = (x - 3)(2x^2 + 7) = 0 \Rightarrow x = 3$. Thus the only critical number is $x = 3$.

(Note: Step 4 could have been done using a graphing calculator.)

Step 5: Apply the First Derivative Test.

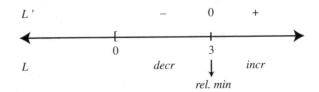

Step 6: Since $x = 3$ is the only relative minimum point in the interval, it is the absolute minimum.

Step 7: At $x = 3$, $Z = \sqrt{(3 - 19)^2 + (3^2 - 2(3) + 1)^2} = \sqrt{(-16)^2 + (4)^2}$
$= \sqrt{272} = \sqrt{16}\sqrt{17} = 4\sqrt{17}$. Thus, the shortest distance is $4\sqrt{17}$.

- Simplify numeric or algebraic expressions only if the question asks you to do so.

Area and Volume Problems

Example—Area Problem

The graph of $y = -\frac{1}{2}x + 2$ encloses a region with the x-axis and y-axis in the first quadrant. A rectangle in the enclosed region has a vertex at the origin and the opposite vertex on the graph of $y = -\frac{1}{2}x + 2$. Find the dimensions of the rectangle so that its area is a maximum.

Solution:

Step 1: Draw a diagram. (See Figure 4.2-2.)

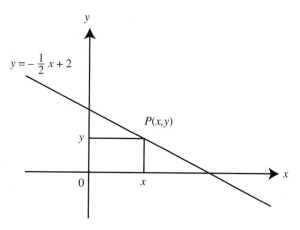

Figure 4.2-2

Step 2: Let $P(x, y)$ be the vertex of the rectangle on the graph of $y = -\frac{1}{2}x + 2$.

Step 3: Thus the area of the rectangle is:

$$A = xy \text{ or } A = x\left(-\frac{1}{2}x + 2\right) = -\frac{1}{2}x^2 + 2x$$

The domain of A is [0, 4].

Step 4: Differentiate.

$$\frac{dA}{dx} = -x + 2$$

Step 5: $\dfrac{dA}{dx}$ is defined for all real numbers.

Set $\dfrac{dA}{dx} = 0 \Rightarrow -x + 2 = 0; x = 2$

$A(x)$ has one critical number $x = 2$.

Step 6: Apply Second Derivative Test

$\dfrac{d^2A}{dx^2} = -1 \Rightarrow A(x)$ has a relative maximum point at $x = 2; A(2) = 2$.

Since $x = 2$ is the only relative maximum, it is the absolute maximum. (Note at the endpoints: $A(0) = 0$ & $A(4) = 0$.)

Step 7: At $x = 2, y = -\dfrac{1}{2}(2) + 2 = 1$.

Therefore the length of the rectangle is 2, and its width is 1.

Example—Volume Problem (with calculator)

If an open box is to be made using a square sheet of tin, 20 inches by 20 inches, by cutting a square from each corner and folding the sides up, find the length of a side of the square being cut so that the box will have a maximum volume.

Solution:

Step 1: Draw a diagram. (See Figure 4.2-3.)

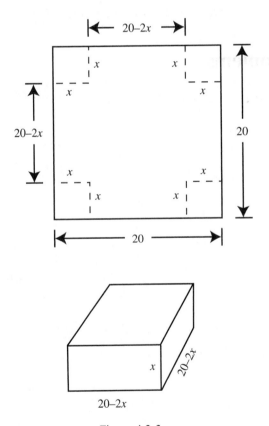

Figure 4.2-3

Step 2: Let x be the length of a side of the square to be cut from each corner.

Step 3: The volume of the box is $V(x) = x(20 - 2x)(20 - 2x)$.
The domain of V is $[0, 10]$.

Step 4: Differentiate $V(x)$.

Enter $d(x * (20 - 2x) * (20 - 2x), x)$ and we have $4(x - 10)(3x - 10)$.

Step 5: $V'(x)$ is defined for all real numbers:

Set $V'(x) = 0$ by entering: solve $(4(x - 10)(3x - 10) = 0, x)$, and obtain $x = 10$ or $x = \dfrac{10}{3}$. The critical numbers of $V(x)$ are $x = 10$ and $x = \dfrac{10}{3}$.

$V(10) = 0$ and $V\left(\dfrac{10}{3}\right) = 592.59$. Since $V(10) = 0$, you need to test only $x = \dfrac{10}{3}$.

Step 6: Using the Second Derivative Test, $d(x * (20 - 2x) * (20 - 2x), x, 2)|x = \dfrac{10}{3}$ and obtain -80. Thus, $V\left(\dfrac{10}{3}\right)$ is a relative maximum. And since it is the only relative maximum on the interval, it is the absolute maximum. (Note at the other endpoint $x = 0$, $V(0) = 0$.)

Step 7: Therefore, the length of a side of the square to be cut is $x = \dfrac{10}{3}$.

> • The formula for the average value of a function f from $x = a$ to $x = b$ is
> $$\frac{1}{b-a}\int_a^b f(x)dx.$$

Business Problems

Summary of Formulas:

1. $P = R - C$: Profit = Revenue − Cost
2. $R = xp$: Revenue = (Units Sold)(Price Per Unit)
3. $\overline{C} = \dfrac{C}{x}$, Average Cost $= \dfrac{\text{Total Cost}}{\text{Units produced/Sold}}$
4. $\dfrac{dR}{dx}$ = Marginal Revenue ≈ Revenue from selling one more unit
5. $\dfrac{dP}{dx}$ = Marginal Profit ≈ Profit from selling one more unit
6. $\dfrac{dC}{dx}$ = Marginal Cost ≈ Cost of producing one more unit

Example 1

Given the cost function $C(x) = 100 + 8x + 0.1x^2$,
(a) find the marginal cost when $x = 50$; and (b) find the marginal profit at $x = 50$, if the price per unit is $20.

Solution:

(a) Marginal cost is $C'(x)$. Enter $d(100 + 8x + 0.1x^2, x)|x = 50$ and obtain $18.

(b) Marginal profit is $P'(x)$
$P = R - C$
$P = 20x - (100 + 8x + 0.1x^2)$. Enter $d(20x - (100 + 8x + 0.1x{\wedge}2, x)|x = 50$ obtain 2.

• Carry all decimal places and round only at the final answer. Round to 3 decimal places unless the question indicates otherwise.

Example 2

Given the cost function $C(x) = 500 + 3x + 0.01x^2$ and the demand function (the price function) $p(x) = 10$, find the number of units produced in order to have maximum profit.

Solution:

Step 1: Write an equation.
Profit = Revenue − Cost
$P = R - C$
Revenue = (Units Sold)(Price Per Unit)
$R = xp(x) = x(10) = 10x$
$P = 10x - (500 + 3x + 0.01x^2)$

Step 2: Differentiate.
Enter $d(10x - (500 + 3x + 0.01x\wedge2, x)$ and obtain $7 - 0.02x$.

Step 3: Find critical numbers.
Set $7 - 0.02x = 0 \Rightarrow x = 350$.
Critical number is $x = 350$.

Step 4: Apply Second Derivative Test.
Enter $d(10x - (500 + 3x + 0.01x\wedge2), x, 2)|x = 350$ and obtain -0.02.
Since $x = 350$ is the only relative maximum, it is the absolute maximum.

Step 5: Write a Solution
Thus, producing 350 units will lead to maximum profit.

4.3 RAPID REVIEW

1. Find the instantaneous rate of change at $x = 5$ of the function $f(x) = \sqrt{2x - 1}$.

   Answer: $f(x) = \sqrt{2x - 1} = (2x - 1)^{1/2}$
   $$f'(x) = \frac{1}{2}(2x - 1)^{-1/2}(2) = (2x - 1)^{-1/2}$$
   $$f'(5) = \frac{1}{3}$$

2. If h is the diameter of a circle and h is increasing at a constant rate of 0.1 cm/sec, find the rate of change of the area of the circle when the diameter is 4 cm.

   Answer: $A = \pi r^2 = \pi \left(\frac{h}{2}\right)^2 = \frac{1}{4}\pi h^2$
   $$\frac{dA}{dt} = \frac{1}{2}\pi h \frac{dh}{dt} = \frac{1}{2}\pi(4)(0.1) = 0.2\pi \text{ cm}^2/\text{sec}.$$

3. The radius of a sphere is increasing at a constant rate of 2 inches per minute. In terms of the surface area, what is the rate of change of the volume of the sphere?

   Answer: $V = \frac{4}{3}\pi r^3$; $\frac{dV}{dt} = 4\pi r^2 \frac{dr}{dt}$ since $S = \pi r^2$, $\frac{dV}{dt} = 28 \text{ in.}^3/\text{min}.$

4. Using your calculator, find the shortest distance between the point (4, 0) and the line $y = x$. (See Figure 4.3-1.)

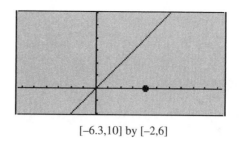

[−6.3,10] by [−2,6]

Figure 4.3-1

Answer:

$S = \sqrt{(x-4)^2 + (y-0)^2} = \sqrt{(x-4)^2 + x^2}$
Enter $y1 = ((x-4)^\wedge 2 + x^\wedge 2)^\wedge(.5)$ and $y2 = d(y1(x), x)$
Use the Zero function for $y2$ and obtain $x = 2$. Use the Value function for $y1$ at $x = 2$ and obtain $y1 = 2.82843$. Thus the shortest distance is approximately 2.828.

4.4 PRACTICE PROBLEMS

Part A—The use of a calculator is not allowed.

1. A spherical balloon is being inflated. Find the volume of the balloon at the instant when the rate of increase of the surface area is eight times the rate of increase of the radius of the sphere.

2. A 13-foot ladder is leaning against a wall. If the top of the ladder is sliding down the wall at 2 ft/sec, how fast is the bottom of the ladder moving away from the wall, when the top of the ladder is 5 feet from the ground? (See Figure 4.4-1.)

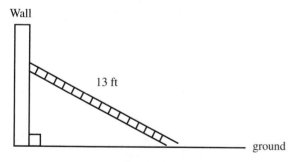

Figure 4.4-1

3. Air is being pumped into a spherical balloon at the rate of 100 cm³/sec. How fast is the diameter increasing when the radius is 5 cm?

4. A man 5 feet tall is walking away from a streetlight hung 20 feet from the ground at the rate of 6 ft/sec. How fast is his shadow lengthening?

5. A water tank in the shape of an inverted cone has an altitude of 18 feet and a base radius of 12 feet. If the tank is full and the water is drained at the rate of 4 ft³/min, how fast is the water level dropping when the water level is 6 feet high?

6. Two cars leave an intersection at the same time. The first car is going due east at the rate of 40 mph and the second is going due south at the rate of 30 mph. How fast is the distance between the two cars increasing when the first car is 120 miles from the intersection?

7. If the perimeter of an isosceles triangle is 18 cm, find the maximum area of the triangle.

8. Find a number in the interval (0, 2) such that the sum of the number and its reciprocal is the absolute minimum.

9. An open box is to be made using a cardboard 8 cm by 15 cm by cutting a square from each corner and folding the sides up. Find the length of a side of the square being cut so that the box will have a maximum volume.

10. What is the shortest distance between the point $\left(2, -\dfrac{1}{2}\right)$ and the parabola $y = -x^2$?

11. If the cost function is $C(x) = 3x^2 + 5x + 12$, find the value of x such that the average cost is a minimum.

12. A man with 200 meters of fence plans to enclose a rectangular piece of land using a river on one side and a fence on the other three sides. Find the maximum area that the man can obtain.

Part B—Calculators are allowed.

13. A trough is 10 meters long and 4 meters wide. (See Figure 4.4-2.) The two sides of the trough are equilateral triangles. Water is pumped into the trough at 1 m³/min. How fast is the water level rising when the water is 2 meters high?

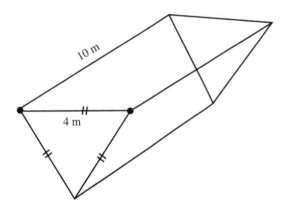

Figure 4.4-2

14. A rocket is sent vertically up in the air with the position function $s = 100t^2$ where s is measured in meters and t in seconds. A camera 3000 m away is recording the rocket. Find the rate of change of the angle of elevation of the camera 5 sec after the rocket went up.

15. A plane lifts off from a runway at an angle of 20°. If the speed of the plane is 300 mph, how fast is the plane gaining altitude?

16. Two water containers are being used. (See Figure 4.4-3.)

    One container is in the form of an inverted right circular cone with a height of 10 feet and a radius at the base of 4 feet. The other container is a right circular cylinder with a radius of 6 feet and a height of 8 feet. If water is being drained from the conical container into the cylindrical container at the rate of 15 ft³/min, how fast is the water level falling in

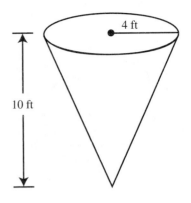

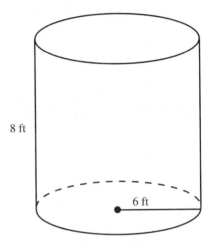

Figure 4.4-3

the conical tank when the water level in the conical tank is 5 feet high? How fast is the water level rising in the cylindrical container?

17. The wall of a building has a parallel fence that is 6 feet high and 8 feet from the wall. What is the length of the shortest ladder that passes over the fence and leans on the wall? (See Figure 4.4-4.)

18. Given the cost function $C(x) = 2500 + 0.02x + 0.004x^2$, find the product level such that the average cost per unit is a minimum.

19. Find the maximum area of a rectangle inscribed in an ellipse whose equation is $4x^2 + 25y^2 = 100$.

20. A right triangle is in the first quadrant with a vertex at the origin and the other two vertices on the x- and y-axes. If the hypotenuse passes through the point (0.5, 4), find the vertices of the triangle so that the length of the hypotenuse is the shortest.

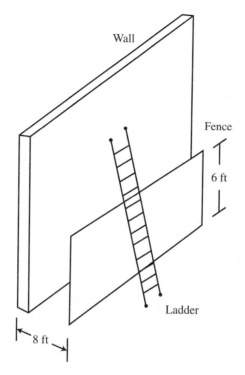

Figure 4.4-4

4.5 CUMULATIVE REVIEW PROBLEMS

"Calculator" indicates that calculators are permitted.

21. If $y = \sin^2(\cos(6x - 1))$, find $\dfrac{dy}{dx}$.

22. Evaluate $\displaystyle\lim_{x \to \infty} \frac{100/x}{-4 + x + x^2}$.

23. The graph of f' is shown in Figure 4.5-1. Find where the function f: (a) has its relative extrema or absolute extrema; (b) is increasing or decreasing; (c) has its point(s) of inflection; (d) is concave upward or downward; and (e) if $f(3) = -2$, draw a possible sketch of f. (See Figure 4.5-1.)

24. (Calculator) At what value(s) of x does the tangent to the curve $x^2 + y^2 = 36$ have a slope of -1.

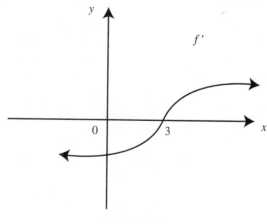

Figure 4.5-1

25. (Calculator) Find the shortest distance between the point $(1, 0)$ and the curve $y = x^3$.

4.6 SOLUTIONS TO PRACTICE PROBLEMS

Part A—No calculators.

1. Volume: $V = \frac{4}{3}\pi r^3$; Surface Area: $S = 4\pi r^2$

   $\frac{dS}{dt} = 8\pi r \frac{dr}{dt}$

   Since $\frac{dS}{dt} = 8\frac{dr}{dt}$, $8\frac{dr}{dt} = 8\pi r \frac{dr}{dt} \Rightarrow 8 = 8\pi r$

   or $r = \frac{1}{\pi}$.

   At $r = \frac{1}{\pi}$, $V = \frac{4}{3}\pi \left(\frac{1}{\pi}\right)^3 = \frac{4}{3\pi^2}$ cubic units.

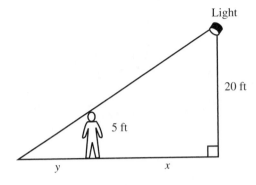

Figure 4.6-1

2. Pythagorean Theorem yields $x^2 + y^2 = (13)^2$.

   Differentiate: $2x\frac{dx}{dt} + 2y\frac{dy}{dt} = 0 \Rightarrow \frac{dy}{dt} = \frac{-x}{y}\frac{dx}{dt}$

   At $x = 5$, $(5)^2 + y^2 = 13^2 \Rightarrow y = \pm 12$, since $y > 0$, $y = 12$.

   Therefore, $\frac{dy}{dt} = -\frac{5}{12}(-2)$ ft/sec $= \frac{5}{6}$ ft/sec.
   The ladder is moving away from the wall at $\frac{5}{6}$ ft/sec when the top of the ladder is 5 feet from the ground.

3. Volume of a sphere is $V = \frac{4}{3}\pi r^3$.

   Differentiate: $\frac{dV}{dt} = \left(\frac{4}{3}\right)(3)\pi r^2 \frac{dr}{dt} = 4\pi r^2 \frac{dr}{dt}$.

   Substitute: $100 = 4\pi(5)^2 \frac{dr}{dt} \Rightarrow \frac{dr}{dt} = \frac{1}{\pi}$ cm/sec.

   Let x be the diameter. Since $x = 2r$, $\frac{dx}{dt} = 2\frac{dr}{dt}$.

   Thus $\frac{dx}{dt}\Big|_{r=5} = 2\left(\frac{1}{\pi}\right)$ cm/sec $= \frac{2}{\pi}$ cm/sec. The diameter is increasing at $\frac{2}{\pi}$ cm/sec when the radius is 5 cm.

4. See Figure 4.6-1. Using similar triangles, with y the length of the shadow you have:

   $\frac{5}{20} = \frac{y}{y+x} \Rightarrow 20y = 5y + 5x \Rightarrow$

   $15y = 5x$ or $y = \frac{x}{3}$.

   Differentiate:

   $\frac{dy}{dt} = \frac{1}{3}\frac{dx}{dt} \Rightarrow \frac{dy}{dt} = \frac{1}{3}(6)$

   $= 2$ ft/sec.

5. See Figure 4.6-2. Volume of a cone $V = \frac{1}{3}\pi r^2 h$.

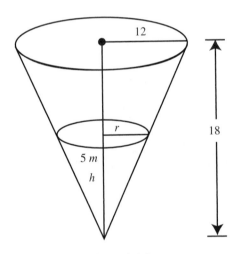

Figure 4.6-2

   Using similar triangles, you have
   $\frac{12}{18} = \frac{r}{h} \Rightarrow 2h = 3r$ or $r = \frac{2}{3}h$, thus reducing

   the equation to $V = \frac{1}{3}\pi \left(\frac{2}{3}h\right)^2 (h) = \frac{4\pi}{27}h^3$.

   Differentiate: $\frac{dV}{dt} = \frac{4}{9}\pi h^2 \frac{dh}{dt}$. Substituting

   known values: $-4 = \frac{4\pi}{9}(6)^2 \frac{dh}{dt} \Rightarrow -4 = 16\pi\frac{dh}{dt}$

   or $\frac{dh}{dt} = -\frac{1}{4\pi}$ ft/min. The water level is

   dropping at $\frac{1}{4\pi}$ ft/min when $h = 6$ ft.

6. See Figure 4.6-3.

   Step 1. Using the Pythagorean Theorem, you have $x^2 + y^2 = z^2$. You also have $\dfrac{dx}{dt} = 40$ and $\dfrac{dy}{dt} = 30$.

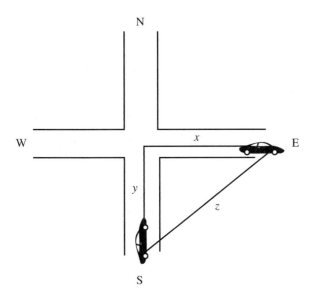

Figure 4.6-3

   Step 2. Differentiate: $2x\dfrac{dx}{dt} + 2y\dfrac{dy}{dt} = 2z\dfrac{dz}{dt}$. At $x = 120$, both cars have traveled 3 hours and thus, $y = 3(30) = 90$. By the Pythagorean Theorem, $(120)^2 + (90)^2 = z^2 \Rightarrow z = 150$.

   Step 3. Substitute all known values into the equation:
   $$2(120)(40) + 2(90)(30) = 2(150)\dfrac{dz}{dt}$$
   Thus $\dfrac{dz}{dt} = 50$ mph.

   Step 4. The distance between the two cars is increasing at 50 mph at $x = 120$.

7. See Figure 4.6-4.

   Step 1. Applying the Pythagorean Theorem, you have $x^2 = y^2 + (9 - x)^2 \Rightarrow y^2 = x^2 - (9 - x)^2 = x^2 - (81 - 18x + x^2) = 18x - 81 = 9(2x - 9)$, or $y = \pm\sqrt{9(2x - 9)} = \pm 3\sqrt{(2x - 9)}$ since $y > 0$, $y = 3\sqrt{(2x - 9)}$. The area of the triangle $A = \dfrac{1}{2}(3\sqrt{2x - 9})(18 - 2x) = (3\sqrt{2x - 9})(9 - x) = 3(2x - 9)^{1/2}(9 - x)$.

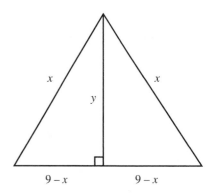

Figure 4.6-4

   Step 2. $\dfrac{dA}{dx} = \dfrac{3}{2}(2x - 9)^{-1/2}(2)(9 - x) + (-1)(3)(2x - 9)^{1/2}$.
   $$= \dfrac{3(9 - x) - 3(2x - 9)}{\sqrt{2x - 9}}$$
   $$= \dfrac{54 - 9x}{\sqrt{2x - 9}}$$

   Step 3. Set $\dfrac{dA}{dx} = 0 \Rightarrow 54 - 9x = 0$; $x = 6$
   $\dfrac{dA}{dx}$ is undefined at $x = \dfrac{9}{2}$. The critical numbers are $\dfrac{9}{2}$ and 6.

   Step 4. First Derivative Test

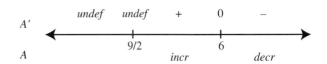

   Thus at $x = 6$, the area A is a relative maximum.
   $$A(6) = \left(\dfrac{1}{2}\right)(3)\left(\sqrt{2(6) - 9}\right)(9 - 6) = 9\sqrt{3}.$$

   Step 5. Check endpoints. The domain of A is $[9/2, 9]$ $A(9/2) = 0$; and $A(9) = 0$. Therefore, the maximum area of an isosceles triangle with the perimeter of 18 cm is $(9\sqrt{3})/2$ cm^2. (Note, at $x = 6$, the triangle is an equilateral triangle.)

8. See Figure 4.6-5.

   Step 1. Let x be the number and $\dfrac{1}{x}$ be its reciprocal.

   Step 2. $s = x + \dfrac{1}{x}$ with $0 < x < 2$.

   Step 3. $\dfrac{ds}{dx} = 1 + (-1)x^{-2} = 1 - \dfrac{1}{x^2}$

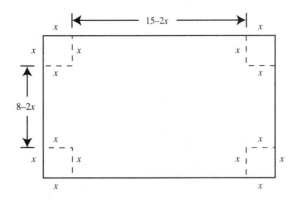

Figure 4.6-5

Step 4. Set $\dfrac{ds}{dx} = 0 \Rightarrow 1 - \dfrac{1}{x^2} = 0 \Rightarrow x = \pm 1$, since the domain is $(0, 2)$, thus $x = 1$. $\dfrac{ds}{dx}$ is defined for all x in $(0, 2)$. Critical number is $x = 1$.

Step 5. Second Derivative Test: $\dfrac{d^2s}{dx^2} = \dfrac{2}{x^3}$ and $\dfrac{d^2s}{dx^2}\Big|_{x=1} = 2.$

Thus at $x = 1$, s is a relative minimum. Since it is the only relative extremum, thus, at $x = 1$, it is the absolute minimum.

9. See Figure 4.6-5.

Step 1. Volume $V = x(8 - 2x)(15 - 2x)$ with $0 \le x \le 4$.

Step 2. Differentiate: Rewrite as $V = 4x^3 - 46x^2 + 120x$ $\dfrac{dV}{dx} = 12x^2 - 92x + 120$

Step 3. Set $V = 0 \Rightarrow 12x^2 - 92x + 120 = 0$ $\Rightarrow 3x^2 - 23x + 30 = 0$. Using the quadratic formula, you have $x = 6$ or $x = \dfrac{5}{3}$. And $\dfrac{dV}{dx}$ is defined for all real numbers.

Step 4. Second Derivative Test.

$$\dfrac{d^2V}{dx^2} = 24x - 92; \ \dfrac{d^2V}{dx^2}\Big|_{x=6}$$

$$= 52 \text{ and } \dfrac{d^2V}{dx^2}\Big|_{x=\frac{5}{3}} = -52$$

Thus at $x = \dfrac{5}{3}$ is a relative maximum.

Step 5. Check endpoints.
At $x = 0$, $V = 0$ and at $x = 4$, $V = 0$.
Therefore, at $x = \dfrac{5}{3}$, V is the absolute maximum.

10. See Figure 4.6-6.

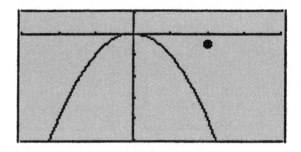

Figure 4.6-6

Step 1. Distance Formula:

$$Z = \sqrt{(x - 2)^2 + \left(y - \left(-\dfrac{1}{2}\right)\right)^2}$$

$$= \sqrt{(x - 2)^2 + \left(-x^2 + \dfrac{1}{2}\right)^2}$$

$$= \sqrt{x^2 - 4x + 4 + x^4 - x^2 + \dfrac{1}{4}}$$

$$= \sqrt{x^4 - 4x + \dfrac{17}{4}}$$

Step 2. Let $S = Z^2$, since S and Z have the same maximums and minimums. $S = x^4 - 4x + \dfrac{17}{4}; \ \dfrac{dS}{dx} = 4x^3 - 4$

Step 3. Set $\dfrac{dS}{dx} = 0$; $x = 1$ and $\dfrac{dS}{dx}$ is defined for all real numbers.

Step 4. Second Derivative Test: $\dfrac{d^2S}{dx^2} = 12x^2$ and $\dfrac{d^2S}{dx^2}\Big|_{x=1} = 12$. Thus at $x = 1$, Z has a minimum and since it is the only relative extremum, it is the absolute minimum.

Step 5. At $x = 1$, $Z = \sqrt{(1)^4 - 4(1) + \dfrac{17}{4}}$ $= \sqrt{\dfrac{5}{4}}.$

Therefore, the shortest distance is $\sqrt{\dfrac{5}{4}}$.

11. Step 1. Average Cost
$$\overline{C} = \frac{C(x)}{x} = \frac{3x^2 + 5x + 12}{x}$$
$$= 3x + 5 + \frac{12}{x}$$

Step 2. $\dfrac{d\overline{C}}{dx} = 3 - 12x^{-2} = 3 - \dfrac{12}{x^2}$

Step 3. Set $\dfrac{d\overline{C}}{dx} = 0 \Rightarrow 3 - \dfrac{12}{x^2} = 0 \Rightarrow 3 = \dfrac{12}{x^2}$
$\Rightarrow x = \pm 2$. Since $x > 0$, $x = 2$ and
$\overline{C}(2) = 17. \dfrac{d\overline{C}}{dx}$ is undefined at $x = 0$
which is not in the domain.

Step 4. Second Derivative Test:
$$\frac{d^2\overline{C}}{dx^2} = \frac{24}{x^3} \text{ and } \left.\frac{d^2\overline{C}}{dx^2}\right|_{x=2} = 3$$
Thus at $x = 2$, the average cost is a minimum.

12. See Figure 4.6-7.

Step 1. Area $A = x(200 - 2x) = 200x - 2x^2$
with $0 \le x \le 100$

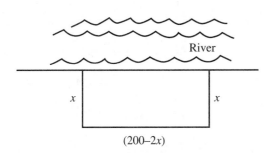

$(200-2x)$

Figure 4.6-7

Step 2. $A'(x) = 200 - 4x$

Step 3. Set $A'(x) = 0 \Rightarrow 200 - 4x = 0$; $x = 50$

Step 4. Second Derivative Test
$A''(x) = -4$; thus at $x = 50$, the area is a relative maximum.
$A(50) = 5000 \text{ m}^2$.

Step 5. Check endpoints
$A(0) = 0$ and $A(100) = 0$; therefore at $x = 50$, the area is the absolute maximum and 5000 m^2 is the maximum area.

13. Step 1. Let h be the height of the trough and 4 be a side of one of the two equilateral triangles. Thus, in a 30–60 right triangle, $h = 2\sqrt{3}$.

Step 2. Volume $V = $ (area of the triangle) $\cdot$ 10
$$= \left[\frac{1}{2}(b)\left(\frac{2}{\sqrt{3}}h\right)\right]10 = \frac{10}{\sqrt{3}}h^2$$

Step 3. Differentiate with respect to t.
$$\frac{dV}{dt} = \left(\frac{10}{\sqrt{3}}\right)(2)h\frac{dh}{dt}$$

Step 4. Substitute known values
$$1 = \frac{20}{\sqrt{3}}(2)\frac{dh}{dt}; \quad \frac{dh}{dt} = \frac{\sqrt{3}}{40} \text{ m/min.}$$

The water level is raising $\dfrac{\sqrt{3}}{40}$ m/min
when the water level is 2 m high.

14. See Figure 4.6-8.

Step 1. $\tan \theta = S/3000$

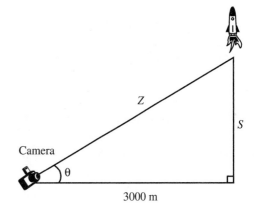

Figure 4.6-8

Step 2. Differentiate with respect to t.
$$\sec^2 \theta \frac{d\theta}{dt} = \frac{1}{3000}\frac{dS}{dt}; \frac{d\theta}{dt}$$
$$= \frac{1}{3000}\left(\frac{1}{\sec^2 \theta}\right)\frac{dS}{dt}$$
$$= \frac{1}{3000}\left(\frac{1}{\sec^2 \theta}\right)(200t)$$

Step 3. At $t = 5$; $S = 100(5)^2 = 2500$;
Thus
$Z^2 = (3000)^2 + (2500)^2 = 15{,}250{,}000$.
Therefore $Z = \pm 500\sqrt{61}$, since $Z > 0$,
$Z = 500\sqrt{61}$. Substitute known values into the question:

$$\frac{d\theta}{dt} = \frac{1}{3000}\left(\frac{1}{\dfrac{500\sqrt{61}}{3000}}\right)^2 (1000),$$

since $\sec\theta = \dfrac{Z}{3000}$.

$\dfrac{d\theta}{dt} = 0.197$ radian/sec. The angle of elevation is changing at 0.197 radian/sec, 5 seconds after lift off.

15. See Figure 4.6-9.

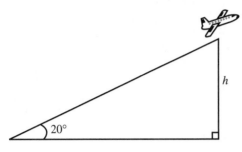

Figure 4.6-9

$\text{Sin } 20° = \dfrac{h}{300t}$

$h = (\sin 20°)300t$; $\dfrac{dh}{dt} = (\sin 20°)(300) \approx$ 102.606 mph. The plane is gaining altitude at 102.606 mph.

16. $V_{\text{cone}} = \dfrac{1}{3}\pi r^2 h$

Similar triangles: $\dfrac{4}{10} = \dfrac{r}{h} \Rightarrow 5r = 2h$ or $r = \dfrac{2h}{5}$

$V_{\text{cone}} = \dfrac{1}{3}\pi\left(\dfrac{2h}{5}\right)^2 h = \dfrac{4\pi}{75}h^3$;

$\dfrac{dV}{dt} = \dfrac{4\pi}{75}(3)h^2\dfrac{dh}{dt}$.

Substitute known values: $-15 = \dfrac{4\pi}{25}(5)^2\dfrac{dh}{dt}$;

$-15 = 4\pi\dfrac{dh}{dt}$; $\dfrac{dh}{dt} = \dfrac{-15}{4\pi} \approx -1.19$ ft/min.

The water level in the cone is falling at $\dfrac{-15}{4\pi}$ ft/min ≈ -1.19 ft/ min when the water level is 5 feet high.

$V_{\text{cylinder}} = \pi R^2 H = \pi(6)^2 H = 36\pi H$.

$\dfrac{dV}{dt} = 36\pi\dfrac{dH}{dt}$; $\dfrac{dH}{dt} = \dfrac{1}{36\pi}\dfrac{dV}{dt}$; $\dfrac{dH}{dt}$

$= \dfrac{1}{36\pi}(15) = \dfrac{5}{12\pi}$ ft/min

≈ 0.1326 ft/min or 1.592 in/min

The water level in the cylinder is rising at $\dfrac{5}{12\pi}$ ft/min $= 0.133$ ft/min.

17. Step 1. Let x be the distance of the foot of the ladder from the higher wall. Let y be the height of the point where the ladder touches the higher wall. The slope of the ladder is $m = \dfrac{y-6}{0-8}$ or $m = \dfrac{6-0}{8-x}$. Thus

$\dfrac{y-6}{-8} = \dfrac{6}{8-x} \Rightarrow (y-6)(8-x) = -48$

$\Rightarrow 8y - xy - 48 + 6x = -48$

$\Rightarrow y(8-x) = -6x \Rightarrow y = \dfrac{-6x}{8-x}$

Step 2. Phythagorean Theorem:

$l^2 = x^2 + y^2 = x^2 + \left(\dfrac{-6x}{8-x}\right)^2$

Since $l > 0$, $l = \sqrt{x^2 + \left(\dfrac{-6x}{8-x}\right)^2}$, $x > 8$

Step 3. Enter

$y_1 = \sqrt{\{x^\wedge 2 + [(-6 * x)/(8-x)]^\wedge 2\}}$.

The graph of y_1 is continuous on the interval $x > 8$. Use the minimum function of the calculator and obtain $x = 14.604$; $y = 17.42$. Thus the minimum value of l is 19.731 or the shortest ladder is approximately 19.731 feet.

18. Step 1. Average Cost $\overline{C} = \dfrac{C}{x}$; thus $\overline{C}(x)$

$= \dfrac{2500 + 0.02x + 0.004x^2}{x}$

$= \dfrac{2500}{x} + 0.02 + .004x$

Step 2. Enter $y_1 = \dfrac{2500}{x} + .02 + .004 * x$

Step 3. Use the Minimum function in the calculator and obtain $x = 790.6$.

Step 4. Verify the result with First Derivative Test. Enter $y_2 = d(2500/x + .02 + 004x, x)$; Use the Zero function and obtain $x = 790.6$. Thus $\dfrac{d\overline{C}}{dx} = 0$; at $x = 790.6$.

Apply the First Derivative Test:

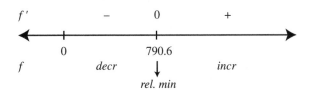

Thus the minimum average cost per unit occurs at $x = 790.6$ (The graph of the average cost function is shown in Figure 4.6-10.)

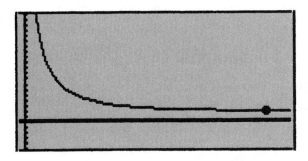

Figure 4.6-10

19. See Figure 4.6-11.

Step 1. Area $A = (2x)(2y); 0 \le x \le 5$ and $0 \le y \le 2$.

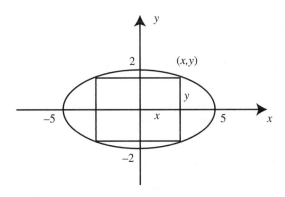

Figure 4.6-11

Step 2. $4x^2 + 25y^2 = 100; 25y^2 = 100 - 4x^2$

$$y^2 = \frac{100 - 4x^2}{25} \Rightarrow y = \pm\sqrt{\frac{100 - 4x^2}{25}}$$

Since $y \ge 0$

$$y = \sqrt{\frac{100 - 4x^2}{25}} = \sqrt{\frac{100 - 4x^2}{5}}$$

Step 3. $A = (2x)\left(\dfrac{2}{5}\right)\left(\sqrt{100 - 4x^2}\right)$

$$= \frac{4x}{5}\sqrt{100 - 4x^2}$$

Step 4. Enter $y_1 = \dfrac{4x}{5}\sqrt{100 - 4x^2}$

Use the Maximum function and obtain $x = 3.536$ and $y_1 = 20$.

Step 5. Verify the result with the First Derivative Test.

Enter $y_2 = d\left(\dfrac{4x}{5}\sqrt{100 - 4x^2}, x\right)$. Use Zero function and obtain $x = 3.536$. Note that:

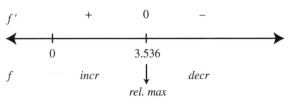

The function f has only one relative extremum. Thus it is the absolute extremum. Therefore, at $x = 3.536$, the area is 20 and the area is the absolute maxima.

20. See Figure 4.6-12.

Step 1. Distance formula: $l^2 = x^2 + y^2; x > 0.5$ and $y > 4$

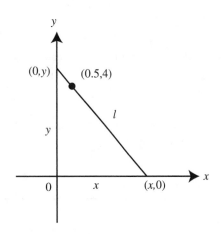

Figure 4.6-12

Step 2. The slope of the hypotenuse:

$$m = \frac{y - 4}{0 - 0.5} = \frac{-4}{x - 0.5}$$

$$\Rightarrow (y - 4)(x - 0.5) = 2$$

$$\Rightarrow xy - 0.5y - 4x + 2 = 2$$

$$y(x - 0.5) = 4x$$

$$y = \frac{4x}{x - 0.5}$$

Step 3. $l^2 = x^2 + \left(\dfrac{4x}{x - 0.5}\right)^2$;

$$l = \pm\sqrt{x^2 + \left(\dfrac{4x}{x - 0.5}\right)^2}$$

Since $l > 0, l = \sqrt{x^2 + \left(\dfrac{4x}{x - 0.5}\right)^2}$

Step 4. Enter $y_1 = \sqrt{x^2 + \left(\dfrac{4x}{x - 0.5}\right)^2}$ and use the minimum function of the calculator and obtain $x = 2.5$.

Step 5. Apply the First Derivative Test. Enter $y_2 = d(y_1(x), x)$ and use the zero function and obtain $x = 2.5$. Note that:

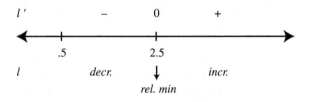

Since f has only one relative extremum, it is the absolute extrenum.

Step 6. Thus at $x = 2.5$, the length of the hypotenuse is the shortest. At $x = 2.5$, $y = \dfrac{4(2.5)}{2.5 - 0.5} = 5$. The vertices of the triangle are $(0, 0), (2.5, 0)$ and $(0, 5)$.

4.7 SOLUTIONS TO CUMULATIVE REVIEW PROBLEMS

21. Rewrite: $y = [\sin(\cos(6x - 1))]^2$

Thus $\dfrac{dy}{dx} = 2\left[\sin\left(\cos\left(6x - 1\right)\right)\right]\left[\cos\left(\cos\left(6x - 1\right)\right)\right]$

$$\times\left[-\sin(6x - 1)\right](6)$$

$$= -12\sin(6x - 1)[\sin(\cos(6x - 1))]$$

$$\times\left[\cos(\cos(6x - 1))\right]$$

22. As $x \to \infty$, the numerator $\dfrac{100}{x}$ approaches 0 and the denominator increases without bound (i.e., ∞). Thus the $\displaystyle\lim_{x \to \infty} \dfrac{100/x}{-4 + x + x^2} = 0$.

23. (a) Summarize the information of f' on a number line.

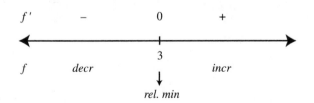

Since f has only one relative extremum, it is the absolute extremum. Thus at $x = 3$, it is an absolute minimum.

(b) The function f is decreasing on the interval $(-\infty, 3)$ and increasing on $(3, \infty)$.

(c)

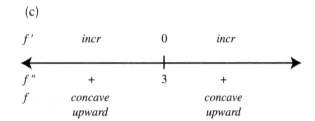

no change of concavity $\Rightarrow$ no point of inflection.

(d) The function f is concave upward for the entire domain $(-\infty, \infty)$.

(e) Possible sketch of the graph for $f(x)$. See Figure 4.7-1.

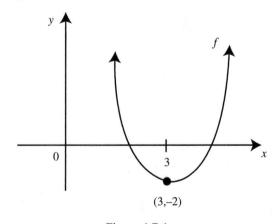

Figure 4.7-1

24. (Calculator) See Figure 4.7-2.

Step 1. Differentiate
$$2x + 2y\frac{dy}{dx} = 0 \Rightarrow \frac{dy}{dx} = -\frac{x}{y}$$

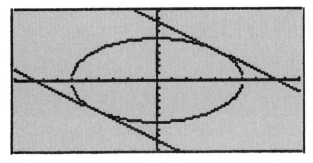

Figure 4.7-2

Step 2. Set $\frac{dy}{dx} = -1 \Rightarrow \frac{-x}{y} = -1 \Rightarrow y = x$

Step 3. Solve for
$y: x^2 + y^2 = 36 \Rightarrow y^2 = 36 - x^2;$
$y = \pm\sqrt{36 - x^2}$

Step 4. Thus, $y = x \Rightarrow \pm\sqrt{36 - x^2} = x \Rightarrow$
$36 - x^2 = x^2 \Rightarrow 36 = 2x^2$ or $x = \pm 3\sqrt{2}$

25. (Calculator) See Figure 4.7-3.

Step 1. Distance formula:
$$z = \sqrt{(x-1)^2 + (x^3)^2} = \sqrt{(x-1)^2 + x^6}$$

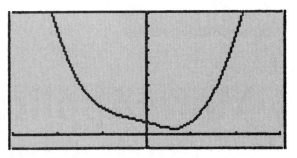

Figure 4.7-3

Step 2. Enter $y_1 = \sqrt{((x-1)^2 + x^6)}$. Use the Minimum function of the calculator and obtain $x = .65052$ and $y_1 = .44488$. Verify the result with the First Deriative Test. Enter $y_2 = d(y_1(x), x)$ and use the Zero Function and obtain $x = .65052$.

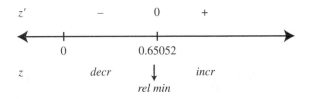

Thus the shortest distance is approximately 0.445.

Chapter 5

More Applications of Derivatives

5.1 TANGENT AND NORMAL LINES

Main Concepts: *Tangent Lines, Normal Lines*

Tangent Lines

If the function y is differentiable at $x = a$, then the slope of the tangent line to the graph of y at $x = a$ is given as $m_{(\text{tangent at } x = a)} = \left. \dfrac{dy}{dx} \right|_{x=a}$.

Types of Tangent Lines:

Horizontal Tangents $\left(\dfrac{dy}{dx} = 0 \right)$. (See Figure 5.1-1.)

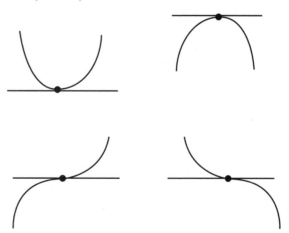

Figure 5.1-1

Vertical Tangents $\left(\dfrac{dy}{dx} \text{ does not exist but } \dfrac{dx}{dy} = 0\right)$. (See Figure 5.1-2.)

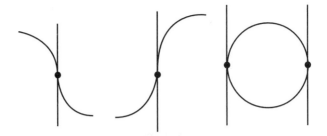

Figure 5.1-2

Parallel Tangents $\left(\left.\dfrac{dy}{dx}\right|_{x=a} = \left.\dfrac{dx}{dy}\right|_{x=c}\right)$. (See Figure 5.1-3.)

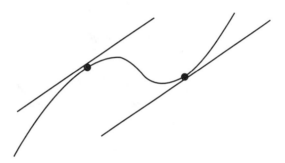

Figure 5.1-3

Example 1

Write an equation of the line tangent to the graph of $y = -3 \sin 2x$ at $x = \dfrac{\pi}{2}$. (See Figure 5.1-4.)

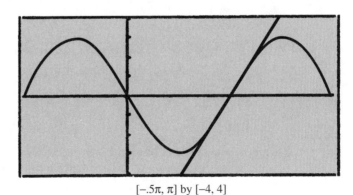

$[-.5\pi, \pi]$ by $[-4, 4]$

Figure 5.1-4

$y = -3 \sin 2x; \quad \dfrac{dy}{dx} = -3[\cos(2x)]2 = -6 \cos(2x)$

Slope of tangent $\left(\text{at } x = \dfrac{\pi}{2}\right): \left.\dfrac{dy}{dx}\right|_{x=\pi/2} = -6 \cos [2(\pi/2)] = -6 \cos \pi = 6.$

Point of tangency: At $x = \dfrac{\pi}{2}$, $y = -3 \sin(2x) = -3 \sin[2(\pi/2)] = -3 \sin(\pi) = 0.$

Therefore $\left(\dfrac{\pi}{2}, 0\right)$ is the point of tangency.

Equation of Tangent: $y - 0 = 6(x - \pi/2)$ or $y = 6x - 3\pi$.

Example 2

If the line $y = 6x + a$ is tangent to the graph of $y = 2x^3$, find the value(s) of a.

Solution:

$y = 2x^3; \dfrac{dy}{dx} = 6x^2$. (See Figure 5.1-5.)

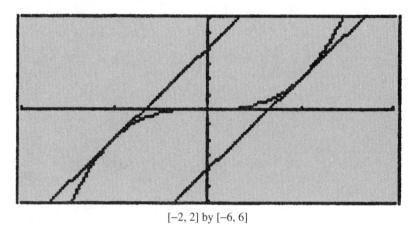

[−2, 2] by [−6, 6]

Figure 5.1-5

The slope of the line $y = 6x + a$ is 6.

Since $y = 6x + a$ is tangent to the graph of $y = 2x^3$, thus $\dfrac{dy}{dx} = 6$ for some values of x.

Set $6x^2 = 6 \Rightarrow x^2 = 1$ or $x = \pm 1$.

At $x = -1$, $y = 2x^3 = 2(-1)^3 = -2$; $(-1, -2)$ is a tangent point. Thus, $y = 6x + a \Rightarrow -2 = 6(-1) + a$ or $a = 4$.

At $x = 1$, $y = 2x^3 = 2(1)^3 = 2$; $(1, 2)$ is a tangent point.

Thus $y = 6x + a \Rightarrow 2 = 6(1) + a$ or $a = -4$.

Therefore, $a = \pm 4$.

Example 3

Find the coordinates of each point on the graph of $y^2 - x^2 - 6x + 7 = 0$ at which the tangent line is vertical. Write an equation of each vertical tangent. (See Figure 5.1-6.)

Step 1: Find $\dfrac{dy}{dx}$

$$y^2 - x^2 - 6x + 7 = 0$$

$$2y\dfrac{dy}{dx} - 2x - 6 = 0$$

$$\dfrac{dy}{dx} = \dfrac{2x + 6}{2y} = \dfrac{x + 3}{y}$$

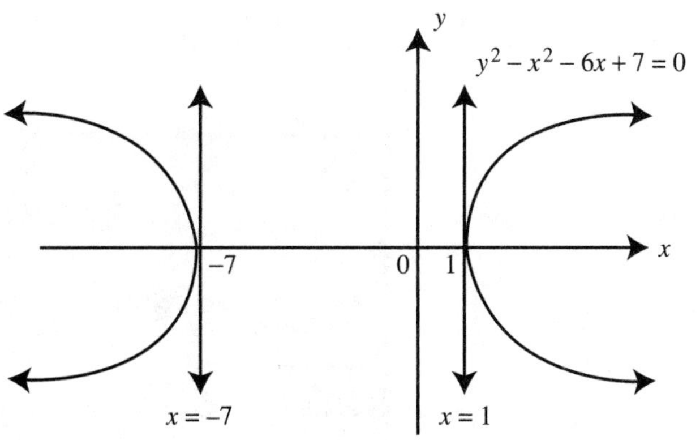

$$y^2 - x^2 - 6x + 7 = 0$$

$x = -7$ $x = 1$

Figure 5.1-6

Step 2: Find $\dfrac{dx}{dy}$

Vertical tangent $\Rightarrow \dfrac{dx}{dy} = 0$

$$\frac{dx}{dy} = \frac{1}{dy/dx} = \frac{1}{(x+3)/y} = \frac{y}{x+3}$$

Set $\dfrac{dx}{dy} = 0 \Rightarrow y = 0$.

Step 3: Find points of tangency.
At $y = 0$, $y^2 - x^2 - 6x + 7 = 0$ becomes $-x^2 - 6x + 7 = 0 \Rightarrow x^2 + 6x - 7 = 0$
$\Rightarrow (x+7)(x-1) = 0 \Rightarrow x = -7$ or $x = 1$.
Thus the points of tangency are $(-7, 0)$ and $(1, 0)$

Step 4: Write equation for vertical tangents:
$x = -7$ and $x = 1$.

Example 4

Find all points on the graph of $y = |xe^x|$ at which the graph has a horizontal tangent.

Step 1: Find $\dfrac{dy}{dx}$

$$y = |xe^x| = \begin{cases} xe^x & \text{if } x \geq 0 \\ -xe^x & \text{if } x < 0 \end{cases}$$

$$\frac{dy}{dx} = \begin{cases} e^x + xe^x & \text{if } x \geq 0 \\ -e^x - xe^x & \text{if } x < 0 \end{cases}$$

Step 2: Find the x-coordinate of points of tangency.
Horizontal Tangent $\Rightarrow \dfrac{dy}{dx} = 0$

If $x \geq 0$, set $e^x + xe^x = 0 \Rightarrow e^x(1 + x) = 0 \Rightarrow x = -1$ but $x \geq 0$, therefore, no solution.

If $x < 0$, set $-e^x - xe^x = 0 \Rightarrow -e^x(1 + x) = 0 \Rightarrow x = -1$.

Step 3: Find points of tangency.
At $x = -1$, $y = -xe^x = -(-1)e^{-1} = \dfrac{1}{e}$.

Thus at the point $(-1, 1/e)$, the graph has a horizontal tangent. (See Figure 5.1-7.)

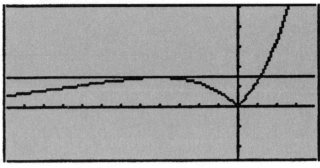

[−3, 1] by [−0.5, 1.25]

Figure 5.1-7

Example 5

Using your calculator, find the value(s) of x to the nearest hundredth at which the slope of the line tangent to the graph of $y = 2 \ln(x^2 + 3)$ is equal to $-\dfrac{1}{2}$. (See Figures 5.1-8 and 5.1-9.)

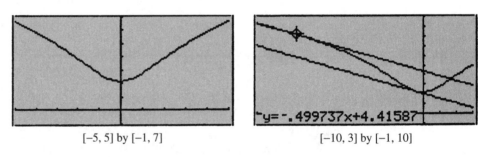

[−5, 5] by [−1, 7]   [−10, 3] by [−1, 10]

Figure 5.1-8   **Figure 5.1-9**

Step 1: Enter $y_1 = 2 * \ln(x^\wedge 2 + 3)$

Step 2: Enter $y_2 = d(y_1(x), x)$ and enter $y_3 = -\dfrac{1}{2}$

Step 3: Using the Intersection function of the calculator for y_2 and y_3, you obtain $x = -7.61$ or $x = -0.39$.

Example 6

Using your calculator, find the value(s) of x at which the graphs of $y = 2x^2$ and $y = e^x$ have parallel tangents.

Step 1: Find $\dfrac{dy}{dx}$ for both $y = 2x^2$ and $y = e^x$

$$y = 2x^2; \frac{dy}{dx} = 4x$$

$$y = e^x; \frac{dy}{dx} = e^x$$

Step 2: Find the x-coordinate of the points of tangency. Parallel tangents $\Rightarrow$ slopes are equal.

Set $4x = e^x \Rightarrow 4x - e^x = 0$

Using the Solve function of the calculator, enter Solve $(4x - e\hat{\ }(x) = 0, \ x)$ and obtain $x = 2.15$ and $x = 0.36$.

- Watch out for different units of measure, e.g., the radius, r, is 2 feet, find $\dfrac{dr}{dt}$ in inches per second.

Normal Lines

The normal line to the graph of f at the point (x_1, y_1) is the line perpendicular to the tangent line at (x_1, y_1). (See Figure 5.1-10.)

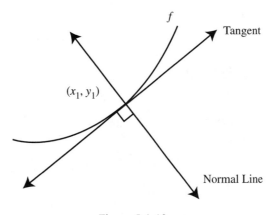

Figure 5.1-10

Note that the slope of the normal line and the slope of the tangent line at any point on the curve are negative reciprocals provided that both slopes exist.

$(m_{\text{normal line}})(m_{\text{tangent line}}) = -1.$

Special Cases: See Figure 5.1-11.

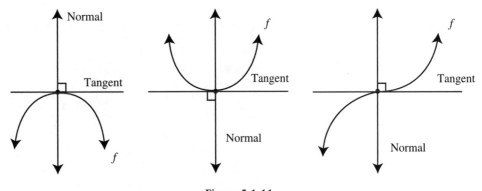

Figure 5.1-11

At these points, $m_{\text{tangent}} = 0$; but m_{normal} does not exist.
See Figure 5.1-12.

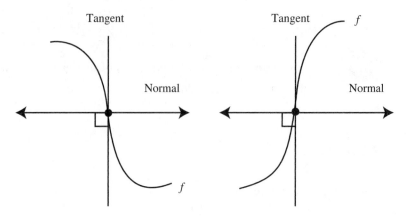

Figure 5.1-12

At these points, m_{tangent} does not exist; however $m_{\text{normal}} = 0$.

Example 1

Write an equation for each normal to the graph of $y = 2 \sin x$ for $0 \le x \le 2\pi$ that has a slope of $\dfrac{1}{2}$.

Step 1: Find m_{tangent}

$$y = 2 \sin x; \quad \frac{dy}{dx} = 2 \cos x$$

Step 2: Find m_{normal}

$$m_{\text{normal}} = -\frac{1}{m_{\text{tangent}}} = -\frac{1}{2 \cos x}$$

Set $m_{\text{normal}} = \dfrac{1}{2} \Rightarrow -\dfrac{1}{2 \cos x} = \dfrac{1}{2} \Rightarrow \cos x = -1$

$\Rightarrow x = \cos^{-1}(-1)$ or $x = \pi$. (See Figure 5.1-13.)

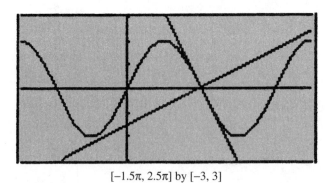

$[-1.5\pi, 2.5\pi]$ by $[-3, 3]$

Figure 5.1-13

Step 3: Write equation of normal line.

At $x = \pi$, $y = 2 \sin x = 2(0) = 0$; $(\pi, 0)$

Since $m = \dfrac{1}{2}$, equation of normal is:

$y - 0 = \dfrac{1}{2}(x - \pi)$ or $y = \dfrac{1}{2}x - \dfrac{\pi}{2}$.

Example 2

Find the point on the graph of $y = \ln x$ such that the normal line at this point is parallel to the line $y = -ex - 1$.

Step 1: Find m_{tangent}

$y = \ln x$; $\dfrac{dy}{dx} = \dfrac{1}{x}$

Step 2: Find m_{normal}

$m_{\text{normal}} = \dfrac{-1}{m_{\text{tangent}}} = \dfrac{-1}{1/x} = -x$

slope of $y = -ex - 1$ is $-e$.
Since normal is parallel to the line $y = -ex - 1$, set $m_{\text{normal}} = -e \Rightarrow -x = -e$ or $x = e$.

Step 3: Find point on graph. At $x = e$, $y = \ln x = \ln e = l$. Thus the point of the graph of $y = \ln x$ at which the normal is parallel to $y = -ex - 1$ is $(e, 1)$. (See Figure 5.1-14.)

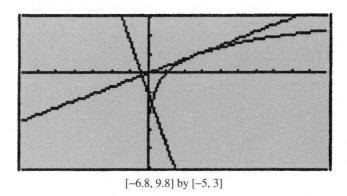

[−6.8, 9.8] by [−5, 3]

Figure 5.1-14

Example 3

Given the curve $y = \dfrac{1}{x}$: (a) write an equation of the normal to the curve $y = \dfrac{1}{x}$ at the point $(2, 1/2)$ and (b) does this normal intersect the curve at any other point? If yes, find the point.

Step 1: Find m_{tangent}

$y = \dfrac{1}{x}$; $\dfrac{dy}{dx} = (-1)(x^{-2}) = -\dfrac{1}{x^2}$

Step 2: Find m_{normal}

$m_{\text{normal}} = \dfrac{-1}{m_{\text{tangent}}} = \dfrac{-1}{-1/x^2} = x^2$

At $(2, 1/2)$, $m_{\text{normal}} = 2^2 = 4$.

Step 3: Write equation of normal

$m_{\text{normal}} = 4;\ (2, 1/2)$

Equation of normal: $y - \dfrac{1}{2} = 4(x - 2)$, or $y = 4x - \dfrac{15}{2}$

Step 4: Find other points of intersection.

$$y = \frac{1}{x};\ y = 4x - \frac{15}{2}$$

Using the Intersection function of your calculator, enter $y_1 = \dfrac{1}{x}$ and $y_2 = 4x - \dfrac{15}{2}$ and obtain $x = -0.125$ and $y = -8$. Thus, the normal line intersects the graph of $y = \dfrac{1}{x}$ at the point $(-0.125, -8)$ as well.

> • Remember that $\displaystyle\int 1\,dx = x + c$ and $\dfrac{d}{dx}(1) = 0$.

5.2 LINEAR APPROXIMATIONS

Main Concepts: *Tangent Line Approximation, Estimating the nth Root of a Number, Estimating the Value of a Trigonometric Function of an Angle*

Tangent Line Approximation

An equation of the tangent line to a curve at the point $(a, f(a))$ is:
$y = f(a) + f'(a)(x - a)$ providing that f is differentiable at a. See Figure 5.2-1.

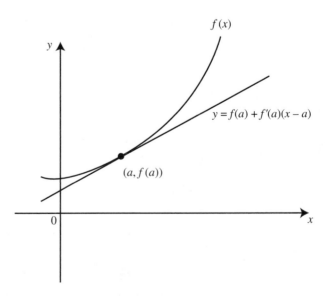

Figure 5.2-1

Tangent Line Approximation (or Linear Approximation):

Since the curve of $f(x)$ and the tangent line are close to each other for points near $x = a$, $f(x) \approx f(a) + f'(a)(x - a)$

Example 1

Write an equation of the tangent line to $f(x) = x^3$ at $(2, 8)$. Use the tangent line to find the approximate values of $f(1.9)$ and $f(2.01)$.

Differentiate $f(x)$: $f'(x) = 3x^2$; $f'(2) = 3(2)^2 = 12$. Since f is differentiable at $x = 2$, an equation of the tangent at $x = 2$ is:

$$y = f(2) + f'(2)(x - 2)$$

$$y = (2)^3 + 12(x - 2) = 8 + 12x - 24 = 12x - 16$$

$$f(1.9) \approx 12(1.9) - 16 = 6.8$$

$$f(2.01) \approx 12(2.01) - 16 = 8.12. \text{ (See Figure 5.2-2.)}$$

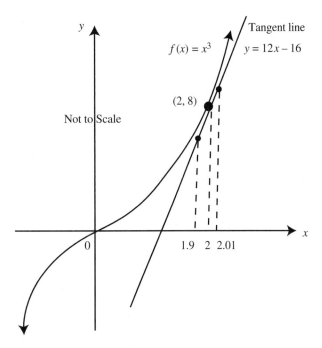

Figure 5.2-2

Example 2

If f is a differentiable function and $f(2) = 6$ and $f'(2) = -\dfrac{1}{2}$, find the approximate value of $f(2.1)$.

Using tangent line approximation, you have

(a) $f(2) = 6 \Rightarrow$ the point of tangency is $(2, 6)$

(b) $f'(2) = -\dfrac{1}{2} \Rightarrow$ the slope of the tangent at $x = 2$ is $m = -\dfrac{1}{2}$.

(c) the equation of the tangent is $y - 6 = -\dfrac{1}{2}(x - 2)$ or $y = -\dfrac{1}{2}x + 7$

(d) thus $f(2.1) \approx -\dfrac{1}{2}(2.1) + 7 \approx 5.95$.

Example 3

The slope of a function at any point (x, y) is $-\dfrac{x+1}{y}$. The point $(3, 2)$ is on the graph of f. (a) Write an equation of the line tangent to the graph of f at $x = 3$. (b) Use the tangent line in part (a) to approximate $f(3.1)$.

(a) Let $y = f(x)$, then $\dfrac{dy}{dx} = -\dfrac{x+1}{y}$

$\left.\dfrac{dy}{dx}\right|_{x=3,\ y=2} = -\dfrac{3+1}{2} = -2.$

Equation of tangent: $y - 2 = -2(x - 3)$ or $y = -2x + 8$

(b) $f(3.1) \approx -2(3.1) + 8 \approx 1.8.$

Estimating the *n*th Root of a Number

Another way of expressing the tangent line approximation is:
$f(a + \Delta x) \approx f(a) + f'(a)\Delta x$; where Δx is a relatively small value.

Example 1

Find the approximation value of $\sqrt{50}$ using linear approximation.

Using $f(a + \Delta x) \approx f(a) + f'(a)\Delta x$, let $f(x) = \sqrt{x}$; $a = 49$ and $\Delta x = 1$.

Thus $f(49 + 1) \approx f(49) + f'(49)(1) \approx \sqrt{49} + \dfrac{1}{2}(49)^{-1/2}(1) \approx 7 + \dfrac{1}{14} \approx 7.0714.$

Example 2

Find the approximate value of $\sqrt[3]{62}$ using linear approximation.

Let $f(x) = x^{1/3}$, $a = 64$, $\Delta x = -2$. Since $f'(x) = \dfrac{1}{3}x^{-2/3} = \dfrac{1}{3x^{2/3}}$ and

$f'(64) = \dfrac{1}{3(64)^{2/3}} = \dfrac{1}{48}$, you can use $f(a + \Delta x) \approx f(a) + f'(a)\Delta x$. Thus

$f(62) = f(64 - 2) \approx f(64) + f'(64)(-2) \approx 4 + \dfrac{1}{48}(-2) \approx 3.958.$

- Use calculus notations and not calculator syntax, e.g., write $\displaystyle\int x^2 dx$ and not $\displaystyle\int (x^\wedge 2,\ x)$.

Estimating the Value of a Trigonometric Function of an Angle

Example

Approximate the value of $\sin 31°$.

Note: You must express the angle measurement in radians before applying linear approximations. $30° = \dfrac{\pi}{6}$ radians and $1° = \dfrac{\pi}{180}$ radians.

Let $f(x) = \sin x$, $a = \dfrac{\pi}{6}$ and $\Delta x = \dfrac{\pi}{180}$.

Since $f'(x) = \cos x$ and $f'\left(\frac{\pi}{6}\right) = \cos\left(\frac{\pi}{6}\right) = \frac{\sqrt{3}}{2}$, you can use linear approximations:

$$f\left(\frac{\pi}{6} + \frac{\pi}{180}\right) \approx f\left(\frac{\pi}{6}\right) + f'\left(\frac{\pi}{6}\right)\left(\frac{\pi}{180}\right)$$

$$\approx \sin\frac{\pi}{6} + \left[\cos\left(\frac{\pi}{6}\right)\right]\left(\frac{\pi}{180}\right)$$

$$\approx \frac{1}{2} + \frac{\sqrt{3}}{2}\left(\frac{\pi}{180}\right) = 0.515.$$

5.3 MOTION ALONG A LINE

Main Concepts: *Instantaneous Velocity and Acceleration, Vertical Motion, Horizontal Motion*

Instantaneous Velocity and Acceleration

Position Function: $s(t)$

Instantaneous Velocity: $v(t) = s'(t) = \dfrac{ds}{dt}$

If particle is moving to the right $\rightarrow$, then $v(t) > 0$.
If particle is moving to the left $\leftarrow$, then $v(t) < 0$.

Acceleration: $a(t) = v'(t) = \dfrac{dv}{dt}$ or $a(t) = s''(t) = \dfrac{d^2s}{dt^2}$

Instantaneous speed: $|v(t)|$

Example 1

The position function of a particle moving on a straight line is $s(t) = 2t^3 - 10t^2 + 5$. Find (a) the position, (b) instantaneous velocity, (c) acceleration and (d) speed of the particle at $t = 1$.

Solution:
(a) $s(1) = 2(1)^3 - 10(1)^2 + 5 = -3$
(b) $v(t) = s'(t) = 6t^2 - 20t$
   $v(1) = 6(1)^2 - 20(1) = -14$
(c) $a(t) = v'(t) = 12t - 20$
   $a(1) = 12(1) - 20 = -8$
(d) Speed $= |v(t)| = |v(1)| = 14$

Example 2

The velocity function of a moving particle is $v(t) = \dfrac{t^3}{3} - 4t^2 + 16t - 64$ for $0 \le t \le 7$.

What are the minimum and maximum acceleration of the particle on $0 \le t \le 7$?

$$v(t) = \frac{t^3}{3} - 4t^2 + 16t - 64$$

$$a(t) = v'(t) = t^2 - 8t + 16$$

See Figure 5.3-1. The graph of $a(t)$ indicates that:

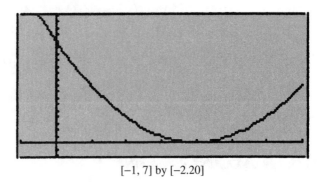

[−1, 7] by [−2.20]

Figure 5.3-1

(1) The minimum acceleration occurs at $t = 4$ and $a(4) = 0$.
(2) The maximum acceleration occurs at $t = 0$ and $a(0) = 16$.

Example 3

The graph of the velocity function is shown in Figure 5.3-2.

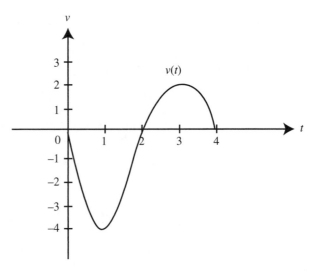

Figure 5.3-2

(a) When is the acceleration 0?
(b) When is the particle moving to the right?
(c) When is the speed the greatest?

Solution:
(a) $a(t) = v'(t)$ and $v'(t)$ is the slope of tangent to the graph of v. At $t = 1$ and $t = 3$, the slope of the tangent is 0.
(b) For $2 < t < 4, v(t) > 0$. Thus the particle is moving to the right during $2 < t < 4$.
(c) Speed $= |v(t)|$ at $t = 1$, $v(t) = -4$.
Thus speed at $t = 1$ is $|-4| = 4$ which is the greatest speed for $0 \le t \le 4$.

• Use only the four specified capabilities of your calculator to get your answer: plotting graph, finding zeros, calculating numerical derivatives, and evaluating definite integrals. All other built-in capabilities can only be used to *check* your solution.

Vertical Motion

Example

From a 400-foot tower, a bowling ball is dropped. The position function of the bowling ball $s(t) = -16t^2 + 400$, $t \geq 0$ is in seconds. Find:

(a) the instantaneous velocity of the ball at $t = 2$ s
(b) the average velocity for the first 3 s
(c) when the ball will hit the ground

Solution:

(a) $v(t) = s'(t) = -32t$
    $v(2) = 32(2) = -64$ ft/s

(b) Average velocity $= \dfrac{s(3) - s(0)}{3 - 0} = \dfrac{(-16(3)^2 + 400) - (0 + 400)}{3} = -48$ ft/s

(c) When the ball hits the ground, $s(t) = 0$.
    Thus set $s(t) = 0 \Rightarrow -16t^2 + 400 = 0$; $16t^2 = 400$; $t = \pm 5$
    Since $t \geq 0$, $t = 5$. The ball hits the ground at $t = 5$ s.

• Remember that the volume of a sphere is $v = \dfrac{4}{3}\pi r^3$ and the surface area is $s = 4\pi r^2$. Note that $v' = s$.

Horizontal Motion

Example

The position function of a particle moving on a straight line is $s(t) = t^3 - 6t^2 + 9t - 1$, $t \geq 0$. Describe the motion of the particle.

Step 1: Find $v(t)$ and $a(t)$. $v(t) = 3t^2 - 12t + 9$

$\qquad a(t) = 6t - 12$

Step 2: Set $v(t)$ and $a(t) = 0$.

$\qquad$ Set $v(t) = 0 \Rightarrow 3t^2 - 12t + 9 = 0 \Rightarrow 3(t^2 - 4t + 3) = 0$

$\qquad\qquad \Rightarrow 3(t - 1)(t - 3) = 0$ or $t = 1$ or $t = 3$.

$\qquad$ Set $a(t) = 0 \Rightarrow 6t - 12 = 0 \Rightarrow 6(t - 2) = 0$ or $t = 2$.

Step 3: Determine the directions of motion. See Figure 5.3-3.

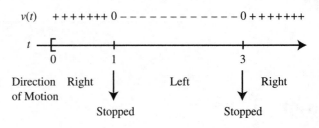

Figure 5.3-3

Step 4: Determine acceleration. See Figure 5.3-4.

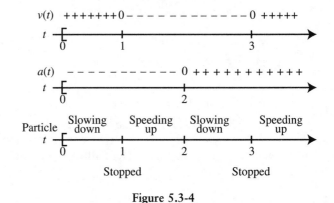

Figure 5.3-4

Step 5: Draw the motion of the particle. See Figure 5.3-5.
$s(0) = -1$, $s(1) = 3$, $s(2) = 1$ and $s(3) = -1$

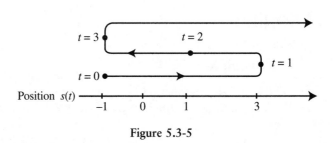

Figure 5.3-5

At $t = 0$, the particle is at -1 and moving to the right. It slows down and stops at $t = 1$ and at $t = 3$. It reverses direction (moving to the left) and speeds up until it reaches 1 at $t = 2$. It continues moving left but slows down and it stops at -1 at $t = 3$. Then it reverses direction (moving to the right) again and speeds up indefinitely. (Note: "Speeding up" is defined as when $|v(t)|$ increases and "slowing down" is defined as when $|v(t)|$ decreases.)

BC 5.4 PARAMETRIC, POLAR, AND VECTOR DERIVATIVES

Main Concepts: *Derivatives of Parametric Equations; Position, Speed, and Acceleration; Derivatives of Polar Equations; Velocity and Acceleration of Vector Functions*

Derivatives of Parametric Equations

If a function is defined parametrically, you can differentiate both $x(t)$ and $y(t)$ with respect to t, and then $\dfrac{dy}{dx} = \dfrac{dy}{dt} \div \dfrac{dx}{dt} = \dfrac{dy}{dt} \cdot \dfrac{dt}{dx}$.

Example 1

A curve is defined by $x(t) = t^2 - 3t$ and $y(t) = 5\cos t$. Find $\dfrac{dy}{dx}$.

Step 1: Differentiate $x(t)$ and $y(t)$ with respect to t. $\dfrac{dx}{dt} = 2t - 3$ and $\dfrac{dy}{dt} = -5\sin t$

Step 2: $\dfrac{dy}{dx} = \dfrac{dy}{dt} \cdot \dfrac{dt}{dx} = \dfrac{-5\sin t}{2t - 3}$

Example 2

A function is defined by $x(t) = 5t - 2$ and $y(t) = 9 - t^2$ when $-5 \le t \le 5$. Find the equation of any horizontal tangent lines to the curve.

Step 1: Differentiate $x(t)$ and $y(t)$ with respect to t. $\dfrac{dx}{dt} = 5$ and $\dfrac{dy}{dt} = -2t$

Step 2: $\dfrac{dy}{dx} = \dfrac{dy}{dt} \cdot \dfrac{dt}{dx} = \dfrac{-2t}{5}$

Step 3: In order for the tangent line to be horizontal, $\dfrac{dy}{dx}$ must be equal to zero, therefore $t = 0$, $x = -2$ and $y = 9$.

Step 4: The equation of the horizontal tangent line at $(-2, 9)$ is $y = 9$.

Example 3

A curve is defined by $x(t) = t^2 - 5t + 2$ and $y(t) = \dfrac{1}{(t+2)^2}$ for $0 \le t \le 3$. Find the equation of the tangent line to the curve when $t = 1$.

Step 1: $\dfrac{dx}{dt} = 2t - 5$ and $\dfrac{dy}{dt} = \dfrac{-2}{(t+2)^3}$

Step 2: $\dfrac{dy}{dx} = \dfrac{-2}{(t+2)^3} \cdot \dfrac{1}{(2t-5)}$

Step 3: At $t = 1$, $m = \dfrac{-2}{(3)^3} \cdot \dfrac{1}{(-3)} = \dfrac{2}{81}$, $x = 1 - 5 + 2 = -2$, and $y = \dfrac{1}{(3)^2} = \dfrac{1}{9}$.

Step 4: The equation of the tangent line is $y - \dfrac{1}{9} = \dfrac{2}{81}(x + 2)$ or $y = \dfrac{2}{81}x + \dfrac{13}{81}$.

Position, Speed, and Acceleration

When the motion of a particle is defined parametrically, its position is given by $(x(t), y(t))$. The speed of the particle is $\sqrt{\left(\dfrac{dx}{dt}\right)^2 + \left(\dfrac{dy}{dt}\right)^2}$ and its acceleration is given by the vector $\left\langle \dfrac{d^2x}{dt^2}, \dfrac{d^2y}{dt^2} \right\rangle$.

Example 1

Find the speed and acceleration of a particle whose motion is defined by $x = 3t$ and $y = 9t - 3t^2$ when $t = 2$.

Step 1: Differentiate $\dfrac{dx}{dt} = 3$ and $\dfrac{dy}{dt} = 9 - 6t$. When $t = 2$, $\dfrac{dx}{dt} = 3$ and $\dfrac{dy}{dt} = -3$.

Step 2: Calculate the speed. $\sqrt{(3)^2 + (-3)^2} = \sqrt{18} = 3\sqrt{2}$.

Step 3: Determine second derivatives. $\dfrac{d^2x}{dt^2} = 0$ and $\dfrac{d^2y}{dt^2} = -6$. The acceleration vector is $\langle 0, -6 \rangle$.

Example 2

A particle moves along the curve $y = \dfrac{1}{2}x^2 - \dfrac{1}{4}\ln x$ so that $x = \dfrac{1}{2}t^2$ and $t > 0$. Find the speed of the particle when $t = 1$.

Step 1: Substitute $x = \dfrac{1}{2}t^2$ in $y = \dfrac{1}{2}x^2 - \dfrac{1}{4}\ln x$ to find

$$y(t) = \frac{1}{2}\left(\frac{1}{2}t^2\right)^2 - \frac{1}{4}\ln\left(\frac{1}{2}t^2\right) = \frac{1}{8}t^4 - \frac{1}{4}\ln\left(\frac{1}{2}t^2\right)$$

$$= \frac{1}{8}t^4 - \frac{1}{4}(-\ln 2 + 2\ln t)$$

$$= \frac{t^4}{8} + \frac{\ln 2}{4} - \frac{\ln t}{2}$$

Step 2: $\dfrac{dx}{dt} = t$ and $\dfrac{dy}{dt} = \dfrac{t^3}{2} - \dfrac{1}{2t}$. Evaluated at $t = 1$, $\dfrac{dx}{dt} = 1$ and $\dfrac{dy}{dt} = 0$.

Step 3: The speed of the particle is $\sqrt{(1)^2 + (0)^2} = 1$.

Derivatives of Polar Equations

For polar representations, remember that $r = f(\theta)$, so $x = r\cos\theta = f(\theta)\cos\theta$ and $y = r\sin\theta = f(\theta)\sin\theta$. Differentiating with respect to θ requires the product rule. $\dfrac{dx}{d\theta} = -r\sin\theta + \cos\theta\dfrac{dr}{d\theta}$ and $\dfrac{dy}{d\theta} = r\cos\theta + \sin\theta\dfrac{dr}{d\theta}$. Dividing $\dfrac{dy}{d\theta}$ by $\dfrac{dx}{d\theta}$ gives $\dfrac{dy}{dx} = \dfrac{r\cos\theta + \sin\theta\, dr/d\theta}{-r\sin\theta + \cos\theta\, dr/d\theta}$.

Example

Find the equation of the tangent line to the curve $r = 2 + 2\sin\theta$ when $\theta = \dfrac{\pi}{4}$

Step 1: $\dfrac{dr}{d\theta} = 2\cos\theta$

Step 2: $\dfrac{dx}{d\theta} = -(2 + 2\sin\theta)\sin\theta + \cos\theta(2\cos\theta) = 2(\cos^2\theta - \sin^2\theta - \sin\theta)$. By the Pythagorean identity $2(\cos^2\theta - \sin^2\theta - \sin\theta) = 2(1 - \sin^2\theta - \sin^2\theta - \sin\theta) = 2(1 - \sin\theta - 2\sin^2\theta) = 2(1 - 2\sin\theta)(1 + \sin\theta)$. Also, $\dfrac{dy}{d\theta} = (2 + 2\sin\theta)\cos\theta + \sin\theta(2\cos\theta) = 2\cos\theta(1 + 2\sin\theta)$

Step 3: $\dfrac{dy}{dx} = \dfrac{2\cos\theta(1 + 2\sin\theta)}{2(1 - 2\sin\theta)(1 + \sin\theta)} = \dfrac{\cos\theta(1 + 2\sin\theta)}{(1 - 2\sin\theta)(1 + \sin\theta)}$

Step 4: When $\theta = \dfrac{\pi}{4}$, $\dfrac{dy}{dx} = \dfrac{\cos\dfrac{\pi}{4}\left(1 + 2\sin\dfrac{\pi}{4}\right)}{\left(1 - 2\sin\dfrac{\pi}{4}\right)\left(1 + \sin\dfrac{\pi}{4}\right)}$. Evaluating,

$$\dfrac{dy}{dx} = \dfrac{\dfrac{\sqrt{2}}{2}(1 + \sqrt{2})}{(1 - \sqrt{2})\left(1 + \dfrac{\sqrt{2}}{2}\right)} = -1 - \sqrt{2}$$

Step 5: When $\theta = \dfrac{\pi}{4}$, $r = 2 + \sqrt{2}$, so $x = r\cos\theta = \left(2 + \sqrt{2}\right)\dfrac{\sqrt{2}}{2} = \sqrt{2} + 1$ and $y = r\sin\theta = \left(2 + \sqrt{2}\right)\dfrac{\sqrt{2}}{2} = \sqrt{2} + 1$

Step 6: The equation of the tangent line is $y - \left(\sqrt{2} + 1\right) = \left(-1 - \sqrt{2}\right)\left(x - \left(\sqrt{2} + 1\right)\right)$ or $y = \left(-1 - \sqrt{2}\right)x + 4 + 3\sqrt{2}$

Velocity and Acceleration of Vector Functions

A vector-valued function assigns a vector to each element in a domain of real numbers. If $r = \langle x, y \rangle$ is a vector-valued function, $\lim\limits_{t \to c} r$ exists only if $\lim\limits_{t \to c} x(t)$ and $\lim\limits_{t \to c} y(t)$ exist. $\lim\limits_{t \to c} r = \left\langle \lim\limits_{t \to c} x(t), \lim\limits_{t \to c} y(t) \right\rangle = \lim\limits_{t \to c} x(t)i + \lim\limits_{t \to c} y(t)j$. A vector-valued function is continuous at c if its component functions are continuous at c. The derivative of a vector-valued function is $\dfrac{dr}{dt} = \left\langle \dfrac{dx}{dt}, \dfrac{dy}{dt} \right\rangle = i\dfrac{dx}{dt} + j\dfrac{dy}{dt}$.

If $r = \langle x, y \rangle$ is a vector-valued function that represent the path of an object in the plane, and x and y are both functions of a variable t, $x = f(t)$ and $y = g(t)$, then the velocity of the object is $v = \dfrac{dr}{dt} = i\dfrac{dx}{dt} + j\dfrac{dy}{dt} = \left\langle \dfrac{dx}{dt}, \dfrac{dy}{dt} \right\rangle$. Speed is the magnitude of velocity so $|v| = \sqrt{\left(\dfrac{dx}{dt}\right)^2 + \left(\dfrac{dy}{dt}\right)^2}$. The direction of v is along the tangent to the path. The acceleration vector is $\left\langle \dfrac{d^2x}{dt^2}, \dfrac{d^2y}{dt^2} \right\rangle$ and the magnitude of acceleration is

$|a| = \sqrt{\left(\dfrac{d^2x}{dt^2}\right)^2 + \left(\dfrac{d^2y}{dt^2}\right)^2}$. The vector T tangent to the path at t is $T(t) = \dfrac{r'(t)}{\|r'(t)\|}$

and the normal vector at t is $N(t) = \dfrac{T'(t)}{\|T'(t)\|}$.

Example 1

The position function $r = \langle t^3, t^2 \rangle = t^3 i + t^2 j$ describes the path of an object moving in the plane. Find the velocity and acceleration of the object at the point $(8, 4)$.

Step 1: The velocity $v = \left\langle \dfrac{dx}{dt}, \dfrac{dy}{dt} \right\rangle = \langle 3t^2, 2t \rangle$. At the point $(8, 4)$, $t = 2$. Evaluating

at $t = 2$, the velocity $v = \langle 12, 4 \rangle$. The speed $|v| = \sqrt{144 + 16} \approx 12.649$.

Step 2: The acceleration vector $\left\langle \dfrac{d^2x}{dt^2}, \dfrac{d^2y}{dt^2} \right\rangle = \langle 6t, 2 \rangle$. Evaluated at $t = 2$, the acceleration is $\langle 12, 2 \rangle$. The magnitude of the acceleration is $|a| = \sqrt{144 + 4} \approx 12.166$.

Example 2

The left field fence in Boston's Fenway Park, nicknamed the Green Monster, is 37 feet high and 310 feet from home plate. If a ball is hit 3 feet above the ground and leaves the bat at an angle of $\dfrac{\pi}{4}$, write a vector-valued function for the path of the ball and use the function to determine the minimum speed at which the ball must leave the bat to be a home run. At that speed, what is the maximum height the ball attains?

Step 1: The horizontal component of the ball's motion, the motion in the "x" direction, is $x = s \cdot \cos \dfrac{\pi}{4} \cdot t = \dfrac{s\sqrt{2}}{2}t$. The vertical component follows the parabolic motion model $y = 3 + s \cdot \sin \dfrac{\pi}{4} t - \dfrac{1}{2}gt^2$, where g is the acceleration due to gravity. The path of the ball can be represented by the vector-valued function $r = \left\langle \dfrac{s \cdot \sqrt{2}}{2}t, 3 + \dfrac{s \cdot \sqrt{2}}{2}t - 16t^2 \right\rangle$.

Step 2: In order for the ball to clear the fence, its height must be greater than 37 feet when its distance from the plate is 310 feet. $\dfrac{s \cdot \sqrt{2}}{2}t = 310$, solved for t, gives $t = \dfrac{620}{s \cdot \sqrt{2}}$ seconds. At this time, $3 + \dfrac{s \cdot \sqrt{2}}{2}t - 16t^2 = 3 + \dfrac{s\sqrt{2}}{2}\left(\dfrac{620}{s\sqrt{2}}\right)$ $- 16\left(\dfrac{620}{s\sqrt{2}}\right)^2$, and this value must exceed 37 feet. Setting $3 + \dfrac{s \cdot \sqrt{2}}{2}\left(\dfrac{620}{s \cdot \sqrt{2}}\right)$ $- 16\left(\dfrac{620}{s \cdot \sqrt{2}}\right)^2 = 37$ and solving gives $s \approx 105.556$. The ball must leave the bat at 105.556 feet per second in order to clear the wall.

Step 3: Since $r = \left\langle \dfrac{s \cdot \sqrt{2}}{2}t, 3 + \dfrac{s \cdot \sqrt{2}}{2}t - 16t^2 \right\rangle$, the derivative is $r' = \left\langle \dfrac{s \cdot \sqrt{2}}{2}, \right.$ $\left. 3 + \dfrac{s \cdot \sqrt{2}}{2} - 32t \right\rangle$, and the ball will attain its maximum height when the vertical component $3 + \dfrac{s \cdot \sqrt{2}}{2} - 32t$ is equal to zero. Since $s \approx 105.556$,

$3 + \dfrac{105.556 \cdot \sqrt{2}}{2} - 32t = 0$ produces $t \approx 2.462$ seconds. For that value

of t, $r = \left\langle \dfrac{105.556\sqrt{2}}{2}(2.462), 3 + \dfrac{105.556\sqrt{2}}{2}(2.462) - 16(2.462)^2 \right\rangle \approx$

$\langle 183.762, 89.779 \rangle$. The ball will reach a maximum height of 89.779 feet, when it is 183.762 feet from home plate.

Example 3

Find the velocity, acceleration, tangent and normal vectors for an object on a path defined by the vector-valued function $r(t) = \langle e^t \cos t, e^t \sin t \rangle$ when $t = \dfrac{\pi}{2}$.

Step 1: $v(t) = r'(t) = \langle e^t(\cos t - \sin t), e^t(\sin t + \cos t) \rangle$. When evaluated at $t = \dfrac{\pi}{2}$, $v\left(\dfrac{\pi}{2}\right) = \langle -e^{\pi/2}, e^{\pi/2} \rangle \approx \langle -4.810, 4.810 \rangle$. The velocity vector is $\langle -4.810, 4.810 \rangle$.

Step 2: $a(t) = \langle -2e^t \sin t, 2e^t \cos t \rangle$. Evaluated at $t = \dfrac{\pi}{2}$, this is $a\left(\dfrac{\pi}{2}\right) = \langle -2e^{\pi/2}, 0 \rangle \approx \langle -9.621, 0 \rangle$

Step 3: The tangent vector is given by $T(t) = \dfrac{r'(t)}{\|r'(t)\|}$. Since $r'(t) = \langle -e^t \sin t + e^t \cos t, e^t \sin t + e^t \cos t \rangle$, the tangent vector becomes

$T(t) = \dfrac{\langle -e^t \sin t + e^t \cos t, e^t \sin t + e^t \cos t \rangle}{\sqrt{(-e^t \sin t + e^t \cos t)^2 + (e^t \sin t + e^t \cos t)^2}}$ which simplifies to

$T(t) = \left\langle \dfrac{\cos t - \sin t}{\sqrt{2}}, \dfrac{\sin t + \cos t}{\sqrt{2}} \right\rangle$. When $t = \dfrac{\pi}{2}$, the tangent vector is

$\left\langle \dfrac{-1}{\sqrt{2}}, \dfrac{1}{\sqrt{2}} \right\rangle = \left\langle \dfrac{-\sqrt{2}}{2}, \dfrac{\sqrt{2}}{2} \right\rangle$.

Step 4: The normal vector $N(t) = \dfrac{T'(t)}{\|T'(t)\|} = \dfrac{\left\langle \dfrac{\cos t - \sin t}{\sqrt{2}}, \dfrac{\sin t + \cos t}{\sqrt{2}} \right\rangle'}{\left\| \left\langle \dfrac{\cos t - \sin t}{\sqrt{2}}, \dfrac{\sin t + \cos t}{\sqrt{2}} \right\rangle \right\|} =$

$\left\langle \dfrac{-\cos t - \sin t}{\sqrt{2}}, \dfrac{\cos t - \sin t}{\sqrt{2}} \right\rangle$. At $t = \dfrac{\pi}{2}, N(t) = \left\langle \dfrac{-1}{\sqrt{2}}, \dfrac{1}{\sqrt{2}} \right\rangle = \left\langle \dfrac{-\sqrt{2}}{2}, \dfrac{\sqrt{2}}{2} \right\rangle$.

Check $T\left(\dfrac{\pi}{2}\right) \cdot N\left(\dfrac{\pi}{2}\right) = \dfrac{-\sqrt{2}}{2} \cdot \dfrac{-\sqrt{2}}{2} + \dfrac{\sqrt{2}}{2} \cdot \dfrac{-\sqrt{2}}{2} = \dfrac{1}{2} - \dfrac{1}{2} = 0$ to be certain the tangent and normal vectors are orthogonal.

5.5 RAPID REVIEW

1. Write an equation of the normal line to the graph $y = e^x$ at $x = 0$.

   Answer: $\dfrac{dy}{dx}\Big|_{x=0}$ $e^x = e^x|_{x=0} = e^0 = 1 \Rightarrow m_{\text{normal}} = -1$.

   At $x = 0, y = e^0 = 1 \Rightarrow$ you have the point $(0, 1)$.

   Equation of normal $y - 1 = -1(x - 0)$ or $y = -x + 1$.

2. Using your calculator, find the values of x at which the function $y = -x^2 + 3x$ and $y = \ln x$ have parallel tangents.

   Answer: $y = -x^2 + 3x \Rightarrow \dfrac{dy}{dx} = -2x + 3$

   $y = \ln x \Rightarrow \dfrac{dy}{dx} = \dfrac{1}{x}$

   Set $-2x + 3 = \dfrac{1}{x}$. Using the Solve function on your calculator, enter

   Solve $\left(-2x + 3 = \dfrac{1}{x}, x\right)$ and obtain $x = 1$ or $x = \dfrac{1}{2}$.

3. Find the linear approximation of $f(x) = x^3$ at $x = 1$ and use the equation to find $f(1.1)$.

   Answer: $f(1) = 1 \Rightarrow (1, 1)$ is on the tangent line and $f'(x) = 3x^2 \Rightarrow f'(1) = 3$.

   $y - 1 = 3(x - 1)$ or $y = 3x - 2$

   $f(1.1) \approx 3(1.1) - 2 \approx 1.3$.

4. See Figure 5.5-1.

   (a) When is the acceleration zero? (b) Is the particle moving to the right or left?

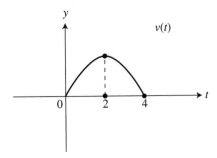

Figure 5.5-1

   Answer: (a) $a(t) = v'(t)$ and $v'(t)$ is the slope of the tangent. Thus, $a(t) = 0$ at $t = 2$.

   (b) Since $v(t) \geq 0$, the particle is moving to the right.

5. Find the maximum acceleration of the particle whose velocity function is $v(t) = t^2 + 3$ on the interval $0 \leq t \leq 4$.

   Answer: $a(t) = v'(t) = 2(t)$ on the interval $0 \leq t \leq 4$, $a(t)$ has its maximum value at $t = 4$. Thus $a(t) = 8$. The maximum acceleration is 8.

6. Find the slope of the tangent to the curve defined by $x = 3t - 5$, $y = t^2 - 9$ when $t = 3$.

   Answer: $\dfrac{dx}{dt} = 3$ and $\dfrac{dy}{dt} = 2t|_{t=3} = 6$, so $\dfrac{dy}{dx} = \dfrac{dy}{dt} \div \dfrac{dx}{dt} = \dfrac{6}{3} = 2$

7. Find the slope of the tangent line to the graph of $r = -3 \cos \theta$.

   Answer: $\dfrac{dr}{d\theta} = 3 \sin \theta$. Since $x = r \cos \theta$, $\dfrac{dx}{d\theta} = -r \sin \theta + \cos \theta \dfrac{dr}{d\theta} = 3 \cos \theta \sin \theta$

   $+ 3 \cos \theta \sin \theta = 6 \sin \theta \cos \theta = 3 \sin 2\theta$. Since $y = r \sin \theta$, $\dfrac{dy}{d\theta} = r \cos \theta$

   $+ \sin \theta \dfrac{dr}{d\theta} = -3 \cos^2 \theta + 3 \sin^2 \theta = -3(\cos^2 \theta - \sin^2 \theta) = -3 \cos 2\theta$.

   $\dfrac{dy}{dx} = \dfrac{dy}{d\theta} \cdot \dfrac{d\theta}{dx} = \dfrac{-3 \cos 2\theta}{3 \sin 2\theta} = -\cot 2\theta$

8. Find $\dfrac{dr}{dt}$ for the vector function $r(t) = 3ti - 2tj = \langle 3t, -2t \rangle$.

   Answer: $\dfrac{dx}{dt} = 3$ and $\dfrac{dy}{dt} = -2$ so $\dfrac{dr}{dt} = \langle 3, -2 \rangle$.

5.6 PRACTICE PROBLEMS

Part A—The use of a calculator is not allowed.

1. Find the linear approximation of $f(x) = (1 + x)^{1/4}$ at $x = 0$ and use the equation to approximate $f(0.1)$.

2. Find the approximate value of $\sqrt[3]{28}$ using linear approximation.

3. Find the approximate value of $\cos 46°$ using linear approximation.

4. Find the point on the graph of $y = |x^3|$ such that the tangent at the point is parallel to the line $y - 12x = 3$.

5. Write an equation of the normal to the graph of $y = e^x$ at $x = \ln 2$.

6. If the line $y - 2x = b$ is tangent to the graph $y = -x^2 + 4$, find the value of b.

7. If the position function of a particle is $s(t) = \dfrac{t^3}{3} - 3t^2 + 4$, find the velocity and position of particle when its acceleration is 0.

8. The graph in Figure 5.6-1 represents the distance in feet covered by a moving particle in t seconds. Draw a sketch of the corresponding velocity function.

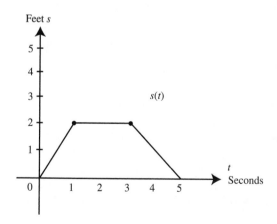

Figure 5.6-1

9. The position function of a moving particle is shown in Figure 5.6-2. For which value(s)

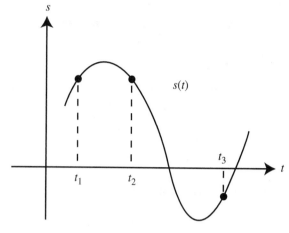

Figure 5.6-2

of $t(t_1, t_2, t_3)$ is:

(a) the particle moving to the left?
(b) the acceleration negative?
(c) the particle moving to the right and slowing down?

10. The velocity function of a particle is shown in Figure 5.6-3.

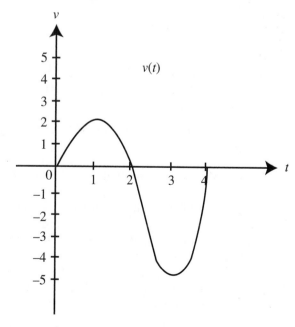

Figure 5.6-3

(a) When does the particle reverse direction?

(b) When is the acceleration 0?

(c) When is the speed the greatest?

11. A ball is dropped from the top of a 640-foot building. The position function of the ball is $s(t) = -16t^2 + 640$, where t is measured in seconds and $s(t)$ is in feet. Find:

(a) The position of the ball after 4 seconds.

(b) The instantaneous velocity of the ball at $t = 4$.

(c) The average velocity for the first 4 seconds.

(d) When the ball will hit the ground.

(e) The speed of the ball when it hits the ground.

12. The graph of the position function of a moving particle is shown. See Figure 5.6-4.

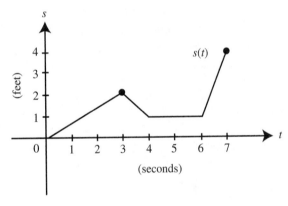

Figure 5.6-4

(a) What is the particle's position at $t = 5$?

(b) When is the particle moving to the left?

(c) When is the particle standing still?

(d) When does the particle have the greatest speed?

Part B—Calculators are permitted.

13. The position function of a particle moving on a line is $s(t) = t^3 - 3t^2 + 1$, $t \geq 0$ where t is measured in seconds and s in meters. Describe the motion of the particle.

14. Find the linear approximation of $f(x) = \sin x$ at $x = \pi$. Use the equation to find the approximate value of $f\left(\dfrac{181\pi}{180}\right)$.

15. Find the linear approximation of $f(x) = \ln(1 + x)$ at $x = 2$.

16. Find the coordinates of each point on the graph of $y^2 = 4 - 4x^2$ at which the tangent line is vertical. Write an equation of each vertical tangent.

17. Find the value(s) of x at which the graphs of $y = \ln x$ and $y = x^2 + 3$ have parallel tangents.

18. The position functions of two moving particles are $s_1(t) = \ln t$ and $s_2(t) = \sin t$ and the domain of both functions is $1 \leq t \leq 8$. Find the values of t such that the velocities of the two particles are the same.

19. The position function of a moving particle on a line is $s(t) = \sin(t)$ for $0 \leq t \leq 2\pi$. Describe the motion of the particle.

20. A coin is dropped from the top of a tower and hits the ground 10.2 seconds later. The position function is given as $s(t) = -16t^2 - v_0 t + s_0$, where s is measured in feet, t in seconds and v_0 is the initial velocity and s_0 is the initial position. Find the approximate height of the building to the nearest foot.

21. Find the equation of the tangent line to the curve defined by $x = \cos t - 1$, $y = \sin t + t$ at the point where $x = \dfrac{-1}{2}$.

22. An object moves on a path defined by $x = e^{2t} + t$ and $y = 1 + e^t$. Find the speed of the object and its acceleration vector with $t = 2$.

23. Find the slope of the tangent line to the curve $r = 3 \sin 4\theta$ at $\theta = \dfrac{5\pi}{6}$.

24. The position of an object is given by $\left\langle 30t, 25 \sin \dfrac{t}{3} \right\rangle$. Find the velocity and acceleration vectors, and determine when the magnitude of the acceleration is equal to 2.

25. Find the tangent vector to the path defined by $r = \langle \ln t, \ln(t + 4) \rangle$ at the point where $t = 4$.

5.7 CUMULATIVE REVIEW PROBLEMS

"Calculator" indicates that calculators are permitted.

26. Find $\dfrac{dy}{dx}$ if $y = x \sin^{-1}(2x)$.

27. Given $f(x) = x^3 - 3x^2 + 3x - 1$ and the point $(1, 2)$ is on the graph of $f^{-1}(x)$. Find the slope of the tangent line to the graph of $f^{-1}(x)$ at $(1, 2)$.

28. Evaluate $\displaystyle\lim_{x \to 100} \dfrac{x - 100}{\sqrt{x} - 10}$.

29. A function f is continuous on the interval $(-1, 8)$ with $f(0) = 0$, $f(2) = 3$, and $f(8) = 1/2$ and the following properties:

Intervals	$(-1, 2)$	$x = 2$	$(2, 5)$	$x = 5$	$(5, 8)$
f'	+	0	−	−	−
f''	−	−	−	0	+

(a) Find the intervals on which f is increasing or decreasing.
(b) Find where f has its absolute extrema.
(c) Find where f has the points of inflection.
(d) Find the intervals on which f is concave upward or downward.
(e) Sketch a possible graph of f.

30. The graph of the velocity function of a moving particle for $0 \le t \le 8$ is shown in Figure 5.7-1. Using the graph:

(a) Estimate the acceleration when $v(t) = 3$ ft/s.

(b) Find the time when the acceleration is a minimum.

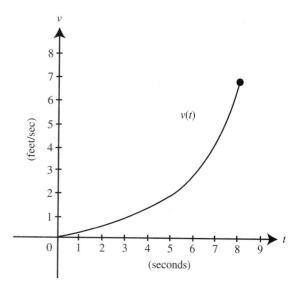

Figure 5.7-1

31. Find the Cartesian equation for the curve defined by $r = 4 \cos \theta$.

32. The motion of an object is modeled by $x = 5 \sin t$, $y = 1 - \cos t$. Find the y-coordinate of the object at the moment when its x-coordinate is 5.

33. Calculate $4u - 3v$ if $u = \langle 6, -1 \rangle$ and $v = \langle -4, 3 \rangle$.

34. Determine the symmetry, if any, of the graph of $r = 2 \sin(4\theta)$.

35. Find the magnitude of the vector $3i + 4j$.

5.8 SOLUTIONS TO PRACTICE PROBLEMS

Part A—No calculators.

1. Equation of tangent line:

$$y = f(a) + f'(a)(x - a)$$
$$f'(x) = \frac{1}{4}(1 + x)^{-3/4}(1) = \frac{1}{4}(1 + x)^{-3/4}$$
$$f'(0) = \frac{1}{4} \text{ and } f(0) = 1;$$

Thus, $y = 1 + \dfrac{1}{4}(x - 0) = 1 + \dfrac{1}{4}x$.

$$f(0.1) = 1 + \frac{1}{4}(0.1) = 1.025.$$

2. $f(a + \Delta x) \approx f(a) + f'(a)\Delta x$

Let $f(x) = \sqrt[3]{x}$ and $f(28) = f(27 + 1)$.

Then $f'(x) = \dfrac{1}{3}(x)^{-2/3}$ and

$f'(27) = \dfrac{1}{27}$ and $f(27) = 3$.

$f(27 + 1) \approx f(27) + f'(27)(1) \approx 3 + \left(\dfrac{1}{27}\right)(1)$

$\approx 3.\overline{037}$.

3. $f(a + \Delta x) \approx f(a) + f'(a)\,\Delta x$. Convert to radians:

$\dfrac{46}{180} = \dfrac{a}{\pi} \Rightarrow a = \dfrac{23\pi}{90}$ and $1° = \dfrac{\pi}{180}$; $45° = \dfrac{\pi}{4}$.

Let $f(x) = \cos x$ and

$f(45°) = f\left(\dfrac{\pi}{4}\right) = \cos\left(\dfrac{\pi}{4}\right) = \dfrac{\sqrt{2}}{2}$

Then $f'(x) = -\sin x$ and

$f'(45°) = f'\left(\dfrac{\pi}{4}\right) = -\dfrac{\sqrt{2}}{2}$

$f(46°) = f\left(\dfrac{23\pi}{90}\right) = f\left(\dfrac{\pi}{4} + \dfrac{\pi}{180}\right)$

$f\left(\dfrac{\pi}{4} + \dfrac{\pi}{180}\right) \approx f\left(\dfrac{\pi}{4}\right) + f'\left(\dfrac{\pi}{4}\right)\left(\dfrac{\pi}{180}\right) \approx$

$\dfrac{\sqrt{2}}{2} - \left(\dfrac{\sqrt{2}}{2}\right)\left(\dfrac{\pi}{180}\right) \approx \dfrac{\sqrt{2}}{2} - \dfrac{\pi\sqrt{2}}{360}$

4. Step 1: Find m_{tangent}

$y = |x^3| = \begin{cases} x^3 & \text{if } x \geq 0 \\ -x^3 & \text{if } x < 0 \end{cases}$

$\dfrac{dy}{dx} = \begin{cases} 3x^2 & \text{if } x > 0 \\ -3x^2 & \text{if } x < 0 \end{cases}$

Step 2: Set $m_{\text{tangent}} =$ slope of line $y - 12x = 3$.
Since $y - 12x = 3 \Rightarrow y = 12x + 3$,
then $m = 12$.
Set $3x^2 = 12 \Rightarrow x = \pm 2$ since $x \geq 0$,
$x = 2$.
Set $-3x^2 = 12 \Rightarrow x^2 = -4$. Thus ϕ.

Step 3: Find the point on the curve. (See Figure 5.8-1.)
At $x = 2$, $y = x^3 = 2^3 = 8$.
Thus the point is $(2, 8)$.

5. Step 1: Find m_{tangent}

$y = e^x$; $\dfrac{dy}{dx} = e^x$

$\dfrac{dy}{dx}\bigg|_{x=\ln 2} = e^{\ln 2} = 2$

Step 2: Find m_{normal}

At $x = \ln 2$, $m_{\text{normal}} = \dfrac{-1}{m_{\text{tangent}}} = -\dfrac{1}{2}$

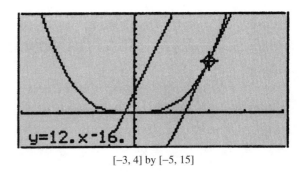

y=12.x-16.

[−3, 4] by [−5, 15]

Figure 5.8-1

Step 3: Write equation of normal
At $x = \ln 2$, $y = e^x = e^{\ln 2} = 2$. Thus the point of tangency is $(\ln 2, 2)$.
The equation of normal:

$y - 2 = -\dfrac{1}{2}(x - \ln 2)$ or

$y = -\dfrac{1}{2}(x - \ln 2) + 2$.

6. Step 1: Find m_{tangent}

$y = -x^2 + 4$; $\dfrac{dy}{dx} = -2x$.

Step 2: Find the slope of line $y - 2x = b$
$y - 2x = b \Rightarrow y = 2x + b$ or $m = 2$.

Step 3: Find point of tangency. Set $m_{\text{tangent}} =$ slope of line $y - 2x = b \Rightarrow -2x = 2 \Rightarrow$
$x = -1$.
At $x = -1$, $y = -x^2 + 4 = -(-1)^2 + 4 = 3$; $(-1, 3)$.

Step 4: Find b.
Since the line $y - 2x = b$ passes through the point $(-1, 3)$, thus $3 - 2(-1) = b$ or $b = 5$.

7. $v(t) = s'(t) = t^2 - 6t$; $a(t) = v'(t) = s''(t) = 2t - 6$
Set $a(t) = 0 \Rightarrow 2t - 6 = 0$ or $t = 3$.
$v(3) = (3)^2 - 6(3) = -9$;
$s(3) = \dfrac{(3)^3}{3} - 3(3)^2 + 4 = -14$.

8. On the interval $(0, 1)$, the slope of the line segment is 2. Thus the velocity $v(t) = 2$ ft/s. On $(1, 3)$, $v(t) = 0$ and on $(3, 5)$, $v(t) = -1$. See Figure 5.8-2.

9. (a) At $t = t_2$, the slope of the tangent is negative. Thus, the particle is moving to the left.

(b) At $t = t_1$, and at $t = t_2$, the curve is concave downward $\Rightarrow \dfrac{d^2s}{dt^2} =$ acceleration is negative.

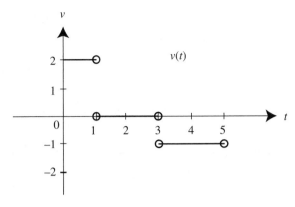

Figure 5.8-2

(c) At $t = t_1$, the slope >0 and thus the particle is moving to the right. The curve is concave downward $\Rightarrow$ the particle is slowing down.

10. (a) At $t = 2$, $v(t)$ changes from positive to negative, and thus the particle reverses its direction.

   (b) At $t = 1$, and at $t = 3$, the slope of the tangent to the curve is 0. Thus the acceleration is 0.

   (c) At $t = 3$, speed is equal to $|-5| = 5$ and 5 is the greatest speed.

11. (a) $s(4) = -16(4)^2 + 640 = 384$ ft

   (b) $v(t) = s'(t) = -32t$
   $v(4) = -32(4)$ ft/s $= -128$ ft/s

   (c) Average Velocity $= \dfrac{s(4) - s(0)}{4 - 0}$
   $= \dfrac{384 - 640}{4} = -64$ ft/s

   (d) Set $s(t) = 0 \Rightarrow -16t^2 + 640 = 0 \Rightarrow 16t^2$
   $= 640$ or $t = \pm 2\sqrt{10}$.
   Since $t \geq 0$, $t = +2\sqrt{10}$ or $t \approx 6.32$ s.

   (e) $|v(2\sqrt{10})| = |-32(2\sqrt{10})| = |-64\sqrt{10}|$ ft/s
   or ≈ 202.39 ft/s

12. (a) At $t = 5$, $s(t) = 1$

   (b) For $3 < t < 4$, $s(t)$ decreases. Thus, the particle moves to the left when $3 < t < 4$.

   (c) When $4 < t < 6$, the particle stays at 1.

   (d) When $6 < t < 7$, speed $= 2$ ft/s, the greatest speed, which occurs where s has the greatest slope.

Part B—Calculators are permitted.

13. Step 1: $v(t) = 3t^2 - 6t$
   $a(t) = 6t - 6$

Step 2: Set $v(t) = 0 \Rightarrow 3t^2 - 6t = 0 \Rightarrow 3t(t - 2)$
   $= 0$, or $t = 0$ or $t = 2$
   Set $a(t) = 0 \Rightarrow 6t - 6 = 0$ or $t = 1$.

Step 3: Determine the directions of motion. See Figure 5.8-3.

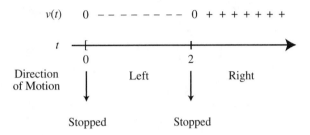

Figure 5.8-3

Step 4: Determine acceleration. See Figure 5.8-4.

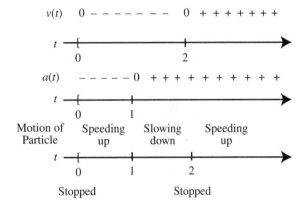

Figure 5.8-4

Step 5: Draw the motion of the particle. See Figure 5.8-5. $s(0) = 1$, $s(1) = -1$ and $s(2) = -3$.

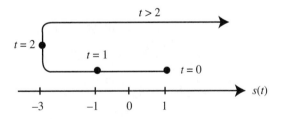

Figure 5.8-5

The particle is initially at 1 ($t = 0$). It moves to the left speeding up until $t = 1$, when it reaches -1. Then it continues moving to the left, but slowing down until $t = 2$ at -3. The particle reverses direction, moving to the right and speeding up indefinitely.

14. Linear approximation:
$y = f(a) + f'(a)(x - a)$ $a = \pi$
$f(x) = \sin x$ and $f(\pi) = \sin \pi = 0$
$f'(x) = \cos x$ and $f'(\pi) = \cos \pi = -1$.
Thus $y = 0 + (-1)(x - \pi)$ or $y = -x + \pi$.
$f\left(\dfrac{181\pi}{180}\right)$ is approximately:

$y = -\left(\dfrac{181\pi}{180}\right) + \pi = \dfrac{-\pi}{180}$ or ≈ -0.0175.

15. $y = f(a) + f'(a)(x - a)$
$f(x) = \ln(1 + x)$ and $f(2) = \ln(1 + 2) = \ln 3$

$f'(x) = \dfrac{1}{1 + x}$ and $f'(2) = \dfrac{1}{1 + 2} = \dfrac{1}{3}$.

Thus $y = \ln 3 + \dfrac{1}{3}(x - 2)$.

16. Step 1: Find $\dfrac{dy}{dx}$.

$y^2 = 4 - 4x^2$

$2y\dfrac{dy}{dx} = -8x \Rightarrow \dfrac{dy}{dx} = \dfrac{-4x}{y}$

Step 2: Find $\dfrac{dx}{dy}$.

$\dfrac{dx}{dy} = \dfrac{1}{dy/dx} = \dfrac{1}{-4x/y} = \dfrac{-y}{4x}$

Set $\dfrac{dx}{dy} = 0 \Rightarrow \dfrac{-y}{4x} = 0$ or $y = 0$.

Step 3: Find points of tangency
At $y = 0, y^2 = 4 - 4x^2$ becomes
$0 = 4 - 4x^2 \Rightarrow x = \pm 1$.
Thus points of tangency are $(1, 0)$ and $(-1, 0)$

Step 4: Write equations of vertical tangents
$x = 1$ and $x = -1$.

17. Step 1: Find $\dfrac{dy}{dx}$ for $y = \ln x$ and $y = x^2 + 3$

$y = \ln x; \dfrac{dy}{dx} = \dfrac{1}{x}$

$y = x^2 + 3; \dfrac{dy}{dx} = 2x$

Step 2: Find the x-coordinate of point(s) of tangency.
Parallel tangents $\Rightarrow$ slopes are equal
Set $\dfrac{1}{x} = 2x$.

Using the Solve function of your calculator, enter solve $\left(\dfrac{1}{x} = 2x, x\right)$ and

obtain $x = \dfrac{\sqrt{2}}{2}$ or $x = \dfrac{-\sqrt{2}}{2}$. Since for

$y = \ln x, x > 0, x = \dfrac{\sqrt{2}}{2}$.

18. $s_1(t) = \ln t$ and $s_1'(t) = \dfrac{1}{t}; 1 \le t \le 8$
$s_2(t) = \sin(t)$ and $s_2'(t) = \cos(t); 1 \le t \le 8$
Enter $y_1 = \dfrac{1}{x}$ and $y_2 = \cos(x)$. Use the
Intersection function of the calculator and
obtain $t = 4.917$ and $t = 7.724$.

19. Step 1: $s(t) = \sin t$
$v(t) = \cos t$
$a(t) = -\sin t$

Step 2: Set $v(t) = 0 \Rightarrow \cos t = 0$; $t = \dfrac{\pi}{2}$ and $\dfrac{3\pi}{2}$.
Set $a(t) = 0 \Rightarrow -\sin t = 0$; $t = \pi$
and 2π.

Step 3: Determine the directions of motion.
See Figure 5.8-6.

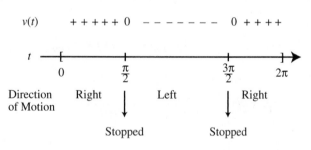

Figure 5.8-6

Step 4: Determine acceleration. See
Figure 5.8-7.

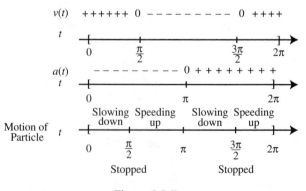

Figure 5.8-7

Step 5: Draw the motion of the particle. See Figure 5.8-8.

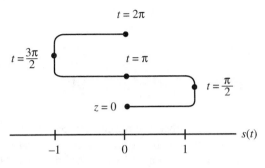

Figure 5.8-8

The particle is initially at 0, $s(0) = 0$. It moves to the right but slows down to a stop at 1 when $t = \dfrac{\pi}{2}$, $s\left(\dfrac{\pi}{2}\right) = 1$. It then turns and moves to the left speeding up until it reaches 0, when $t = \pi$, $s(\pi) = 0$ and continues to the left but slowing down to a stop at -1 when $t = \dfrac{3\pi}{2}$, $s\left(\dfrac{3\pi}{2}\right) = -1$. It then turns around again, moving to the right, speeding up to 0 when $t = 2\pi$, $s(2\pi) = 0$.

20. $s(t) = -16t^2 + v_0 t + s_0$.
s_0 = height of building and $v_0 = 0$.
Thus $s(t) = -16t^2 + s_0$.
When the coin hits the ground, $s(t) = 0$, $t = 10.2$. Thus, set $s(t) = 0 \Rightarrow -16t^2 + s_0 = 0 \Rightarrow -16(10.2)^2 + s_0 = 0$
$s_0 = 1664.64$ ft. The building is approximately 1665 ft tall.

21. When $x = \dfrac{-1}{2} = \cos t - 1$, $\cos t = \dfrac{1}{2}$, and $t = \dfrac{\pi}{3}$, and so $y = \sin\left(\dfrac{\pi}{3}\right) + \dfrac{\pi}{3} = \dfrac{\sqrt{3}}{2} + \dfrac{\pi}{3}$. Find $\dfrac{dx}{dt} = -\sin t$ and $\dfrac{dy}{dt} = \cos t + 1$, and divide to find $\dfrac{dy}{dx} = \dfrac{\cos t + 1}{-\sin t}$. Evaluate at $t = \dfrac{\pi}{3}$ to find the slope $m = \dfrac{\cos(\pi/3) + 1}{-\sin(\pi/3)} = \dfrac{3/2}{-\sqrt{3}/2} = -\sqrt{3}$. Therefore the equation of the tangent line is $y - \left(\dfrac{\sqrt{3}}{2} + \dfrac{\pi}{3}\right) = -\sqrt{3}\left(x + \dfrac{1}{2}\right)$ or simplifying, $y = -\sqrt{3}x + \dfrac{\pi}{3}$.

22. Differentiate to find $\dfrac{dx}{dt} = 2e^{2t} + 1$ and $\dfrac{dy}{dt} = e^t$.
The speed of the object is $\sqrt{(2e^{2t} + 1)^2 + (e^t)^2}$
$= \sqrt{4e^{4t} + 5e^{2t} + 1}$. When $t = 2$,

$\sqrt{4e^{4t} + 5e^{2t} + 1} \approx 110.444$. Find second derivatives $\dfrac{d^2x}{dt^2} = 4e^{2t}$ and $\dfrac{d^2y}{dt^2} = e^t$ and evaluate at $t = 2$, to find the acceleration vector $\langle 4e^{2t}, e^t \rangle \approx \langle 218.393, 7.389 \rangle$.

23. Since $x = r\cos\theta$ and $y = \sin\theta$,
$\dfrac{dy}{dx} = \dfrac{dy}{d\theta} \div \dfrac{dx}{d\theta} = \dfrac{r\cos\theta + \sin\theta\,(dr/d\theta)}{-r\sin\theta + \cos\theta\,(dr/d\theta)}$.
Find $\dfrac{dr}{d\theta} = 12\cos 4\theta$ and substitute.
$\dfrac{dy}{dx} = \dfrac{(3\sin 4\theta)\cos\theta + \sin\theta\,(12\cos 4\theta)}{-(3\sin 4\theta)\sin\theta + \cos\theta\,(12\cos 4\theta)}$. When $\theta = \dfrac{5\pi}{6}$, $4\theta = \dfrac{10\pi}{3} = \dfrac{4\pi}{3} + 2\pi$ so the functions of $\dfrac{10\pi}{3}$ are equal to those of $\dfrac{4\pi}{3}$. Evaluate $\dfrac{dy}{dx} = \dfrac{(3\sin(4\pi/3))\cos(5\pi/6) + \sin(5\pi/6)(12\cos(4\pi/3))}{-(3\sin(4\pi/3))\sin(5\pi/6) + \cos(5\pi/6)(12\cos(4\pi/3))}$
at
$= \dfrac{\left(3\left(-\sqrt{3}/2\right)\right)\left(-\sqrt{3}/2\right) + (1/2)(12(-1/2))}{-\left(3\left(-\sqrt{3}/2\right)\right)(1/2) + \left(-\sqrt{3}/2\right)(12(-1/2))}$
$= \dfrac{(9/4) - 3}{(3\sqrt{3}/4) + 3\sqrt{3}} = -\dfrac{\sqrt{3}}{15}$. The slope of the tangent line is $\dfrac{-\sqrt{3}}{15}$.

24. If the position of the object is given by $\left\langle 30t, 25\sin\dfrac{t}{3} \right\rangle$, then the velocity vector is $\left\langle \dfrac{dx}{dt}, \dfrac{dy}{dt} \right\rangle = \left\langle 30, \dfrac{25}{3}\cos\dfrac{t}{3} \right\rangle$, and the acceleration vector is $\left\langle \dfrac{d^2x}{dt^2}, \dfrac{d^2y}{dt^2} \right\rangle = \left\langle 0, \dfrac{-25}{9}\sin\dfrac{t}{3} \right\rangle$. The magnitude of the acceleration is equal to $\left| \dfrac{-25}{9}\sin\dfrac{t}{3} \right| = 2$ when $\sin\dfrac{t}{3} = \dfrac{18}{25}$. Solve to find $t = 3\sin^{-1}\dfrac{18}{25} \approx 2.411$.

25. If $r = \langle \ln t, \ln(t + 4) \rangle$, then $\dfrac{dr}{dt} = \left\langle \dfrac{1}{t}, \dfrac{1}{t + 4} \right\rangle$.
Evaluate at $t = 4$ for $\dfrac{dr}{dt} = \left\langle \dfrac{1}{4}, \dfrac{1}{8} \right\rangle$, then find $\left\| \dfrac{dr}{dt} \right\| = \sqrt{\left(\dfrac{1}{t}\right)^2 + \left(\dfrac{1}{t + 4}\right)^2}$, which, at $t = 4$,

is equal to $\left\| \dfrac{dr}{dt} \right\| = \sqrt{\dfrac{1}{16} + \dfrac{1}{64}} = \dfrac{\sqrt{5}}{8}$. The

tangent vector is $T = \dfrac{\langle 1/t, 1/(t+4) \rangle}{\sqrt{5}/8}$

$= \left\langle \dfrac{8}{t\sqrt{5}}, \dfrac{8}{(t+4)\sqrt{5}} \right\rangle$. When $t = 4$,

$T = \left\langle \dfrac{2\sqrt{5}}{5}, \dfrac{\sqrt{5}}{5} \right\rangle$.

5.9 SOLUTIONS TO CUMULATIVE REVIEW PROBLEMS

26. Using product rule, let $u = x$; $v = \sin^{-1}(2x)$.

$$\dfrac{dy}{dx} = (1)\sin^{-1}(2x) + \dfrac{1}{\sqrt{1 - (2x)^2}}(2)(x)$$

$$= \sin^{-1}(2x) + \dfrac{2x}{\sqrt{1 - 4x^2}}$$

27. Let $y = f(x) \Rightarrow y = x^3 - 3x^2 + 3x - 1$. To find $f^{-1}(x)$, switch x and y: $x = y^3 - 3y^2 + 3y - 1$.

$$\dfrac{dx}{dy} = 3y^2 - 6y + 3;$$

$$\dfrac{dy}{dx} = \dfrac{1}{dx/dy} = \dfrac{1}{3y^2 - 6y + 3}$$

$$\dfrac{dy}{dx}\bigg|_{y=2} = \dfrac{1}{3(2)^2 - 6(2) + 3} = \dfrac{1}{3}$$

28. Substituting $x = 100$ into the expression $\dfrac{x - 100}{\sqrt{x} - 10}$ would lead to $\dfrac{0}{0}$. Multiply both numerator and denominator by the conjugate of the denominator $(\sqrt{x} + 10)$:

$$\lim_{x \to 100} \dfrac{(x - 100)}{(\sqrt{x} - 10)} \cdot \dfrac{(\sqrt{x} + 10)}{(\sqrt{x} + 10)} =$$

$$\lim_{x \to 100} \dfrac{(x - 100)(\sqrt{x} + 10)}{(x - 100)}$$

$$\lim_{x \to 100} (\sqrt{x} + 10) = 10 + 10 = 20.$$

An alternative solution is to factor the numerator:

$$\lim_{x \to 10} \dfrac{(\sqrt{x} - 10)(\sqrt{x} + 10)}{(\sqrt{x} - 10)} = 20.$$

29. (a) $f' > 0$ on $(-1, 2)$, f is increasing on $(-1, 2)$ $f' < 0$ on $(2, 8)$, f is decreasing on $(2, 8)$.

(b) At $x = 2$, $f' = 0$ and $f'' < 0$, thus at $x = 2$, f has a relative maximum. Since it is the only relative extremum on the interval,

it is an absolute maximum. Since f is a continuous function on a closed interval and at its endpoints $f(-1) < 0$ and $f(8) = 1/2$, thus f has an absolute minimum at $x = -1$.

(c) At $x = 5$, f has a change of concavity and f' exists at $x = 5$.

(d) $f'' < 0$ on $(-1, 5)$, f is concave downward on $(-1, 5)$.
$f'' > 0$ on $(5, 8)$, f is concave upward on $(5, 8)$.

(e) A possible graph of f is given in Figure 5.9-1.

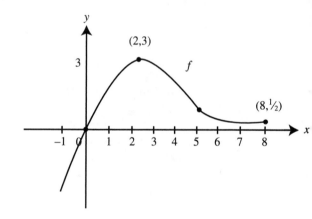

Figure 5.9-1

30. (a) $v(t) = 3$ ft/s at $t = 6$. The tangent line to the graph of $v(t)$ at $t = 6$ has a slope of approximately $m = 1$. (The tangent passes through the points $(8, 5)$ and $(6, 3)$; thus $m = 1$.) Therefore the acceleration is 1 ft/s^2.

(b) The acceleration is a minimum at $t = 0$, since the slope of the tangent to the curve of $v(t)$ is the smallest at $t = 0$.

31. To convert $r = 4\cos\theta$ to a Cartesian representation, recall that $r = \sqrt{x^2 + y^2}$, and $\tan\theta = \dfrac{y}{x}$. Then $\sqrt{x^2 + y^2} = 4\cos\left(\tan^{-1}\dfrac{y}{x}\right)$.
Since $\cos\left(\tan^{-1}\dfrac{y}{x}\right) = \dfrac{x}{\sqrt{x^2 + y^2}}$, the equation becomes $\sqrt{x^2 + y^2} = \dfrac{4x}{\sqrt{x^2 + y^2}}$. Multiply

through by $\sqrt{x^2 + y^2}$ to produce $x^2 + y^2 = 4x$. Completing the square produces $(x - 2)^2 + y^2 = 4$.

32. When $x = 5 \sin t = 5$, $t = \dfrac{\pi}{2}$, so $y = 1 - \cos \dfrac{\pi}{2} = 1$.

33. If $u = \langle 6, -1 \rangle$ and $v = \langle -4, 3 \rangle$, $4\langle 6, -1 \rangle - 3\langle -4, 3 \rangle = \langle 24, -4 \rangle + \langle 12, -9 \rangle = \langle 36, -13 \rangle$.

34. Replace θ with $-\theta$. $2 \sin(-4\theta) = -2 \sin(4\theta) \neq 2 \sin(4\theta)$, so the graph is not symmetric

about the polar axis. Replace θ with $\pi - \theta$.
$2 \sin(4(\pi - \theta)) = 2 \sin(4\pi - 4\theta)$
$= 2 [\sin 4\pi \cos 4\theta - \sin 4\theta \cos 4\pi] = -2 \sin 4\theta$
$\neq 2 \sin 4\theta$ so the graph is not symmetric
about the line $x = \dfrac{\pi}{2}$. Replace θ with $\theta + \pi$.

$2 \sin(4(\theta + \pi)) = 2 \sin(4\theta + 4\pi) =$
$2 [\sin 4\theta \cos 4\pi + \cos 4\theta \sin 4\pi] = 2 \sin 4\theta$
so the graph is symmetric about the pole.

35. The magnitude of the vector $3i + 4j$ is
$\| 3i + 4j \| = \sqrt{3^2 + 4^2} = 5$.

Chapter 6

Integration

6.1 EVALUATING BASIC INTEGRALS

Main Concepts: *Antiderivatives and Integration Formulas, Evaluating Integrals*

• Answer all parts of a question from Section II even if you think your answer to an earlier part of the question might not be correct. Also, if you don't know the answer to part one of a question, and you need it to answer part two, just make it up and continue.

Antiderivatives and Integration Formulas

Definition: A function F is an antiderivative of another function f if $F'(x) = f(x)$ for all x in some open interval. Any two antiderivatives of f differ by an additive constant C. We denote the set of antiderivatives of f by $\int f(x)dx$, called the indefinite integral of f.

Integration Rules:

1. $\displaystyle\int f(x)dx = F(x) + C \Leftrightarrow F'(x) = f(x)$

2. $\displaystyle\int af(x)dx = a\int f(x)dx$

3. $\displaystyle\int -f(x)dx = -\int f(x)dx$

4. $\displaystyle\int \left[f(x) \pm g(x)\right] dx = \int f(x)dx \pm \int g(x)dx$

Differentiation Formulas: *Integration Formulas:*

1. $\dfrac{d}{dx}(x) = 1$ 1. $\displaystyle\int 1dx = x + c$

2. $\dfrac{d}{dx}(ax) = a$ 2. $\displaystyle\int a\, dx = ax + c$

Differentiation Formulas (cont.):   *Integration Formulas (cont.):*

3. $\dfrac{d}{dx}(x^n) = nx^{n-1}$

3. $\displaystyle\int x^n dx = \dfrac{x^{n+1}}{n+1} + c, n \neq -1$

4. $\dfrac{d}{dx}(\cos x) = -\sin x$

4. $\displaystyle\int \sin x\, dx = -\cos x + c$

5. $\dfrac{d}{dx}(\sin x) = \cos x$

5. $\displaystyle\int \cos x\, dx = \sin x + c$

6. $\dfrac{d}{dx}(\tan x) = \sec^2 x$

6. $\displaystyle\int \sec^2 x\, dx = \tan x + c$

7. $\dfrac{d}{dx}(\cot x) = -\csc^2 x$

7. $\displaystyle\int \csc^2 x\, dx = -\cot x + c$

8. $\dfrac{d}{dx}(\sec x) = \sec x\ \tan x$

8. $\displaystyle\int \sec x(\tan x)\, dx = \sec x + c$

9. $\dfrac{d}{dx}(\csc x) = -\csc x(\cot x)$

9. $\displaystyle\int \csc x(\cot x)\, dx = -\csc x + c$

10. $\dfrac{d}{dx}(\ln x) = \dfrac{1}{x}$

10. $\displaystyle\int \dfrac{1}{x}dx = \ln|x| + c$

11. $\dfrac{d}{dx}(e^x) = e^x$

11. $\displaystyle\int e^x dx = e^x + c$

12. $\dfrac{d}{dx}(a^x) = (\ln a)a^x$

12. $\displaystyle\int a^x dx = \dfrac{a^x}{\ln a} + c\ a > 0, a \neq 1$

13. $\dfrac{d}{dx}(\sin^{-1} x) = \dfrac{1}{\sqrt{1-x^2}}$

13. $\displaystyle\int \dfrac{1}{\sqrt{1-x^2}}dx = \sin^{-1} x + c$

14. $\dfrac{d}{dx}(\tan^{-1} x) = \dfrac{1}{1+x^2}$

14. $\displaystyle\int \dfrac{1}{1+x^2}dx = \tan^{-1} x + c$

15. $\dfrac{d}{dx}(\sec^{-1}x) = \dfrac{1}{|x|\sqrt{x^2-1}}$

15. $\displaystyle\int \dfrac{1}{|x|\sqrt{x^2-1}}dx = \sec^{-1}x + c$

More Integration Formulas:

16. $\displaystyle\int \tan x\, dx = \ln|\sec x| + c \text{ or } -\ln|\cos x| + c$

17. $\displaystyle\int \cot x\, dx = \ln|\sin x| + c \text{ or } -\ln|\csc x| + c$

18. $\displaystyle\int \sec x\, dx = \ln|\sec x + \tan x| + c$

19. $\displaystyle\int \csc x\, dx = \ln|\csc x - \cot x| + c$

20. $\displaystyle\int \ln x\, dx = x \ln|x| - x + c$

21. $\displaystyle\int \dfrac{1}{\sqrt{a^2-x^2}}dx = \sin^{-1}\left(\dfrac{x}{a}\right) + c$

22. $\displaystyle\int \dfrac{1}{\sqrt{a^2+x^2}}dx = \dfrac{1}{a}\tan^{-1}\left(\dfrac{x}{a}\right) + c$

23. $\displaystyle\int \dfrac{1}{x\sqrt{x^2+a^2}}dx = \dfrac{1}{a}\sec^{-1}\left|\dfrac{x}{a}\right| + c \text{ or } \dfrac{1}{a}\cos^{-1}\left|\dfrac{a}{x}\right| + c$

24. $\displaystyle\int \sin^2 x\, dx = \dfrac{x}{2} - \dfrac{\sin(2x)}{4} + c.$ Note: $\sin^2 x = \dfrac{1-\cos 2x}{2}$

Note: After evaluating an integral, always check the result by taking the derivative of the answer (i.e., taking the derivative of the antiderivative).

- Remember that the volume of a right-circular cone is $v = \dfrac{1}{3}\pi r^2 h$ where r is the radius of the base and h is the height of the cone.

Evaluating Integrals

Integral	Rewrite	Antiderivative
$\int x^3\,dx$		$\dfrac{x^4}{4} + c$
$\int dx$	$\int 1\,dx$	$x + c$
$\int 5\,dx$		$5x + c$
$\int \sqrt{x}\,dx$	$\int x^{1/2}\,dx$	$\dfrac{x^{3/2}}{3/2} + c$ or $\dfrac{2x^{3/2}}{3} + c$
$\int x^{5/2}\,dx$		$\dfrac{x^{7/2}}{7/2} + c$ or $\dfrac{2x^{7/2}}{7} + c$
$\int \dfrac{1}{x^2}\,dx$	$\int x^{-2}\,dx$	$\dfrac{x^{-1}}{-1} + c$ or $\dfrac{-1}{x} + c$
$\int \dfrac{1}{\sqrt[3]{x^2}}\,dx$	$\int \dfrac{1}{x^{2/3}}\,dx = \int x^{-2/3}\,dx$	$\dfrac{x^{1/3}}{1/3} + c$ or $3\sqrt[3]{x} + c$
$\int \dfrac{x+1}{x}\,dx$	$\int \left(1 + \dfrac{1}{x}\right)\,dx$	$x + \ln|x| + c$
$\int x(x^5 + 1)\,dx$	$\int (x^6 + x)\,dx$	$\dfrac{x^7}{7} + \dfrac{x^2}{2} + c$

Example 1

Evaluate $\int (x^5 - 6x^2 + x - 1)\,dx$

Applying the formula $\int x^n\,dx = \dfrac{x^{n+1}}{n+1} + c$, $n \neq 1$.

$\int (x^5 - 6x^2 + x - 1)\,dx = \dfrac{x^6}{6} - 2x^3 + \dfrac{x^2}{2} - x + c$

Example 2

Evaluate $\int \left(\sqrt{x} + \dfrac{1}{x^3}\right)\,dx$

Rewrite $\int \left(\sqrt{x} + \dfrac{1}{x^3}\right)\,dx$ as $\int \left(x^{1/2} + x^{-3}\right)\,dx = \dfrac{x^{3/2}}{3/2} + \dfrac{x^{-2}}{-2} + c$

$$= \dfrac{2}{3}x^{3/2} - \dfrac{1}{2x^2} + c$$

Example 3

If $\dfrac{dy}{dx} = 3x^2 + 2$, and the point $(0, -1)$ lies on the graph of y, find y.

Since $\dfrac{dy}{dx} = 3x^2 + 2$, then y is an antiderivative of $\dfrac{dy}{dx}$. Thus

$y = \displaystyle\int \left(3x^2 + 2\right) dx = x^3 + 2x + c$. The point $(0, -1)$ is on the graph of y.

Thus $y = x^3 + 2x + c$ becomes $-1 = 0^3 + 2(0) + c$ or $c = -1$. Therefore, $y = x^3 + 2x - 1$.

Example 4

Evaluate $\displaystyle\int \left(1 - \dfrac{1}{\sqrt[3]{x^4}}\right) dx$

Rewrite as $\displaystyle\int \left(1 - \dfrac{1}{x^{4/3}}\right) dx = \int \left(1 - x^{-4/3}\right) dx$

$$= x - \dfrac{x^{-1/3}}{-1/3} + c = x + \dfrac{3}{\sqrt[3]{x}} + c$$

Example 5

Evaluate $\displaystyle\int \dfrac{3x^2 + x - 1}{x^2} dx$

Rewrite as $\displaystyle\int \left(3 + \dfrac{1}{x} - \dfrac{1}{x^2}\right) dx = \int \left(3 + \dfrac{1}{x} - x^{-2}\right) dx$

$$= 3x + \ln|x| - \dfrac{x^{-1}}{-1} + c = 3x + \ln|x| + \dfrac{1}{x} + c$$

Example 6

Evaluate $\displaystyle\int \sqrt{x}\left(x^2 - 3\right) dx$

Rewrite $\displaystyle\int x^{1/2}\left(x^2 - 3\right) dx = \int \left(x^{5/2} - 3x^{1/2}\right) dx$

$$= \dfrac{x^{7/2}}{7/2} - \dfrac{3x^{3/2}}{3/2} + c = \dfrac{2}{7}x^{7/2} - 2\sqrt{x^3} + c$$

Example 7

Evaluate $\displaystyle\int \left(x^3 - 4\sin x\right) dx$

$\displaystyle\int \left(x^3 - 4\sin x\right) dx = \dfrac{x^4}{4} + 4\cos x + c$

Example 8

Evaluate $\displaystyle\int \left(4\cos x - \cot x\right) dx$

$\displaystyle\int \left(4\cos x - \cot x\right) dx = 4\sin x - \ln|\sin x| + c$

Example 9

Evaluate $\displaystyle\int \dfrac{\sin x - 1}{\cos} dx$

Rewrite $\displaystyle\int \left(\dfrac{\sin x}{\cos x} - \dfrac{1}{\cos x}\right) dx = \int (\tan x - \sec x)\, dx = \int \tan x\, dx - \int \sec x\, dx$

$$= \ln|\sec x| - \ln|\sec x + \tan x| + c = \ln\left|\dfrac{\sec x}{\sec x + \tan x}\right| + c$$

$$\text{or } -\ln|\sin x + 1| + c$$

Example 10

Evaluate $\int \dfrac{e^{2x}}{e^x} dx$

Rewrite the integral as $\int e^x dx = e^x + c$

Example 11

Evaluate $\int \dfrac{3}{1+x^2} dx$

Rewrite as $3 \int \dfrac{1}{1+x^2} dx = 3\tan^{-1} x + c$

Example 12

Evaluate $\int \dfrac{1}{\sqrt{9-x^2}} dx$

Rewrite as $\int \dfrac{1}{\sqrt{3^2 - x^2}} dx = \sin^{-1}\left(\dfrac{x}{3}\right) + c$

Example 13

Evaluate $\int 7^x dx$

$\int 7^x dx = \dfrac{7^x}{\ln 7} + c$

Reminder: You can always check the result by taking the derivative of the answer.

- Be familiar with the instructions for the different parts of the exam before the day of exam. Visit the College Board website at: *www.collegeboard.com/ap* for more information.

6.2 INTEGRATION BY U-SUBSTITUTION

Main Concepts: *The U-Substitution Method, U-Substitution and Algebraic Functions, U-Substitution and Trigonometric Functions, U-Substitution and Inverse Trigonometric, U-Substitution and Logarithmic and Exponential Functions*

The U-Substitution Method

The Chain Rule for Differentiation:

$$\frac{d}{dx}F(g(x)) = f(g(x))g'(x) \quad \text{where } F' = f$$

The Integral of a Composite Function:

If $f(g(x))$ and f' are continuous and $F' = f$, then

$$\int f(g(x))g'(x)dx = F(g(x)) + c$$

Making a U-Substitution:

Let $u = g(x)$; then $du = g'(x)dx$

$$\int f(g(x))g''(x)dx = \int f(u)du = F(u) + c = F(g(x)) + c$$

!

Procedure for Making a U-Substitution:

Steps:
1. Given $f(g(x))$; Let $u = g(x)$
2. Differentiate: $du = g'(x)dx$
3. Rewrite the integral in terms of u.
4. Evaluate the integral.
5. Replace u by $g(x)$.
6. Check your result by taking the derivative of the answer.

U-Substitution and Algebraic Functions

Another Form of the Integral of a Composite Function:

If f is a differentiable function, then

$$\int (f(x))^n f'(x)dx = \frac{(f(x))^{n+1}}{n+1} + c, n \neq -1$$

Making a U-Substitution:

Let $u = f(x)$; then $du = f'(x)dx$

$$\int (f(x))^n f'(x)dx = \int u^n du = \frac{u^{n+1}}{n+1} + c = \frac{(f(x))^{n+1}}{n+1} + c, n \neq -1$$

Example 1

Evaluate $\int x(x+1)^{10}dx$

Step 1. Let $u = x + 1$; then $x = u - 1$.

Step 2. Differentiate: $du = dx$.

Step 3. Rewrite: $\int (u-1)u^{10}du = \int \left(u^{11} - u^{10} \right)du$

Step 4. Integrate: $\dfrac{u^{12}}{12} - \dfrac{u^{11}}{11} + c$

Step 5. Replace u: $\dfrac{(x+1)^{12}}{12} - \dfrac{(x+1)^{11}}{11} + c$

Step 6. Differentiate and Check: $\dfrac{12(x+1)^{11}}{12} - \dfrac{11(x+1)^{10}}{11} = (x+1)^{11} - (x+1)^{10}$

$$= (x+1)^{10}(x+1-1) = (x+1)^{10}x \text{ or } x(x+1)^{10}$$

Example 2

Evaluate $\int x\sqrt{x-2}\,dx$

Step 1. Let $u = x - 2$; then $x = u + 2$

Step 2. Differentiate $du = dx$

Step 3. Rewrite: $\int (u+2)\sqrt{u}\,du = \int (u+2)u^{1/2}\,du = \int \left(u^{3/2} + 2u^{1/2}\right)du$

Step 4. Integrate: $\dfrac{u^{5/2}}{5/2} + \dfrac{2u^{3/2}}{3/2} + c$

Step 5. Replace: $\dfrac{2(x-2)^{5/2}}{5} + \dfrac{4(x-2)^{3/2}}{3} + c$

Step 6. Differentiate and Check: $\left(\dfrac{5}{2}\right)\dfrac{2(x-2)^{3/2}}{5} + \left(\dfrac{3}{2}\right)\dfrac{4(x-2)^{1/2}}{3}$

$$= (x-2)^{3/2} + 2(x-2)^{1/2}$$

$$= (x-2)^{1/2}[(x-2)+2] = (x-2)^{1/2}\,x \text{ or } x\sqrt{x-2}$$

Example 3

Evaluate $\int (2x-5)^{2/3}\,dx$

Step 1. Let $u = 2x - 5$

Step 2. Differentiate $du = 2dx \Rightarrow \dfrac{du}{2} = dx$

Step 3. Rewrite: $\int u^{2/3}\dfrac{du}{2} = \dfrac{1}{2}\int u^{2/3}\,du$

Step 4. Integrate: $\dfrac{1}{2}\left(\dfrac{u^{5/3}}{5/3}\right) + c = \dfrac{3u^{5/3}}{10} + c$

Step 5. Replace u: $\dfrac{3(2x-5)^{5/3}}{10} + c$

Step 6. Differentiate and Check: $\left(\dfrac{3}{10}\right)\left(\dfrac{5}{3}\right)(2x-5)^{2/3}(2) = (2x-5)^{2/3}$

Example 4

Evaluate $\int \dfrac{x^2}{(x^3-8)^5}\,dx$

Step 1. Let $u = x^3 - 8$

Step 2. Differentiate: $du = 3x^2\,dx \Rightarrow \dfrac{du}{3} = x^2\,dx$

Step 3. Rewrite: $\int \dfrac{1}{u^5}\dfrac{du}{3} = \dfrac{1}{3}\int \dfrac{1}{u^5}\,du = \dfrac{1}{3}\int u^{-5}\,du$

Step 4. Integrate: $\dfrac{1}{3}\left(\dfrac{u^{-4}}{-4}\right) + c$

Step 5. Replace u: $\dfrac{1}{-12}\left(x^3-8\right)^{-4}+c$ or $\dfrac{-1}{12\left(x^3-8\right)^4}+c$

Step 6. Differentiate and Check: $\left(-\dfrac{1}{12}\right)(-4)\left(x^3-8\right)^{-5}\left(3x^2\right)=\dfrac{x^2}{\left(x^3-8\right)^5}$

U-Substitution and Trigonometric Functions

Example 1

Evaluate $\displaystyle\int \sin 4x\,dx$

Step 1. Let $u=4x$

Step 2. Differentiate: $du=4\,dx$ or $\dfrac{du}{4}=dx$

Step 3. Rewrite: $\displaystyle\int \sin u\,\dfrac{du}{4}=\dfrac{1}{4}\int \sin u\,du$

Step 4. Integrate: $\dfrac{1}{4}(-\cos u)+c=-\dfrac{1}{4}\cos u+c$

Step 5. Replace u: $-\dfrac{1}{4}\cos(4x)+c$

Step 6. Differentiate and Check: $\left(-\dfrac{1}{4}\right)(-\sin 4x)(4)=\sin 4x$

Example 2

Evaluate $\displaystyle\int 3\left(\sec^2 x\right)\sqrt{\tan x}\,dx$

Step 1. Let $u=\tan x$

Step 2. Differentiate: $du=\sec^2 x\,dx$

Step 3. Rewrite: $3\displaystyle\int (\tan x)^{1/2}\sec^2 x\,dx=3\int u^{1/2}\,du$

Step 4. Integrate $3\dfrac{u^{3/2}}{3/2}+c=2u^{3/2}+c$

Step 5. Replace u: $2(\tan x)^{3/2}+c$ or $2\tan^{3/2}x+c$

Step 6. Differentiate and Check: $(2)\left(\dfrac{3}{2}\right)\left(\tan^{1/2}x\right)\left(\sec^2 x\right)=3\left(\sec^2 x\right)\sqrt{\tan x}$

Example 3

Evaluate $\displaystyle\int 2x^2\cos\left(x^3\right)dx$

Step 1. Let $u=x^3$

Step 2. Differentiate $du=3x^2\,dx\Rightarrow\dfrac{du}{3}=x^2\,dx$

Step 3. Rewrite: $2\displaystyle\int \left[\cos(x^3)\right]x^2\,dx=2\int \cos u\,\dfrac{du}{3}=\dfrac{2}{3}\int \cos u\,du$

Step 4. Integrate: $\dfrac{2}{3}\sin u+c$

Step 5. Replace u: $\frac{2}{3}\sin\left(x^3\right) + c$

Step 6. Differentiate and Check: $\frac{2}{3}\left[\cos\left(x^3\right)\right]3x^2 = 2x^2\cos\left(x^3\right)$

- Remember that the area of a semi-circle is $\frac{1}{2}\pi r^2$. Don't forget the $\frac{1}{2}$. If the cross sections of a solid are semi-circles, the integral for the volume of the solid will involve $\left(\frac{1}{2}\right)^2$ which is $\frac{1}{4}$.

U-Substitution and Inverse Trigonometric Functions

Example 1

Evaluate $\displaystyle\int \frac{dx}{\sqrt{9 - 4x^2}}$

Step 1. Let $u = 2x$

Step 2. Differentiate $du = 2x$; $\dfrac{du}{2} = dx$

Step 3. Rewrite: $\displaystyle\int \frac{1}{\sqrt{9 - u^2}}\frac{du}{2} = \frac{1}{2}\int \frac{du}{\sqrt{3^2 - u^2}}$

Step 4. Integrate: $\dfrac{1}{2}\sin^{-1}\left(\dfrac{u}{3}\right) + c$

Step 5. Replace u: $\dfrac{1}{2}\sin^{-1}\left(\dfrac{2x}{3}\right) + c$

Step 6. Differentiate and Check: $\dfrac{1}{2}\dfrac{1}{\sqrt{1 - (2x/3)^2}}\cdot\dfrac{2}{3} = \dfrac{1}{3}\dfrac{1}{\sqrt{1 - 4x^2/9}}$

$= \dfrac{1}{\sqrt{9}}\dfrac{1}{\sqrt{1 - 4x^2/9}} = \dfrac{1}{\sqrt{9\left(1 - 4x^2/9\right)}}$

$= \dfrac{1}{\sqrt{9 - 4x^2}}$

Example 2

Evaluate $\displaystyle\int \frac{1}{x^2 + 2x + 5}dx$

Step 1. Rewrite: $\displaystyle\int \frac{1}{\left(x^2 + 2x + 1\right) + 4} = \int \frac{1}{(x + 1)^2 + 2^2}dx = \int \frac{1}{2^2 + (x + 1)^2}dx$

Let $u = x + 1$

Step 2. Differentiate: $du = dx$

Step 3. Rewrite: $\displaystyle\int \frac{1}{2^2 + u^2}du$

Step 4. Integrate: $\dfrac{1}{2}\tan^{-1}\left(\dfrac{u}{2}\right) + c$

Step 5. Replace u: $\dfrac{1}{2}\tan^{-1}\left(\dfrac{x+1}{2}\right)+c$

Step 6. Differentiate and Check: $\left(\dfrac{1}{2}\right)\dfrac{1\,(1/2)}{1+[(x+1)/2]^2}=\left(\dfrac{1}{4}\right)\dfrac{1}{1+(x+1)^2/4}$

$$\left(\dfrac{1}{4}\right)\dfrac{4}{4+(x+1)^2}=\dfrac{1}{x^2+2x+5}.$$

- If the problem gives you that the diameter of a sphere is 6 and you are using formulas such as $v=\dfrac{4}{3}\pi r^3$ or $s=4\pi r^2$, don't forget that $r=3$.

U-Substitution and Logarithmic and Exponential Functions

Example 1

Evaluate $\displaystyle\int\dfrac{x^3}{x^4-1}dx$

Step 1. Let $u=x^4-1$

Step 2. Differentiate: $du=4x^3dx\Rightarrow\dfrac{du}{4}=x^3dx$

Step 3. Rewrite: $\displaystyle\int\dfrac{1}{u}\dfrac{du}{4}=\dfrac{1}{4}\int\dfrac{1}{u}du$

Step 4. Integrate: $\dfrac{1}{4}\ln|u|+c$

Step 5. Replace u: $\dfrac{1}{4}\ln\left|x^4-1\right|+c$

Step 6. Differentiate and Check: $\left(\dfrac{1}{4}\right)\dfrac{1}{x^4-1}\left(4x^3\right)=\dfrac{x^3}{x^4-1}.$

Example 2

Evaluate $\displaystyle\int\dfrac{\sin x}{\cos x+1}dx$

Step 1. Let $u=\cos x+1$

Step 2. Differentiate: $du=-\sin x\,dx\Rightarrow-du=\sin x\,dx$

Step 3. Rewrite: $\displaystyle\int\dfrac{-du}{u}=-\int\dfrac{du}{u}$

Step 4. Integrate: $-\ln|u|+c$

Step 5. Replace u: $-\ln|\cos x+1|+c$

Step 6. Differentiate and Check: $-\left(\dfrac{1}{\cos x+1}\right)(-\sin x)=\dfrac{\sin x}{\cos x+1}$

Example 3

Evaluate $\displaystyle\int\dfrac{x^2+3}{x-1}dx$

Step 1. Rewrite $\dfrac{x^2+3}{x-1} = x+1+\dfrac{4}{x-1}$; by dividing (x^2+3) by $(x-1)$.

$$\int \frac{x^2+3}{x-1}dx = \int \left(x+1+\frac{4}{x-1}\right)dx = \int (x+1)\,dx + \int \frac{4}{x-1}dx$$

$$= \frac{x^2}{2}+x+4\int \frac{1}{x-1}dx$$

Let $u = x-1$.

Step 2. Differentiate: $du = dx$

Step 3. Rewrite: $4\int \dfrac{1}{u}\,du$

Step 4. Integrate: $4\ln|u| + c$

Step 5. Replace u: $4\ln|x-1| + c$

$$\int \frac{x^2+3}{x-1}dx = \frac{x^2}{2}+x+4\ln|x-1|+c$$

Step 6. Differentiate and Check:

$$\frac{2x}{2}+1+4\left(\frac{1}{x-1}\right)+c = x+1+\frac{4}{x-1} = \frac{x^2+3}{x-1}$$

Example 4

Evaluate $\displaystyle\int \frac{\ln x}{3x}\,dx$

Step 1. Let $u = \ln x$

Step 2. Differentiate: $du = \dfrac{1}{x}dx$

Step 3. Rewrite: $\dfrac{1}{3}\displaystyle\int u\,dx$

Step 4. Integrate $\left(\dfrac{1}{3}\right)\dfrac{u^2}{2}+c = \dfrac{1}{6}u^2+c$

Step 5. Replace u: $\dfrac{1}{6}(\ln x)^2 + c$

Step 6. Differentiate and Check: $\dfrac{1}{6}(2)(\ln x)\left(\dfrac{1}{x}\right) = \dfrac{\ln x}{3x}$.

Example 5

Evaluate $\displaystyle\int e^{(2x-5)}\,dx$

Step 1. Let $u = 2x-5$

Step 2. Differentiate: $du = 2dx \Rightarrow \dfrac{du}{2} = dx$

Step 3. Rewrite: $\displaystyle\int e^u\left(\frac{du}{2}\right) = \frac{1}{2}\int e^u\,du$

Step 4. Integrate: $\dfrac{1}{2}e^u + c$

Step 5. Replace u: $\dfrac{1}{2}e^{(2x-5)} + c$

Step 6. Differentiate and Check: $\dfrac{1}{2}e^{2x-5}(2) = e^{2x-5}$.

Example 6

Evaluate $\displaystyle\int \dfrac{e^x}{e^x+1}\,dx$

Step 1. Let $u = e^x + 1$

Step 2. Differentiate: $du = e^x\,dx$

Step 3. Rewrite: $\displaystyle\int \dfrac{1}{u}\,du$

Step 4. Integrate: $\ln|u| + c$

Step 5. Replace u: $\ln|e^x + 1| + c$

Step 6. Differentiate and Check: $\dfrac{1}{e^x + 1} \cdot e^x = \dfrac{e^x}{e^x + 1}$.

Example 7

Evaluate $\displaystyle\int xe^{3x^2}\,dx$

Step 1. Let $u = 3x^2$

Step 2. Differentiate: $du = 6x\,dx \Rightarrow \dfrac{du}{6} = x\,dx$

Step 3. Rewrite: $\displaystyle\int e^u\dfrac{du}{6} = \dfrac{1}{6}\int e^u\,du$

Step 4. Integrate: $\dfrac{1}{6}e^u + c$

Step 5. Replace u: $\dfrac{1}{6}e^{3x^2} + c$

Step 6. Differentiate and Check: $\dfrac{1}{6}\left(e^{3x^2}\right)(6x) = xe^{3x^2}$.

Example 8

Evaluate $\displaystyle\int 5^{(2x)}\,dx$

Step 1. Let $u = 2x$

Step 2. Differentiate: $du = 2dx \Rightarrow \dfrac{du}{2} = dx$

Step. 3. Rewrite: $\displaystyle\int 5^u\dfrac{du}{2} = \dfrac{1}{2}\int 5^u\,du$

Step 4. Integrate: $\dfrac{1}{2}(5^u)/\ln 5 + c = 5^u/2\ln 5 + c$

Step 5. Replace u: $\dfrac{5^{2x}}{2\ln 5} + c$

Step 6. Differentiate and Check: $(5^{2x})(2)\ln 5/2\ln 5 = 5^{2x}$.

Example 9

Evaluate $\int x^3 5^{(x^4)} dx$

Step 1. Let $u = x^4$

Step 2. Differentiate: $du = 4x^3 dx \Rightarrow \dfrac{du}{4} = x^3 dx$

Step 3. Rewrite: $\int 5^u \dfrac{du}{4} = \dfrac{1}{4} \int 5^u du$

Step 4. Integrate: $\dfrac{1}{4}(5^u)/\ln 5 + c$

Step 5. Replace u: $\dfrac{5^{x^4}}{4 \ln 5} + c$

Step 6. Differentiate and Check: $5^{(x^4)} (4x^3) \ln 5/4 \ln 5 = x^3 5^{(x^4)}$.

Example 10

Evaluate $\int (\sin \pi x)\, e^{\cos \pi x} dx$

Step 1. Let $u = \cos \pi x$

Step 2. Differentiate: $du = -\pi \sin \pi x\, dx; -\dfrac{du}{\pi} = \sin \pi x\, dx$

Step 3. Rewrite: $\int e^u \left(\dfrac{-du}{\pi} \right) = -\dfrac{1}{\pi} \int e^u du$

Step 4. Integrate: $-\dfrac{1}{\pi} e^u + c$

Step 5. Replace u: $-\dfrac{1}{\pi} e^{\cos \pi x} + c$

Step 6: Differentiate and Check: $-\dfrac{1}{\pi}(e^{\cos \pi x})(-\sin \pi x)\pi = (\sin \pi x) e^{\cos \pi x}$.

6.3   TECHNIQUES OF INTEGRATION

Main Concepts: *Integration by Parts, Integration by Partial Fractions*

Integration by Parts

According to the product rule for differentiation $\dfrac{d}{dx}(uv) = u\dfrac{dv}{dx} + v\dfrac{du}{dx}$. Integrating tells us that $uv = \int u\dfrac{dv}{dx} + \int v\dfrac{du}{dx}$, and therefore $\int u\dfrac{dv}{dx} = uv - \int v\dfrac{du}{dx}$. To integrate a product, careful identification of one factor as u and the other as $\dfrac{dv}{dx}$ allows the application of this rule for integration by parts. Choice of one factor to be u (and therefore the other to be dv) is simpler if you remember the mnemonic **LIPET**. Each letter in the acronym represents a type of function: Logarithmic, Inverse trigonometric, Polynomial, Exponential, and Trigonometric. As you consider integrating by parts, assign the factor that falls earlier in the *LIPET* list as *u*, and the other as *dv*.

Example 1

$$\int xe^{-x}dx$$

Step 1: Identify $u = x$ and $dv = e^{-x}dx$ since x is a Polynomial, which comes before Exponential in *LIPET*.

Step 2: Differentiate $du = dx$ and integrate $v = -e^{-x}$

Step 3: $\int xe^{-x}dx = xe^{-x} - \int -e^{-x}dx = xe^{-x} - e^{-x} + C$

Example 2

$$\int x \sin 4x \, dx$$

Step 1: Identify $u = x$ and $dv = \sin 4x \, dx$ since Polynomial, which comes before Trigonometric in *LIPET*.

Step 2: Differentiate $du = dx$ and integrate $v = \dfrac{-1}{4} \cos 4x$

Step 3: $\int x \sin 4x \, dx = \dfrac{-x}{4} \cos 4x + \dfrac{1}{4} \int \cos 4x \, dx = \dfrac{-x}{4} \cos 4x + \dfrac{1}{16} \sin 4x + C$

Integration by Partial Fractions

A rational function with a factorable denominator can be integrated by decomposing the integrand into a sum of simpler fractions. Each linear factor of the denominator becomes the denominator of one of the partial fractions.

Example 1

$$\int \frac{dx}{x^2 + 3x - 4}$$

Step 1: Factor the denominator. $\int \dfrac{dx}{x^2 + 3x - 4} = \int \dfrac{dx}{(x+4)(x-1)}$

Step 2: Let A and B represent the numerators of the partial fractions
$$\frac{1}{(x+4)(x-1)} = \frac{A}{x+4} + \frac{B}{x-1}$$

Step 3: The algorithm for adding fractions tells us that $A(x-1) + B(x+4) = 1$ so $Ax + Bx = 0$ and $-A + 4B = 1$. Solving gives us $A = -0.2$ and $B = 0.2$.

Step 4: $\int \dfrac{dx}{x^2 + 3x - 4} = \int \dfrac{-0.2}{x+4}dx + \int \dfrac{0.2}{x-1}dx = -0.2 \ln(x+4) + 0.2 \ln(x-1) + C$

Example 2

$$\int \frac{x^5 + 2x^2 + 1}{x^3 - x}dx$$

Step 1: Use long division to rewrite $\int \dfrac{x^5 + 2x^2 + 1}{x^3 - x} dx =$

$$\int \left(x^2 + 1 + \frac{2x^2 + x + 1}{x^3 - x} \right) dx = \int (x^2 + 1) dx + \int \frac{2x^2 + x + 1}{x^3 - x} dx$$

Step 2: Factor the denominator. $\dfrac{2x^2 + x + 1}{x^3 - x} = \dfrac{2x^2 + x + 1}{x(x + 1)(x - 1)}$

Step 3: Let A, B, and C represent the numerators of the partial fractions.
$$\frac{2x^2 + x + 1}{x(x + 1)(x - 1)} = \frac{A}{x} + \frac{B}{x + 1} + \frac{c}{x - 1}$$

Step 4: $2x^2 + x + 1 = A(x + 1)(x - 1) + Bx(x - 1) + Cx(x + 1)$, therefore $Ax^2 + Bx^2 + Cx^2 = 2x^2$, $Cx - Bx = x$, and $-A = 1$. Solving gives $A = -1$, $B = 1$, and $C = 2$.

Step 5: $\displaystyle \int \frac{x^5 + 2x^2 + 1}{x^3 - x} dx = \int (x^2 + 1) dx + \int \frac{-1}{x} dx + \int \frac{1}{x + 1} dx + \int \frac{2}{x - 1} dx$

$$= \frac{x^3}{3} + x - \ln x + \ln(x + 1) + 2\ln(x - 1) + C$$

6.4 RAPID REVIEW

1. Evaluate $\int \dfrac{1}{x^2} dx$.

   Answer: Rewrite as $\int x^{-2} dx = \dfrac{x^{-1}}{-1} + c = -\dfrac{1}{x} + c$.

2. Evaluate $\int \dfrac{x^3 - 1}{x} dx$.

   Answer: Rewrite as $\int \left(x^2 - \dfrac{1}{x} \right) dx = \dfrac{x^3}{3} - \ln |x| + c$.

3. Evaluate $\int x\sqrt{x^2 - 1}\, dx$.

   Answer: Rewrite as $\int x(x^2 - 1)^{1/2} dx$. Let $u = x^2 - 1$.

   Thus $\dfrac{du}{2} = x\, dx \Rightarrow \dfrac{1}{2} \int u^{1/2} du = \dfrac{1 u^{3/2}}{2^{3/2}} + c = \dfrac{1}{3}(x^2 - 1)^{3/2} + c$.

4. Evaluate $\int \sin x\, dx$.

   Answer: $-\cos x + c$.

5. Evaluate $\int \cos(2x) dx$.

   Answer: Let $u = 2x$ and obtain $\dfrac{1}{2} \sin 2x + c$.

6. Evaluate $\int \dfrac{\ln x}{x} dx$.

   Answer: Let $u = \ln x$; $du = \dfrac{1}{x} dx$ and obtain $\dfrac{(\ln x)^2}{2} + c$.

7. Evaluate $\int x e^{x^2} dx$.

   Answer: Let $u = x^2$; $\dfrac{du}{2} = x\, dx$ and obtain $\dfrac{e^{x^2}}{2} + c$.

8. $\int x\cos x\,dx$

Answer: Let $u = x, du = dx, dv = \cos x dx,$ and $v = \sin x,$ then

$$\int x\cos x\,dx = x\sin x - \int \sin x\,dx = x\sin x + \cos x + C$$

9. $\int \dfrac{5}{(x+3)(x-7)}dx$

Answer: $\int \dfrac{5}{(x+3)(x-7)}dx = \int \left(\dfrac{-1/2}{x+3} + \dfrac{1/2}{x-7}\right)dx$

$$= -\dfrac{1}{2}\ln|x+3| + \dfrac{1}{2}\ln|x-7| + C = \dfrac{1}{2}\ln\left|\dfrac{x-7}{x+3}\right| + C$$

6.5 PRACTICE PROBLEMS

Evaluate the following integrals in problems 1 to 20. No calculators allowed. (However, you may use calculators to check your results.)

1. $\int (x^5 + 3x^2 - x + 1)dx$

2. $\int \left(\sqrt{x} - \dfrac{1}{x^2}\right)dx$

3. $\int x^3(x^4 - 10)^5\,dx$

4. $\int x^3\sqrt{x^2 + 1}\,dx$

5. $\int \dfrac{x^2 + 5}{\sqrt{x - 1}}\,dx$

6. $\int \tan\left(\dfrac{x}{2}\right)dx$

7. $\int x\csc^2(x^2)dx$

8. $\int \dfrac{\sin x}{\cos^3 x}dx$

9. $\int \dfrac{1}{x^2 + 2x + 10}dx$

10. $\int \dfrac{1}{x^2}\sec^2\left(\dfrac{1}{x}\right)dx$

11. $\int (e^{2x})(e^{4x})dx$

12. $\int \dfrac{1}{x\ln x}dx$

13. $\int \ln(e^{5x+1})dx$

14. $\int \dfrac{e^{4x} - 1}{e^x}dx$

15. $\int (9 - x^2)\sqrt{x}\,dx$

16. $\int \sqrt{x}\left(1 + x^{3/2}\right)^4 dx$

17. If $\dfrac{dy}{dx} = e^x + 2$ and the point $(0, 6)$ is on the graph of y, find y.

18. $\int -3e^x \sin(e^x)dx$

19. $\int \dfrac{e^x - e^{-x}}{e^x + e^{-x}}dx$

20. If $f(x)$ is the antiderivative of $\dfrac{1}{x}$ and $f(1) = 5$, find $f(e)$.

 21. $\int x^2\sqrt{1 - x}\,dx$

22. $\int 3x^2 \sin x\,dx$

23. $\int \dfrac{x\,dx}{x^2 - 3x - 4}$

24. $\int \dfrac{dx}{x^2 + x}$

25. $\int \dfrac{\ln x}{(x + 5)^2}\,dx$

6.6 CUMULATIVE REVIEW PROBLEMS

"Calculator" indicates that calculators are permitted.

26. The graph of the velocity function of a moving particle for $0 \le t \le 10$ is shown in Figure 6.6-1.

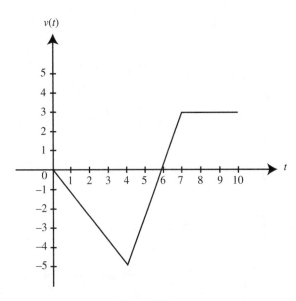

Figure 6.6-1

(a) At what value of t is the speed of the particle the greatest?

(b) At what time is the particle moving to the right?

27. Air is pumped into a spherical balloon, whose maximum radius is 10 meters. For what value of r is the rate of increase of the volume a hundred times that of the radius?

28. Evaluate $\displaystyle\int \frac{\ln^3(x)}{x}\,dx$

29. (Calculator) The function f is continuous and differentiable on $(0, 2)$ with $f''(x) > 0$ for all x in the interval $(0, 2)$. Some of the points on the graph are shown below.

x	0	0.5	1	1.5	2
$f(x)$	1	1.25	2	3.25	5

Which of the following is the best approximation for $f'(1)$?

(a) $f'(1) < 2$
(b) $0.5 < f'(1) < 1$
(c) $1.5 < f'(1) < 2.5$
(d) $2.5 < f'(1) < 3.5$
(e) $f'(1) > 2$

30. The graph of the function f'' on the interval $[1, 8]$ is shown in Figure 6.6-2. At what value(s) of t on the open interval $(1, 8)$, if any, does the graph of the function f':

(a) have a point of inflection?
(b) have a relative maximum or minimum?
(c) concave upward?

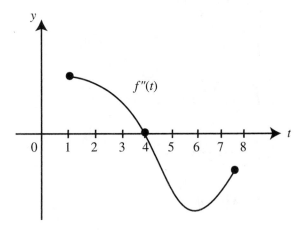

Figure 6.6-2

 31. Evaluate $\displaystyle\lim_{x \to -2} \frac{x^2 - x - 6}{x^2 - 4}$

32. If the position of an object is given by $x = 4\sin(\pi t)$, $y = t^2 - 3t + 1$, find the position of the object at $t = 2$.

33. Find the slope of the tangent line to the curve $r = 3\cos\theta$ when $\theta = \dfrac{\pi}{4}$.

6.7 SOLUTIONS TO PRACTICE PROBLEMS

No calculators except for verifying your results.

1. $\dfrac{x^6}{6} + x^3 - \dfrac{x^2}{2} + x + c$

2. Rewrite:

$$\int (x^{1/2} - x^{-2})dx = \frac{x^{3/2}}{3/2} - \frac{x^{-1}}{-1} + c$$

$$= \frac{2x^{3/2}}{3} + \frac{1}{x} + c$$

3. Let $u = x^4 - 10 \; du = 4x^3 dx$ or $\dfrac{du}{4} = x^3 dx$.

Rewrite: $\displaystyle\int u^5 \frac{du}{4} = \frac{1}{4}\int u^5 du = \left(\frac{1}{4}\right)\frac{u^6}{6} + c$

$$= \frac{(x^4 - 10)^6}{24} + c$$

4. Let $u = x^2 + 1 \Rightarrow (u - 1) = x^2$ and $du = 2x \, dx$
   or $\dfrac{du}{2} = x \, dx$

Rewrite: $\displaystyle\int x^2\sqrt{x^2 + 1}(x \, dx) = \int (u - 1)\sqrt{u} \, \frac{du}{2}$

$$= \frac{1}{2}\int (u - 1)u^{1/2} du = \frac{1}{2}\int (u^{3/2} - u^{1/2})du$$

$$= \frac{1}{2}\left(\frac{u^{5/2}}{5/2} - \frac{u^{3/2}}{3/2}\right) + c = \frac{u^{5/2}}{5} - \frac{u^{3/2}}{3} + c$$

$$= \frac{(x^2 + 1)^{5/2}}{5} - \frac{(x^2 + 1)^{3/2}}{3} + c$$

5. Let $u = x - 1; du = dx$ and $(u + 1) = x$
   Rewrite:

$$\int \frac{(u + 1)^2 + 5}{\sqrt{u}} du = \int \frac{u^2 + 2u + 6}{u^{1/2}} du$$

$$= \int \left(u^{3/2} + 2u^{1/2} + 6u^{-1/2}\right) du$$

$$= \frac{u^{5/2}}{5/2} + \frac{2u^{3/2}}{3/2} + \frac{6u^{1/2}}{1/2} + c$$

$$= \frac{2(x - 1)^{5/2}}{5} + \frac{4(x - 1)^{3/2}}{3}$$
$$+ 12(x - 1)^{1/2} + c$$

6. Let $u = \dfrac{x}{2}; du = \dfrac{1}{2}dx$ or $2du = dx$

Rewrite: $\displaystyle\int \tan u(2 \, du) = 2\int \tan u \, du$

$$= -2\ln|\cos u| + c$$

$$= -2\ln\left|\cos \frac{x}{2}\right| + c$$

7. Let $u = x^2; du = 2x \, dx$ or $\dfrac{du}{2} = x \, dx$

Rewrite: $\displaystyle\int \csc^2 u\frac{du}{2} = \frac{1}{2}\int \csc^2 u \, du$

$$= -\frac{1}{2}\cot u + c$$

$$= -\frac{1}{2}\cot(x^2) + c$$

8. Let $u = \cos x; du = -\sin x \, dx$ or $-du = \sin x \, dx$

Rewrite: $\displaystyle\int \frac{-du}{u^3} = -\int \frac{-du}{u^3} = -\frac{u^{-2}}{-2} + c$

$$= \frac{1}{2\cos^2 x} + c$$

9. Rewrite $\displaystyle\int \frac{1}{(x^2 + 2x + 1) + 9}dx$

$$= \int \frac{1}{(x + 1)^2 + 3^2}dx$$

Let $u = x + 1; du = dx$

Rewrite $\displaystyle\int \frac{1}{u^2 + 3^2}du = \frac{1}{3}\tan^{-1}\left(\frac{u}{3}\right) + c$

$$= \frac{1}{3}\tan^{-1}\left(\frac{x + 1}{3}\right) + c$$

10. Let $u = \dfrac{1}{x}; du = \dfrac{-1}{x^2}dx$ or $-du = \dfrac{1}{x^2}dx$

Rewrite $\displaystyle\int \sec^2 u(-du) = -\int \sec^2 u \, du$

$$= -\tan u + c$$

$$= -\tan\left(\frac{1}{x}\right) + c$$

11. Rewrite $\displaystyle\int e^{(2x + 4x)}dx = \int e^{6x}dx$

Let $u = 6x; du = 6 \, dx$ or $\dfrac{du}{6} = dx$

Rewrite $\displaystyle\int e^u \frac{du}{6} = \frac{1}{6}\int e^u du = \frac{1}{6}e^u + c$

$$= \frac{1}{6}e^{6x} + c$$

12. Let $u = \ln x; du = \dfrac{1}{x}dx$

Rewrite $\displaystyle\int \frac{1}{u}du = \ln|u| + c = \ln|\ln x| + c$

13. Since e^x and $\ln x$ are inverse functions:

$$\int \ln\left(e^{5x + 1}\right) dx = \int (5x + 1)dx$$

$$= \frac{5x^2}{2} + x + c$$

14. Rewrite $\int \left(\dfrac{e^{4x}}{e^x} - \dfrac{1}{e^x}\right) dx = \int \left(e^{3x} - e^{-x}\right) dx$

$$= \int e^{3x} dx - \int e^{-x} dx$$

Let $u = 3x$; $du = 3dx$;

$$\int e^{3x} dx = \int e^u \left(\dfrac{du}{3}\right) = \dfrac{1}{3} e^u + c_1$$

$$= \dfrac{1}{3} e^{3x} + c$$

Let $v = -x$; $dv = -dx$;

$$\int e^{-x} dx = \int e^v (-dv) = e^v + c_2$$

$$= -e^{-x} + c_2$$

Thus $\int e^{3x} dx - \int e^{-x} dx = \dfrac{1}{3} e^{3x} + e^{-x} + c$

Note: c_1 and c_2 are arbitrary constants, and thus $c_1 + c_2 = c$.

15. Rewrite $\int (9 - x^2) x^{1/2} dx = \int \left(9x^{1/2} - x^{5/2}\right) dx$

$$= \dfrac{9x^{3/2}}{3/2} - \dfrac{x^{7/2}}{7/2} + c$$

$$= 6x^{3/2} - \dfrac{2x^{7/2}}{7} + c$$

16. Let $u = 1 + x^{3/2}$; $du = \dfrac{3}{2} x^{1/2} dx$ or

$$\dfrac{2}{3} du = x^{1/2} dx = \sqrt{x}\, dx$$

Rewrite:

$$\int u^4 \left(\dfrac{2}{3} du\right) = \dfrac{2}{3} \int u^4 du = \dfrac{2}{3} \left(\dfrac{u^5}{5}\right) + c$$

$$= \dfrac{2 \left(1 + x^{3/2}\right)^5}{15} + c$$

17. Since $\dfrac{dy}{dx} = e^x + 2$, then $y = \int (e^x + 2) dx$

$$= e^x + 2x + c.$$

The point $(0, 6)$ is on the graph of y. Thus, $6 = e^0 + 2(0) + c \Rightarrow 6 = 1 + c$ or $c = 5$. Therefore, $y = e^x + 2x + 5$.

18. Let $u = e^x$; $du = e^x dx$
Rewrite: $-3 \int \sin(u) du = -3(-\cos u) + c$
$$= 3 \cos (e^x) + c$$

19. Let $u = e^x + e^{-x}$; $du = (e^x - e^{-x}) dx$
Rewrite: $\int \dfrac{1}{u} du = \ln |u| + c = \ln |e^x + e^{-x}| + c$

or $= \ln \left|e^x + \dfrac{1}{e^x}\right| + c$

$$= \ln \left|\dfrac{e^{2x} + 1}{e^x}\right| + c$$

$$= \ln |e^{2x} + 1| - \ln |e^x| + c$$
$$= \ln |e^{2x} + 1| - x + c$$

20. Since $f(x)$ is the antiderivative of $\dfrac{1}{x}$,

$$f(x) = \int \dfrac{1}{x} d = \ln |x| + c.$$

Given $f(1) = 5$; thus $\ln (1) + c = 5 \Rightarrow 0 + c = 5$ or $c = 5$.

Thus, $f(x) = \ln |x| + 5$ and $f(e) = \ln (e) + 5 = 1 + 5 = 6$.

BC 21. Integrate $\int x^2 \sqrt{1 - x}\, dx$ by parts.

Let $u = x^2$, $du = 2x\, dx$, $dv = \sqrt{1 - x}\, dx$, and $v = -\dfrac{2}{3} (1 - x)^{3/2}$. Then $\int x^2 \sqrt{1 - x}\, dx =$

$-\dfrac{2}{3} x^2 (1 - x)^{3/2} + \dfrac{4}{3} \int x(1 - x)^{3/2} dx$. Use parts again with $u = x$, $du = dx$, $dv = (1 - x)^{3/2} dx$,

and $v = -\dfrac{2}{5} (1 - x)^{5/2}$ so that $\int x^2 \sqrt{1 - x}\, dx$

$$= -\dfrac{2}{3} x^2 (1 - x)^{3/2} + \dfrac{4}{3} \left[-\dfrac{2}{5} (1 - x)^{5/2} - \right.$$

$\left. \int -\dfrac{2}{5} (1 - x)^{5/2} dx\right]$. Integrate for

$-\dfrac{2}{3} x^2 (1 - x)^{3/2} - \dfrac{8}{15} x(1-x)^{5/2} - \dfrac{16}{105} x(1-x)^{7/2}$

and simplify to

$$-\dfrac{2}{105} (1 - x)^{3/2} \left[15x^2 + 12x + 8\right]$$

22. For $\int 3x^2 \sin x\, dx$, use integration by parts with $u = 3x^2$, $du = 6x\, dx$, $dv = \sin x\, dx$, and $v = -\cos x$. $\int 3x^2 \sin x\, dx = -3x^2 \cos x +$

$\left[6x \sin x - \int 6 \sin x\, dx\right] = -3x^2 \cos x +$

$6x \sin x + 6 \cos x$

23. Factor the denominator so that

$$\int \dfrac{x\, dx}{x^2 - 3x - 4} = \int \dfrac{x\, dx}{(x - 4)(x + 1)}.$$

Use a partial fraction decomposition,

$\dfrac{x}{(x - 4)(x + 1)} = \dfrac{A}{(x - 4)} + \dfrac{B}{(x + 1)}$ which

implies $Ax + A + Bx - 4B = x$. Solve $A + B = 1$ and $A - 4B$ to find $A = \dfrac{4}{5}$ and $B = \dfrac{1}{5}$. Integrate

$$\int \dfrac{x\, dx}{x^2 - 3x - 4} = \int \dfrac{4/5}{x - 4} dx + \int \dfrac{1/5}{x + 1} dx$$

$$= \dfrac{4}{5} \ln |x - 4| + \dfrac{1}{5} \ln |x + 1| + c$$

$$= \dfrac{1}{5} \ln \left|(x - 4)^4 (x + 1)\right| + c$$

24. Factor $\int \dfrac{dx}{x^2 + x} = \int \dfrac{dx}{x(x+1)}$ and use partial fractions. If $\dfrac{1}{x(x+1)} = \dfrac{A}{x} + \dfrac{B}{(x+1)}$,

$Ax + A + Bx = 1$ and $A = -B = 1$. $\int \dfrac{dx}{x^2 + x}$

$= \int \dfrac{dx}{x} + \int \dfrac{-dx}{x+1} = \ln |x| - \ln |x+1| + c$

$= \ln \left| \dfrac{x}{x+1} \right| + c$

25. Begin with integration by parts, using $u = \ln x$,

$du = \dfrac{dx}{x}$, $dv = \dfrac{dx}{(x+5)^2}$, and $v = \dfrac{-1}{x+5}$.

Then $\int \dfrac{\ln x}{(x+5)^2} dx = \dfrac{-\ln |x|}{x+5} + \int \dfrac{dx}{x(x+5)}$.

Use partial fractions to decompose $\dfrac{1}{x(x+5)}$

$= \dfrac{A}{x} + \dfrac{B}{(x+5)}$. Solve to find $A = -B = \dfrac{1}{5}$.

Then $\int \dfrac{\ln x}{(x+5)^2} dx = \dfrac{-\ln |x|}{x+5} - \dfrac{1}{5} \ln |x| +$

$\dfrac{1}{5} \ln |x+5| = \dfrac{-\ln |x|}{x+5} + \dfrac{1}{5} \ln \left| \dfrac{x+5}{x} \right|$

6.8 SOLUTIONS TO CUMULATIVE REVIEW PROBLEMS

"Calculator" indicates that calculators are permitted.

26. (a) At $t = 4$, speed is 5 which is the greatest on $0 \le t \le 10$.

    (b) The particle is moving to the right when $6 < t < 10$.

27. $V = \dfrac{4}{3}\pi r^3$; $\dfrac{dV}{dt} = \left(\dfrac{4}{3}\right)(3)\pi r^2 \dfrac{dr}{dt} = 4\pi r^2 \dfrac{dr}{dt}$

    If $\dfrac{dV}{dt} = 100 \dfrac{dr}{dt}$, then $100 \dfrac{dr}{dt} = 4\pi r^2 \dfrac{dr}{dt} \Rightarrow 100$

    $= 4\pi r^2$ or $r = \pm \sqrt{\dfrac{25}{\pi}} = \pm \dfrac{5}{\sqrt{\pi}}$.

    Since $r \ge 0$, $r = \dfrac{5}{\sqrt{\pi}}$ meters.

28. Let $u = \ln x$; $du = \dfrac{1}{x} dx$

    Rewrite: $\int u^3 du = \dfrac{u^4}{4} + c = \dfrac{(\ln x)^4}{4} + c$

    $= \dfrac{\ln^4 (x)}{4} + c$

29. Label given points as A, B, C, D, and E. Since $f''(x) > 0 \Rightarrow f$ is concave upward for all x in the interval $[0, 2]$.
    Thus $m_{\overline{BC}} < f'(x) < m_{\overline{CD}}$; $m_{\overline{BC}}$
    $= 1.5$ and $m_{\overline{CD}} = 2.5$
    Therefore $1.5 < f'(1) < 2.5$, choice (c). See Figure 6.8-1.

30. (a) f'' is decreasing on $[1, 6) \Rightarrow f''' < 0 \Rightarrow f'$ is concave downward on $[1, 6)$ and f'' is increasing on $(6, 8] \Rightarrow f'$ is concave upward on $(6, 8]$. Thus, at $x = 6$, f' has a change of concavity. Since f'' exists at $x = 6$ (which

implies there is a tangent to the curve of f' at $x = 6$), f' has a point of inflection at $x = 6$.

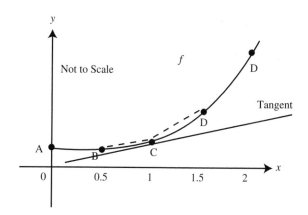

Figure 6.8-1

(b) $f'' > 0$ on $[1, 4] \Rightarrow f'$ is increasing and $f'' < 0$ on $(4, 8] \Rightarrow f'$ is decreasing. Thus at $x = 4$, f' has a relative maximum at $x = 4$. There is no relative minimum.

(c) f'' is increasing on $[6, 8] \Rightarrow f''' > 0 \Rightarrow f'$ is concave upward on $[6, 8]$.

31. $\displaystyle\lim_{x \to -2} \dfrac{x^2 - x - 6}{x^2 - 4} \to \dfrac{0}{0}$

    $= \displaystyle\lim_{x \to -2} \dfrac{2x - 1}{2x} = \dfrac{5}{4}$

32. At $t = 2$, $x = 4\sin(2\pi) = 0$, and $y = 2^2 - 3 \cdot 2 + 1 = -1$, so the position of the object at $t = 2$ is $(0, -1)$.

33. To find the slope of the tangent line to the curve $r = 3\cos\theta$ when $\theta = \dfrac{\pi}{4}$, begin with $x = r\cos\theta$ and $y = r\sin\theta$, and find $\dfrac{dx}{d\theta}$ and $\dfrac{dy}{d\theta}$. $x = 3\cos^2\theta$ so $\dfrac{dx}{d\theta} = -6\cos\theta\sin\theta$.

$y = 3\cos\theta\sin\theta$ so $\dfrac{dy}{d\theta} = 3\cos^2\theta - 3\sin^2\theta$.

Then the slope of the tangent line is

$$\frac{dy}{dx} = \frac{dy}{d\theta}\cdot\frac{d\theta}{dx} = \frac{3\cos^2\theta - 3\sin^2\theta}{-6\cos\theta\sin\theta} = \frac{\cos 2\theta}{-\sin 2\theta}.$$

Evaluate at $\theta = \dfrac{\pi}{4}$ to get $\dfrac{\cos\left(\pi/2\right)}{-\sin\left(\pi/2\right)} = \dfrac{0}{-1} = 0$.

The slope of the tangent line is zero, indicating that the tangent is horizontal.

Chapter 7

Definite Integrals

7.1 RIEMANN SUMS AND DEFINITE INTEGRALS

Main Concepts: *Sigma Notation, Definition of a Riemann Sum, Definition of a Definite Integral, and Properties of Definite Integrals*

Sigma Notation or Summation Notation

$$\sum_{i=1}^{n} a_1 + a_2 + a_3 + \cdots + a_n$$

where i is the index of summation, l is the lower limit and n is the upper limit of summation. (Note: The lower limit may be any non-negative integer $\leq n$.)

Examples

$$\sum_{i=5}^{7} i^2 = 5^2 + 6^2 + 7^2$$

$$\sum_{k=0}^{3} 2k = 2(0) + 2(1) + 2(2) + 2(3)$$

$$\sum_{i=-1}^{3} (2i + 1) = -1 + 1 + 3 + 5 + 7$$

$$\sum_{k=1}^{4} (-1)^k (k) = -1 + 2 - 3 + 4$$

Summation Formulas

If n is a positive integer, then:

1. $\displaystyle\sum_{i=1}^{n} a = an$

2. $\displaystyle\sum_{i=1}^{n} i = \frac{n(n+1)}{2}$

3. $\displaystyle\sum_{i=1}^{n} i^2 = \frac{n(n+1)(2n+1)}{6}$

4. $\displaystyle\sum_{i=1}^{n} i^3 = \frac{n^2(n+1)^2}{4}$

5. $\displaystyle\sum_{i=1}^{n} i^4 = \frac{n(n+1)(6n^3+9n^2+n-1)}{30}$

Example

Evaluate $\displaystyle\sum_{i=1}^{n} \frac{i(i+1)}{n}$

Rewrite $\displaystyle\sum_{i=1}^{n} \frac{i(i+1)}{n}$ as $\displaystyle\frac{1}{n}\sum_{i=1}^{n}(i^2+i) = \frac{1}{n}\left(\sum_{i=1}^{n} i^2 + \sum_{i=1}^{n} i\right)$

$$= \frac{1}{n}\left(\frac{n(n+1)(2n+1)}{6} + \frac{n(n+1)}{2}\right)$$

$$= \frac{1}{n}\left[\frac{n(n+1)(2n+1)+3n(n+1)}{6}\right] = \frac{(n+1)(2n+1)+3(n+1)}{6}$$

$$= \frac{(n+1)[(2n+1)+3]}{6} = \frac{(n+1)(2n+4)}{6}$$

$$= \frac{(n+1)(n+2)}{3}$$

(Note: This question has not appeared in an AP Calculus AB Exam in recent years.)

- Remember: in exponential growth/decay problems, the formulas are $\dfrac{dy}{dx} = ky$ and $y = y_0 e^{kt}$.

Definition of a Riemann Sum

Let f be defined on $[a, b]$ and x_i's be points on $[a, b]$ such that $x_0 = a$, $x_n = b$ and $a < x_1 < x_2 < x_3 \ldots < x_{n-1} < b$. The points a, x_1, x_2, $x_3, \ldots x_{n+1}$, b form a partition of f denoted as Δ on $[a, b]$. Let Δx_i be the length of the ith interval $[x_{i-1}, x_i]$ and c_i be any point in the ith interval. Then the Riemann sum of f for the partition is $\displaystyle\sum_{i=1}^{n} f(c_i)\Delta x_i$.

Example 1

Let f be a continuous function defined on [0, 12] as shown below.

x	0	2	4	6	8	10	12
$f(x)$	3	7	19	39	67	103	147

Find the Riemann sum for $f(x)$ over [0, 12] with 3 subdivisions of equal length and the midpoints of the intervals as c_i's.

Length of an interval $\Delta x_i = \dfrac{12 - 0}{3} = 4$. (See Figure 7.1-1.)

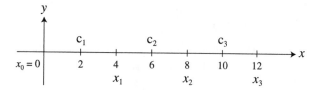

Figure 7.1-1

$$\text{Riemann sum} = \sum_{i=1}^{3} f(c_i)\Delta x_i = f(c_1)\Delta x_1 + f(c_2)\Delta x_2 + f(c_3)\Delta x_3$$

$$= 7(4) + 39(4) + 103(4) = 596$$

The Riemann sum is 596.

Example 2

Find the Riemann sum for $f(x) = x^3 + 1$ over the interval [0, 4] using 4 subdivisions of equal length and the midpoints of the intervals as c_i's. (See Figure 7.1-2.)

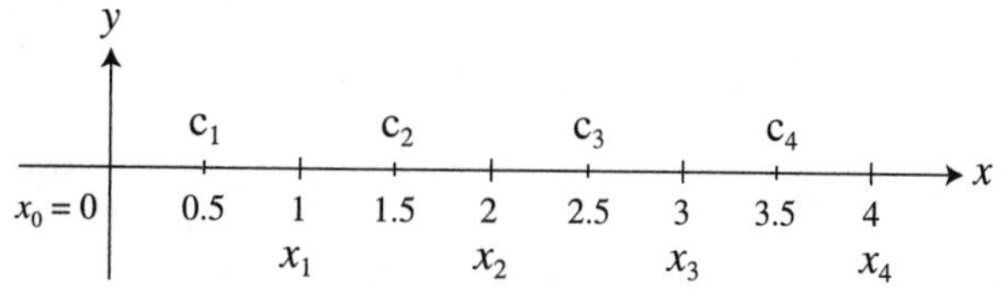

Figure 7.1-2

Length of an interval $\Delta x_i = \dfrac{b - a}{n} = \dfrac{4 - 0}{4} = 1$; $c_i = 0.5 + (i - 1) = i - 0.5$

$$\text{Riemann sum} = \sum_{i=1}^{4} f(c_i)\Delta x_i = \sum_{i=1}^{4} \left[(i - 0.5)^3 + 1\right]1$$

$$= \sum_{i=1}^{4} (i - 0.5)^3 + 1$$

Enter $\sum \left((1 - 0.5)^3 + 1, i, 1, 4\right) = 66$.

The Riemann sum is 66.

Definition of a Definite Integral

Let f be defined on $[a, b]$ with the Riemann sum for f over $[a, b]$ written as $\sum_{i=1}^{n} f(c_i)\Delta x_i$.

If max Δx_i is the length of the largest subinterval in the partition and the $\lim_{\max \Delta x_i \to 0} \sum_{i=1}^{n} f(c_i)\Delta x_i$ exists, then the limit is denoted by:

$$\lim_{\max \Delta x_i \to 0} \sum_{i=1}^{n} f(c_i)\Delta x_i = \int_a^b f(x)dx$$

$\int_a^b f(x)dx$ is the definite integral of f from a to b.

Example 1

Use a midpoint Riemann sum with three subdivisions of equal length to find the approximate value of $\int_0^6 x^2 dx$.

$$\Delta x = \frac{6 - 0}{3} = 2, \ f(x) = x^2$$

midpoints are $x = 1, 3,$ and 5.

$$\int_0^6 x^2 dx \approx f(1)\Delta x + f(3)\Delta x + f(5)\Delta x = 1(2) + 9(2) + 25(2)$$

$$\approx 70.$$

Example 2

Using the limit of the Riemann sum, find $\int_1^5 3x\,dx$.

Using n subintervals of equal lengths, the length of an interval

$$\Delta x_i = \frac{5 - 1}{n} = \frac{4}{n}; x_i = 1 + \left(\frac{4}{n}\right)i$$

$$\int_1^5 3x\,dx = \lim_{\max \Delta x_i \to 0} \sum_{i=1}^{n} f(c_i)\Delta x_i$$

Let $c_i = x_i$; $\max \Delta x_i \to 0 \Rightarrow n \to \infty$

$$\int_1^5 3x\,dx = \lim_{n \to \infty} \sum_{i=1}^{n} f\left(1 + \frac{4i}{n}\right)\left(\frac{4}{n}\right) = \lim_{n \to \infty} \sum_{i=1}^{n} 3\left(1 + \frac{4i}{n}\right)\left(\frac{4}{n}\right)$$

$$= \lim_{n \to \infty} \frac{12}{n} \sum_{i=1}^{n} \left(1 + \frac{4i}{n}\right) = \lim_{n \to \infty} \frac{12}{n}\left(n + \frac{4}{n}\left[n\left(\frac{n+1}{2}\right)\right]\right)$$

$$= \lim_{n \to \infty} \frac{12}{n}(n + 2(n + 1)) = \lim_{n \to \infty} \frac{12}{n}(3n + 2) = \lim_{n \to \infty}\left(36 + \frac{24}{n}\right) = 36$$

Thus $\int_1^5 3x\,dx = 36$.

(Note: This question has not appeared in an AP Calculus AB Exam in recent years.)

Properties of Definite Integrals

1. If f is defined on $[a, b]$, and the limit $\displaystyle\lim_{\max \Delta x_i \to 0} \sum_{i=1}^{n} f(x_i)\Delta x_i$ exists, then f is integrable on $[a, b]$.

2. If f is continuous on $[a, b]$, then f is integrable on $[a, b]$.

   If $f(x)$, $g(x)$, and $h(x)$ are integrable on $[a, b]$, then

3. $\displaystyle\int_{a}^{a} f(x)dx = 0$

4. $\displaystyle\int_{a}^{b} f(x)dx = -\int_{b}^{a} f(x)$

5. $\displaystyle\int_{a}^{b} cf(x)dx = c\int_{a}^{b} f(x)dx$ when c is a constant.

6. $\displaystyle\int_{a}^{b} [f(x) \pm g(x)]\,dx = \int_{a}^{b} f(x)dx \pm \int_{a}^{b} g(x)dx$

7. $\displaystyle\int_{a}^{b} f(x)dx \geq 0$ provided $f(x) \geq 0$ on $[a, b]$

8. $\displaystyle\int_{a}^{b} f(x)dx \geq \int_{a}^{b} g(x)dx$ provided $f(x) \geq g(x)$ on $[a, b]$

9. $\displaystyle\left|\int_{a}^{b} f(x)dx\right| \leq \int_{a}^{b} |f(x)|\,dx$

10. $\displaystyle\int_{a}^{b} g(x)dx \leq \int_{a}^{b} f(x)dx \leq \int_{a}^{b} h(x)dx$; provided $g(x) \leq f(x) \leq h(x)$ on $[a, b]$

11. $m(b - a) \leq \displaystyle\int_{a}^{b} f(x)dx \leq M(b - a)$; provided $m \leq f(x) \leq M$ on $[a, b]$

12. $\displaystyle\int_{a}^{c} f(x)dx = \int_{a}^{b} f(x)dx + \int_{b}^{c} f(x)dx$; provided $f(x)$ is integrable on an interval containing a, b, c.

Examples

1. $\displaystyle\int_{\pi}^{\pi} \cos x\,dx = 0$

2. $\displaystyle\int_{1}^{5} x^4 dx = -\int_{5}^{1} x^4 dx$

3. $\displaystyle\int_{-2}^{7} 5x^2 dx = 5\int_{-2}^{7} x^2 dx$

4. $\displaystyle\int_{0}^{4} \left(x^3 - 2x + 1\right)dx = \int_{0}^{4} x^3 dx - 2\int_{0}^{4} x\,dx + \int_{0}^{4} 1\,dx$

5. $\displaystyle\int_{1}^{5} \sqrt{x}\,dx = \int_{1}^{3} \sqrt{x}\,dx + \int_{3}^{5} \sqrt{x}\,dx$

   Note: Or $\displaystyle\int_{1}^{3} \sqrt{x}\,dx = \int_{1}^{5} \sqrt{x}\,dx + \int_{5}^{3} \sqrt{x}\,dx$

   $\displaystyle\int_{a}^{c} = \int_{a}^{b} + \int_{b}^{c}$ a, b, c do not have to be arranged from smallest to largest.

The remaining properties are best illustrated in terms of the area under the curve of the function as discussed in the next section.

> • Don't forget that $\int_0^{-3} f(x)dx = -\int_{-3}^0 f(x)dx.$

7.2 FUNDAMENTAL THEOREMS OF CALCULUS

Main Concepts: *The First Fundamental Theorem of Calculus, The Second Fundamental Theorem of Calculus*

First Fundamental Theorem of Calculus

If f is continuous on $[a, b]$ and F is an antiderivative of f on $[a, b]$, then

$$\int_a^b f(x)dx = F(b) - F(a).$$

Note $F(b) - F(a)$ is often denoted as $F(x)\Big]_a^b$.

Example 1

Evaluate $\int_0^2 \left(4x^3 + x - 1\right)dx$

$$\int_0^2 \left(4x^3 + x - 1\right)dx = \frac{4x^4}{4} + \frac{x^2}{2} - x\Bigg]_0^2 = x^4 + \frac{x^2}{2} - x\Bigg]_0^2$$

$$= \left(2^4 + \frac{2^2}{2} - 2\right) - (0) = 16$$

Example 2

Evaluate $\int_{-\pi}^{\pi} \sin x\,dx$

$$\int_{-\pi}^{\pi} \sin x\,dx = -\cos x\Bigg]_{-\pi}^{\pi} = [-\cos \pi] - [-\cos(-\pi)]$$

$$= [-(-1)] - [-(-1)] = (1) - (1) = 0$$

Example 3

If $\int_{-2}^k (4x + 1)dx = 30$, $k > 0$, find k.

$$\int_{-2}^k (4x + 1)dx = 2x^2 + x\Bigg]_{-2}^k = \left(2k^2 + k\right) - \left(2(-2)^2 - 2\right)$$

$$= 2k^2 + k - 6$$

Set $2k^2 + k - 6 = 30 \Rightarrow 2k^2 + k - 36 = 0$

$$\Rightarrow (2k + 9)(k - 4) = 0 \text{ or } k = -\frac{9}{2} \text{ or } k = 4.$$

Since $k > 0$, $k = 4$.

Example 4

If $f'(x) = g(x)$ and g is a continuous function for all real values of x, express $\int_2^5 g(3x)dx$ in terms of f.

Let $u = 3x$; $du = 3dx$ or $\dfrac{du}{3} = dx$

$$\int g(3x)dx = \int g(u)\frac{du}{3} = \frac{1}{3}\int g(u)du = \frac{1}{3}f(u) + c$$

$$= \frac{1}{3}f(3x) + c$$

$$\int_2^5 g(3x)dx = \frac{1}{3}f(3x)\Big]_2^5 = \frac{1}{3}f(3(5)) - \frac{1}{3}f(3(2))$$

$$= \frac{1}{3}f(15) - \frac{1}{3}f(6).$$

Example 5

Evaluate $\displaystyle\int_0^4 \frac{1}{x-1}dx$

Cannot evaluate using the First Fundamental Theorem of Calculus since $f(x) = \dfrac{1}{x-1}$ is discontinuous at $x = 1$.

Example 6

Using a graphing calculator, evaluate $\displaystyle\int_{-2}^2 \sqrt{4 - x^2}dx$.

Using a TI-89 graphing calculator, enter $\int \left(\sqrt{(4 - x^{\wedge}2)},\ x,\ -2,\ 2\right)$ and obtain 2π.

Second Fundamental Theorem of Calculus

If f is continuous on $[a, b]$ and $F(x) = \displaystyle\int_a^x f(t)dt$, then $F'(x) = f(x)$ at every point x in $[a, b]$.

Example 1

Evaluate $\displaystyle\int_{\pi/4}^\pi \cos(2t)\, dt$

Let $u = 2t$; $du = 2dt$ or $\dfrac{du}{2} = dt$

$$\int \cos(2t)dt = \int \cos u \frac{du}{2} = \frac{1}{2}\int \cos u\, du$$

$$= \frac{1}{2}\sin u + c = \frac{1}{2}\sin(2t) + c$$

$$\int_{\pi/4}^{x} \cos(2t)dt = \frac{1}{2}\sin(2t)\Big]_{\pi/4}^{x}$$

$$= \frac{1}{2}\sin(2x) - \frac{1}{2}\sin\left(2\left(\frac{\pi}{4}\right)\right)$$

$$= \frac{1}{2}\sin(2x) - \frac{1}{2}\sin\left(\frac{\pi}{2}\right)$$

$$= \frac{1}{2}\sin(2x) - \frac{1}{2}$$

Example 2

If $h(x) = \int_{3}^{x} \sqrt{t+1}\, dt$ find $h'(8)$.

$h'(x) = \sqrt{x+1};\ h'(8) = \sqrt{8+1} = 3.$

Example 3

Find $\frac{dy}{dx}$; if $y = \int_{1}^{2x} \frac{1}{t^3} dt.$

Let $u = 2x$; then $\frac{dy}{dx} = 2.$

Rewrite: $y = \int_{1}^{u} \frac{1}{t^3} dt$

$$\frac{dy}{dx} = \frac{dy}{du} \cdot \frac{du}{dx} = \frac{1}{u^3} \cdot (2) = \frac{1}{(2x)^3} \cdot 2 = \frac{1}{4x^3}$$

Example 4

Find $\frac{dy}{dx}$; if $y = \int_{x^2}^{1} \sin t\, dt.$

Rewrite: $y = -\int_{1}^{x^2} \sin t\, dt$

Let $u = x^2$; then $\frac{du}{dx} = 2x$

Rewrite: $y = -\int_{1}^{u} \sin t\, dt$

$$\frac{dy}{dx} = \frac{dy}{du} \cdot \frac{du}{dx} = (-\sin u)2x = (-\sin x^2)2x$$

$$= -2x\sin(x^2)$$

Example 5

Find $\dfrac{dy}{dx}$; if $y = \displaystyle\int_x^{x^2} \sqrt{e^t + 1}\, dt$.

$$y = \int_x^0 \sqrt{e^t + 1}\, dt + \int_0^{x^2} \sqrt{e^t + 1}\, dt = -\int_0^x \sqrt{e^t + 1}\, dt + \int_0^{x^2} \sqrt{e^t + 1}\, dt$$

$$= \int_0^{x^2} \sqrt{e^t + 1}\, dt - \int_0^x \sqrt{e^t + 1}\, dt$$

Since $y = \displaystyle\int_0^{x^2} \sqrt{e^t + 1}\, dt - \int_0^x \sqrt{e^t + 1}\, dt$

$$\frac{dy}{dx} = \left(\frac{d}{dx}\int_0^{x^2} \sqrt{e^t + 1}\, dt\right) - \left(\frac{d}{dx}\int_0^x \sqrt{e^t + 1}\, dt\right)$$

$$= \left(\sqrt{e^{x^2} + 1}\right)\frac{d}{dx}(x^2) - \left(\sqrt{e^x + 1}\right)$$

$$= 2x\sqrt{e^{x^2} + 1} - \sqrt{e^x + 1}$$

Example 6

$F(x) = \displaystyle\int_1^x (t^2 - 4)dt$, integrate to find $F(x)$ and then differentiate to find $F'(x)$.

$$F(x) = \frac{t^3}{3} - 4t\Big]_1^x = \left(\frac{x^3}{3} - 4x\right) - \left(\frac{1^3}{3} - 4(1)\right)$$

$$= \frac{x^3}{3} - 4x + \frac{11}{3}$$

$$F'(x) = 3\left(\frac{x^2}{3}\right) - 4 = x^2 - 4.$$

7.3 EVALUATING DEFINITE INTEGRALS

Main Concepts: *Definite Integrals Involving Algebraic Functions;*
Definite Integrals Involving Absolute Volume; Definite
Integrals Involving Trigonometric, Logarithmic, and
Exponential Functions; Definite Integrals Involving
Odd and Even Functions

• If the problem asks you to determine the concavity of f' (not f), you need to know if f'' is increasing or decreasing or if f''' is positive or negative.

Definite Integrals Involving Algebraic Functions

Example 1

Evaluate: $\displaystyle\int_1^4 \frac{x^3 - 8}{\sqrt{x}}\,dx$

Rewrite: $\displaystyle\int_1^4 \frac{x^3 - 8}{\sqrt{x}}\,dx = \int_1^4 \left(x^{5/2} - 8x^{-1/2}\right)dx$

$$= \frac{x^{7/2}}{7/2} - \frac{8x^{1/2}}{1/2}\Bigg]_1^4 = \frac{2x^{7/2}}{7} - 16x^{1/2}\Bigg]_1^4$$

$$= \left(\frac{2(4)^{7/2}}{7} - 16(4)^{1/2}\right) - \left(\frac{2(1)^{7/2}}{7} - 16(1)^{1/2}\right) = \frac{142}{7}$$

Verify your result with a calculator.

Example 2

Evaluate: $\displaystyle\int_0^2 x(x^2 - 1)^7\,dx$

Begin by evaluating the indefinite integral $\displaystyle\int x(x^2 - 1)^7\,dx$.

Let $u = x^2 - 1;\ du = 2x\,dx$ or $\dfrac{du}{2} = x\,dx$

Rewrite: $\displaystyle\int \frac{u^7\,du}{2} = \frac{1}{2}\int u^7\,du = \frac{1}{2}\left(\frac{u^8}{8}\right) + c = \frac{u^8}{16} + c = \frac{(x^2 - 1)^8}{16} + c$

Thus the definite integral $\displaystyle\int_0^2 x(x^2 - 1)^7\,dx = \frac{(x^2 - 1)^8}{16}\Bigg]_0^2$

$$= \frac{(2^2 - 1)^8}{16} - \frac{(0^2 - 1)^8}{16} = \frac{3^8}{16} - \frac{(-1)^8}{16} = \frac{3^8 - 1}{16} = 410$$

Verify your result with a calculator.

Example 3

Evaluate $\displaystyle\int_{-8}^{-1}\left(\sqrt[3]{y} + \frac{1}{\sqrt[3]{y}}\right)dy$

Rewrite: $\displaystyle\int_{-8}^{-1}\left(y^{1/3} + \frac{1}{y^{1/3}}\right)dy = \int_{-8}^{-1}\left(y^{1/3} + y^{-1/3}\right)dy$

$$= \frac{y^{4/3}}{4/3} + \frac{y^{2/3}}{2/3}\Bigg]_{-8}^{-1} = \frac{3y^{4/3}}{4} + \frac{3y^{2/3}}{2}\Bigg]_{-8}^{-1}$$

$$= \left(\frac{3(-1)^{4/3}}{4} + \frac{3(-1)^{2/3}}{2}\right) - \left(\frac{3(-8)^{4/3}}{4} + \frac{3(-8)^{2/3}}{2}\right)$$

$$= \left(\frac{3}{4} + \frac{3}{2}\right) - (12 + 6) = \frac{-63}{4}$$

Verify your result with a calculator.

Definite Integrals Involving Absolute Value

Example 1

Evaluate: $\displaystyle\int_{1}^{4} |3x - 6|\,dx$

Set $3x - 6 = 0$; $x = 2$; Thus $|3x - 6| = \begin{cases} 3x - 6 & \text{if } x \geq 2 \\ -(3x - 6) & \text{if } x < 2 \end{cases}$

Rewrite Integral:

$$\int_{1}^{4} |3x - 6|\,dx = \int_{1}^{2} -(3x - 6)\,dx + \int_{2}^{4} (3x - 6)\,dx$$

$$= \left[\frac{-3x^2}{2} + 6x\right]_{1}^{2} + \left[\frac{3x^2}{2} - 6x\right]_{2}^{4}$$

$$= \left(\frac{-3(2)^2}{2} - 6(2)\right) - \left(\frac{-3(1)^2}{2} - 6(1)\right)$$

$$+ \left(\frac{3(4)^2}{2} - 6(4)\right) - \left(\frac{3(2)^2}{2} - 6(2)\right)$$

$$= (-6 + 12) - \left(-\frac{3}{2} + 6\right) + (24 - 24) - (6 - 12)$$

$$= 6 - 4\frac{1}{2} + 0 + 6 = \frac{15}{2}$$

Verify your result with a calculator.

Example 2

Evaluate $\displaystyle\int_{0}^{4} |x^2 - 4|\,dx$

Set $x^2 - 4 = 0$; $x = \pm 2$

Thus $|x^2 - 4| = \begin{cases} x^2 - 4 & \text{if } x \geq 2 \text{ or } x \leq -2 \\ -(x^2 - 4) & \text{if } -2 < x < 2 \end{cases}$

Thus $\displaystyle\int_{0}^{4} |x^2 - 4|\,dx = \int_{0}^{2} -(x^2 - 4)\,dx + \int_{2}^{4} (x^2 - 4)\,dx$

$$= \left[\frac{-x^3}{3} + 4x\right]_{0}^{2} + \left[\frac{x^3}{3} - 4x\right]_{2}^{4}$$

$$= \left(\frac{-2^3}{3} + 4(2)\right) - (0) + \left(\frac{4^3}{3} - 4(4)\right) - \left(\frac{2^3}{3} - 4(2)\right)$$

$$= \left(\frac{-8}{3} + 8\right) + \left(\frac{64}{3} - 16\right) - \left(\frac{8}{3} - 8\right) = 16$$

Verify your result with a calculator.

> • You are not required to clear the memories in your calculator for the exam.

Definite Integrals Involving Trigonometric, Logarithmic, and Exponential Functions

Example 1

Evaluate $\displaystyle\int_0^x (x + \sin x)dx$

Rewrite: $\displaystyle\int_0^x (x + \sin x)dx = \frac{x^2}{2} - \cos x \Big]_0^\pi = \left(\frac{\pi^2}{2} - \cos \pi\right) - (0 - \cos 0)$

$$= \frac{\pi^2}{2} + 1 + 1 = \frac{\pi^2}{2} + 2$$

Verify your result with a calculator.

Example 2

Evaluate $\displaystyle\int_{\pi/4}^{\pi/2} \csc^2(3t)dt$

Let $u = 3t; du = 3dt$ or $\dfrac{du}{3} = dt$

Rewrite the indefinite integral : $\displaystyle\int \csc^2 u \frac{du}{3} = -\frac{1}{3}\cot u + c$

$$= -\frac{1}{3}\cot(3t) + c$$

$$\int_{\pi/4}^{\pi/2} \csc^2(3t)dt = -\frac{1}{3}\cot(3t)\Big]_{\pi/4}^{\pi/2}$$

$$= -\frac{1}{3}\left[\cot\left(\frac{3\pi}{2}\right) - \cot\left(\frac{3\pi}{4}\right)\right]$$

$$= -\frac{1}{3}[0 - (-1)] = -\frac{1}{3}$$

Verify your result with a calculator.

Example 3

Evaluate: $\displaystyle\int_1^e \frac{\ln t}{t}dt$

Let $u = \ln t,\ du = \dfrac{1}{t}dt$

Rewrite: $\int \dfrac{\ln t}{t} dt = \int u\, du = \dfrac{u^2}{2} + c = \dfrac{(\ln t)^2}{2} + c$

$\displaystyle\int_1^e \dfrac{\ln t}{t} dt = \dfrac{(\ln t)^2}{2}\Bigg]_1^e = \dfrac{(\ln e)^2}{2} - \dfrac{(\ln 1)^2}{2}$

$$= \dfrac{1}{2} - 0 = \dfrac{1}{2}$$

Verify your result with a calculator.

Example 4

Evaluate: $\displaystyle\int_{-1}^2 xe^{(x^2+1)} dx$

Let $u = x^2 + 1; du = 2x\, dx$ or $\dfrac{dx}{2} = x\, dx$

$\displaystyle\int xe^{(x^2+1)} dx = \int e^u \dfrac{du}{2} = \dfrac{1}{2}e^u + c = \dfrac{1}{2}e^{(x^2+2)} + c$

Rewrite: $\displaystyle\int_{-1}^2 xe^{(x^2+1)} dx = \dfrac{1}{2}e^{(x^2+1)}\Bigg]_{-1}^2 = \dfrac{1}{2}e^5 - \dfrac{1}{2}e^2 = \dfrac{1}{2}e^2\left(e^3 - 1\right)$

Verify your result with a calculator.

Definite Integrals Involving Odd and Even Functions

If f is an even function, that is, $f(-x) = f(x)$, and is continuous on $[-a, a]$, then

$$\int_{-a}^a f(x)dx = 2\int_0^a f(x)dx$$

If f is an odd function, that is, $F(x) = -f(-x)$, and is continuous on $[-a, a]$ then

$$\int_{-a}^a f(x)dx = 0$$

Example 1

Evaluate: $\displaystyle\int_{-\pi/2}^{\pi/2} \cos x\, dx$

Since $f(x) = \cos x$ is an even function,

$\displaystyle\int_{-\pi/2}^{\pi/2} \cos x\, dx = 2\int_0^{\pi/2} \cos x\, dx = 2\,[\sin x]_0^{\pi/2} = 2\left[\sin\left(\dfrac{\pi}{2}\right) - \sin(0)\right]$

$$= 2(1 - 0) = 2$$

Verify your result with a calculator.

Example 2

Evaluate: $\displaystyle\int_{-3}^3 \left(x^4 - x^2\right)dx$

Since $f(x) = x^4 - x^2$ is an even function, i.e., $f(-x) = f(x)$, thus

$$\int_{-3}^{3} \left(x^4 - x^2\right)dx = 2\int_{0}^{3} \left(x^4 - x^2\right)dx = 2\left[\frac{x^5}{5} - \frac{x^3}{3}\right]_{0}^{3}$$

$$= 2\left[\left(\frac{3^5}{5} - \frac{3^3}{3}\right) - 0\right] = \frac{396}{5}$$

Verify your result with a calculator.

Example 3

Evaluate: $\displaystyle\int_{-\pi}^{\pi} \sin x \, dx$

Since $f(x) = \sin x$ is an odd function, i.e., $f(-x) = -f(x)$, thus

$$\int_{-\pi}^{\pi} \sin x \, dx = 0$$

Verify your result algebraically.

$$\int_{-\pi}^{\pi} \sin x \, dx = -\cos x\Big]_{-\pi}^{\pi} = (-\cos \pi) - [-\cos(-\pi)]$$

$$= [-(-1)] - [-(1)] = (1) - (1) = 0$$

You can also verify the result with a calculator.

Example 4

If $\displaystyle\int_{-k}^{k} f(x)dx = 2\int_{0}^{k} f(x)dx$ for all values of k, then which of the following could be the graph of f? See Figure 7.3-1.

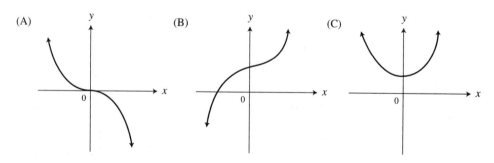

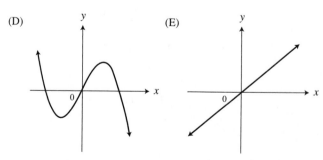

Figure 7.3-1

$$\int_{-k}^{k} f(x)dx = \int_{-k}^{0} f(x)dx + \int_{0}^{k} f(x)dx$$

Since $\int_{-k}^{k} f(x)dx = 2\int_{0}^{k} f(x)dx$, then $\int_{0}^{k} f(x)dx = \int_{-k}^{0} f(x)dx$

Thus f is an even function. Choice (c).

7.4 IMPROPER INTEGRALS

Main Concepts: *Infinite Intervals of Integration, Infinite Discontinuities*

Infinite Intervals of Integration

Improper integrals are integrals with infinite intervals of integration or infinite discontinuities within the interval of integration. For infinite intervals of integration,

$\int_{a}^{\infty} f(x)dx = \lim\limits_{l \to \infty} \int_{a}^{l} f(x)dx$ and $\int_{-\infty}^{b} f(x)dx = \lim\limits_{l \to -\infty} \int_{1}^{b} f(x)dx$. If the limit exists, the integral converges. $\int_{-\infty}^{\infty} f(x)dx = \int_{-\infty}^{c} f(x)dx + \int_{c}^{\infty} f(x)dx$ for some value c.

Example 1

Evaluate $\int_{1}^{\infty} \frac{1}{x} dx$

$\int_{1}^{\infty} \frac{1}{x} dx = \lim\limits_{k \to \infty} \int_{1}^{k} \frac{1}{x} dx = \lim\limits_{k \to \infty} \left[\ln x\right]_{1}^{k} = \lim\limits_{k \to \infty} \left(\ln k\right) = \infty$ so the integral diverges.

Example 2

Evaluate $\int_{-\infty}^{\infty} xe^{-x^2} dx$.

Since both limits of integration are infinite, consider the sum of the improper integrals $\int_{-\infty}^{\infty} xe^{-x^2} dx = \int_{-\infty}^{0} xe^{-x^2} dx + \int_{0}^{\infty} xe^{-x^2} dx$. This sum is the sum of the limits $\lim\limits_{k \to -\infty} \int_{k}^{0} xe^{-x^2} dx + \lim\limits_{c \to \infty} \int_{0}^{c} xe^{-x^2} dx = \lim\limits_{k \to -\infty} \left[-\frac{1}{2}e^{-x^2}\right]_{k}^{0} + \lim\limits_{c \to \infty} \left[-\frac{1}{2}e^{-x^2}\right]_{0}^{c}$

$= \lim\limits_{k \to -\infty} \left[-\frac{1}{2} + \frac{1}{2}e^{-k^2}\right] + \lim\limits_{c \to \infty} \left[-\frac{1}{2}e^{-c^2} + \frac{1}{2}\right] = -\frac{1}{2} + \frac{1}{2} = 0$. Since the limit exists, the integral converges and $\int_{-\infty}^{\infty} xe^{-x^2} dx = 0$.

Infinite Discontinuities

If the function has an infinite discontinuity at one of the limits of integration, then $\int_{a}^{b} f(x)dx = \lim\limits_{l \to b^-} \int_{a}^{l} f(x)dx$ or $\int_{a}^{b} f(x)dx = \lim\limits_{l \to a^+} \int_{l}^{b} f(x)dx$. If an infinite discontinuity occurs at $x = c$ within the interval of integration (a, b), then the integral can be broken into sections at the discontinuity and the sum of the two improper integrals can be found. $\int_a^b f(x)dx = \int_a^c f(x)dx + \int_c^b f(x)dx = \lim_{k \to c^-} \int_a^k f(x)dx + \lim_{l \to c^+} \int_l^b f(x)dx$

Example:

Evaluate $\int_0^{\pi/2} \dfrac{\cos x}{\sqrt{1 - \sin x}} dx$

Since $f(x) = \dfrac{\cos x}{\sqrt{1 - \sin x}}$ has an infinite discontinuity at $x = \dfrac{\pi}{2}$, the integral is improper. Evaluate $\int_0^{\pi/2} \dfrac{\cos x}{\sqrt{1 - \sin x}} dx = \lim_{k \to \pi/2^-} \int_0^k \dfrac{\cos x}{\sqrt{1 - \sin x}} dx = \lim_{k \to \pi/2^-} \left[-2\sqrt{1 - \sin x} \right]_0^k = \lim_{k \to \pi/2^-} \left[-2\sqrt{1 - \sin k} + 2 \right] = 2$. Since the limit exists, $\int_0^{\pi/2} \dfrac{\cos x}{\sqrt{1 - \sin x}} dx = 2$.

7.5 RAPID REVIEW

1. Evaluate: $\int_{\pi/2}^x \cos t \, dt$.

   Answer: $\sin t]_{x/2}^x = \sin x - \sin\left(\dfrac{\pi}{2}\right) = \sin x - 1$.

2. Evaluate $\int_0^1 \dfrac{1}{x + 1} dx$

   Answer: $\ln(x + 1)]_0^1 = \ln 2 - \ln 1 = \ln 2$.

3. If $G(x) = \int_0^x (2t + 1)^{3/2} dt$, find $G'(4)$.

   Answer: $G'(x) = (2x + 1)^{3/2}$ and $G'(4) = 9^{3/2} = 27$.

4. If $\int_1^k 2x \, dx = 8$, find k.

   Answer: $x^2]_1^k = 8 \Rightarrow k^2 - 1 = 8 \Rightarrow k = \pm 3$.

5. If $G(x)$ is a antiderivative of $(e^x + 1)$ and $G(0) = 0$, find $G(1)$.

   Answer: $G(x) = e^x + x + c$
   $G(0) = e^0 + 0 + c = 0 \Rightarrow c = -1$.
   $G(1) = e^1 + 1 - 1 = e$.

6. If $G'(x) = g(x)$, express $\int_0^2 g(4x)dx$ in terms of $G(x)$.

   Answer: Let $u = 4x$; $\dfrac{du}{4} = dx$.
   $\int g(u)\dfrac{du}{4} = \dfrac{1}{4}G(u)$. Thus $\int_0^2 g(4x)dx = \dfrac{1}{4}G(4x)\Big]_0^2 = \dfrac{1}{4}[G(8) - G(0)]$.

7. $\int_1^\infty \dfrac{dx}{x^2}$

   Answer: $\int_1^\infty \dfrac{dx}{x^2} = \lim_{n \to \infty} \int_1^n \dfrac{dx}{x^2} = \lim_{n \to \infty} \left[\dfrac{-1}{x}\right]_1^n = \lim_{n \to \infty} \left[\dfrac{-1}{n} + 1\right] = 1$.

8. $\int_0^1 \dfrac{dx}{\sqrt{x}}$

Answer: $\int_0^1 \dfrac{dx}{\sqrt{x}} = \lim_{k \to 0^+} \int_k^1 \dfrac{dx}{\sqrt{x}} = \lim_{k \to 0^+} \left[2\sqrt{x}\right]_k^1 = \lim_{k \to 0^+} \left[2 - 2\sqrt{k}\right] = 2.$

7.6 PRACTICE PROBLEMS

Part A—The use of a calculator is not allowed.

Evaluate the following definite integrals.

1. $\int_{-1}^0 (1 + x - x^3)dx$

2. $\int_6^{11} (x - 2)^{1/2}\, dx$

3. $\int_1^3 \dfrac{t}{t+1}\, dt$

4. $\int_0^6 |x - 3|dx$

5. If $\int_0^k (6x - 1)dx = 4$, find k.

6. $\int_0^\pi \dfrac{\sin x}{\sqrt{1 + \cos x}}dx$

7. If $f'(x) = g(x)$ and g is a continuous function for all real values of x, express $\int_1^2 g(4x)dx$ in terms of f.

8. $\int_{\ln 2}^{\ln 3} 10e^x\, dx$

9. $\int_e^{e^2} \dfrac{1}{t+3}dt$

10. If $f(x) = \int_{-\pi/4}^x \tan^2(t)dt$, find $f'\left(\dfrac{\pi}{6}\right)$

11. $\int_{-1}^1 4xe^{x^2}\, dx$

12. $\int_{-\pi}^\pi \left(\cos x = x^2\right)dx$

Part B—Calculators are permitted.

13. Find k if $\int_0^2 \left(x^3 + k\right)dx = 10$

14. Evaluate $\int_{-1.2}^{3.3} 2\theta \cos\theta\, d\theta$ to the nearest 100th.

15. If $y = \int_1^{x^3} \sqrt{t^2 + 1}\, dt$, find $\dfrac{dy}{dx}$.

16. Use a midpoint Riemann sum with four subdivisions of equal length to find the approximate value of $\int_0^2 \left(x^3 + 1\right)dx$.

17. Given $\int_{-2}^2 g(x)dx = 8$ and $\int_0^2 g(x)dx = 3$ find

    (a) $\int_{-2}^0 g(x)dx$

    (b) $\int_2^{-2} g(x)dx$

    (c) $\int_0^{-2} 5g(x)dx$

    (d) $\int_{-2}^2 2g(x)dx$

18. Evaluate: $\int_0^{1/2} \dfrac{dx}{\sqrt{1 - x^2}}$

19. Find $\dfrac{dy}{dx}$ if $y = \int_{\cos x}^{\sin x} (2t + 1)dt$

20. Let f be a continuous function defined on [0, 30] with selected values as shown below:

x	0	5	10	15	20	25	30
$f(x)$	1.4	2.6	3.4	4.1	4.7	5.2	5.7

Use a midpoint Riemann sum with three subdivisions of equal length to find the approximate value of $\int_0^{30} f(x)\,dx$.

21. $\int_0^{\infty} e^{-x}\,dx$

22. $\int_{-\infty}^0 \dfrac{dx}{(4-x)^2}$

23. $\int_0^1 \ln x\,dx$

24. $\int_{-2}^2 \dfrac{dx}{\sqrt{4-x^2}}$

25. $\int_{-1}^8 \dfrac{dx}{\sqrt[3]{x}}$

7.7 CUMULATIVE REVIEW PROBLEMS

26. Evaluate $\displaystyle\lim_{x \to -\infty} \dfrac{\sqrt{x^2-4}}{3x-9}$

27. Find $\dfrac{dy}{dx}$ at $x=3$ if $y=\ln|x^2-4|$.

28. The graph of f', the derivative of f, $-6 \le x \le 8$ is shown in Figure 7.7-1.

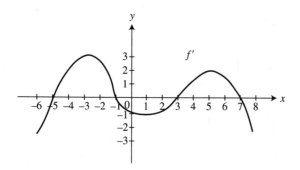

Figure 7.7-1

(a) Find all values of x such that f attains a relative maximum or a relative minimum.
(b) Find all values of x such that f is concave upward.
(c) Find all values of x such that f has a change of concavity.

29. (Calculator) Given the equation $9x^2 + 4y^2 - 18x + 16y = 11$, find the points on the graph

where the equation has a vertical or horizontal tangent.

30. (Calculator) Two corridors, one 6 feet wide and another 10 feet wide meet at a corner. See Figure 7.7-2. What is the maximum length of a pipe of negligible thickness that can be carried horizontally around the corner?

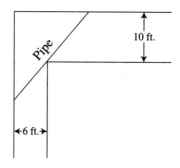

Figure 7.7-2

31. Evaluate $\displaystyle\lim_{x \to -1} \dfrac{1 + \cos \pi x}{x^2 - 1}$

32. Determine the speed of an object moving along the path described by $x = 3 - 2t^2$, $y = t^2 + 1$ when $t = \dfrac{1}{2}$.

33. $\displaystyle\int 2x\sqrt{x+3}\,dx$

7.8 SOLUTIONS TO PRACTICE PROBLEMS

Part A—No calculators.

1. $\displaystyle\int_{-1}^0 \left(1 + x - x^3\right)dx = x + \dfrac{x^2}{2} - \dfrac{x^4}{4}\bigg]_{-1}^0$

$= 0 - \left[(-1) + \dfrac{(-1)^2}{2} - \dfrac{(-1)^4}{4}\right]$

$= \dfrac{3}{4}$

2. Let $u = x - 2 \ du = dx$

$$\int (x-2)^{1/2} dx = \int u^{1/2} du = \frac{2u^{1/2}}{3} + c$$

$$= \frac{2}{3}(x-2)^{3/2} + c$$

Thus $\int_6^{11} (x-2)^{1/2} dx = \frac{2}{3} (x-2)^{3/2} \Big]_6^{11}$

$$= \frac{2}{3} \Big[(11-2)^{3/2}$$

$$-(6-2)^{3/2} \Big]$$

$$= \frac{2}{3}(27-8) = \frac{38}{3}$$

3. Let $u = t + 1$; $du = dt$ and $t = u - 1$.

Rewrite: $\int \frac{t}{t+1} dt = \int \frac{u-1}{u} du$

$$= \int \left(1 - \frac{1}{u} \right) du$$

$$= u - \ln|u| + c$$

$$= t + 1 - \ln|t+1| + c$$

$$\int_1^3 \frac{t}{t+1} dt = \big[t + 1 - \ln|t+1| \big]_1^3$$

$$= \big[(3) + 1 - \ln|3+1| \big]$$

$$- \big((1) + 1 - \ln|1+1| \big)$$

$$= 4 - \ln 4 - 2 + \ln 2$$

$$= 2 - \ln 4 + \ln 2$$

$$= 2 - \ln(2)^2 + \ln 2$$

$$= 2 - 2\ln 2 + \ln 2$$

$$= 2 - \ln 2$$

4. Set $x - 3 = 0$; $x = 3$.

$$|x-3| = \begin{cases} (x-3) \text{ if } x \geq 3 \\ -(x-3) \text{ if } x < 3 \end{cases}$$

$$\int_0^6 |x-3| \, dx = \int_0^3 -(x-3) dx + \int_3^6 (x-3) dx$$

$$= \left[\frac{-x^2}{2} + 3x \right]_0^3 + \left[\frac{x^2}{2} - 3x \right]_3^6$$

$$= \left(-\frac{(3)^2}{2} + 3(3) \right) - 0$$

$$+ \left(\frac{6^2}{2} - 3(6) \right) - \left(\frac{3^2}{2} - 3(3) \right)$$

$$= \frac{9}{2} + \frac{9}{2} = 9$$

5. $\int_0^k (6x-1) dx = 3x^2 - x \Big]_0^k = 3k^2 - k$

Set $3k^2 - k = 4 \Rightarrow 3k^2 - k - 4 = 0$

$$\Rightarrow (3k-4)(k+1) = 0$$

$$\Rightarrow k = \frac{4}{3} \text{ or } k = -1$$

Verify your results by evaluating

$$\int_0^{4/3} (6x-1) dx \text{ and } \int_0^{-1} (6x-1) dx.$$

6. Let $u = 1 + \cos x$; $du = -\sin x \, dx$ or $-du = \sin x \, dx$.

$$\int \frac{\sin x}{\sqrt{1+\cos x}} dx = \int \frac{-1}{\sqrt{u}} (du) = -\int \frac{1}{u^{1/2}} du$$

$$= -\int u^{-1/2} du = -\frac{u^{1/2}}{1/2} + c$$

$$= -2u^{1/2} + c$$

$$= -2(1+\cos x)^{1/2} + c$$

$$\int_0^\pi \frac{\sin x}{\sqrt{1+\cos x}} dx = -2(1+\cos x)^{1/2} \Big]_0^\pi$$

$$= -2 \Big[(1+\cos \pi)^{1/2}$$

$$-(1+\cos 0)^{1/2} \Big]$$

$$= -2 \Big[0 - 2^{1/2} \Big] = 2\sqrt{2}$$

7. Let $u = 4x$; $du = 4 \, dx$ or $\frac{du}{4} = dx$

$$\int g(4x) dx = \int g(u) \frac{du}{4} = \frac{1}{4} \int g(u) du$$

$$= \frac{1}{4} f(u) + c$$

$$= \frac{1}{4} f(4x) + c$$

$$\int_1^2 g(4x)dx = \frac{1}{4} f(4x)]_1^2$$

$$= \frac{1}{4}f(4(2)) - \frac{1}{4}(f(4(1)))$$

$$= \frac{1}{4}f(8) - \frac{1}{4}f(4).$$

8. $\displaystyle\int_{\ln 2}^{\ln 3} 10e^x dx = 10e^x]_{\ln 2}^{\ln 3} = 10\left[\left(e^{\ln 3}\right) - \left(e^{\ln 2}\right)\right]$

$$= 10(3 - 2) = 10.$$

9. Let $u = t + 3$; $du = dt$.

$$\int \frac{1}{t+3}dt = \int \frac{1}{u}du = \ln|u| + c$$

$$= \ln|t+3| + c$$

$$\int_e^{e^2} \frac{1}{t+3}dt = \ln|t+3|]_e^{e^2}$$

$$= \ln(e^2 + 3) - \ln(e + 3)$$

$$= \ln\left(\frac{e^2 + 3}{e + 3}\right)$$

10. $f'(x) = \tan^2 x$;

$$f'\left(\frac{\pi}{6}\right) = \tan^2\left(\frac{\pi}{6}\right) = \left(\frac{1}{\sqrt{3}}\right)^2 = \frac{1}{3}.$$

11. Let $u = x^2$; $du = 2x\,dx$ or $\dfrac{du}{2} = x\,dx$

$$\int 4xe^{x^2} dx = 4\int e^u \left(\frac{du}{2}\right)$$

$$= 2\int e^u du = 2e^u + c = 2e^{x^2} + c$$

$$\int_{-1}^1 4xe^{x^2} dx = 2e^{x^2}]_{-1}^1$$

$$= 2\left[e^{(1)^2} - e^{(-1)^2}\right] = 2(e - e) = 0$$

Note that $f(x) = 4xe^{x^2}$ is an odd function. Thus $\displaystyle\int_{-a}^a f(x)dx = 0.$

12. $\displaystyle\int_{-\pi}^{\pi} \left(\cos x - x^2\right)dx = \sin x - \frac{x^3}{3}\Bigg]_{-\pi}^{\pi}$

$$= \left(\sin\pi - \frac{\pi^3}{3}\right)$$

$$- \left(\sin(-\pi) - \frac{(-\pi)^3}{3}\right)$$

$$= -\frac{\pi^3}{3} - \left(0 - \frac{-\pi^3}{3}\right)$$

$$= -\frac{2\pi^3}{3}$$

Note that $f(x) = \cos x - x^2$ is an even function. Thus you could have written

$$\int_{-\pi}^{\pi} \left(\cos x - x^2\right)dx = 2\int_0^{\pi} \left(\cos x - x^2\right)dx \text{ and}$$

obtain the same result.

Part B—Calculators are permitted.

13. $\displaystyle\int_0^2 \left(x^3 + k\right)dx = \frac{x^4}{4} + kx]_0^2 = \left(\frac{2^4}{4} + k(2)\right) - 0$

$$= 4 + 2k$$

Set $4 + 2k = 10$ and $k = 3$.

14. Enter $\displaystyle\int (2x * \cos(x), x, -1.2, 3.1)$ and obtain $-4.70208 \approx -4.702.$

15. $\dfrac{d}{dx}\left(\displaystyle\int_1^{x^3} \sqrt{t^2 + 1}\,dt\right) = \sqrt{(x^3)^2 + 1}\dfrac{d}{dx}\left(x^3\right)$

$$= 3x^2\sqrt{x^6 + 1}.$$

16. $\Delta x = \dfrac{8 - 0}{4} = 2$ Midpoints are $x = 1, 3, 5,$ and 7.

$$\int_0^{12} \left(x^3 + 1\right)dx = (1^3 + 1)(2) + (3^3 + 1)(2)$$

$$+ (5^3 + 1)(2) + (7^3 + 1)(2)$$

$$= (2)(2) + (28)(2) + (126)(1)$$

$$+ (344)(2) = 874$$

17. (a) $\displaystyle\int_{-2}^0 g(x)dx + \int_0^2 g(x)dx = \int_{-2}^2 g(x)dx$

$$\int_{-2}^0 g(x)dx + 3 = 8. \text{ Thus } \int_{-2}^0 g(x)dx = 5$$

(b) $\displaystyle\int_2^{-2} g(x)dx = -\int_{-2}^2 g(x)dx = -8$

(c) $\displaystyle\int_0^{-2} 5g(x)dx = 5\int_0^{-2} g(x)dx$

$$= 5\left(-\int_{-2}^0 g(x)dx\right)$$

$$= 5(-5) = -25$$

(d) $\displaystyle\int_{-2}^2 2g(x)dx = 2\int_{-2}^2 g(x)dx = 2(8) = 16$

18. $\displaystyle\int_0^{1/2} \frac{dx}{\sqrt{1-x^2}} = \sin^{-1}(x)\Big]_0^{1/2}$

$\displaystyle= \sin^{-1}\left(\frac{1}{2}\right) - \sin^{-1}(0)$

$\displaystyle= \frac{\pi}{6} - 0 = \frac{\pi}{6}$

19. $\displaystyle\int_{\cos x}^{\sin x} (2t+1)dt = \int_0^{\sin x} (2t+1)dt$

$\displaystyle\quad - \int_0^{\cos x} (2t+1)dt$

$\displaystyle\frac{dy}{dx} = \frac{d}{dx}\left(\int_{\cos x}^{\sin x} (2t+1)dt = (2\sin x + 1)\frac{d}{dx}\right.$

$\displaystyle\left.\sin x - (2\cos x + 1)\frac{d}{dx}(\cos x)\right)$

$\displaystyle= (2\sin x + 1)\cos x - (2\cos x + 1)(-\sin x)$

$\displaystyle= 2\sin x\cos x + \cos x + 2\sin x\cos x + \sin x$

$\displaystyle= 4\sin x\cos x + \cos x + \sin x.$

20. $\displaystyle\Delta x = \frac{30-0}{3} = 10$

Midpoints are $x = 5, 15,$ and $25.$

$\displaystyle\int_0^{30} f(x)dx = [f(5)]\,10 + [f(15)]\,10 + [f(25)]\,10$

$\displaystyle= (2.6)(10) + (4.1)(10) + (5.2)10$

$\displaystyle= 119.$

21. $\displaystyle\int_0^\infty e^{-x}dx = \lim_{k\to\infty}\int_0^k e^{-x}dx = \lim_{k\to\infty}\left(-e^{-x}\big|_0^k\right)$

$\displaystyle= \lim_{k\to\infty}\left(-e^{-k}+1\right) = 1$

22. $\displaystyle\int_{-\infty}^0 \frac{dx}{(4-x)^2} = \lim_{k\to-\infty}\int_k^0 \frac{dx}{(4-x)^2}$

$\displaystyle= \lim_{k\to-\infty}\frac{-1}{4-x}\Big|_k^0 = \lim_{k\to-\infty}\left(\frac{-1}{4-k}+\frac{1}{4}\right) = \frac{1}{4}$

23. $\displaystyle\int_0^1 \ln x\,dx = \lim_{k\to 0^+}\int_k^1 \ln x\,dx$

$\displaystyle= \lim_{k\to 0^+}\big[x\ln x - x\big]_k^1 = \lim_{k\to 0^+}(-1 - k\ln k + k) = -1$

24. $\displaystyle\int_{-2}^2 \frac{dx}{\sqrt{4-x^2}} = 2\int_0^2 \frac{dx}{\sqrt{4-x^2}}$

$\displaystyle= 2\lim_{k\to 2}\int_0^k \frac{dx}{\sqrt{4-x^2}}$

$\displaystyle= 2\lim_{k\to 2}\left[\sin^{-1}\left(\frac{x}{2}\right)\right]_0^k$

$\displaystyle= 2\lim_{k\to 2}\left[\sin^{-1}\left(\frac{k}{2}\right) - \sin^{-1}(0)\right]$

$\displaystyle= 2\lim_{k\to 2^-}\left[\sin^{-1}\left(\frac{k}{2}\right)\right] = 2\left(\frac{\pi}{2}\right) = \pi$

25. $\displaystyle\int_{-1}^8 \frac{dx}{\sqrt[3]{x}} = \int_{-1}^0 \frac{dx}{\sqrt[3]{x}} + \int_0^8 \frac{dx}{\sqrt[3]{x}}$

$\displaystyle= \lim_{k\to 0^-}\int_{-1}^k \frac{dx}{\sqrt[3]{x}} + \lim_{k\to 0^+}\int_k^8 \frac{dx}{\sqrt[3]{x}}$

$\displaystyle= \lim_{k\to 0^-}\left[\frac{3}{2}x^{2/3}\right]_{-1}^k + \lim_{k\to 0^-}\left[\frac{3}{2}x^{2/3}\right]_k^8$

$\displaystyle= \lim_{k\to 0^-}\left[\frac{3}{2}k^{2/3} - \frac{3}{2}\right]$

$\displaystyle+ \lim_{k\to 0^+}\left[\frac{12}{2} - \frac{3}{2}k^{2/3}\right] = \frac{9}{2}$

7.9 SOLUTIONS TO CUMULATIVE REVIEW PROBLEMS

26. As $x \to -\infty,\; x = -\sqrt{x^2}$

$\displaystyle\lim_{x\to-\infty}\frac{\sqrt{x^2-4}}{3x-9} = \lim_{x\to-\infty}\frac{\sqrt{x^2-4}/-\sqrt{x^2}}{(3x-9)/x}$

$\displaystyle= \lim_{x\to-\infty}\frac{-\sqrt{(x^2-4)/x^2}}{3-(9/x)}$

$\displaystyle= \lim_{x\to-\infty}\frac{-\sqrt{1-(4/x)^2}}{3-9/x}$

$\displaystyle= \frac{-\sqrt{1-0}}{3-0} = -\frac{1}{3}.$

27. $\displaystyle y = \ln|x^2-4|,\; \frac{dy}{dx} = \frac{1}{(x^2-4)}(2x)$

$\displaystyle\frac{dy}{dx}\Big|_{x=3} = \frac{2(3)}{(3^2-4)} = \frac{6}{5}$

28. (a) See Figure 7.9-1.
    The function f has a relative minimum at $x = -5$ and $x = 3$, and f has a relative maximum at $x = -1$ and $x = 7$.

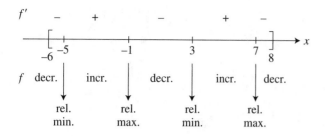

Figure 7.9-1

(b) See Figure 7.9-2.

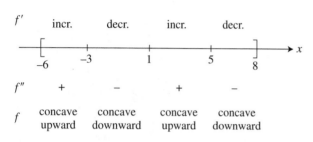

Figure 7.9-2

The function f is concave upward on intervals $(-6, -3)$ and $(1, 5)$.

(c) A change of concavity occurs at $x = -3$, $x = 1$, and $x = 5$.

29. (Calculator) Differentiate both sides of $9x^2 + 4y^2 - 18x + 16y = 11$.

$$18x + 8y\frac{dy}{dx} - 18 + 16\frac{dy}{dx} = 0$$

$$8y\frac{dy}{dx} + 16\frac{dy}{dx} = -18x + 18$$

$$\frac{dy}{dx}(8y + 16) = -18x + 18$$

$$\frac{dy}{dx} = \frac{-18x + 18}{8y + 16}$$

Horizontal tangent $\Rightarrow \dfrac{dy}{dx} = 0$

Set $\dfrac{dy}{dx} = 0 \Rightarrow -18x + 18 = 0$ or $x = 1$

At $x = 1, 9 + 4y^2 - 18 + 16y = 11$

$$4y^2 + 16y - 20 = 0$$

Using a calculator, enter solve $(4y^2 + 16y - 20 = 0, y)$; obtaining $y = -5$ or $y = 1$.

Thus each of the points at $(1, 1)$ and $(1, -5)$ the graph has a horizontal tangent at each point.

Vertical tangent $\Rightarrow \dfrac{dy}{dx}$ is undefined.

Set $8y + 16 = 0 \Rightarrow y = -2$.

At $y = -2, 9x^2 + 16 - 18x - 32 = 11$

$$9x^2 - 18x - 27 = 0$$

Enter solve $(9x^2 - 8x - 27 = 0, x)$ and obtain $x = 3$ or $x = -1$.

Thus at each of the points $(3, -2)$ and $(-1, -2)$, the graph has a vertical tangent. See Figure 7.9-3.

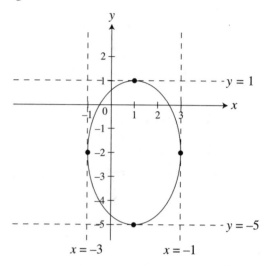

Figure 7.9-3

30. (Calculator)

Step 1. See Figure 7.9-4. Let $P = x + y$ where P is the length of the pipe and x and y are as shown. The minimum value of P is the maximum length of the pipe to be able to turn in the corner. By similar triangles, $\dfrac{y}{10} = \dfrac{x}{\sqrt{x^2 - 36}}$ and thus

$$y = \frac{10x}{\sqrt{x^2 - 36}}, \quad x > 6$$

$$P = x + y = x + \frac{10x}{\sqrt{x^2 - 36}}$$

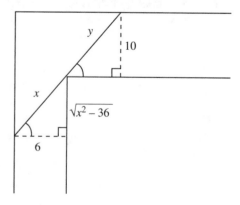

Figure 7.9-4

Step 2. Find the minimum value of P.
Enter $y_t = x + 10 * x/(\sqrt{(x^2 - 36)})$.
Use the Minimum function of the calculator and obtain the minimum point (9.306, 22.388).

Step 3. Verify with the First Derivative Test.
Enter $y_2 = (y_1(x), x)$ and observe.
(See Figure 7.9-5.)

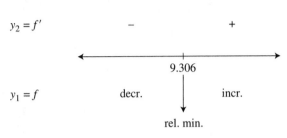

Figure 7.9-5

Step 4. Check endpoints.
The domain of x is $(6, \infty)$
Since at $x = 9.306$ is the only relative extremum, it is the absolute minimum.

Thus the maximum length of the pipe is 22.388 feet.

31. $\displaystyle\lim_{x \to -1} \frac{1 + \cos \pi x}{x^2 - 1} = \lim_{x \to -1} \frac{-\pi \sin \pi x}{2x} = \frac{0}{-2} = 0$

32. $\dfrac{dx}{dt} = -4t$ $\dfrac{dy}{dt} = 2t$. The speed of the object is
$\sqrt{(-4t)^2 + (2t)^2} = \sqrt{16t^2 + 4t^2} = \sqrt{20t^2} = 2t\sqrt{5}$. Evaluated at $t = \dfrac{1}{2}$, the speed is
$2\left(\dfrac{1}{2}\right)\sqrt{5} = \sqrt{5}$.

33. Integrate $\displaystyle\int 2x\sqrt{x + 3}\,dx$ by parts with $u = 2x$,
$du = 2dx$, $dv = \sqrt{x + 4}\,dx$, and $v = \dfrac{2}{3}(x + 4)^{3/2}$.
Then $\displaystyle\int 2x\sqrt{x + 3}\,dx = \dfrac{4}{3}x(x + 3)^{3/2} - \dfrac{4}{3}\int (x + 3)^{3/2}\,dx = \dfrac{4}{3}x(x + 3)^{3/2} - \dfrac{4}{3}\left(\dfrac{2}{5}(x + 3)^{5/2}\right)$. Simplifying this expression,
we get $\displaystyle\int 2x\sqrt{x + 3}\,dx = \dfrac{4}{5}(x + 3)^{3/2}(x - 2)$.

Chapter 8

Areas and Volumes

8.1 THE FUNCTION $F(x) = \int_a^x f(t)\,dt$

The Second Fundamental Theorem of Calculus defines

$$F(x) = \int_a^x f(t)\,dt$$

and states that if f is continuous on $[a, b]$, then $F'(x) = f(x)$ for every point x in $[a, b]$.

If $f \geq 0$, then $F \geq 0$. $F(x)$ can be interpreted geometrically as the area under the curve of f from $t = a$ to $t = x$. (See Figure 8.1-1.)

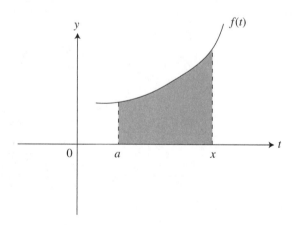

Figure 8.1-1

If $f < 0$, $F < 0$, $F(x)$ can be treated as the negative value of the area between the curve of f and the t-axis from $t = a$ to $t = x$. (See Figure 8.1-2.)

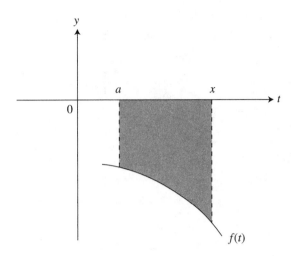

Figure 8.1-2

Example 1

If $f(x) = \displaystyle\int_0^x 2\cos t\, dt$ for $0 \le x \le 2\pi$, find the value(s) of x where f has a local minimum.

Method 1: Since $f(x) = \displaystyle\int_0^x 2\cos t\, dt$, $f'(x) = 2\cos x$.

Set $f'(x) = 0$; $2\cos x = 0$, $x = \dfrac{\pi}{2}$ or $\dfrac{3\pi}{2}$.

$f''(x) = -2\sin x$ and $f''\left(\dfrac{\pi}{2}\right) = -2$ and $f''\left(\dfrac{3\pi}{2}\right) = 2$

Thus at $x = \dfrac{3\pi}{2}$, f has a local minimum.

Method 2: You can solve this problem geometrically by using area. See Figure 8.1-3.

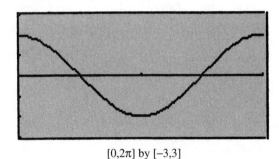

$[0, 2\pi]$ by $[-3, 3]$

Figure 8.1-3

The area "under the curve" is above the t-axis on $[0, \pi/2]$ and below the x-axis on $[\pi/2, 3\pi/2]$. Thus the local minimum occurs at $3\pi/2$.

Example 2

Let $p(x) = \displaystyle\int_0^x f(t)dt$ and the graph of f is shown in Figure 8.1-4.

(a) Evaluate: $p(0)$, $p(1)$, $p(4)$
(b) Evaluate: $p(5)$, $p(7)$, $p(8)$

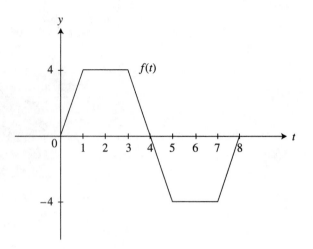

Figure 8.1-4

(c) At what value of t does p have a maximum value?
(d) On what interval(s) is p decreasing?
(e) Draw a sketch of the graph of p.

Solution:

(a) $p(0) = \displaystyle\int_0^0 f(t)dt = 0$

$p(1) = \displaystyle\int_0^1 f(t)dt = \dfrac{(1)(4)}{2} = 2$

$p(4) = \displaystyle\int_0^4 f(t)dt = \dfrac{1}{2}(2+4)(4) = 12$

(Note: $f(t)$ forms a trapezoid from $t = 0$ to $t = 4$.)

(b) $p(5) = \displaystyle\int_0^5 f(t)dt = \int_0^4 f(t)dt + \int_4^5 f(t)dt$

$\quad\quad = 12 - \dfrac{(1)(4)}{2} = 10$

$p(7) = \displaystyle\int_0^7 f(t)dt = \int_0^4 f(t)dt + \int_4^5 f(t)dt + \int_5^7 f(t)dt$

$\quad\quad = 12 - 2 - (2)(4) = 2$

$p(8) = \displaystyle\int_0^8 f(t)dt = \int_0^4 f(t)dt + \int_4^8 f(t)dt$

$\quad\quad = 12 - 12 = 0$

(c) Since $f \geq 0$ on the interval $[0, 4]$, p attains a maximum at $t = 4$.

(d) Since $f(t)$ is below the x-axis from $t = 4$ to $t = 8$, if $x > 4$,

$$\int_0^x f(t)dt = \int_0^4 f(t)dt + \int_4^x f(t)dt \text{ where } \int_4^x f(t)dt < 0.$$

Thus p is decreasing on the interval $(4, 8)$.

(e) $p(x) = \int_0^x f(t)dt$. See Figure 8.1-5 for a sketch.

x	0	1	2	3	4	5	6	7	8
$p(x)$	0	2	6	10	12	10	6	2	0

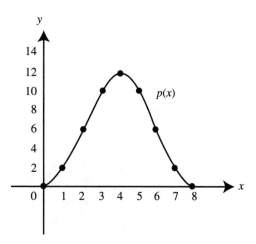

Figure 8.1-5

• Remember differentiability implies continuity, but the converse is not true, i.e., continuity does not imply differentiability, e.g., as in the case of a cusp or a corner.

Example 3

The position function of a moving particle on a coordinate axis is:

$$s = \int_0^t f(x)dx \text{ feet.}$$

The function f is a differentiable function and its graph is shown below in Figure 8.1-6.

(a) What is the particle's velocity at $t = 4$?
(b) What is the particle's position at $t = 3$?
(c) When is the acceleration zero?
(d) When is the particle moving to the right?
(e) At $t = 8$, is the particle on the right side or left side of the origin?

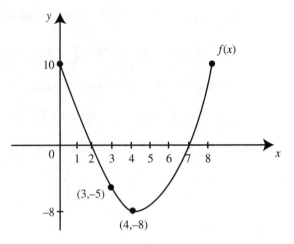

Figure 8.1-6

Solution:

(a) Since $s = \displaystyle\int_0^t f(x)dx$, then $v(t) = s'(t) = f(t)$
    Thus $v(4) = -8$ *ft.*

(b) $s(3) = \displaystyle\int_0^3 f(x)dx = \int_0^2 f(x)dx + \int_2^3 f(x)dx = \frac{1}{2}(10)(2) - \frac{1}{2}(1)(5) = \frac{15}{2}$

(c) $a(t) = v'(t)$. Since $v'(t) = f'(t)$, $v'(t) = 0$ at $t = 4$. Thus $a(4) = 0$.

(d) The particle is moving to the right when $v(t) > 0$. Thus the particle is moving to the right on intervals $(0, 2)$ and $(7, 8)$

(e) The area of f below the x-axis from $x = 2$ to $x = 7$ is larger than the area of f above the x-axis from $x = 0$ to $x = 2$ and $x = 7$ to $x = 8$. Thus $\displaystyle\int_0^8 f(x)dx < 0$ and the particle is on the left side of the origin.

> • Don't forget that $(fg)' = f'g + g'f$ and *not* $f'g'$. However, $\lim(fg) = (\lim f)(\lim g)$

8.2 APPROXIMATING THE AREA UNDER A CURVE

Main Concepts: *Rectangular Approximations, Trapezoidal Approximations*

Rectangular Approximations

If $f \geq 0$, the area under the curve of f can be approximated using three common types of rectangles: left-endpoint rectangles, right-endpoint rectangles, or midpoint rectangles. (See Figure 8.2-1.)

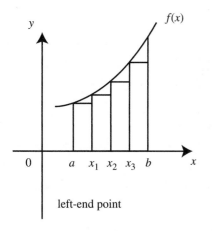 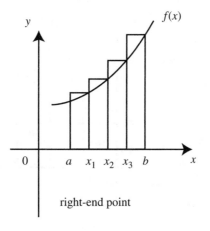 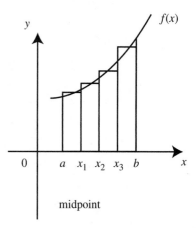

left-end point — right-end point — midpoint

Figure 8.2-1

The area under the curve using n rectangles of equal length is approximately:

$$\sum_{i=1}^{n} \left(\text{area of rectangle}\right) = \begin{cases} \sum_{i=1}^{n} f(x_{i-1})\Delta x \text{ left-endpoint rectangles} \\ \sum_{i=1}^{n} f(x_i)\Delta x \text{ right-endpoint rectangles} \\ \sum_{i=1}^{n} f\left(\dfrac{x_i + x_{i-1}}{2}\right)\Delta x \text{ midpoint rectangles} \end{cases}$$

where $\Delta x = \dfrac{b - a}{n}$ and $a = x_0 < x_1 < x_2 < \cdots < x_n = b$

If f is increasing on $[a, b]$, then left-endpoint rectangles are inscribed rectangles and the right-endpoint rectangles are circumscribed rectangles. If f is decreasing on $[a, b]$, then left-endpoint rectangles are circumscribed rectangles and the right-endpoint rectangles are inscribed. Furthermore,

$$\sum_{i=1}^{n} \text{inscribed rectangle} \leq \text{area under the curve} \leq \sum_{i=1}^{n} \text{circumscribed rectangle.}$$

Example 1

Find the approximate area under the curve of $f(x) = x^2 + 1$ from $x = 0$ to $x = 2$, using 4 left-endpoint rectangles of equal length. (See Figure 8.2-2.)

Let Δx_i be the length of ith rectangle. The length $\Delta x_i = \dfrac{2 - 0}{4} = \dfrac{1}{2}; x_{i-1} = \dfrac{1}{2}(i - 1)$

Area under the curve $\approx \sum_{i=1}^{4} f(x_{i-1})\Delta x_i = \sum_{i=1}^{4} \left(\left(\dfrac{1}{2}(i - 1)\right)^2 + 1\right)\left(\dfrac{1}{2}\right)$

Enter $\sum \left(\left((0.5(x - 1))^2 + 1\right) * 0.5, x, 1, 4\right)$ and obtain 3.75.

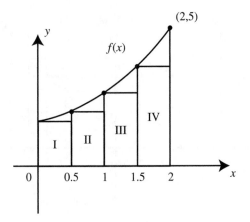

Figure 8.2-2

Or, find the area of each rectangle:

$$\text{Area of Rect}_I = (f(0))\Delta x_1 = (1)\left(\frac{1}{2}\right) = \frac{1}{2}$$

$$\text{Area of Rect}_{II} = f(0.5)\Delta x_2 = ((0.5)^2 + 1)\left(\frac{1}{2}\right) = 0.625$$

$$\text{Area of Rect}_{III} = f(1)\Delta x_3 = (1^2 + 1)\left(\frac{1}{2}\right) = 1$$

$$\text{Area of Rect}_{IV} = f(1.5)\Delta x_4 = (1.5^2 + 1)\left(\frac{1}{2}\right) = 1.625$$

Area of $(\text{Rect}_I + \text{Rect}_{II} + \text{Rect}_{III} + \text{Rect}_{IV}) = 3.75$

Thus the approximate area under the curve of $f(x)$ is 3.75.

Example 2

Find the approximate area under the curve of $f(x) = \sqrt{x}$ from $x = 4$ to $x = 9$ using 5 right-endpoint rectangles. (See Figure 8.2-3.)

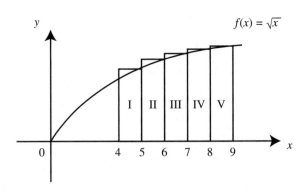

Figure 8.2-3

Let Δx_i be the length of ith rectangle. The length $\Delta x_i = \dfrac{9-4}{5} = 1$; $x_i = 4 + (1)i = 4 + i$

$$\text{Area of Rect}_\text{I} = f(x_1)\Delta x_1 = f(5)(1) = \sqrt{5}$$

$$\text{Area of Rect}_\text{II} = f(x_2)\Delta x_2 = f(6)(1) = \sqrt{6}$$

$$\text{Area of Rect}_\text{III} = f(x_3)\Delta x_3 = f(7)(1) = \sqrt{7}$$

$$\text{Area of Rect}_\text{IV} = f(x_4)\Delta x_4 = f(8)(1) = \sqrt{8}$$

$$\text{Area of Rect}_\text{V} = f(x_5)\Delta x_5 = f(9)(1) = \sqrt{9} = 3$$

$$\sum_{i=1}^{5}(\text{Area of Rect}_\text{I}) = \sqrt{5} + \sqrt{6} + \sqrt{7} + \sqrt{8} + 3 = 13.160$$

Or, using $\sum$ notation:

$$\sum_{i=1}^{5} f(x_i)\,\Delta x_i = \sum_{i=1}^{5} f(4+i)(1) = \sum_{i=1}^{5} \sqrt{4+1}$$

Enter $\sum\left(\sqrt{(4+x)}, x, 1, 5\right)$ and obtain 13.160

Thus the area under the curve is approximately 13.160.

Example 3

The function f is continuous on $[1, 9]$ and $f > 0$. Selected values of f are given below:

x	1	2	3	4	5	6	7	8	9
$f(x)$	1	1.41	1.73	2	2.37	2.45	2.65	2.83	3

Using 4 midpoint rectangles, approximate the area under the curve of f for $x = 1$ to $x = 9$. (See Figure 8.2-4.)

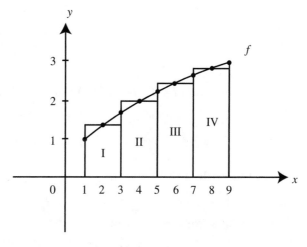

Figure 8.2-4

Let Δx_i be the length of ith rectangle. The length $\Delta x_i = \dfrac{9-1}{4} = 2$.

Area of $\text{Rect}_\text{I} = f(2)(2) = (1.41)2 = 2.82$

Area of $\text{Rect}_\text{II} = f(4)(2) = (2)2 = 4$

Area of $\text{Rect}_\text{III} = f(6)(2) = (2.45)2 = 4.90$

Area of $\text{Rect}_\text{IV} = f(8)(2) = (2.83)2 = 5.66$

Area of $(\text{Rect}_\text{I} + \text{Rect}_\text{II} + \text{Rect}_\text{III} + \text{Rect}_\text{IV}) = 2.82 + 4 + 4.90 + 5.66 = 17.38$.

Thus the area under the curve is approximately 17.38.

Trapezoidal Approximations

Another method of approximating the area under a curve is to use trapezoids. See Figure 8.2-5.

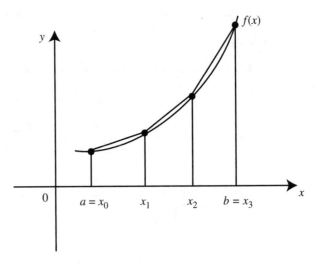

Figure 8.2-5

Formula for Trapezoidal Approximation:

If f is continuous, the area under the curve of f from $x = a$ to $x = b$ is:

$$\text{Area} \approx \frac{b-a}{2n}\left[f(x_0) + 2f(x_1) + 2f(x_2) + \cdots + 2f(x_{n-1}) + f(x_n)\right]$$

Example 1

Find the approximate area under the curve of $f(x) = \cos\left(\dfrac{x}{2}\right)$ from $x = 0$ to $x = \pi$, using 4 trapezoids. (See Figure 8.2-6.)

Since $n = 4$, $\Delta x = \dfrac{\pi - 0}{4} = \dfrac{\pi}{4}$

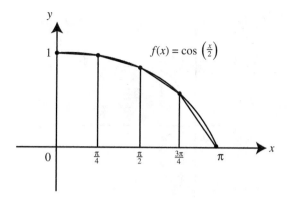

Figure 8.2-6

Area under the curve:

$$\approx \frac{\pi}{4} \cdot \frac{1}{2}\left[\cos(0) + 2\cos\left(\frac{\pi/4}{2}\right) + 2\cos\left(\frac{\pi/2}{2}\right) + 2\cos\left(\frac{3\pi/4}{2}\right) + \cos\left(\frac{\pi}{2}\right)\right]$$

$$\approx \frac{\pi}{8}\left[\cos(0) + 2\cos\left(\frac{\pi}{8}\right) + 2\cos\left(\frac{\pi}{4}\right) + 2\cos\left(\frac{3\pi}{8}\right) + \cos\left(\frac{\pi}{2}\right)\right]$$

$$\approx \frac{\pi}{8}\left[1 + 2(.9239) + 2\left(\frac{\sqrt{2}}{2}\right) + 2(.3827) + 0\right] \approx 1.9743$$

- When using a graphing calculator in solving a problem, you are required to write the setup that leads to the answer. For example, if you are finding the volume of a solid, you must write the definite integral and then use the calculator to compute the numerical value, e.g.,
  Volume $= \pi \int_0^3 (5x)^2 dx = 225\pi$. Simply indicating the answer without writing the integral would get you only one point for the answer. And you will not get full credit for the problem.

8.3   AREA AND DEFINITE INTEGRALS

Main Concepts: *Area under a Curve Area between Two Curves*

Area Under a Curve

If $y = f(x)$ is continuous and non-negative on $[a, b]$, then the area under the curve of f from a to b is:

$$\text{Area} = \int_a^b f(x)dx$$

If f is continuous and $f < 0$ on $[a, b]$, then the area under the curve from a to b is:

$$\text{Area} = -\int_a^b f(x)dx. \text{ See Figure 8.3-1.}$$

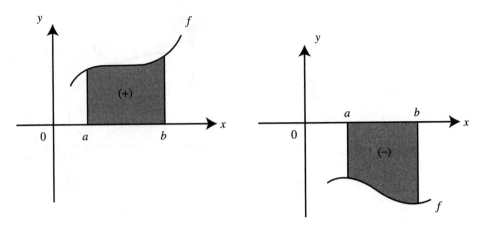

Figure 8.3-1

If $x = g(y)$ is continuous and non-negative on $[c, d]$, then the area under the curve of g from c to d is:

$$\text{Area} \int_c^d g(y)dy. \text{ See Figure 8.3-2.}$$

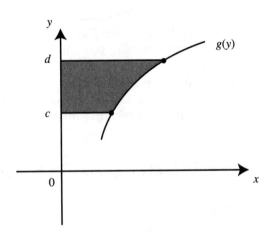

Figure 8.3-2

Example 1

Find the area under the curve of $f(x) = (x - 1)^3$ from $x = 0$ to $x = 2$.

Step 1. Sketch the graph of $f(x)$. See Figure 8.3-3.

Step 2. Set up integrals:

$$\text{Area} = \left| \int_0^1 f(x)dx \right| + \int_1^2 f(x)dx.$$

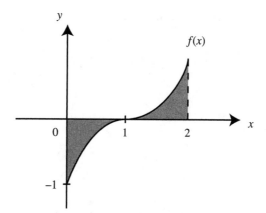

Figure 8.3-3

Step 3. Evaluate integrals:

$$\left| \int_0^1 (x-1)^3 dx \right| = \left| \frac{(x-1)^4}{4} \right]_0^1 \right| = \left| -\frac{1}{4} \right| = \frac{1}{4}$$

$$\int_1^2 (x-1)^3 dx = \frac{(x-1)^4}{4} \right]_1^2 = \frac{1}{4}$$

Thus the total area is $\dfrac{1}{4} + \dfrac{1}{4} = \dfrac{1}{2}$.

Another solution is to find the area using a calculator.

Enter $\int \left(abs\left((x-1)^{\wedge}3 \right), x, 0, 2 \right)$ and obtain $\dfrac{1}{2}$.

Example 2

Find the area of the region bounded by the graph of $f(x) = x^2 - 1$, the lines $x = -2$ and $x = 2$ and the x-axis.

Step 1. Sketch the graph of $f(x)$. See Figure 8.3-4.

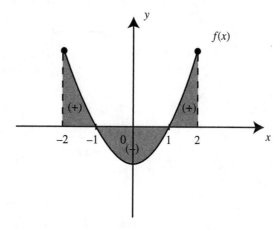

Figure 8.3-4

Step 2. Set up integrals.

$$\text{Area} = \int_{-2}^{-1} f(x)dx + \left| \int_{-1}^{1} f(x)dx \right| + \int_{1}^{2} f(x)dx.$$

Step 3. Evaluate the integrals:

$$\int_{-2}^{-1} \left(x^2 - 1\right) dx = \frac{x^3}{3} - x \Big]_{-2}^{-1} = \frac{2}{3} - \left(-\frac{2}{3}\right) = \frac{4}{3}$$

$$\left| \int_{-1}^{1} \left(x^2 - 1\right) dx \right| = \left| \frac{x^3}{3} - x \Big]_{-1}^{1} \right| = \left| -\frac{2}{3} - \left(\frac{2}{3}\right) \right| = \left| -\frac{4}{3} \right| = \frac{4}{3}$$

$$\int_{1}^{2} \left(x^2 - 1\right) dx = \frac{x^3}{3} - x \Big]_{1}^{2} = \frac{2}{3} - \left(-\frac{2}{3}\right) = \frac{4}{3}$$

Thus the total area $= \frac{4}{3} + \frac{4}{3} + \frac{4}{3} = 4.$

Note: Since $f(x) = x^2 - 1$ is an even function, you can use the symmetry of the graph and set area $= 2 \left(\left| \int_{0}^{1} f(x)dx \right| + \int_{1}^{2} f(x)dx \right).$

An alternate solution is to find the area using a calculator.

Enter $\int (abs(x^2 - 1), x, -2, 2)$ and obtain 4.

Example 3

Find the area of the region bounded by $x = y^2$, $y = -1$, and $y = 3$. See Figure 8.3-5.

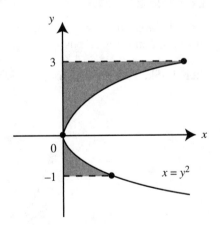

Figure 8.3-5

$$\text{Area} = \int_{-1}^{3} y^2 dy = \frac{y^3}{3} \Big]_{-1}^{3} = \frac{3^3}{3} - \frac{(-1)^3}{3} = \frac{28}{3}.$$

Example 4

Using a calculator, find the area bounded by $f(x) = x^3 + x^2 - 6x$ and the x-axis. See Figure 8.3-6.

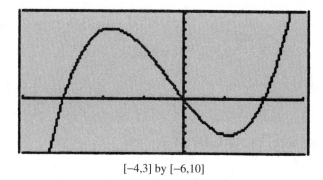

[−4,3] by [−6,10]

Figure 8.3-6

Step 1. Enter $y_1 = x^\wedge 3 + x^\wedge 2 - 6x$

Step 2. Enter $\int \left(abs\left(x^\wedge 3 + x^\wedge 2 - 6 * x\right), x, -3, 2\right)$ and obtain 21.083.

Example 5

The area under the curve $y = e^x$ from $x = 0$ to $x = k$ is 1. Find the value of k.

$\text{Area} = \int_0^k e^x dx = e^x]_0^k = e^k - e^0 = e^k - 1 \Rightarrow e^k = 2$. Take ln of both sides:

$\ln(e^k) = \ln 2; \ k = \ln 2$.

Example 6

The region bounded by the x-axis, and the graph of $y = \sin x$ between $x = 0$ and $x = \pi$ is divided into 2 regions by the line $x = k$. If the area of the region for $0 \le x \le k$ is twice the area of the region $k \le x \le \pi$, find k. (See Figure 8.3-7.)

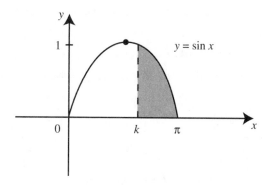

Figure 8.3-7

$$\int_0^k \sin x \, dx = 2 \int_k^\pi \sin x \, dx$$

$$-\cos x]_0^k = 2 \left[-\cos x\right]_k^\pi$$

$$-\cos k - (-\cos(0)) = 2\left(-\cos \pi - (-\cos k)\right)$$

$$-\cos k + 1 = 2(1 + \cos k)$$

$$-\cos k + 1 = 2 + 2\cos k$$

$$-3\cos k = 1$$

$$\cos k = -\frac{1}{3}$$

$$k = \arccos\left(-\frac{1}{3}\right) = 1.91063.$$

Area Between Two Curves

Area Bounded by Two Curves: See Figure 8.3-8.

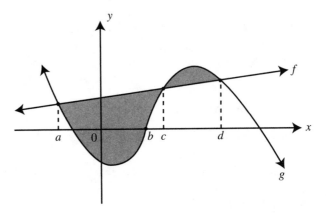

Figure 8.3-8

$$\text{Area} = \int_{a}^{c} \left[f(x) - g(x)\right] dx + \int_{c}^{d} \left[g(x) - f(x)\right] dx$$

Note: Area $= \int_{a}^{d} \left(\text{upper curve} - \text{lower curve}\right) dx$

Example 1

Find the area of the region bounded by the graphs of $f(x) = x^3$ and $g(x) = x$. (See Figure 8.3-9.)

Step 1. Sketch the graphs of $f(x)$ and $g(x)$.

Step 2. Find the points of intersection.

Set $f(x) = g(x)$

$x^3 = x$

$x(x^2 - 1) = 0$

$x(x - 1)(x + 1) = 0$

$x = 0, 1, \text{ and } -1$

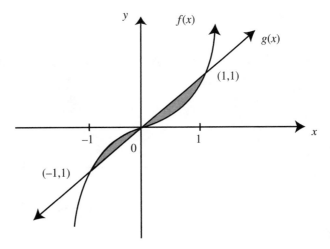

Figure 8.3-9

Step 3. Set up integrals.

$$\text{Area} = \int_{-1}^{0} \left(f(x) - g(x)\right)dx + \int_{0}^{1} \left(g(x) - f(x)\right)dx$$

$$= \int_{-1}^{0} \left(x^3 - x\right)dx + \int_{0}^{1} \left(x - x^3\right)dx$$

$$= \left[\frac{x^4}{4} - \frac{x^2}{2}\right]_{-1}^{0} + \left[\frac{x^2}{2} - \frac{x^4}{4}\right]_{0}^{1}$$

$$= 0 - \left(\frac{(-1)^4}{4} - \frac{(-1)^2}{2}\right) + \left(\frac{1^2}{2} - \frac{1^4}{4}\right) - 0$$

$$= -\left(-\frac{1}{4}\right) + \frac{1}{4} = \frac{1}{2}.$$

Note: You can use the symmetry of the graphs and let area $= 2\int_{0}^{1} \left(x - x^3\right)dx$. An alternate solution is to find the area using a calculator.

$$\text{Enter } \int \left(abs\left(x^\wedge 3 - x\right), x, -1, 1\right) \text{ and obtain } \frac{1}{2}.$$

Example 2

Find the area of the region bounded by the curve $y = e^x$, the y-axis and the line $y = e^2$.

Step 1. Sketch a graph. See Figure 8.3-10.

Step 2. Find the point of intersection. Set $e^2 = e^x \Rightarrow x = 2$.

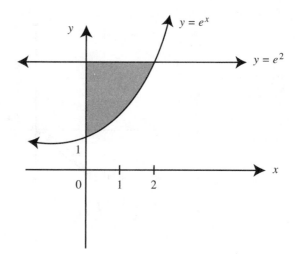

Figure 8.3-10

Step 3. Set up an integral:

$$\text{Area} = \int_0^2 (e^2 - e^x)dx = (e^2)x - e^x\Big]_0^2$$

$$= (2e^2 - e^2) - (0 - e^0)$$

$$= e^2 + 1.$$

Or using a calculator, enter $\int \left((e^\wedge 2 - e^\wedge x), x, 0, 2\right)$ and obtain $(e^2 + 1)$.

Example 3

Using a calculator, find the area of the region bounded by $y = \sin x$ and $y = \dfrac{x}{2}$ between $0 \le x \le \pi$.

Step 1. Sketch a graph. See Figure 8.3-11.

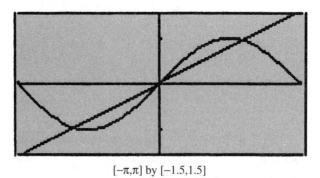

$[-\pi,\pi]$ by $[-1.5,1.5]$

Figure 8.3-11

Step 2. Find the points of intersection.
Using the Intersection function of the calculator, the intersection points are $x = 0$ and $x = 1.89549$.

Step 3. Enter nInt($\sin(x) - .5x$, x, 0, 1.89549) and obtain $0.420798 \approx 0.421$.

(Note: You could also use the $\int$ function on your calculator and get the same result.)

Example 4

Find the area of the region bounded by the curve $xy = 1$ and the lines $y = -5$, $x = e$ and $x = e^3$.

Step 1. Sketch a graph. See Figure 8.3-12.

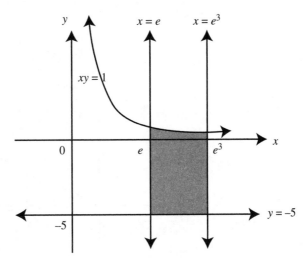

Figure 8.3-12

Step 2. Set up an integral:

$$\text{Area} = \int_e^{e^3} \left(\frac{1}{x} - (-5) \right) dx$$

Step 3. Evaluate the integral:

$$\text{Area} = \int_e^{e^3} \left(\frac{1}{x} - (-5) \right) dx = \int_e^{e^3} \left(\frac{1}{x} + 5 \right) dx$$

$$= \ln |x| + 5x]_e^{e^3} = \left[\ln(e^3) + 5(e^3) \right] - \left[\ln(e) + 5(e) \right]$$

$$= 3 + 5e^3 - 1 - 5e = 2 - 5e + 5e^3.$$

• Remember: if $f' > 0$, then f is increasing and if $f'' > 0$ then the graph of f is concave upward.

8.4 VOLUMES AND DEFINITE INTEGRALS

Main Concepts: *Solids with Known Cross Sections, The Disc Method, The Washer Method*

Solids with Known Cross Sections

If $A(x)$ is the area of a cross section of a solid and $A(x)$ is continuous on $[a, b]$, then the volume of the solid from $x = a$ to $x = b$ is:

$$V = \int_a^b A(x)dx$$

See Figure 8.4-1.

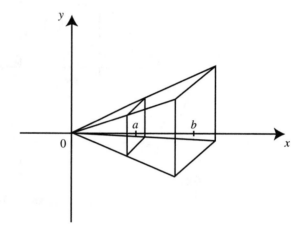

Figure 8.4-1

Note: A cross section of a solid is perpendicular to the height of the solid.

Example 1

The base of a solid is the region enclosed by the ellipse $\dfrac{x^2}{4} + \dfrac{y^2}{25} = 1$. The cross sections are perpendicular to the x-axis and are isosceles right triangles whose hypotenuse are on the ellipse. Find the volume of the solid. See Figure 8.4-2.

Step 1. Find the area of a cross section $A(x)$.

Pythagorean Theorem: $a^2 + a^2 = (2y)^2$

$$2a^2 = 4y^2$$

$$a = \sqrt{2}y, \ a > 0$$

$$A(x) = \frac{1}{2}a^2 = \frac{1}{2}\left(\sqrt{2}y\right)^2 = y^2$$

Since $\dfrac{x^2}{4} + \dfrac{y^2}{25} = 1, \dfrac{y^2}{25} = 1 - \dfrac{x^2}{4}$ or $y^2 = 25 - \dfrac{25x^2}{4}$

$$A(x) = 25 - \frac{25x^2}{4}$$

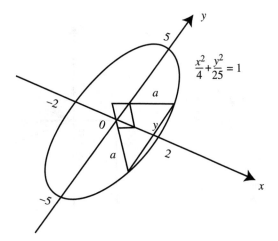

Figure 8.4-2

Step 2. Set up an integral.

$$V = \int_{-2}^{2} \left(25 - \frac{25x^2}{4} \right) dx$$

Step 3. Evaluate the integral.

$$V = \int_{-2}^{2} \left(25 - \frac{25x^2}{4} \right) dx = 25x - \frac{25}{12}x^3 \Big]_{-2}^{2}$$

$$= \left(25(2) - \frac{25}{12}(2)^3 \right) - \left(25(-2) - \frac{25}{12}(-2)^3 \right)$$

$$= \frac{100}{3} - \left(-\frac{100}{3} \right) = \frac{200}{3}.$$

The volume of the solid is $\frac{200}{3}$.

Verify your result with a graphing calculator.

Example 2

Find the volume of a pyramid whose base is a square with a side of 6 feet long, and a height of 10 feet. See Figure 8.4-3.

Step 1. Find the area of a cross section $A(x)$. Note each cross section is a square of side 2s.

Similar triangles: $\dfrac{x}{s} = \dfrac{10}{3} \Rightarrow s = \dfrac{3x}{10}$

$A(x) = (2s)^2 = 4s^2 = 4\left(\dfrac{3x}{10} \right)^2 = \dfrac{9x^2}{25}$

Step 2. Set up an integral.

$$V = \int_{0}^{10} \frac{9x^2}{25} dx$$

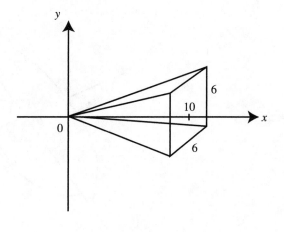

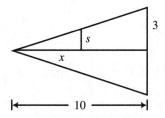

Figure 8.4-3

Step 3. Evaluate the integral.

$$V = \int_0^{10} \frac{9x^2}{25} dx = \frac{3x^3}{25}\bigg]_0^{10} = \frac{3(10)^3}{25} - 0 = 120.$$

The volume of the pyramid is 120 ft^3.

Example 3

The base of a solid is the region enclosed by a triangle whose vertices are (0, 0), (4, 0) and (0, 2). The cross sections are semicircles perpendicular to the x-axis. Using a calculator, find the volume of the solid. (See Figure 8.4-4.)

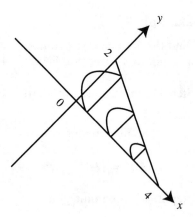

Figure 8.4-4

Step 1. Find the area of a cross section. Equation of the line passing through (0, 2) and (4, 0):

$$y = mx + b; \; m = \frac{0 - 2}{4 - 0} = -\frac{1}{2}; \; b = 2$$

$$y = -\frac{1}{2}x + 2$$

Area of semicircle $= \frac{1}{2}\pi r^2; \; r = \frac{1}{2}y = \frac{1}{2}\left(-\frac{1}{2}x + 2\right) = -\frac{1}{4}x + 1$

$$A(x) = \frac{1}{2}\pi\left(\frac{y}{2}\right)^2 = \frac{\pi}{2}\left(-\frac{1}{4}x + 1\right)^2.$$

Step 2. Set up an integral.

$$V = \int_0^4 A(x)\,dx = \int_0^4 \frac{\pi}{2}\left(-\frac{1}{4}x + 1\right)^2 dx$$

Step 3. Evaluate the integral.

Enter $\int\left(\left(\frac{\pi}{2}\right) * (-.25x + 1)^\wedge 2, \; x, \; 0, \; 4\right)$ and obtain 2.0944.

Thus the volume of the solid is 2.094.

> • Remember: if $f' < 0$, then f is decreasing and if $f'' < 0$ then the graph of f is concave downward.

The Disc Method

The volume of a solid of revolution using discs:
Revolving about the x-axis:

$$V = \pi \int_a^b \left(f(x)\right)^2 dx, \; f(x) = \text{radius}$$

Revolving about the y-axis:

$$V = \pi \int_c^d (g(y))^2 \, dy, \; g(y) = \text{radius}$$

See Figure 8.4-5.
Revolving about a line $y = k$:

$$V = \pi \int_a^b \left(f(x) - k\right)^2 dx, \; \text{where } |f(x) - k| = \text{radius}$$

Revolving about the a line $x = h$:

$$V = \pi \int_c^d \left(g(y) - h\right)^2 dy, \; \text{where } |g(y) - h| = \text{radius}$$

See Figure 8.4-6 on page 247.

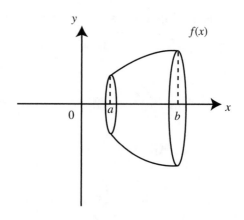

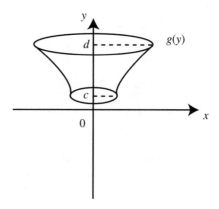

Figure 8.4-5

Example 1

Find the volume of the solid generated by revolving the region bounded by the graph of $f(x) = \sqrt{x-1}$, the x-axis and the line $x = 5$ about the x-axis.

Step 1. Draw a sketch. See Figure 8.4-7.

Step 2. Determine the radius of a disc from a cross section.

$$r = f(x) = \sqrt{x-1}$$

Step 3. Set up an integral.

$$V = \pi \int_1^5 (f(x))^2 dx = \pi \int_1^5 \left(\sqrt{x-1}\right)^2 dx$$

Step 4. Evaluate the integral.

$$V = \pi \int_1^5 \left(\sqrt{x-1}\right)^2 dx = \pi\,[(x-1)]_1^5 = \pi \left[\frac{x^2}{2} - x\right]_1^5$$

$$= \pi \left(\left(\frac{5^2}{2} - 5\right) - \left(\frac{1^2}{2} - 1\right)\right) = 8\pi$$

Verify your result with a calculator.

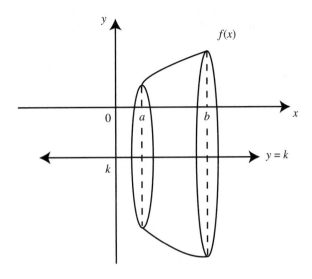

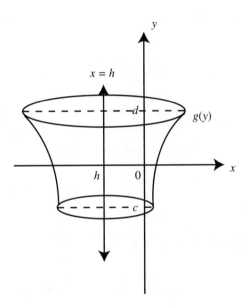

Figure 8.4-6

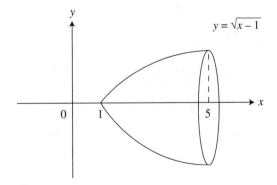

Figure 8.4-7

Example 2

Find the volume of the solid generated by revolving about the x-axis the region bounded by the graph of $y = \sqrt{\cos x}$ where $0 \leq x \leq \frac{\pi}{2}$, the x-axis and the y-axis.

Step 1. Draw a sketch. See Figure 8.4-8.

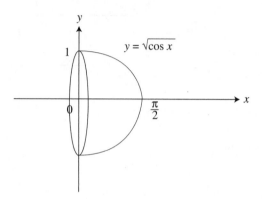

Figure 8.4-8

Step 2. Determine the radius from a cross section.

$$r = f(x) = \sqrt{\cos x}$$

Step 3. Set up an integral.

$$V = \pi \int_0^{\pi/2} \left(\sqrt{\cos x}\right)^2 dx = \pi \int_0^{\pi/2} \cos x\, dx.$$

Step 4. Evaluate the integral.

$$V = \pi \int_0^{\pi/2} \cos x\, dx = \pi \left[\sin x\right]_0^{\pi/2} = \pi \left(\sin\left(\frac{\pi}{2}\right) - \sin 0\right) = \pi.$$

Thus the volume of the solid is π.
Verify your result with a calculator.

Example 3

Find the volume of the solid generated by revolving about the y-axis the region in the first quadrant bounded by the graph of $y = x^2$, the y-axis, and the line $y = 6$.

Step 1. Draw a sketch. See Figure 8.4-9.

Step 2. Determine the radius from a cross section.

$$y = x^2 \Rightarrow x = \pm\sqrt{y}$$

$x = \sqrt{y}$ is the part of the curve involved in the region.

$$r = x = \sqrt{y}$$

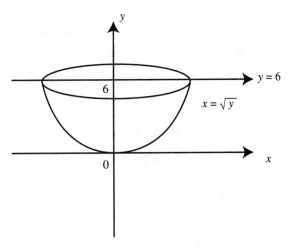

Figure 8.4-9

Step 3. Set up an integral.

$$V = \pi \int_0^6 x^2 dy = \pi \int_0^6 \left(\sqrt{y}\right)^2 dy = \pi \int_0^6 y\, dy.$$

Step 4. Evaluate the integral.

$$V = \pi \int_0^6 y\, dy = \pi \left[\frac{y^2}{2}\right]_0^6 = 18\pi.$$

The volume of the solid is 18π.
Verify your result with a calculator.

Example 4

Using a calculator, find the volume of the solid generated by revolving the region bounded by the graph of $y = x^2 + 4$, the line $y = 8$ about the line $y = 8$.

Step 1. Draw a sketch. See Figure 8.4-10.

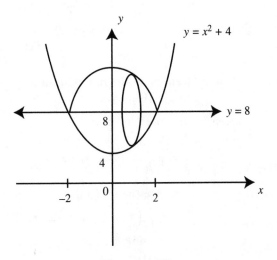

Figure 8.4-10

Step 2. Determine the radius from a cross section.
$$r = 8 - y = 8 - (x^2 + 4) = 4 - x^2$$

Step 3. Set up an integral.
To find the intersection points, set $8 = x^2 + 4 \Rightarrow x = \pm 2$.
$$V = \pi \int_{-2}^{2} \left(4 - x^2\right)^2 dx$$

Step 4. Evaluate the integral.

Enter $\int \left(\pi \left(4 - x^\wedge 2\right)^\wedge 2,\ x,\ -2,\ 2\right)$ and obtain $\dfrac{512}{15}\pi$.

Thus the volume of the solid is $\dfrac{512}{15}\pi$.
Verify your result with a calculator.

Example 5

Using a calculator, find the volume of the solid generated by revolving the region bounded by the graph of $y = e^x$, the y-axis, the lines $x = \ln 2$ and $y = -3$ about the line $y = -3$.

Step 1. Draw a sketch. See Figure 8.4-11.

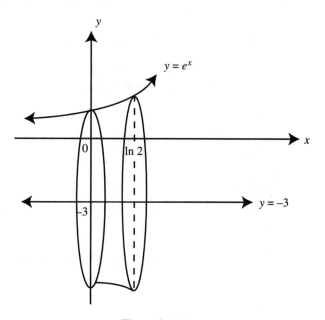

Figure 8.4-11

Step 2. Determine the radius from a cross section.
$$r = y - (-3) = y + 3 = e^x + 3.$$

Step 3. Set up an integral.
$$V = \pi \int_{0}^{\ln 2} \left(e^x + 3\right)^2 dx.$$

Step 4. Evaluate the integral.
Enter $\int \left(\pi \left(e^\wedge(x) + 3\right)^\wedge 2,\ x,\ 0\ \ln(2)\right)$ and obtain $\pi \left(9 \ln 2 + \dfrac{15}{2}\right)$
$= 13.7383\pi$

The volume of the solid is approximately 13.7383π.

> • Remember: if f' is increasing, then $f'' > 0$ and the graph of f is concave upward.

The Washer Method

The volume of a solid (with a hole in the middle) generated by revolving a region bounded by 2 curves:

About the x-axis:

$$V = \pi \int_a^b \left[(f(x))^2 - (g(x))^2 \right] dx; \text{ where } f(x) = \text{outer radius } \& \ g(x) = \text{inner radius}$$

About the y-axis:

$$V = \pi \int_c^d \left[(p(y))^2 - (q(y))^2 \right] dy; \text{ where } p(y) = \text{outer radius } \& \ q(y) = \text{inner radius}$$

About a line $x = h$:

$$V = \pi \int_a^b \left[(f(x) - h)^2 - (g(x) - h)^2 \right] dx$$

About a line $y = k$:

$$V = \pi \int_c^d \left[(p(y) - k)^2 - (q(y) - k)^2 \right] dy$$

Example 1

Using the Washer Method, find the volume of the solid generated by revolving the region bounded by $y = x^3$ and $y = x$ in the first quadrant about the x-axis.

Step 1. Draw a sketch. See Figure 8.4-12.

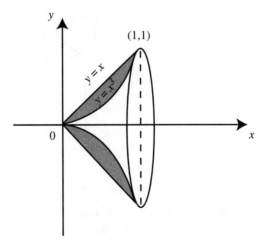

Figure 8.4-12

To find the points of intersection, set $x = x^3 \Rightarrow x^3 - x = 0$ or $x(x^2 - 1) = 0$. or $x = -1, 0, 1$. In the first quadrant $x = 0, 1$.

Step 2. Determine the outer and inner radii of a washer, whose outer radius $= x$; and inner radius $= x^3$.

Step 3. Set up an integral.

$$V = \int_0^1 \left[x^2 - \left(x^3 \right)^2 \right] dx$$

Step 4. Evaluate the integral.

$$V = \int_0^1 \left(x^2 - x^6 \right) dx = \pi \left[\frac{x^3}{3} - \frac{x^7}{7} \right]_0^1$$

$$= \pi \left(\frac{1}{3} - \frac{1}{7} \right) = \frac{4\pi}{21}$$

Verify your result with a calculator.

Example 2

Using the Washer Method and a calculator, find the volume of the solid generated by revolving the region in Example 1 about the line $y = 2$.

Step 1. Draw a sketch. See Figure 8.4-13.

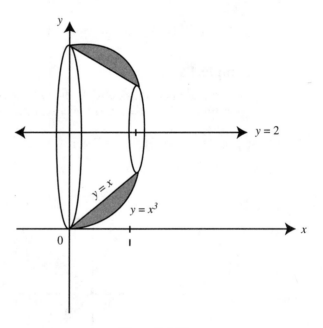

Figure 8.4-13

Step 2. Determine the outer & inner radii of a washer.
The outer radius $= (2 - x^3)$ and inner radius $= (2 - x)$.

Step 3. Set up an integral.

$$V = \pi \int_0^1 \left[\left(2 - x^3 \right)^2 - \left(2 - x \right)^2 \right] dx$$

Step 4. Evaluate the integral.

Enter $\int \left(\pi * \left((2 - x\char`\^3)\char`\^2 - (2 - x)\char`\^2\right), x, 0, 1\right)$ and obtain $\dfrac{17\pi}{21}$.

The volume of the solid is $\dfrac{17\pi}{21}$.

Example 3

Using the Washer Method and a calculator, find the volume of the solid generated by revolving the region bounded by $y = x^2$ and $x = y^2$ about the y-axis.

Step 1. Draw a sketch. See Figure 8.4-14.

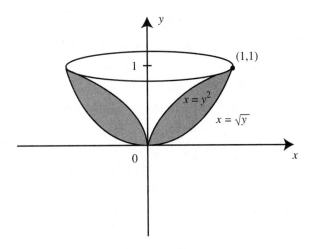

Figure 8.4-14

Intersection points: $y = x^2$; $x = y^2 \Rightarrow y = \pm\sqrt{x}$

Set $x^2 = \sqrt{x} \Rightarrow x^4 = x \Rightarrow x^4 - x = 0 \Rightarrow x(x^3 - 1) = 0 \Rightarrow x = 0$ or $x = 1$

$x = 0, \; y = 0 \; (0, 0)$

$x = 1, \; y = 1 \; (1, 1).$

Step 2. Determine the outer and inner radii of a washer, whose outer radius: $x = \sqrt{y}$ and inner radius: $x = y^2$.

Step 3. Set up an integral.

$$V = \pi \int_0^1 \left((\sqrt{y})^2 - (y^2)^2\right) dy$$

Step 4. Evaluate the integral

Enter $\int \left(\pi * \left((\sqrt{y})\char`\^2 - (y\char`\^2)\char`\^2\right), y, 0, 1\right)$ and obtain $\dfrac{3\pi}{10}$.

The volume of the solid is $\dfrac{3\pi}{10}$.

BC 8.5 INTEGRATION OF PARAMETRIC, POLAR AND VECTOR CURVES

Main Concepts: *Area, Arc Length, and Surface Area for Parametric Curves; Area and Arc Length for Polar Curves; Integration of a Vector-Valued Function*

Area, Arc Length, and Surface Area for Parametric Curves

Area for Parametric Curves

For a curve defined parametrically by $x = f(t)$ and $y = g(t)$, the area bounded by the portion of the curve between $t = \alpha$ and $t = \beta$ is $A = \int_{\alpha}^{\beta} g(t)f'(t)dt$.

Example 1

Find the area bounded by $x = 2\sin t$, $y = 3\sin^2 t$.

Step 1: Determine the limits of integration. The symmetry of the graph allows us to integrate from $t = 0$ to $t = \pi/2$ and multiply by 2.

Step 2: Differentiate $\dfrac{dx}{dt} = 2\cos t$

Step 3: $A = 2\int_{0}^{\pi/2} 3\sin^2 t (2\cos t)dt = 12\int_{0}^{\pi/2}(\sin^2 t \cos t)dt = 4\sin^3 t \Big|_{0}^{\pi/2} = 4$

Arc Length for Parametric Curves

The length of that arc is $L = \int_{\alpha}^{\beta}\sqrt{\left(\dfrac{dx}{dt}\right)^2 + \left(\dfrac{dy}{dt}\right)^2}\,dt$

Example 2

Find the length of the arc defined by $x = e^t \cos t$ and $y = e^t \sin t$ from $t = 0$ to $t = 4$.

Step 1: Differentiate $\dfrac{dx}{dt} = e^t \cos t - e^t \sin t$ and $\dfrac{dy}{dt} = e^t \cos t + e^t \sin t$

Step 2: $L = \int_{0}^{4}\sqrt{(e^t \cos t - e^t \sin t)^2 + (e^t \cos t + e^t \sin t)^2}\,dt$

$L = \int_{0}^{4}\sqrt{2e^{2t}(\cos^2 t + \sin^2 t)}\,dt = \int_{0}^{4}\sqrt{2e^{2t}}\,dt = \sqrt{2}\int_{0}^{4} e^t dt = \sqrt{2}e^t \Big|_{0}^{4}$

$= \sqrt{2}e^4 - \sqrt{2}$

Surface Area for Parametric Curves

The surface area created when that arc is revolved about the x-axis is

$$S = \int_{\alpha}^{\beta} 2\pi y \sqrt{\left(\dfrac{dx}{dt}\right)^2 + \left(\dfrac{dy}{dt}\right)^2}\,dt.$$

Example 3

Find the area of the surface generated by revolving about the x-axis the arc defined by $x = 3 - 2t$ and $y = \sqrt{20 - t^2}$ when $0 \le t \le 4$.

Step 1: Differentiate $\dfrac{dx}{dt} = -2$ and $\dfrac{dy}{dt} = \dfrac{-t}{\sqrt{20 - t^2}}$

Step 2: $S = \displaystyle\int_0^4 2\pi\sqrt{20-t^2}\sqrt{(-2)^2 + \left(\dfrac{-t}{\sqrt{20-t^2}}\right)^2}\, dt = 2\pi\int_0^4 \sqrt{20-t^2}\sqrt{4 + \dfrac{t^2}{20-t^2}}\, dt$

$= 2\pi\displaystyle\int_0^4 \sqrt{20-t^2}\sqrt{\dfrac{80-3t^2}{20-t^2}}\, dt = 2\pi\int_0^4 \sqrt{80-3t^2}\, dt \approx 2\pi(31.7768) \approx 199.6595$

Area and Arc Length for Polar Curves

Area for Polar Curves

If $r = f(\theta)$ is a continuous polar curve on the interval $\alpha \le \theta \le \beta$ and $\alpha < \beta < \alpha + 2\pi$, then the area enclosed by the polar curve is $A = \dfrac{1}{2}\displaystyle\int_\alpha^\beta [f(\theta)]^2\, d\theta = \dfrac{1}{2}\int_\alpha^\beta r^2\, d\theta$.

Example 1

Find the area enclosed by $r = 2 + 2\cos\theta$ on the interval from $\theta = 0$ to $\theta = \pi$.

Step 1: Square $r^2 = 4 + 8\cos\theta + 4\cos^2\theta$

Step 2: $A = \dfrac{1}{2}\displaystyle\int_0^\pi \left(4 + 8\cos\theta + 4\cos^2\theta\right) d\theta = 2\int_0^\pi \left(1 + 2\cos\theta + 2\cos^2\theta\right) d\theta$

$= 2\left[2\theta + 4\sin\theta + 2\left(\dfrac{\theta}{2} + \dfrac{1}{4}\sin 2\theta\right)\right]_0^\pi = 6\theta + 8\sin\theta + \sin 2\theta \,\big|_0^\pi = 6\pi$

Arc Length for Polar Curves

For a polar graph defined on a interval (α, β), if the graph does not retrace itself in that interval and if $\dfrac{dr}{d\theta}$ is continuous, then the length of the arc from $\theta = \alpha$ to $\theta = \beta$ is $L = \displaystyle\int_\alpha^\beta \sqrt{r^2 + \left(\dfrac{dr}{d\theta}\right)^2}\, d\theta$.

Example 2

Find the length of the spiral $r = e^\theta$ from $\theta = 0$ to $\theta = \pi$.

Step 1: Differentiate $\dfrac{dr}{d\theta} = e^\theta$

Step 2: Square $r^2 = e^{2\theta}$

Step 3: $L = \int_0^\pi \sqrt{e^{2\theta} + e^{2\theta}}\,d\theta = \int_0^\pi \sqrt{2e^{2\theta}}\,d\theta = \sqrt{2}\int_0^\pi e^\theta\,d\theta = \sqrt{2}\,e^\theta\Big|_0^\pi$

$\qquad = \sqrt{2}e^\pi - \sqrt{2}$

Integration of a Vector-Valued Function

Integrating a Vector Function

For a vector-valued function $r(t) = \langle x(t), y(t) \rangle$, $\int r(t)\,dt = \int x(t)\,dt \cdot i + \int y(t)\,dt \cdot j$.

Example 1

The acceleration vector of a particle at any time $t \geq 0$ is $a(t) = \langle e^t, e^{2t} \rangle$. If at time $t = 0$, its velocity is $i + j$ and its displacement is 0, find the functions for the position and velocity at any time t.

Step 1: $a(t) = \langle e^t, e^{2t} \rangle$ so $v(t) = \int a(t)\,dt = \int x(t)\,dt \cdot i + \int y(t)\,dt \cdot j$

$v(t) = \int e^t\,dt \cdot i + \int e^{2t}\,dt \cdot j = e^t \cdot i + \dfrac{e^{2t}}{2} \cdot j + C$. Since velocity at $t = 0$

is known to be $i + j$, $i + \dfrac{1}{2} \cdot j + C = i + j$, and $C = \dfrac{1}{2} \cdot j$; therefore, $v(t) =$

$e^t \cdot i + \dfrac{e^{2t}}{2} \cdot j + \dfrac{1}{2} \cdot j = \left\langle e^t, \dfrac{e^{2t} + 1}{2} \right\rangle$.

Step 2: The position function $s(t) = \int v(t)\,dt = \int e^t\,dt \cdot i + \int \dfrac{e^{2t} + 1}{2}\,dt \cdot j$.

$s(t) = \int v(t)\,dt = \int e^t\,dt \cdot i + \int \dfrac{e^{2t} + 1}{2}\,dt \cdot j = e^t \cdot i + \left(\dfrac{e^{2t} + 2t}{4} \right) \cdot j +$

C. Displacement is 0 at $t = 0$, so $i + \dfrac{1}{4} \cdot j + C = 0$ and $C = -i - \dfrac{1}{4} \cdot j$.

The position function $s(t) = e^t \cdot i + \left(\dfrac{e^{2t} + 2t}{4} \right) \cdot j - i - \dfrac{1}{4} \cdot j = (e^t - 1)\,i$

$+ \left(\dfrac{e^{2t} + 2t - 1}{4} \right) j = \left\langle e^t - 1, \dfrac{e^{2t} + 2t - 1}{4} \right\rangle$.

Length of a Vector Curve

The length of a curve defined by the vector-valued function $r(t) = \langle x(t), y(t) \rangle$ traced from $t = a$ to $t = b$ is $s = \int_a^b \|r'(t)\|\,dt$.

Example 2

Find the length of the curve $r(t) = \langle 2\sin t, 5t \rangle$ from $t = 0$ to $t = \pi$.

Step 1: $r'(t) = \langle 2\cos t, 5 \rangle$

Step 2: $\|r'(t)\| = \sqrt{4\cos^2 t + 25}$

Step 3: With the aid of a graphing calculator, the arc length $s = \int_0^\pi \sqrt{4\cos^2 t + 25}\,dt$ can be found to be approximately equal to 16.319 units.

8.6 RAPID REVIEW

1. If $f(x) = \int_0^x g(t)dt$ and the graph of g is shown in Figure 8.6-1. Find $f(3)$.

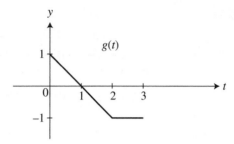

Figure 8.6-1

Answer: $f(3) = \int_0^3 g(t)dt = \int_0^1 g(t)\,dt + \int_1^3 g(t)\,dt$

$= 0.5 - 1.5 = -1.$

2. The function f is continous on $[1, 5]$ and $f > 0$ and selected values of f are given below.

x	1	2	3	4	5
$f(x)$	2	4	6	8	10

Using 2 midpoint rectangles, approximate the area under the curve of f for $x = 1$ to $x = 5$.

Answer: Midpoints are $x = 2$ and $x = 4$ and the width of each rectangle $= \dfrac{5-1}{2} = 2.$

Area $\approx$ Area of Rect$_1$ + Area of Rect$_2$ $\approx 4(2) + 8(2) \approx 24.$

3. Set up an integral to find the area of the regions bounded by the graphs of $y = x^3$ and $y = x$. Do not evaluate the integral.

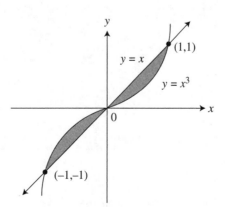

Figure 8.6-2

Answer: Graphs interesect at $x = -1$ and $x = 1$. See Figure 8.6-2.

$$\text{Area} = \int_{-1}^{0} \left(x^3 - x \right) dx + \int_{0}^{1} \left(x - x^3 \right) dx$$

Or, using symmetry, $\text{Area} = 2 \int_{0}^{1} \left(x - x^3 \right) dx.$

4. The base of a solid is the region bounded by the lines $y = x$, $x = 1$, and the x-axis. The cross sections are squares perpendicular to the x-axis. Set up an integral to find the volume of the solid. Do not evaluate the integral.

Answer: Area of cross section $= x^2$

$$\text{Volume of solid} = \pi \int_{0}^{1} x^2 dx.$$

5. Set up an integral to find the volume of a solid generated by revolving about the x-axis the region bounded by the graph of $y = \sin x$, where $0 \leq x \leq \pi$ and the x-axis. Do not evaluate the integral.

Answer: Volume $= \pi \int_{0}^{\pi} (\sin x)^2 dx.$

6. The area under the curve of $y = \dfrac{1}{x}$ from $x = a$ to $x = 5$ is apprximately 0.916 where $1 \leq a < 5$. Using your calculator, find a.

Answer: $\displaystyle\int_{a}^{5} \frac{1}{x} dx = \ln x \big|_{a}^{5} = \ln 5 - \ln a = 0.916$

$\ln a = \ln 5 - 0.916 \approx .693$

$a \approx e^{0.693} \approx 2.$

7. Find the length of the arc defined by $x = t^2$ and $y = 3t^2 - 1$ from $t = 2$ to $t = 5$.

Answer: $\dfrac{dx}{dt} = 2t$ and $\dfrac{dy}{dt} = 6t$. $L = \displaystyle\int_{2}^{5} \sqrt{(2t)^2 + (6t)^2} dt = \int_{2}^{5} \sqrt{40t^2} dt$

$= \displaystyle\int_{2}^{5} 2t\sqrt{10} dt = \left[t^2 \sqrt{10} \right]_{2}^{5} = 25\sqrt{10} - 4\sqrt{10} = 21\sqrt{10}.$

8. Find the area bounded by the $r = 3 + \cos \theta$.

Answer: To trace out the graph completely, without retracing, we need $0 \leq \theta \leq 2\pi$. Then

$$A = \frac{1}{2} \int_{0}^{2\pi} (3 + \cos \theta)^2 d\theta = \frac{1}{2} \int_{0}^{2\pi} \left(9 + 6 \cos \theta + \cos^2 \theta \right) d\theta$$

$$= \frac{1}{2} \left[9\theta + 6 \sin \theta + \frac{1}{2}\theta + \frac{1}{4} \sin 2\theta \right]_{0}^{2\pi} = \frac{1}{2} [(18\pi + \pi) - 0] = \frac{19\pi}{2}.$$

9. Find the area of the surface formed when the curve defined by $x = \sin \theta$, $y = 3 \sin \theta$ on the interval $\dfrac{\pi}{3} \leq \theta \leq \dfrac{\pi}{6}$ is revolved about the x-axis.

Answer: $\dfrac{dx}{d\theta} = \cos\theta$ and $\dfrac{dy}{d\theta} = 3\cos\theta$ so

$$S = \int_{\pi/6}^{\pi/3} 2\pi(3\sin\theta)\sqrt{\cos^2\theta + 9\cos^2\theta}\,d\theta = 6\pi \int_{\pi/6}^{\pi/3} \sin\theta\sqrt{10\cos^2\theta}\,d\theta$$

$$= 3\pi\sqrt{10}\int_{\pi/6}^{\pi/3} 2\sin\theta\cos\theta\,d\theta = 3\pi\sqrt{10}\int_{\pi/6}^{\pi/3} \sin 2\theta\,d\theta = -\frac{3}{2}\pi\sqrt{10}\cos 2\theta\Big]_{\pi/6}^{\pi/3}$$

$$= -\frac{3}{2}\pi\sqrt{10}\left[\left(\cos\frac{2\pi}{3}\right) - \left(\cos\frac{\pi}{3}\right)\right] = -\frac{3}{2}\pi\sqrt{10}\left[\left(-\frac{1}{2}\right) - \left(-\frac{1}{2}\right)\right] = \frac{3}{2}\pi\sqrt{10}$$

10. If $\left\langle \dfrac{dx}{dt}, \dfrac{dy}{dt} \right\rangle = \langle 5 - t^2, 4t - 3\rangle$ and $\langle x_0, y_0\rangle = \langle 0, 0\rangle$, find $\langle x, y\rangle$.

Answer: $x = \displaystyle\int \left(5 - t^2\right)dt = 5t - \frac{t^3}{3} + c_1$ and $y = \displaystyle\int (4t - 3)dt = 2t^2 - 3t + c_2$.

Since $\langle x_0, y_0\rangle = \langle 0, 0\rangle$, $c_1 = c_2 = 0$ so $\langle x, y\rangle = \left\langle 5t - \frac{1}{3}t^3,\ 2t^2 - 3t\right\rangle$.

8.7 PRACTICE PROBLEMS

Part A—The use of a calculator is not allowed.

1. Let $F(x) = \displaystyle\int_0^x f(t)dt$ where the graph of f is given in Figure 8.7-1.

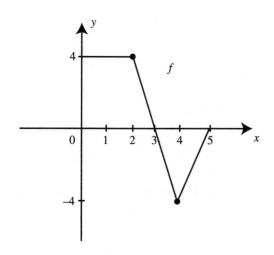

Figure 8.7-1

(a) Evaluate $F(0)$, $F(3)$, and $F(5)$.
(b) On what interval(s) is F decreasing?
(c) At what value of t does F have a maximum value?
(d) On what interval is F concave up?

2. Find the area of the region(s) enclosed by the curve $f(x) = x^3$, the x-axis, and the lines $x = -1$ and $x = 2$.

3. Find the area of the region(s) enclosed by the curve $y = |2x - 6|$, the x-axis, and the lines $x = 0$ and $x = 4$.

4. Find the approximate area under the curve $f(x) = \dfrac{1}{x}$ from $x = 1$ to $x = 5$, using four right-endpoint rectangles of equal lengths.

5. Find the approximate area under the curve $y = x^2 + 1$ from $x = 0$ to $x = 3$, using the Trapezoidal Rule with $n = 3$.

6. Find the area of the region bounded by the graphs $y = \sqrt{x}$, $y = -x$, and $x = 4$.

7. Find the area of the region bounded by the curves $x = y^2$ and $x = 4$.

8. Find the area of the region bounded by the graphs of all four equations: $f(x) = \sin\left(\dfrac{x}{2}\right)$; x-axis; and the lines, $x = \dfrac{\pi}{2}$ and $x = \pi$.

9. Find the volume of the solid obtained by revolving about the x-axis, the region bounded by the graphs of $y = x^2 + 4$, the x-axis, the y-axis, and the lines $x = 3$.

10. The area under the curve $y = \dfrac{1}{x}$ from $x = 1$ to $x = k$ is 1. Find the value of k.

11. Find the volume of the solid obtained by revolving about the y-axis the region bounded by $x = y^2 + 1$, $x = 0$, $y = -1$, and $y = 1$.

12. Let R be the region enclosed by the graph $y = 3x$, the x-axis and the line $x = 4$. The line $x = a$ divides region R into two regions such that when the regions are revolved about the x-axis, the resulting solids have equal volume. Find a.

Part B—Calculators are allowed.

13. Find the volume of the solid obtained by revolving about the x-axis the region bounded by the graphs of $f(x) = x^3$ and $g(x) = x^2$.

14. The base of a solid is a region bounded by the circle $x^2 + y^2 = 4$. The cross of the solid sections are perpendicular to the x-axis and are equilateral triangles. Find the volume of the solid.

15. Find the volume of the solid obtained by revolving about the y-axis, the region bounded by the curves $x = y^2$ and $y = x - 2$.

For Problems 16 thru 19, find the volume of the solid obtained by revolving the region as described below. See Figure 8.7-2.

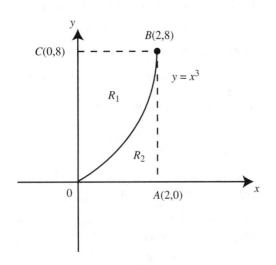

Figure 8.7-2

16. R_1 about the x-axis.

17. R_2 about the y-axis.

18. R_1 about the line $\overleftrightarrow{BC}$.

19. R_2 about the line $\overleftrightarrow{AB}$.

20. The function $f(x)$ is continuous on $[0, 12]$ and the selected values of $f(x)$ are shown below.

x	0	2	4	6	8	10	12
$f(x)$	1	2.24	3	3.61	4.12	4.58	5

Find the approximate area under the curve of f from 0 to 12 using three midpoint rectangles.

21. Find the area bounded by the curve defined by $x = 2 \cos t$ and $y = 3 \sin t$ from $t = 0$ to $t = \pi$.

22. Find the length of the arc of $r = \sin^2 \left(\dfrac{\theta}{2} \right)$ from $\theta = 0$ to $\theta = \pi$.

23. Find the area of the surface formed when the curve defined by $x = e^t \sin t$ and $y = e^t \cos t$ from $t = 0$ to $t = \dfrac{\pi}{2}$ is revolved about the x-axis.

24. Find the area bounded by $r = 2 + 2 \sin \theta$.

25. The acceleration vector for an object is $\langle -e^t, e^t \rangle$. Find the position of the object at $t = 1$ if the initial velocity is $v_0 = \langle 3, 1 \rangle$ and the initial position of the object is at the origin.

8.8 CUMULATIVE REVIEW PROBLEMS

"Calculator" indicates that calculators are permitted.

26. If $\displaystyle\int_{-a}^{a} e^{x^1} \, dx = k$, find $\displaystyle\int_{0}^{a} e^{x^2} \, dx$ in terms of k.

27. A man wishes to pull a log over a 9 foot high garden wall as shown. See Figure 8.8-1. He is pulling at a rate of 2 ft/sec. At what rate is the angle between the rope and the ground

changing when there are 15 feet of rope between the top of the wall and the log?

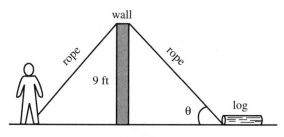

wall

rope

rope

9 ft

θ

log

Figure 8.8-1

28. (Calculator) Find a point on the parabola $y = \frac{1}{2}x^2$ that is closest to the point $(4, 1)$.

29. The velocity function of a particle moving along the x-axis is $v(t) = t\cos(t^2 + 1)$ for $t \geq 0$.

(a) If at $t = 0$, the particle is at the origin, find the position of the particle at $t = 2$.

(b) Is the particle moving to the right or left at $t = 2$?

(c) Find the acceleration of the particle at $t = 2$ and determine if the velocity of the particle is increasing or decreasing. Explain why.

30. (Calculator) given $f(x) = xe^x$ and $g(x) = \cos x$, find:

(a) the area of the region in the first quadrant bounded by the graphs $f(x)$, $g(x)$, and $x = 0$.

(b) The volume obtained by revolving the region in part (a) about the x-axis.

BC

31. Find the slope of the tangent line to the curve defined by $r = 5\cos 2\theta$ at the point where $\theta = \frac{3\pi}{2}$.

32. $\displaystyle\int \frac{2}{x^2 - 4x}dx$

33. $\displaystyle\int_e^\infty \frac{dx}{x}$

8.9 SOLUTIONS TO PRACTICE PROBLEMS

Part A—No calculators.

1. (a) $F(0) = \displaystyle\int_0^0 f(t)dt = 0$.

$F(3) = \displaystyle\int_0^3 f(t)dt = \frac{1}{2}(3 + 2)(4) = 10$.

$F(5) = \displaystyle\int_0^5 f(t)dt = \int_0^3 f(t)dt + \int_3^5 f(t)dt$

$= 10 + (-4) = 6$.

(b) Since $\displaystyle\int_3^5 f(t)dt \leq 0$, F is decreasing on the interval $[3, 5]$.

(c) At $t = 3$, F has a maximum value.

(d) $F'(x) = f(x)$, $F'(x)$ is increasing on $(4, 5)$ which implies $F \leq (x) > 0$. Thus F is concave upwards on $(4, 5)$.

2. See Figure 8.9-1.

$A = \left|\displaystyle\int_{-1}^0 x^3 dx\right| + \int_0^2 x^3 dx$

$= \left|\left[\dfrac{x^4}{4}\right]_{-1}^0\right| + \left[\dfrac{x^4}{4}\right]_0^2$

$= \left|0 - \dfrac{(-1)^4}{4}\right| + \left(\dfrac{2^4}{4} - 0\right)$

$= \dfrac{1}{4} + 4 = \dfrac{17}{4}$

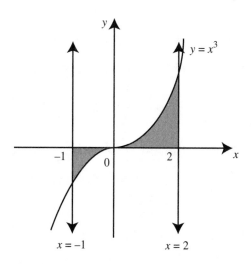

$y = x^3$

$x = -1$

$x = 2$

Figure 8.9-1

3. See Figure 8.9-2.
   Set $2x - 6 = 0$; $x = 3$ and

   $$f(x) = \begin{cases} 2x - 6 & \text{if } x \geq 3 \\ -(2x - 6) & \text{if } x < 3 \end{cases}$$

   $$A = \int_0^3 -(2x - 6)dx + \int_3^4 (2x - 6)dx$$

   $$= \left[-x^2 + 6x\right]_0^3 + \left[x^2 - 6x\right]_3^4 = \left[-(3)^2 + 6(3)\right]$$

   $$- 0 + \left[4^2 + 6(4)\right] - \left[3^2 - 6(3)\right]$$

   $$= 9 + 1 = 10.$$

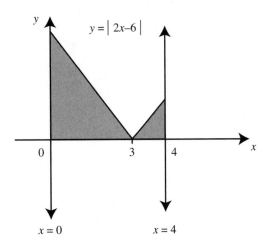

Figure 8.9-2

4. See Figure 8.9-3.
   Length of $\Delta x_1 = \dfrac{5 - 1}{4} = 1$

   $$\text{Area of Rect}_I = f(2)\Delta x_1 = \frac{1}{2}(1) = \frac{1}{2}$$

   $$\text{Area of Rect}_{II} = f(3)\Delta x_2 = \frac{1}{3}(1) = \frac{1}{3}$$

   $$\text{Area of Rect}_{III} = f(4)\Delta x_3 = \frac{1}{4}(1) = \frac{1}{4}$$

   $$\text{Area of Rect}_{IV} = f(5)\Delta x_4 = \frac{1}{5}(1) = \frac{1}{5}$$

   $$\text{Total Area} = \frac{1}{2} + \frac{1}{3} + \frac{1}{4} + \frac{1}{5} = \frac{77}{60}.$$

5. See Figure 8.9-4.

   $$\text{Trapezoid Rule} = \frac{b - a}{2n}\left(f(a) + 2f(x_1)\right)$$

$$+ 2f(x_2) + f(b))$$

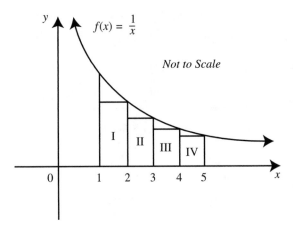

Figure 8.9-3

$$A = \frac{3 - 0}{2(3)}\left(f(0) + 2f(1) + 2f(2) + f(3)\right)$$

$$= \frac{1}{2}(1 + 4 + 10 + 10) = \frac{25}{2}.$$

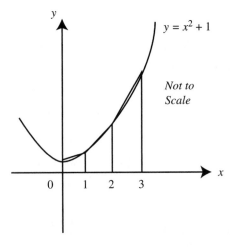

Figure 8.9-4

6. See Figure 8.9-5.

   $$A = \int_0^4 \left(\sqrt{x} - (-x)\right)dx = \int_0^4 \left(x^{1/2} + x\right)dx$$

   $$= \left[\frac{2x^{3/2}}{3} + \frac{x^2}{2}\right]_0^4 = \left(\frac{2(4)^{3/2}}{3} + \frac{4^2}{2}\right) - 0$$

   $$= \frac{16}{3} + 8 = \frac{40}{3}$$

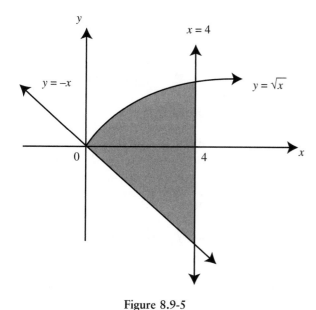

Figure 8.9-5

7. See Figure 8.9-6.
Intersection points: $4 = y^2 \Rightarrow y = \pm 2$

$$A = \int_{-2}^{2} \left(4 - y^2\right) dy = \left[4y - \frac{y^3}{3}\right]_{-2}^{2}$$

$$= \left(4(2) - \frac{2^3}{3}\right) - \left(4(-2) - \frac{(-2)^3}{3}\right)$$

$$= \left(8 - \frac{8}{3}\right) - \left(-8 + \frac{8}{3}\right)$$

$$= \frac{16}{3} + \frac{16}{3} = \frac{32}{3}$$

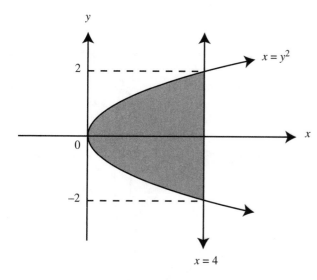

Figure 8.9-6

You can use the symmetry of the region and obtain the area $= 2\int_{-2}^{2} (4 - y^2) dy$. An alternative method is to find the area by setting up an integral with respect to the x-axis and expressing $x = y^2$ as $y = \sqrt{x}$ and $y = -\sqrt{x}$.

8. See Figure 8.9-7.

$$A = \int_{\pi/2}^{\pi} \sin\left(\frac{x}{2}\right) dx$$

Let $u = \frac{x}{2}$ and $du = \frac{dx}{2}$ or $2\,du = dx$.

$$\int \sin\left(\frac{x}{2}\right) dx = \int \sin u (2\,du) = 2\int \sin u\, du =$$
$$-2\cos u + c = -2\cos\left(\frac{x}{2}\right) + c$$

$$A = \int_{\pi/2}^{\pi} \sin\left(\frac{x}{2}\right) dx = \left[-2\cos\left(\frac{x}{2}\right)\right]_{\pi/2}^{\pi}$$

$$= -2\left[\cos\left(\frac{\pi}{2}\right) - \cos\left(\frac{\pi/2}{2}\right)\right]$$

$$= -2\left(\cos\left(\frac{\pi}{2}\right) - \cos\left(\frac{\pi}{4}\right)\right)$$

$$= -2\left(0 - \frac{\sqrt{2}}{2}\right) = \sqrt{2}$$

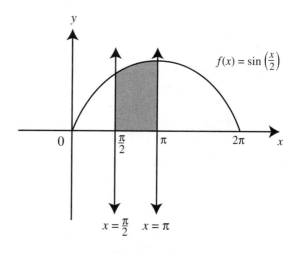

Figure 8.9-7

9. See Figure 8.9-8.
Disc Method:

$$V = \pi\int_{0}^{3} \left(x^2 + 4\right)^2 dx = \pi\int_{0}^{3} \left(x^4 + 8x^2 + 16\right) dx$$

$$= \pi\left[\frac{x^5}{5} + \frac{8x^3}{3} + 16x\right]_{0}^{3}$$

$$= \pi\left[\frac{3^5}{5} + \frac{8(3)^3}{3} + 16(3)\right] - 0 = \frac{843}{5}\pi.$$

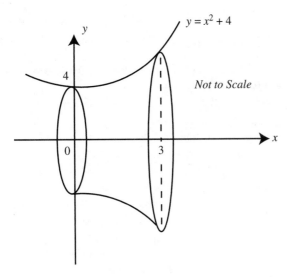

$y = x^2 + 4$

4

Not to Scale

0   3

Figure 8.9-8

10. Area $= \int_1^k \frac{1}{x} dx = \ln x]_1^k = \ln k - \ln 1 = \ln k$
Set $\ln k = 1$. Thus $e^{\ln k} = e^1$ or $k = e$.

11. See Figure 8.9-9.

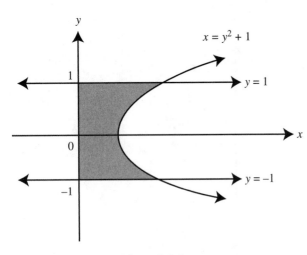

$x = y^2 + 1$

1      $y = 1$

0   x

-1 $y = -1$

Figure 8.9-9

Disc Method:

$$V = \pi \int_{-1}^{1} \left(y^2 + 1 \right)^2 dy$$

$$= \pi \int_{-1}^{1} \left(y^4 + 2y^2 + 1 \right) dy$$

$$= \pi \left[\frac{y^5}{5} + \frac{2y^3}{3} + y \right]_{-1}^{1}$$

$$= \pi \left[\left(\frac{1^5}{5} + \frac{2(1)^3}{3} + 1 \right) \right.$$

$$\left. - \left(\frac{(-1)^5}{5} + \frac{2(-1)^3}{3} + (-1) \right) \right]$$

$$= \pi \left(\frac{28}{15} + \frac{28}{15} \right) = \frac{56\pi}{15}.$$

Note: You can use the symmetry of the region and find the volume by $2\pi \int_0^1 \left(y^2 + 1 \right)^2 dy$.

12. Volume of solid by revolving R:

$$V_R = \int_0^4 \pi (3x)^2 dx = \pi \int_0^4 9x^2 dx = \pi \left[3x^2 \right]_0^4$$

$$= 192\pi.$$

Set $\int_0^4 \pi (3x)^2 dx = \frac{192\pi}{2}$

$$\Rightarrow 3a^3 \pi = 96\pi$$

$$a^3 = 32$$

$$a = (32)^{1/3} = 2 (2)^{2/3}.$$

You can verify your result by evaluating $\int_0^{2(2)^{2/3}} \pi (3x)^2 dx$. The result is 96π.

Part B—Calculators are permitted.

13. See Figure 8.9-10.

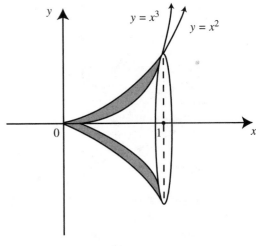

$y = x^3$

$y = x^2$

0   1   x

Figure 8.9-10

Step 1. Washer Method
Points of intersection: Set $x^3 = x^2 \Rightarrow$
$x^3 - x^2 = 0 \Rightarrow x^2(x-1) = 0$ or $x = 1$.
Outer radius $= x^2$; Inner radius $= x^3$.

Step 2. $V = \pi \int_0^1 \left(\left(x^2\right)^2 - \left(x^3\right)^2 \right) dx$

$= \pi \int_0^1 (x^4 - x^6) dx$

Step 3. Enter $\int \left(\pi \left(x^\wedge 4 - x^\wedge 6\right), x, 0, 1 \right)$ and

obtain $\dfrac{2\pi}{35}$.

14. See Figure 8.9-11.

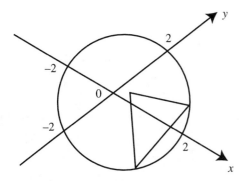

Figure 8.9-11

Step 1. $x^2 + y^2 = 4 \Rightarrow y^2 = 4 - x^2 \Rightarrow$
$y = \pm\sqrt{4 - x^2}$
Let $s = a$ side of an equilateral triangle
$s = 2\sqrt{4 - x^2}$.

Step 2. Area of a cross section:

$$A(x) = \frac{s^2 \sqrt{3}}{4} = \frac{\left(2\sqrt{4 - x^2}\right)^2 \sqrt{3}}{4}.$$

Step 3. $V = \int_{-2}^{2} \left(2\sqrt{4 - x^2}\right)^2 \frac{\sqrt{3}}{4} dx$

$= \int_{-2}^{2} \sqrt{3}(4 - x^\wedge 2) dx$

Step 4. Enter $\int \left(\sqrt{(3)} * (4 - x^2), x, -2, 2 \right)$

and obtain $\dfrac{32\sqrt{3}}{3}$.

15. See Figure 8.9-12.

Step 1. Washer Method
Points of Intersection:
$y = x - 2 \Rightarrow x = y + 2$

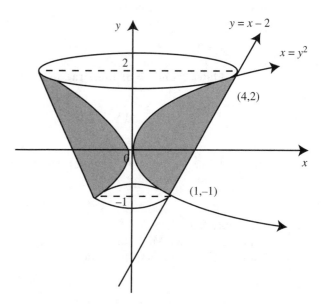

Figure 8.9-12

Set $y^2 = y + 2 \Rightarrow y^2 - y - 2 = 0$
$\Rightarrow (y - 2)(y + 1) = 0$ or $y = -1$ or $y = 2$.
Outer radius $= y + 2$ and inner radius
$= y^2$.

Step 2. $V = \pi \int_{-1}^{2} \left((y + 2)^2 - \left(y^2\right)^2 \right) dy$.

Step 3. Enter $\pi \int \left((y + 2)^2 - y^\wedge 4, -1, 2 \right)$ and

obtain $\dfrac{72}{5}\pi$.

16. See Figure 8.9-13.

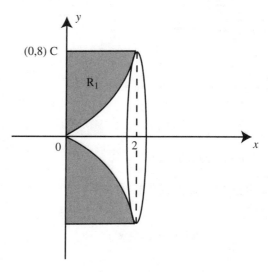

Figure 8.9-13

Step 1: Washer Method

$y = 8$, $y = x^3$

Outer radius $= 8$; Inner radius $= x^3$

$$V = \pi \int_0^2 \left(8^2 - \left(x^3 \right)^2 \right) dx$$

Step 2: Enter $\int \pi \left(8^2 - x^6,\ x,\ 0,\ 2 \right)$ and obtain
$\dfrac{768\pi}{7}$.

17. See Figure 8.9-14.
Using the Washer Method:
Outer radius: $x = 2$ and Inner radius: $x = y^{1/3}$

$$V = \pi \int_0^1 \left(2^2 - \left(y^{1/3} \right)^2 \right) dy$$

Using your calculator, you obtain $V = \dfrac{64\pi}{5}$.

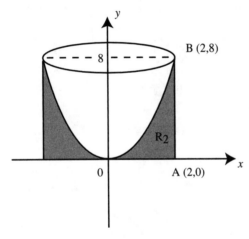

Figure 8.9-14

18. See Figure 8.9-15.

Step 1: Disc Method:

Radius $= (8 - x^3)$

$$V = \pi \int_0^2 \left(8 - x^3 \right)^2 dx$$

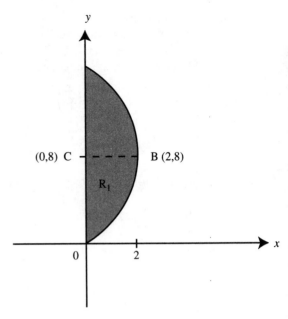

Figure 8.9-15

Step 2: Enter $\int \left(\pi * (8 - x^{\wedge}3)^{\wedge}2,\ x,\ 0,\ 2 \right)$ and
obtain $\dfrac{576\pi}{7}$.

19. See Figure 8.9-16.
Using the Disc Method:

Radius $= 2 - x = \left(2 - y^{1/3} \right)$

$$V = \pi \int_0^8 \left(2 - y^{1/3} \right)^2 dy$$

Using your calculator, you obtain $V = \dfrac{16\pi}{5}$.

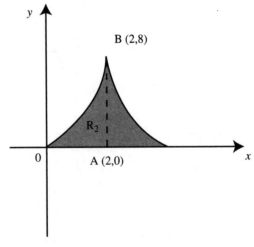

Figure 8.9-16

20. Area $= \sum_{i=1}^{3} f(x_i)\Delta x_i$

$x_i =$ midpoint of the ith interval.

Length of $\Delta x_i = \dfrac{12 - 0}{3} = 4$

Area of Rect$_I = f(2)\Delta x_1 = (2.24)(4) = 8.96$

Area of Rect$_{II} = f(6)\Delta x_2 = (3.61)(4) = 14.44$

Area of Rect$_{III} = f(10)\Delta x_3 = (4.58)(4) = 18.32$

Total Area $= 8.96 + 14.44 + 18.32 = 41.72$.

The area under the curve is approximately 41.72.

21. The area enclosed by the curve is the upper half of an ellipse. Find $\dfrac{dx}{dt} =$

$-2\sin t$. $A = \int_0^{\pi} (3\sin t)(-2\sin t)dt =$

$-6\int_0^{\pi} \sin^2 t\, dt = -6\left[\dfrac{1}{2}t - \dfrac{1}{4}\sin 2t\right]_0^{\pi} = -3\pi$.

The negative simply indicates that the area has been swept from right to left, rather than left to right, and so may be ignored. The area enclosed by the curve is 3π.

22. Differentiate to find $\dfrac{dr}{d\theta} = \sin\left(\dfrac{\theta}{2}\right)\cos\left(\dfrac{\theta}{2}\right)$ and

calculate $r^2 = \sin^4\left(\dfrac{\theta}{2}\right)$ and

$\left(\dfrac{dr}{d\theta}\right)^2 = \sin^2\left(\dfrac{\theta}{2}\right)\cos^2\left(\dfrac{\theta}{2}\right)$. Then the length

of the arc is

$L = \int_0^{\pi} \sqrt{\sin^4\left(\dfrac{\theta}{2}\right) + \sin^2\left(\dfrac{\theta}{2}\right)\cos^2\left(\dfrac{\theta}{2}\right)}\, d\theta$

$= \int_0^{\pi} \sqrt{\sin^2\left(\dfrac{\theta}{2}\right)\left(\sin^2\left(\dfrac{\theta}{2}\right) + \cos^2\left(\dfrac{\theta}{2}\right)\right)}\, d\theta$

$= \int_0^{\pi} \left|\sin\left(\dfrac{\theta}{2}\right)\right| d\theta = -2\cos\left(\dfrac{\theta}{2}\right)\Big|_0^{\pi}$

$= -2\cos\dfrac{\pi}{2} + 2\cos 0 = -2(0) + 2(1) = 2$

23. Find $\dfrac{dx}{dt} = e^t\cos t + e^t\sin t$ and $\dfrac{dy}{dt} = e^t\cos t - e^t\sin t$. Square each derivative.

$\left(\dfrac{dx}{dt}\right)^2 = (e^t\cos t + e^t\sin t)^2$

$= e^{2t}(\cos^2 t + 2\sin t\cos t + \sin^2 t)$

$= e^{2t}(1 + 2\sin t\cos t)$ and $\left(\dfrac{dy}{dt}\right)^2$

$= (e^t\cos t - e^t\sin t)^2$

$= e^{2t}(\cos^2 t - 2\cos t\sin t + \sin^2 t)$

$= e^{2t}(1 - 2\cos t\sin t)$. Then

$S = 2\pi e^t\cos t$

$\times \int_0^{\pi/2} \sqrt{e^{2t}[(1 + 2\sin t\cos t)] + [(1 - 2\sin t\cos t)]}\, dt$

$= 2\pi e^t\cos t \int_0^{\pi/2} e^t\sqrt{2}\, dt$

$= 2\sqrt{2}\pi e^{2t}\cos t\Big|_0^{\pi/2} = -2\sqrt{2}\pi$

24. Square $r^2 = 4 + 8\sin\theta + 4\sin^2\theta$. The area

$A = \dfrac{1}{2}\int_0^{2\pi}\left(4 + 8\sin\theta + 4\sin^2\theta\right)d\theta$

$= \dfrac{1}{2}\left[4\theta - 8\cos\theta + 4\left(\dfrac{1}{2}\theta - \dfrac{1}{4}\sin 2\theta\right)\right]_0^{2\pi}$

$= \left[3\theta - 4\cos\theta - \dfrac{1}{2}\sin 2\theta\right]_0^{2\pi}$

$= (6\pi - 4) - (-4) = 6\pi$

25. The acceleration vector for an object moving in the plane is $\langle -e^t, e^t\rangle$. Find the position of the object at $t = 1$ if the initial velocity is $v_0 = \langle 3, 1\rangle$ and the initial position of the object is at the origin. The acceleration of the object is known to be $a = \langle -e^t, e^t\rangle = -e^t i + e^t j$. Integrate to find the velocity. $v = -e^t i + e^t j + C$ and since the initial velocity is $v_0 = \langle 3, 1\rangle$, $v_0 = -i + j + C = 3i + j$ and $C = 4i$. The velocity vector is $v = -e^t i + e^t j + 4i = (4 - e^t)i + e^t j$. Integrate again to find the position vector $s = (4t - e^t)i + e^t j + C$. The initial position at the origin means that $s_0 = (4.0 - e^0)i + e^0 j + C = -i + j + C = 0$ and therefore $C = i - j$. The position vector $s = (4t - e^t + 1)i + (e^t - 1)j$ can be evaluated at $t = 1$ to find the position as $(3 - e)i + (e - 1)j = \langle 3 - e, e - 1\rangle$.

8.10   SOLUTIONS TO CUMULATIVE REVIEW PROBLEMS

26. (See Figure 8.10-1.)

$$\int_{-a}^{a} e^{x^2} dx = \int_{-a}^{0} e^{x^2} dx + \int_{0}^{a} e^{x^2} dx$$

Since e^{x^2} is an even function, thus

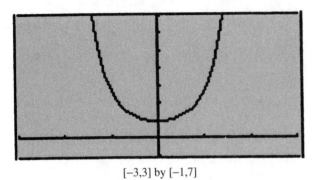

[−3,3] by [−1,7]

Figure 8.10-1

$$\int_{-a}^{a} e^{x^2} dx = \int_{0}^{a} e^{x^2} dx.$$

$$k = 2 \int_{0}^{a} e^{x^2} dx \text{ and } \int_{0}^{a} e^{x^2} dx = \frac{k}{2}.$$

27. See Figure 8.10-2.

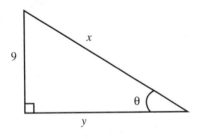

Figure 8.10-2

$$\sin \theta = \frac{9}{x}$$

Differentiate both sides:

$$\cos \theta \frac{d\theta}{dt} = (9)(-x^{-2}) \frac{dx}{dt}$$

When $x = 15$, $9^2 + y^2 = 15^2 \Rightarrow y = 12$

Thus $\cos \theta = \dfrac{12}{15} = \dfrac{4}{5}; \dfrac{dx}{dt} = -2$ ft/sec.

$$\frac{4}{5} \frac{d\theta}{dt} = 9 \left(-\frac{1}{15^2} \right) (-2)$$

$$= \frac{d\theta}{dt} = \frac{18}{15^2} \frac{5}{4} = \frac{1}{10} \text{ radian/sec.}$$

28. See Figure 8.10-3.

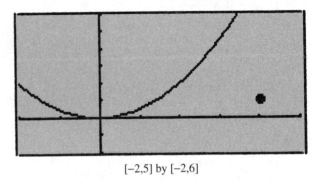

[−2,5] by [−2,6]

Figure 8.10-3

Step 1. Distance Formula

$$L = \sqrt{(x-4)^2 + (y-1)^2}$$

$$= \sqrt{(x-4)^2 + \left(\frac{x^2}{2} - 1 \right)^2}$$

where the domain is all real numbers.

Step 2. Enter $y_1 = \sqrt{((x-4)\text{^}2 + (.5x\text{^}2 - 1)\text{^}2)}$

Enter $y_2 = d(y_1(x), x)$

Step 3. Use the Zero Function and obtain $x = 2$ for y_2.

Step 4. Use the First Derivative Test. (See Figures 8.10-4 and 8.10-5.)
At $x = 2$, L has a relative minimum. Since at $x = 2$, L has the only relative extremum, it is an absolute minimum.

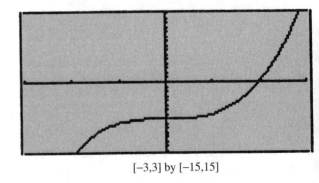

[−3,3] by [−15,15]

Figure 8.10-4

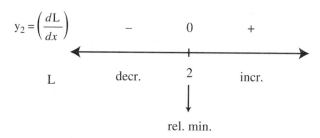

$$y_2 = \left(\frac{d\text{L}}{dx}\right)$$

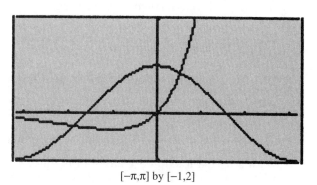

$[-\pi,\pi]$ by $[-1,2]$

Figure 8.10-6

Step 5. At $x = 2$, $y = \frac{1}{2}(x^2) = \frac{1}{2}(2^2) = 2$.

Thus the point on $y = \frac{1}{2}(x^2)$ closest to the point $(4, 1)$ is the point $(2, 2)$.

29. (a) $s(0) = 0$ and

$$s(t) = \int v(t)dt = \int t\cos(t^2 + 1)dt. \text{ Enter}$$

$$\int (x * \cos(x^\wedge 2 + 1), x) \text{ and obtain}$$

$$\frac{\sin(x^2 + 1)}{2}.$$

Thus $s(t) = \dfrac{\sin(t^2 + 1)}{2} + c.$

Since $s(0) = 0 \Rightarrow \dfrac{\sin(0^2 + 1)}{2} + c = 0$

$$\Rightarrow \frac{.841471}{2} + c = 0$$

$$\Rightarrow c = -0.420735 = -0.421$$

$$s(t) = \frac{\sin(t^2 + 1)}{2} - 0.420735$$

$$s(2) = \frac{\sin(2^2 + 1)}{2} - 0.420735$$

$$= -0.900197 \approx -0.900$$

(b) $v(2) = 2\cos(2^2 + 1) = 2\cos(5) = 0.567324$

Since $v(2) > 0$, the particle is moving to the right at $t = 2$.

(c) $a(t) = v'(t)$

Enter $d(x * \cos(x^\wedge 2 + 1), x)|x = 2$ and obtain 7.95506.

Thus, the velocity of the particle is increasing at $t = 2$, since $a(2) > 0$.

30. See Figure 8.10-6

(a) Point of Intersection: Use the Intersection Function of the calculation and obtain $(0.517757, 0.868931)$

$$\text{Area} = \int_0^{0.51775} (\cos x - xe^x)dx$$

Enter $\displaystyle\int (\cos(x) - x * e^\wedge x, x, 0, 0.51775)$

and obtain 0.304261.
The area of the region is approximately 0.304.

(b) **Step 1.** Washer Method:

Outer radius $= \cos x$ and Inner radius $= xe^x$

$$V = \pi \int_0^{0.51775} \left[(\cos x)^2 - (xe^x)^2\right] dx$$

Step 2. Enter

$$\int \left(\pi\left((\cos(x)^\wedge 2) - (x * e^\wedge(x))^\wedge 2\right),\right.$$

$$\left. x, 0.51775\right) \text{ and obtain } 1.16678.$$

The volume of the solid is approximately 1.167.

 31. Convert to a parametric representation with $x = r\cos\theta = 5\cos\theta\cos 2\theta$ and $y = r\sin\theta = 5\cos 2\theta\sin\theta$. Differentiate with respect to θ.

$$\frac{dx}{d\theta} = -5\cos 2\theta\sin\theta - 10\sin 2\theta\sin\theta \text{ and}$$

$$\frac{dy}{d\theta} = 5\cos 2\theta\cos\theta - 10\sin 2\theta\sin\theta. \text{ Divide to}$$

find $\dfrac{dy}{dx} = \dfrac{5\cos 2\theta \sin \theta - 10 \sin 2\theta \sin \theta}{-5 \cos 2\theta \sin \theta - 10 \sin 2\theta \sin \theta}$

$= \dfrac{-\cos 2\theta \cos \theta + 2 \sin 2\theta \sin \theta}{\cos 2\theta \sin \theta + 2 \sin 2\theta \sin \theta}$. Evaluated at

$\theta = \dfrac{3\pi}{2}, \dfrac{dy}{dx} = 0$. The slope of the tangent line is zero, including a horizontal tangent.

32. $\displaystyle\int \dfrac{2}{x^2 - 4x}\,dx = \int \dfrac{2}{x(x-4)}\,dx$ can be integrated with a partial fraction decomposition. Since
$\dfrac{A}{x} + \dfrac{B}{x-4} = \dfrac{2}{x(x-4)}, A = \dfrac{-1}{2}$ and $B = \dfrac{1}{2}$.

Therefore,

$$\int \dfrac{2}{x^2 - 4x}\,dx = \dfrac{-1}{2}\int \dfrac{dx}{x} + \dfrac{1}{2}\int \dfrac{dx}{x-4}$$

$$= \dfrac{-1}{2}\ln|x| + \dfrac{1}{2}\ln|x-4| + C$$

$$= \dfrac{1}{2}\ln\left|\dfrac{x-4}{x}\right| + C.$$

33. $\displaystyle\int_e^\infty \dfrac{dx}{x} = \lim_{k\to\infty}\int_e^k \dfrac{dx}{x} = \lim_{k\to\infty} \ln|x|\Big|_e^k$

$$= \lim_{k\to\infty}\big[\ln|k| - 1\big] = \infty.$$

Chapter 9

More Applications of Definite Integrals

9.1   AVERAGE VALUE OF A FUNCTION

Main Concepts: *Mean Value Theorem for Integrals, Average Value of a Function on* [a, b]

Mean Value Theorem for Integrals

Mean Value Theorem for Integrals:
If f is continuous on $[a, b]$, then there exists a number c in $[a, b]$ such that $\int_a^b f(x)\,dx = f(c)(b - a)$. See Figure 9.1-1.

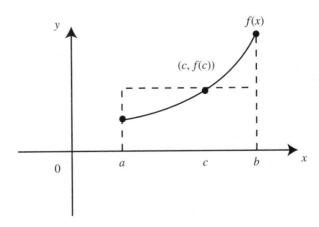

Figure 9.1-1

Example 1

Given $f(x) = \sqrt{x-1}$, verify the hypotheses of the Mean Value Theorem for Integrals for f on $[1, 10]$ and find the value of c as indicated in the theorem.

The function f is continuous for $x \geq 1$, thus:

$$\int_1^{10} \sqrt{x-1}\, dx = f(c)(10 - 1)$$

$$\frac{2(x-1)^{1/2}}{3}\Bigg]_1^{10} = 9f(c)$$

$$\frac{2}{3}\left[(10-1)^{1/2} - 0\right] = 9f(c)$$

$$18 = 9f(c); \; 2 = f(c); \; 2 = \sqrt{c-1}; \; 4 = c - 1$$

$$5 = c$$

Example 2

Given $f(x) = x^2$, verify the hypotheses of the Mean Value Theorem for Integrals for f on $[0, 6]$ and find the value of c as indicated in the theorem.

Since f is a polynomial, it is continuous and differentiable everywhere,

$$\int_0^6 x^2 dx = f(c)(6 - 0)$$

$$\frac{x^3}{3}\Bigg]_0^6 = f(c)6$$

$$72 = 6f(c); \; 12 = f(c); \; 12 = c^2$$

$$c = \pm\sqrt{12} = \pm 2\sqrt{3}\left(\pm 2\sqrt{3} \approx \pm 3.4641\right)$$

Since only $2\sqrt{3}$ is in the interval $[0, 6]$, $c = 2\sqrt{3}$.

> • Remember: if f' is decreasing, then $f'' < 0$ and the graph of f is concave downward.

Average Value of a Function on [a, b]

Average Value of a Function on an Interval:
If f is a continuous function on $[a,b]$, then the Average Value of f on $[a,b]$
$$= \frac{1}{b-a}\int_a^b f(x)dx.$$

Example 1

Find the average value of $y = \sin x$ between $x = 0$ to $x = \pi$.

$$\text{Average value} = \frac{1}{\pi - 0} \int_0^\pi \sin x \, dx$$

$$= \frac{1}{\pi}[-\cos x]_0^\pi = \frac{1}{\pi}[-\cos \pi - (-\cos(0))]$$

$$= \frac{1}{\pi}[1 + 1] = \frac{2}{\pi}.$$

Example 2

The graph of a function f is shown in Figure 9.1-2. Find the average value of f on $[0, 4]$.

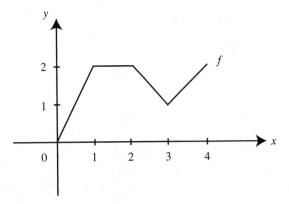

Figure 9.1-2

$$\text{Average value} = \frac{1}{4 - 0} \int_0^4 f(x) \, dx$$

$$= \frac{1}{4}\left(1 + 2 + \frac{3}{2} + \frac{3}{2}\right) = \frac{3}{2}.$$

Example 3

The velocity of a particle moving on a line is $v(t) = 3t^2 - 18t + 24$. Find the average velocity from $t = 1$ to $t = 3$.

$$\text{Average velocity} = \frac{1}{3 - 1} \int_1^3 (3t^2 - 18t + 24) \, dt$$

$$= \frac{1}{2}\left[t^3 - 9t^2 + 24t\right]_1^3$$

$$= \frac{1}{2}\left[\left(3^3 - 9(3^2) + 24(3)\right) - \left(1^3 - 9(1^2) + 24(1)\right)\right]$$

$$= \frac{1}{2}(18 - 16) = \frac{1}{2}(2) = 1.$$

Note: The average velocity for $t = 1$ to $t = 3$ is $\dfrac{s(3) - s(1)}{2}$, which is equivalent to the computations above.

9.2 DISTANCE TRAVELED PROBLEMS

Summary of Formulas:

Position Function: $s(t); s(t) = \int v(t)\, dt$

Velocity: $v(t) = \dfrac{ds}{dt}; v(t) = \int a(t)\, dt$

Acceleration: $a(t) = \dfrac{dv}{dt}$

Speed: $|v(t)|$

Displacement from t_1 to $t_2 = \displaystyle\int_{t_1}^{t_2} v(t)\, dt = s(t_2) - s(t_1)$

Total Distance Traveled from t_1 to $t_2 = \displaystyle\int_{t_1}^{t_2} |v(t)|\, dt$

Example 1

See Figure 9.2-1.

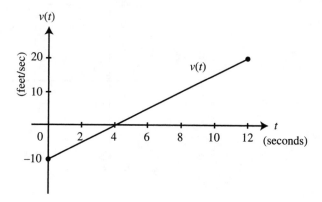

Figure 9.2-1

The graph of the velocity function of a moving particle is shown in Figure 9.2-1. What is the total distance traveled by the particle during $0 \le t \le 12$?

$$\text{Total Distance Traveled} = \left| \int_0^4 v(t)dt \right| + \int_4^{12} v(t)dt$$

$$= \frac{1}{2}(4)(10) + \frac{1}{2}(8)(20) = 20 + 80 = 100 \text{ feet.}$$

Example 2

The velocity function of a moving particle on a coordinate line is $v(t) = t^2 + 3t - 10$ for $0 \le t \le 6$. Find (a) the displacement by the particle during $0 \le t \le 6$, and (b) the total distance traveled during $0 \le t \le 6$.

(a) Displacement $= \displaystyle\int_{t_1}^{t_2} v(t)dt$

$$= \int_0^6 (t^2 + 3t - 10)dt = \frac{t^3}{3} + \frac{3t^2}{2} - 10t \Big]_0^6 = 66.$$

(b) Total Distance Traveled $= \int_{t_1}^{t_2} |v(t)| dt$

$$= \int_0^6 |t^2 + 3t - 10| dt$$

Let $t^2 + 3t - 10 = 0 \Rightarrow (t+5)(t-2) = 0 \Rightarrow t = -5 \text{ or } t = 2$

$$|t^2 + 3t - 10| = \begin{cases} -(t^2 + 3t - 10) & \text{if } 0 \le t \le 2 \\ t^2 + 3t - 10 & \text{if } t > 2 \end{cases}$$

$$\int_0^6 |t^2 + 3t - 10| dt = \int_0^2 -(t^2 + 3t - 10)dt + \int_2^6 (t^2 + 3t - 10)dt$$

$$= \left[\frac{-t^3}{3} - \frac{3t^2}{2} + 10t \right]_0^2 + \left[\frac{t^3}{3} + \frac{3t^2}{2} - 10t \right]_2^6$$

$$= \frac{34}{3} + \frac{232}{3} = \frac{266}{3} \approx 88.667$$

The total distance traveled by the particle is $\frac{266}{3}$ or approximately 88.667.

Example 3

The velocity function of a moving particle on a coordinate line is $v(t) = t^3 - 6t^2 + 11t - 6$. Using a calculator, find (a) the displacement by the particle during $1 \le t \le 4$, and (b) the total distance traveled by the particle during $1 \le t \le 4$.

(a) Displacement $= \int_{t_1}^{t_2} v(t)dt$

$$= \int_1^4 (t^3 - 6t^2 + 11t - 6)dt$$

Enter $\int (x^3 - 6x^2 + 11x - 6, x, 1, 4)$ and obtain $\frac{9}{4}$.

(b) Total Distance Traveled $= \int_{t_1}^{t_2} |v(t)| dt$.

Enter $y_1 = x^3 - 6x^2 + 11x - 6$ and use the Zero Function to obtain x-intercepts at $x = 1, 2, 3$.

$$|v(t)| = \begin{cases} v(t) & \text{if } 1 \le t \le 2 \text{ and } 3 \le t \le 4 \\ -v(t) & \text{if } 2 < t < 3 \end{cases}$$

Total Distance Traveled $\int_1^2 v(t)dt + \int_2^3 -v(t)dt + \int_3^4 v(t)dt$.

Enter $\int (y_1(x), x, 1, 2)$ and obtain $\frac{1}{4}$.

Enter $\int (-y_1(x), x, 2, 3)$ and obtain $\frac{1}{4}$.

Enter $\int (y_1(x), x, 3, 4)$ and obtain $\frac{9}{4}$.

Thus, total distance traveled is $\left(\frac{1}{4} + \frac{1}{4} + \frac{9}{4} \right) = \frac{11}{4}$.

Example 4

The acceleration function of a moving particle on a coordinate line is $a(t) = -4$ and $v_0 = 12$ for $0 \le t \le 8$. Find the total distance traveled by the particle during $0 \le t \le 8$.

$$a(t) = -4$$

$$v(t) = \int a(t)dt = \int -4dt = -4t + c$$

Since $v_0 = 12 \Rightarrow -4(0) + c = 12$ or $c = 12$

Thus $v(t) = -4t + 12$

$$\text{Total Distance Traveled} = \int_0^4 |-4t + 12| \, dt$$

Let $-4t + 12 = 0 \Rightarrow t = 3$.

$$|-4t + 12| = \begin{cases} -4t + 12 & \text{if } 0 \le t \le 3 \\ -(-4t + 12) & \text{if } t > 3 \end{cases}$$

$$\int_0^6 |-4t + 12| \, dt = \int_0^3 |-4t + 12| \, dt + \int_3^6 -(-4t + 12)dt$$

$$= \left[-12t^2 + 12t \right]_0^3 + \left[2t^2 + 12t \right]_3^6$$

$$= 18 + 50 = 68$$

Total distance traveled by the particle is 68.

Example 5

The velocity function of a moving particle on a coordinate line is $v(t) = 3\cos(2t)$ for $0 \le t \le 2\pi$. Using a calculator:

(a) Determine when the particle is moving to the right.

(b) Determine when the particle stops.

(c) The total distance traveled by the particle during $0 \le t \le 2\pi$.

Solution:

(a) The particle is moving to the right when $v(t) > 0$.

Enter $y_1 = 3\cos(2x)$. Obtain $y_1 = 0$ when $t = \dfrac{\pi}{4}, \dfrac{3\pi}{4}, \dfrac{5\pi}{4}$, and $\dfrac{7\pi}{4}$.

The particle is moving to the right when:

$$0 < t < \frac{\pi}{4}, \frac{3\pi}{4} < t < \frac{5\pi}{4}, \frac{7\pi}{4} < t < 2\pi.$$

(b) The particle stops when $v(t) = 0$.

Thus the particle stops at $t = \dfrac{\pi}{4}, \dfrac{3\pi}{4}, \dfrac{5\pi}{4}$, and $\dfrac{7\pi}{4}$.

(c) Total distance traveled $\displaystyle\int_0^{2\pi} |3\cos(2t)| \, dt$

Enter $\displaystyle\int (\text{abs}(3\cos(2x)), x, 0, 2\pi)$ and obtain 12.

The total distance traveled by the particle is 12.

9.3 DEFINITE INTEGRAL AS ACCUMULATED CHANGE

Main Concepts: *Business Problems, Temperature Problems, Leakage Problems, Growth Problems*

Business Problems

$$P(x) = R(x) - C(x) \quad \text{Profit} = \text{Revenue} - \text{Cost}$$
$$R(x) = P(x) \qquad \text{Revenue} = (\text{price})(\text{items sold})$$
$$P'(x) \qquad \text{Marginal Profit}$$
$$R'(x) \qquad \text{Marginal Revenue}$$
$$C'(x) \qquad \text{Marginal Cost}$$

$P'(x), R'(x)$, and $C'(x)$ are the instantaneous rates of change of profit, revenue, and cost respectively.

Example 1

The marginal profit of manufacturing and selling a certain drug is $P'(x) = 100 - 0.005x$.

How much profit should the company expect if it sells 10,000 units of this drug?

$$P(t) = \int_0^1 P'(x)dx$$

$$= \int_0^{10,000} (100 - 0.005x) \, dx = 100x - \frac{0.005x^2}{2} \Big]_0^{10,000}$$

$$= \left(100(10,000) - \frac{0.005}{2}(10,000)^2\right) = 750,000.$$

- If $f''(a) = 0, f$ may or may not have a point of inflection at $x = a$, e.g., as in the function $f(x) = x^4, f''(0) = 0$ *but* at $x = 0, f$ has an absolute minimum.

Example 2

If the marginal cost of producing x units of a commodity is $C'(x) = 5 + 0.4x$,

find (a) the marginal cost when $x = 50$.

   (b) the cost of producing the first 100 units.

Solution:

(a) marginal cost at $x = 50$:

   $$C'(50) = 5 + 0.4(50) = 5 + 20 = 25.$$

(b) cost of producing 100 units:

   $$C(t) = \int_0^1 C'(x)dx$$

   $$= \int_0^{100} (5 + 0.4x)dx$$

$$= 5x + 0.2x^2 \Big]_0^{100}$$

$$= \left(5(100) + 0.2(100)^2\right) - 0 = 2500.$$

Temperature Problems

Example 3

On a certain day, the changes in the temperature in a greenhouse beginning at 12 noon are represented by $f(t) = \sin\left(\dfrac{t}{2}\right)$ degrees Fahrenheit, where t is the number of hours elapsed after 12 noon. If at 12 noon, the temperature is 95°F, find the temperature in the greenhouse at 5 p.m.

Let $F(t)$ represent the temperature of the greenhouse.

$$F(0) = 95°F$$

$$F(t) = 95 + \int_0^5 f(x)\, dx$$

$$F(5) = 95 + \int_0^5 \sin\left(\frac{x}{2}\right) dx$$

$$= 95 + \left[-2\cos\left(\frac{x}{2}\right)\right]_0^5 = 95 + \left[-2\cos\left(\frac{5}{2}\right) - (-2\cos(0))\right]$$

$$= 95 + 3.602 = 98.602$$

The temperature in the greenhouse at 5 p.m. is 98.602°F.

Leakage Problems

Example 4

Water is leaking from a faucet at the rate of $l(t) = 10e^{-0.5t}$ gallons per hour, where t is measured in hours. How many gallons of water will have leaked from the faucet after a 24 hour period?

Let $L(x)$ represent the number of gallons that have leaked after x hours.

$$L(x) = \int_0^x l(t)\, dt = \int_0^{24} 10e^{-0.5t}\, dt$$

Using your calculator, enter $\int\left(10e^{\wedge}(-0.5x), x, 0, 24\right)$ and obtain 19.9999. Thus, the number of gallons of water that have leaked after x hours is approximately 20 gallons.

- You are permitted to use the following 4 built-in capabilities of your calculator to obtain an answer; plotting the graph of a function, finding the zeros of a function, finding the numerical derivative of a function, and evaluating a definite integral. All other capabilities of your calculator can only be used to *check* your answer. For example, you may *not* use the built-in Inflection function of your calculator to find points of inflection. You must use calculus using derivatives and showing change of concavity.

Growth Problems

Example 5

In a farm, the animal population is increasing at a rate which can be approximately represented by $g(t) = 20 + 50 \ln(2 + t)$, where t is measured in years. How much will the animal population increase to the nearest tens between the 3rd and 5th year?

Let $G(x)$ be the increase in animal population after x years.

$$G(x) = \int_0^x g(t)\, dt$$

Thus, the population increase between the 3rd and 5th years

$$= G(5) - G(3)$$

$$= \int_0^5 \left(20 + 50\ln(2+t)\, dt\right) - \int_0^3 \left(20 + 50\ \ln(2+t)dt\right)$$

$$= \int_3^5 \left[20 + 50\ \ln(2+t)\right] dt$$

Enter $\int (20 + 50\ \ln(2+x), x, 3, 5)$ and obtain 218.709.

Thus the animal population will increase by approximately 220 between the 3rd and 5th years.

9.4   DIFFERENTIAL EQUATIONS

Main Concepts: *Exponential Growth/Decay Problems, Separable Differential Equations*

Exponential Growth/Decay Problems

1. If $\dfrac{dy}{dx} = ky$, then the rate of change of y is proportional to y.

2. If y is a differentiable function of t with $y > 0$ $\dfrac{dy}{dx} = ky$, then $y(t) = y_0 e^{kt}$; where y_0 is initial value of y and k is constant. If $k > 0$, then k is a growth constant and if $k < 0$, then k is the decay constant.

Example 1—Population Growth:

If the amount of bacteria in a culture at any time increases at a rate proportional to the amount of bacteria present and there are 500 bacteria after one day and 800 bacteria after the third day:

(a) approximately how many bacteria are there initially, and

(b) approximately how many bacteria are there after 4 days?

Solution:

(a) Since the rate of increase is proportional to the amount of bacteria present, Then:

$\dfrac{dy}{dx} = ky$ where y is the amount of bacteria at any time

Therefore this is an exponential growth/decay model: $y(t) = y_0 e^{kt}$

Step 1. $y(1) = 500$ and $y(3) = 800$
$500 = y_0 e^k$ and $800 = y_0 e^{3k}$

Step 2. $500 = y_0 e^k \Rightarrow y_0 = \dfrac{500}{e^k} = 500 e^{-k}$

Substitute $y_0 = 500 e^{-k}$ into $800 = y_0 e^{3k}$

$800 = (500)(e^{-k})(e^{3k})$

$800 = 500 e^{2k} \Rightarrow \dfrac{8}{5} = e^{2k}$

Take the ln of both sides:

$\ln\left(\dfrac{8}{5}\right) = \ln\left(e^{2k}\right)$

$\ln\left(\dfrac{8}{5}\right) = 2k$

$k = \dfrac{1}{2}\ln\left(\dfrac{8}{5}\right) = \ln\sqrt{\dfrac{8}{5}}$

Step 3. Substitute $k = \dfrac{1}{2}\ln\left(\dfrac{8}{5}\right)$ into one of the equations

$500 = y_0 e^k$

$500 = y_0 e^{\ln\left(\sqrt{\frac{8}{5}}\right)}$

$500 = y_0 \left(\sqrt{\dfrac{8}{5}}\right)$

$y_0 = \dfrac{500}{\sqrt{8/5}} = 125\sqrt{10} \approx 395.285;$
Thus, there are 395 bacteria present initially.

(b) $y_0 = 125\sqrt{10}, k = \ln\sqrt{\dfrac{8}{5}}$

$y(t) = y_0 e^{kt}$

$y(t) = \left(125\sqrt{10}\right) e^{\left(\ln\sqrt{\frac{8}{5}}\right)t} = \left(125\sqrt{10}\right)\left(\dfrac{8}{5}\right)^{(1/2)t}$

$y(4) = \left(125\sqrt{10}\right)\left(\dfrac{8}{5}\right)^{(1/2)4} = \left(125\sqrt{10}\right)\left(\dfrac{8}{5}\right)^{2} = 1011.93$

Thus there are approximately 1011 bacteria present after 4 days.

> • Get a good night's sleep the night before. Have a light breakfast before the exam.

Example 2—Radioactive Decay:

Carbon-14 has a half-life of 5750 years. If initially there are 60 grams of carbon-14, how many grams are left after 3000 years?

Step 1. $y(t) = y_0 e^{kt} = 60e^{kt}$

Since half-life is 5750 years, $30 = 60e^{k(5750)} \Rightarrow \dfrac{1}{2} = e^{5750k}$

$$\ln\left(\frac{1}{2}\right) = \ln\left(e^{5750k}\right)$$

$$-\ln 2 = 5750k$$

$$\frac{-\ln 2}{5750} = k$$

Step 2. $y(t) = y_0 e^{kt}$

$$y(t) = 60e^{\left[\frac{-\ln 2}{5750}\right]}$$

$$y(t) = 60e^{\left[\frac{-\ln 2}{5750}\right](3000)}$$

$$y(3000) \approx 41.7919$$

Thus, there will be approximately 41.792 grams of carbon-14 after 3000 years.

Separable Differential Equations

General Procedure:

1. Separate the variables: $g(y)dy = f(x)dx$
2. Integrate both sides: $\int g(y)dy = \int f(x)dx$
3. Solve for y to get a general solution.
4. Substitute given conditions to get a particular solution.
5. Verify your result by differentiating.

Example 1

Given $\dfrac{dy}{dx} = 4x^3 y^2$ and $y(1) = -\dfrac{1}{2}$, solve the differential equation.

Step 1. Separate the variables: $\dfrac{1}{y^2}dy = 4x^3 dx$

Step 2. Integrate both sides: $\displaystyle\int \frac{1}{y^2}dy = \int 4x^3 dx; \quad -\frac{1}{y} = x^4 + c$

Step 3. General solution: $y = \dfrac{-1}{x^4 + c}$

Step 4. Particular solution: $-\dfrac{1}{2} = \dfrac{-1}{1 + c} \Rightarrow c = 1; y = \dfrac{-1}{x^4 + c}$

Step 5. Verify result by differentiating

$$y = \frac{-1}{x^4 + 1} = (-1)(x^4 + 1)^{-1}$$

$$\frac{dy}{dx} = (-1)(-1)(x^4 + 1)^{-2}(4x^3) = \frac{4x^3}{(x^4 + 1)^2}$$

$$\text{Note}: y = \frac{-1}{x^4 + 1} \text{ implies } y^2 = \frac{1}{(x^4 + 1)^2}$$

$$\text{Thus } \frac{dy}{dx} = \frac{4x^3}{(x^4 + 1)^2} = 4x^3 y^2.$$

- To get your AP grade, you can call 888-0013 in July. There is a charge of about $13.

Example 2

Find a solution of the differentiation equation $\dfrac{dy}{dx} = x \sin(x^2); y(0) = -1$.

Step 1. Separate variables: $dy = x \sin(x^2)dx$

Step 2. Integrate both sides: $\displaystyle\int dy = \int x \sin(x^2)dx; \int dy = y$

$$\text{Let } u = x^2; du = 2x \, dx \text{ or } \frac{du}{2} = x \, dx$$

$$\int x \sin(x^2)dx = \int \sin u \left(\frac{du}{2}\right) = \frac{1}{2}\int \sin u \, du = -\frac{1}{2}\cos u + c$$

$$= -\frac{1}{2}\cos(x^2) + c$$

$$\text{Thus}: y = -\frac{1}{2}\cos(x^2) + c.$$

Step 3. Substitute given condition:

$$y(0) = -1; -1 = -\frac{1}{2}\cos(0) + c; -1 = \frac{-1}{2} + c; -\frac{1}{2} = c$$

$$\text{Thus, } y = -\frac{1}{2}\cos(x^2) - \frac{1}{2}.$$

Step 4. Verify result by differentiating:

$$\frac{dy}{dx} = \frac{1}{2}\left[\sin(x^2)\right](2x) = x \sin(x^2).$$

Example 3

If $\dfrac{d^2y}{dx^2} = 2x + 1$ and at $x = 0$, $y' = -1$, and $y = 3$, find a solution of the differential equation.

Step 1. Rewrite: $\dfrac{d^2y}{dx^2}$ as $\dfrac{dy'}{dx}$; $\dfrac{dy'}{dx} = 2x + 1$.

Step 2. Separate variables: $dy' = (2x + 1)dx$.

Step 3. Integrate both sides: $\displaystyle\int dy' = \int (2x + 1)dx$; $y' = x^2 + x + c_1$.

Step 4. Substitute given condition: At $x = 0$, $y' = -1$; $-1 = 0 + 0 + c_1 \Rightarrow c_1 = -1$.
Thus $y' = x^2 + x - 1$.

Step 5. Rewrite: $y' = \dfrac{dy}{dx}$; $\dfrac{dy}{dx} = x^2 + x - 1$.

Step 6. Separate variables: $dy = (x^2 + x - 1)dx$.

Step 7. Integrate both sides: $\displaystyle\int dy = \int (x^2 + x - 1)dx$
$$y = \frac{x^3}{3} + \frac{x^2}{2} - x + c_2.$$

Step 8. Substitute given condition: At $x = 0$, $y = 3$; $3 = 0 + 0 - 0 + c_2 \Rightarrow c_2 = 3$
Therefore $y = \dfrac{x^3}{3} + \dfrac{x^2}{2} - x + 3$.

Step 9. Verify result by differentiating: $y = \dfrac{x^3}{3} + \dfrac{x^2}{2} - x + 3$

$$\frac{dy}{dx} = x^2 + x - 1; \frac{d^2y}{dx^2} = 2x + 1.$$

Example 4

Find the general solution of the differential equation $\dfrac{dy}{dx} = \dfrac{2xy}{x^2 + 1}$.

Step 1. Separate variables:

$$\frac{dy}{y} = \frac{2x}{x^2 + 1}dx.$$

Step 2. Integrate both sides: $\displaystyle\int \frac{dy}{y} = \int \frac{2x}{x^2 + 1}dx$ (let $u = x^2 + 1$; $du = 2x\,dx$)
$$\ln|y| = \ln(x^2 + 1) + c_1$$

Step 3. General Solution: solve for y

$$e^{\ln|y|} = e^{\ln(x^2+1)+c_1}$$

$$|y| = e^{\ln(x^2+1)} \cdot e^{c_1}; |y| = e^{c_1}(x^2 + 1)$$

$$y = \pm e^{c_1}(x^2 + 1)$$

The general solution is $y = c(x^2 + 1)$.

Step 4. Verify result by differentiating:

$$y = c(x^2 + 1)$$

$$\frac{dy}{dx} = 2cx = 2x\frac{c(x^2 + 1)}{x^2 + 1} = \frac{2xy}{x^2 + 1}.$$

Example 5

Write an equation for the curve that passes through the point (3, 4) and has a slope at any point (x, y) as $\dfrac{dy}{dx} = \dfrac{x^2 + 1}{2y}$.

Step 1. Separate variables: $2y\,dy = (x^2 + 1)dx$.

Step 2. Integrate both sides: $\displaystyle\int 2y\,dy = \int (x^2 + 1)\,dx; y^2 = \dfrac{x^3}{3} + x + c.$

Step 3. Substitute given condition: $4^2 = \dfrac{3^3}{3} + 3 + c \Rightarrow c = 4$

Thus $y^2 = \dfrac{x^3}{3} + x + 4.$

Step 4. Verify the result by differentiating: $2y\dfrac{dy}{dx} = x^2 + 1$

$$\frac{dy}{dx} = \frac{x^2 + 1}{2y}.$$

 9.5 SLOPE FIELDS

Main Concepts: *Slope Fields, Solution of Different Equations*

Slope Fields

A *slope field* (or a *direction field*) for first-order differential equations is a graphic representation of the slopes of a family of curves. It consists of a set of short line segments drawn on a pair of axes. These line segments are the tangents to a family of solution curves for the differential equation at various points. The tangents show the direction in which the solution curves will follow. Slope fields are useful in sketching solution curves without having to solve a differential equation algebraically.

Example 1

If $\dfrac{dy}{dx} = 0.5x$, draw a slope field for the given differential equation.

Step 1: Set up a table of values for $\dfrac{dy}{dx}$ for selected values of x.

x	-4	-3	-2	-1	0	1	2	3	4
$\dfrac{dy}{dx}$	-2	-1.5	-1	-0.5	0	0.5	1	1.5	2

Note that since $\dfrac{dy}{dx} = 0.5x$, the numerical value of $\dfrac{dy}{dx}$ is independent of the value of y. For example, at the points $(1, -1)$, $(1, 0)$, $(1, 1)$, $(1, 2)$, $(1, 3)$ and at all the points whose x-coordinates are 1, the numerical value of $\dfrac{dy}{dx}$ is 0.5 regardless of their y-coordinates. Similarly, for all the points, whose x-coordinates are 2 (e.g., $(2, -1)$, $(2, 0)$, $(2, 3)$, etc.), $\dfrac{dy}{dx} = 1$. Also, remember that $\dfrac{dy}{dx}$ represents the slopes of the tangents lines to the curve at various points. You are now ready to draw these tangents

Step 2: Draw short line segments with the given slopes at the various points. The slope field for the differential equation $\dfrac{dy}{dx} = 0.5x$ is shown in Figure A-1.

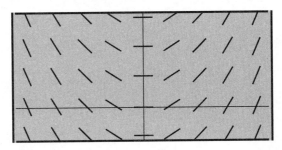

Figure A-1

Example 2

Figure A-2 shows a slope field for one of the differential equations given below. Identify the equation.

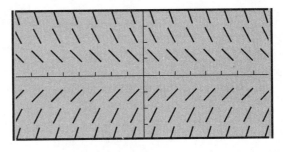

Figure A-2

(a) $\dfrac{dy}{dx} = 2x$ (b) $\dfrac{dy}{dx} = -2x$ (c) $\dfrac{dy}{dx} = y$

(d) $\dfrac{dy}{dx} = -y$ (e) $\dfrac{dy}{dx} = x + y$

Solution:

If you look across horizontally at any row of tangents, you'll notice that the tangents have the same slope. (Points on the same row have the same y-coordinate but different x-coordinates.) Therefore, the numerical value of $\dfrac{dy}{dx}$ (which represents the slope of the tangent) depends solely on the y-coordinate of a point and it is independent of the x-coordinate. Thus, only choice (c) and choice (d) satisfy this condition. Also notice that the tangents have a negative slope when $y > 0$ and have a positive slope when $y < 0$.

Therefore, the correct choice is (c) $\dfrac{dy}{dx} = -y$.

Example 3

A slope field for a differential equation is shown in Figure A-3. Draw a possible graph for the particular solution $y = f(x)$ to the differential equation function, if (a) the initial condition is $f(0) = -2$ and (b) the initial condition is $f(0) = 0$.

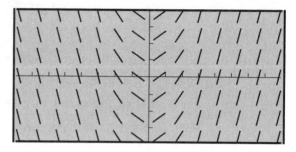

Figure A-3

Solution:

Being by locating the point $(0, -2)$ as given in the initial condition. Follow the flow of the field and sketch the graph of the function. Repeat the same procedure with the point $(0, 0)$. See the curves as shown in Figure A-4.

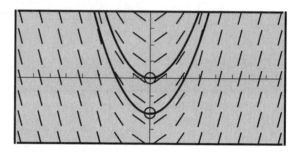

Figure A-4

Example 4

Given the differential equation $\dfrac{dy}{dx} = -xy$.

(a) Draw a slope field for the differential equation at the 15 points indicated on the provided set of axes in Figure A-5.

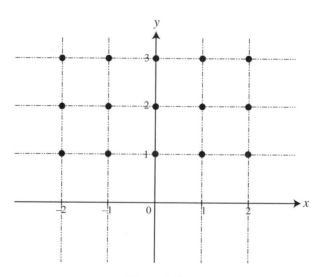

Figure A-5

(b) Sketch a possible graph for the particular solution $y = f(x)$ to the differential equation with the initial condition $f(0) = 3$.

(c) Find, algebraically, the particular solution $y = f(x)$ to the differential equation with the initial condition $f(0) = 3$.

Solution:

(a) Set up a table of values for $\dfrac{dy}{dx}$ at the 15 given points.

	$x = -2$	$x = -1$	$x = 0$	$x = 1$	$x = 2$
$y = 1$	2	1	0	−1	−2
$y = 2$	4	2	0	−2	−4
$y = 3$	6	3	0	−3	−6

Then sketch the tangents at the various points as shown in Figure A-6.

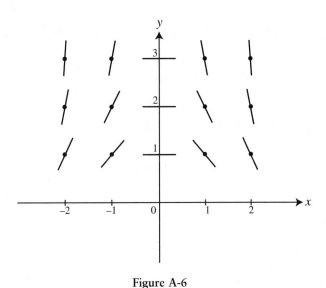

Figure A-6

(b) Locate the point $(0, 3)$ as indicated in the initial condition. Follow the flow of the field and sketch the curve as shown Figure A-7.

(c) Step 1: Rewrite $\dfrac{dy}{dx} = -xy$ as $\dfrac{dy}{y} = -x\,dx.$

Step 2: Integrate both sides $\displaystyle\int \dfrac{dy}{y} = \int -x\,dx$ and obtain $\ln|y| = -\dfrac{x^2}{2} + c$

Step 3: Apply the exponential function to both sides and obtain $e^{\ln|y|} = e^{-\frac{x^2}{2} + c}$

Step 4: Simplify the equation and get $y = \left(e^{\frac{x^2}{2}}\right)(e^c) = \dfrac{e^c}{e^{\frac{x^2}{2}}}$

Let $k = e^c$ and you have $y = \dfrac{k}{e^{\frac{x^2}{2}}}$

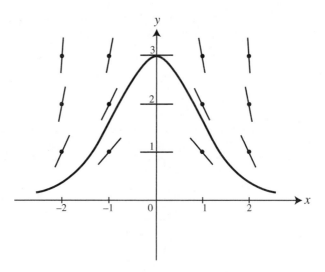

Figure A-7

Step 5: Substitute initial condition $(0, 3)$ and obtain $k = 3$. Thus you have

$$y = \frac{3}{e^{\frac{x^2}{2}}}.$$

9.6 LOGISTIC DIFFERENTIAL EQUATIONS

Main Concepts: *Logistic Growth*

The Logistic Differential Equation

Often population may grow exponentially at first, but eventually slows as it nears a limit, called the carrying capacity. This patten is called logistic growth, and is represented by the differential equation $\frac{dP}{dt} = kP\left(1 - \frac{P}{K}\right)$, in which P is the population, K is the carrying capacity, and k is the proportional constant. The differential equation is separable so $\frac{dP}{dt} = kP\left(1 - \frac{P}{K}\right) \Rightarrow \frac{dP}{dt} = \frac{kP(K - P)}{K} \Rightarrow \frac{K\,dP}{P(K - P)} = k\,dt$. This equation can be integrated using a partial fraction decomposition.

$$\int \frac{K\,dP}{P(K - P)} = \int k\,dt$$

$$\int \left(\frac{1}{P} + \frac{-1}{K - P}\right) dP = \int k\,dt$$

$$\ln|P| - \ln|K - P| = kt + c_1$$

$$\ln\left|\frac{P}{K - P}\right| = kt + c_1$$

Exponention produces $\frac{P}{K - P} = e^{(kt + c_1)} \Rightarrow \frac{P}{K - P} = e^{kt} \cdot e^{c_1} \Rightarrow \frac{P}{K - P} = c_2 e^{kt}$. Solving for P yields $P = c_2 e^{kt}(K - P) \Rightarrow P = c_2 e^{kt}K - c_2 e^{kt}P \Rightarrow P + c_2 e^{kt}P = c_2 e^{kt}K \Rightarrow P(1 + c_2 e^{kt}) = c_2 e^{kt}K \Rightarrow P = \frac{c_2 e^k K}{1 + c_2 e^{kt}}$. Dividing numerator and denominator by c_2e^{kt}, $P(t) = \dfrac{c_2e^{kt}K}{1+c_2e^{kt}} = \dfrac{K}{\dfrac{1}{c_2e^{kt}}+1} = \dfrac{K}{\left(\dfrac{1}{c_2}\right)e^{-kt}+1}$. At $t=0$, $P_0 = \dfrac{K}{\dfrac{1}{c_2}+1}$.

Solving for c_2 yields $P_0\left(\dfrac{1}{c_2}+1\right) = K \Rightarrow \dfrac{1+c_2}{c_2} = \dfrac{K}{P_0} \Rightarrow P_0 + P_0c_2 = Kc_2 \Rightarrow P_0 = Kc_2 - P_0c_2$ or $c_2 = \dfrac{P_0}{K-P_0}$. Let $A = \dfrac{1}{c_2}$, and the solution of this logistic differential equation with initial condition $P(0) = P_0$ is $P(t) = \dfrac{K}{Ae^{-kt}+1}$ where K is the carrying capacity and $A = \dfrac{K-P_0}{P_0}$.

Example 1

The population of the Great Britain was 57.1 million in 2001 and 60.6 million in 2006. Find a logistic model for the growth of the population, assuming a carrying capacity of 100 million. Use the model to predict the population in 2020.

Step 1: Since the carrying capacity is $K = 100$, $\dfrac{dP}{dt} = kP\left(1-\dfrac{P}{100}\right)$.

Step 2: The solution of the differential equation, if $A = \dfrac{k-P_0}{P_0} = \dfrac{100-57.1}{57.1} \approx .7513$ is $P(t) = \dfrac{100}{Ae^{-kt}+1}$ or $P(t) = \dfrac{100}{.7513e^{-kt}+1}$.

Step 3: Take 2006 as $t = 5$, $P(5) = 60.6$. Then $60.6 = \dfrac{100}{.7513e^{-k(5)}+1}$. Solving gives $k \approx 0.0289$ so $P(t) = \dfrac{100}{.7513e^{-0.0289t}+1}$.

Step 4: Since the year 2020 corresponds to $t = 19$, Substitute and evaluate $P(19) = \dfrac{100}{.7513e^{-0.0289(19)}+1} \approx 69.742$. The population of the Great Britain in 2020 is predicted to be approximately 69.742 million.

Example 2

The spread of an infectious disease can often be modeled by a logistic equation with the total exposed population as the carrying capacity. In a community of 2000 individuals, the first case of a new virus is diagnosed on March 31, and by April 10, there are 500 individuals infected. Write a differential equation that models the rate at which the virus spread through the community and determine when 98% of the population will have contracted the virus.

Step 1: The rate of spread is $\dfrac{dP}{dt} = kP\left(1-\dfrac{P}{2000}\right)$.

Step 2: The solution of the differential equation is $P(t) = \dfrac{2000}{Ae^{-kt}+1}$, and with one person exposed, $A = \dfrac{2000-1}{1} = 1999$, or $P(t) = \dfrac{2000}{1999e^{-kt}+1}$.

Step 3: Taking April 10 as day 10, $P(10) = 500 = \dfrac{2000}{1999e^{-k(10)}+1}$. Solving the equation gives $k \approx .6502$, so $P(t) = \dfrac{2000}{1999e^{-.6502t}+1}$.

Step 4: 98% of the population of 2000 is 1960 people. To determine the day when 1960 people are infected, solve $1960 = \dfrac{2000}{1999e^{-.6502t} + 1}$. This gives $t \approx 17.6749$, so the 98% infection rate should be reached by April 18.

9.7 EULER'S METHOD

Main Concepts: *Approximating Solutions of Differential Equations by Euler's Method.*

Euler's Method

Euler's Method provides a means of estimating the numerical solution of differential equation by a series of successive linear approximations. Represent the differential equation by $y' = f'(x, y)$ and the initial condition $y_0 = f(x_0)$, and choose a small value, Δx, as the increment between estimates. Begin with the initial value y_0, and evaluate $y_1 = y_0 + \Delta x \cdot f'(x_0, y_0)$. Continue with $y_2 = y_1 + \Delta x \cdot f'(x_1, y_1)$ and in general, $y_n = y_{n-1} + \Delta x \cdot f'(x_{n-1}, y_{n-1})$.

Example 1

Given the initial value problem $\dfrac{dy}{dt} = \cos 2\pi t$ with $y(0) = 1$, approximate $y(1)$, using five steps.

Step 1: The interval $(0, 1)$ divided into five steps gives us $\Delta t = 0.2$.

Step 2: Create a table showing the iterations.

t	y	$\cos 2\pi t$	$y_n = y_{n-1} + \Delta x \cdot f'(x_{n-1}, y_{n-1})$
0	1	1	$1 + .2(1) = 1.2$
0.2	1.2	.309016	$1.2 + .2(.309016) = 1.261803$
0.4	1.261803	$-.809016$	$1.261803 + .2(-.809016) = 1.1$
0.6	1.1	$-.809016$	$1.1 + .2(-.809016) = 0.938196$
0.8	.938196	.309016	$.938196 + .2(.309016) = 1$
1	1		

Step 3: $y(1) \approx 1$

Example 2

Use Euler's method with a step size of $\Delta x = 0.1$ to compute $y(1)$ if $y(x)$ is the solution of the differential equation $\dfrac{dy}{dx} + 3x^2 y = 6x^2$ with initial condition $y(0) = 3$.

Step 1: For case of evaluation, transform $\dfrac{dy}{dx} + 3x^2 y = 6x^2$ to $\dfrac{dy}{dx} = 3x^2(2 - y)$.

Step 2: Create a table showing the iterations. A simple problem, stored in your calculator and modified with the new differential equation and initial condition, will allow you to generate the table quickly.

x	y	$3x^2(2 - y)$	$y_n = y_{n-1} + \Delta x \cdot f'(x_{n-1}, y_{n-1})$
0	3	0	3
0.1	3	-0.03	2.997
0.2	2.997	-0.11964	2.985036
0.3	2.985036	-0.265959	2.9584400
0.4	2.958440	-0.460051	2.912434
0.5	2.912434	-0.684326	2.844002
0.6	2.844002	-0.911522	2.752850
0.7	2.752850	-1.106689	2.642181
0.8	2.642181	-1.232987	2.518882
0.9	2.518882	-1.260884	2.392793
1	2.392793		

Step 3: $y(1) \approx 2.393$

Example 3

Use Euler's Method to approximate $P(4)$, given $\dfrac{dP}{dt} = .3P\left(1 - \dfrac{P}{20}\right)$ with initial condition $P(0) = 4$. Use an increment of $\Delta t = 0.5$.

t	P	$.3P\left(1 - \dfrac{P}{20}\right)$	$y_n = y_{n-1} + \Delta x \cdot f'(x_{n-1}, y_{n-1})$
0	4	.96	$4 + .5(.96) = 4.48$
0.5	4.48	1.042944	$4.48 + .5(1.042944) = 5.001472$
1	5.001472	1.125220	$5.001472 + .5(1.125220) = 5.564082$
1.5	5.564082	1.204839	$5.564082 + .5(1.204839) = 6.166502$
2	6.166502	1.279564	$6.166502 + .5(1.279564) = 6.806284$
2.5	6.806284	1.347002	$6.806284 + .5(1.347002) = 7.479785$
3	7.479785	1.404727	$7.479785 + .5(1.404727) = 8.182149$
3.5	8.182149	1.450431	$8.182149 + .5(1.450431) = 8.907365$
4	8.907365		

$P(4) \approx 8.907$

9.8 RAPID REVIEW

1. Find the average value of $y = \sin x$ on $[0, \pi]$.

   Answer: Average value $= \dfrac{1}{\pi - 0} \displaystyle\int_0^\pi \sin x \, dx$

   $$= \frac{1}{\pi}\left[-\cos x\right]_0^\pi = \frac{2}{\pi}$$

2. Find the total distance traveled by a particle during $0 \leq t \leq 3$ whose velocity function is shown in Figure 9.8-1.

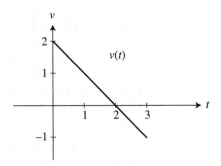

Figure 9.8-1

   Answer: The Total Distance Traveled $= \displaystyle\int_0^2 v(t)dt + \left|\int_2^3 v(t)dt\right|$

   $$= 2 + 0.5 = 2.5$$

3. Oil is leaking from a tank at the rate of $f(t) = 5e^{-0.1t}$ gallons/hour, where t is measured in hours. Write an integral to find the total number of gallons of oil that will have leaked from the tank after 10 hours. Do not evaluate the integral.

   Answer: Total number of gallons leaked $= \displaystyle\int_0^{10} 5e^{-0.1t}\, dt$

4. How much money should Mary invest at 7.5% interest a year compounded continuously, so that Mary will have \$100,000 after 20 years.

   Answer: $y(t) = y_0 e^{kt}$, $k = 0.075$, and $t = 20$. $y(20) = 100{,}000 = y_0 e^{(0.075)(20)}$. Thus you obtain $y_0 \approx 22313$, using a calculator.

5. Given $\dfrac{dy}{dx} = \dfrac{x}{y}$ and $y(1) = 0$, solve the differential equation.

   Answer:
   $$y\,dy = x\,dx \Rightarrow \int y\,dy = \int x\,dx \Rightarrow \frac{y^2}{2} = \frac{x^2}{2} + c \Rightarrow 0 = \frac{1}{2} + c \Rightarrow c = -\frac{1}{2}.$$
   Thus $\dfrac{y^2}{2} = \dfrac{x^2}{2} - \dfrac{1}{2}$ or $y^2 = x^2 - 1$.

6. Identify the differential equation for the slope field shown.

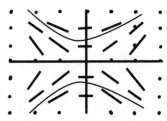

Answer: The slope field suggests a hyperbola of the form $y^2 - x^2 = k$, so $2y\dfrac{dy}{dx} - 2x = 0$ and $\dfrac{dy}{dx} = \dfrac{x}{y}$.

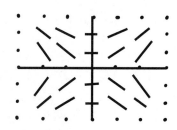

7. Find the solution of the initial value problem $\dfrac{dP}{dt} = .75P\left(1 - \dfrac{P}{2500}\right)$ with $P_0 = 10$.

Answer: $\dfrac{dP}{dt} = .75P\left(1 - \dfrac{P}{2500}\right) \Rightarrow \dfrac{2500}{P(2500 - P)}dP = .75dt$

$\Rightarrow \displaystyle\int \dfrac{2500}{P(2500 - P)}dP = \int .75dt \Rightarrow \int \left(\dfrac{1}{P} + \dfrac{1}{2500 - P}\right)dP = \int .75dt$

$\Rightarrow \ln|P| - \ln|2500 - P| = .75t + c_1 \Rightarrow \ln\left|\dfrac{P}{2500 - P}\right| = .75t + c_1$

$\Rightarrow \dfrac{P}{2500 - P} = c_2 e^{.75t} \Rightarrow P(t) = \dfrac{2500 c_2 e^{.75t}}{1 + c_2 e^{.75t}} \Rightarrow P(0) = \dfrac{2500 c_2}{1 + c_2} = 10$

$\Rightarrow 2500 c_2 = 10 + 10 c_2 \Rightarrow 2490 c_2 = 10 \Rightarrow c_2 = \dfrac{1}{249}$. Therefore,

$P(t) = \dfrac{2500\,(1/249)\,e^{.75t}}{1 + (1/249)\,e^{.75t}} = \dfrac{2500 e^{.75t}}{249 + e^{.75t}}$ so $P(t) = \dfrac{2500}{249 e^{-.75t} + 1}$

8. Use Euler's method with a step size of $\Delta x = 0.5$ to compute $y(2)$ if $y(x)$ is the solution of the differential equation $\dfrac{dy}{dx} = y + xy$ with initial condition $y(0) = 1$.

Answer: $y(0) = 1$; $y(0.5) = 1 + 0.5(1 + 0.1) = 1.5$;
$y(1) = 1.5 + 0.5\,[1.5 + (0.5)(1.5)] = 1.5 + 0.5[2.25] = 2.625$
$y(1.5) = 2.625 + 0.5\,[2.625 + (1)(2.625)] = 2.625 + 2.625 = 5.25$
$y(2) = 5.25 + 0.5\,[5.25 + (1.5)(5.25)] = 5.25 + 0.5[13.125] = 11.8125$

9.9 PRACTICE PROBLEMS

Part A—The use of a calculator is not allowed.

1. Find the value of c as stated in the Mean Value Theorem for Integrals for $f(x) = x^3$ on $[2, 4]$.

2. The graph of f is shown in Figure 9.9-1. Find the average value of f on $[0, 8]$.

3. The position function of a particle moving on a coordinate line is given as $s(t) = t^2 - 6t - 7$, $0 \le t \le 10$. Find the displacement and total distance traveled by the particle from $1 \le t \le 4$.

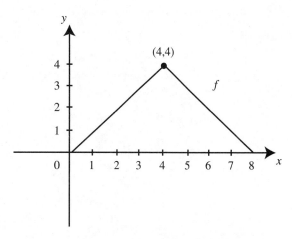

Figure 9.9-1

4. The velocity function of a moving particle on a coordinate line is $v(t) = 2t + 1$ for $0 \leq t \leq 8$. At $t = 1$, its position is -4. Find the position of the particle at $t = 5$.

5. The rate of depreciation for a new piece of equipment at a factory is given as $p(t) = 50t - 600$ for $0 \leq t \leq 10$, where t is measured in years. Find the total loss of value of the equipment over the first 5 years.

6. If the acceleration of a moving particle on a coordinate line is $a(t) = -2$ for $0 \leq t \leq 4$, and the initial velocity $v_0 = 10$, find the total distance traveled by the particle during $0 \leq t \leq 4$.

7. The graph of the velocity function of a moving particle is shown in Figure 9.9-2. What is the total distance traveled by the particle during $0 \leq t \leq 12$?

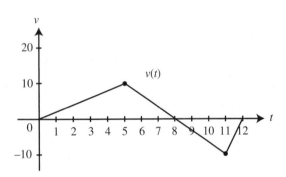

Figure 9.9-2

8. If oil is leaking from a tanker at the rate of $f(t) = 10e^{0.2t}$ gallons per hour where t is measured in hours, how many gallons of oil will have leaked from the tanker after the first 3 hours?

9. The change of temperature of a cup of coffee measured in degrees Fahrenheit in a certain room is represented by the function

$$f(t) = -\cos\left(\frac{t}{4}\right) \text{ for } 0 \leq t \leq 5, \text{ where } t \text{ is}$$

measured in minutes. If the temperature of the coffee is initially 92°F, find its temperature after the first 5 minutes.

10. If the *half-life* of a radioactive element is 4500 years, and initially there are 100 grams of this element, approximately how many grams are left after 5000 years?

11. Find a solution of the differential equation:

$$\frac{dy}{dx} = x\cos(x^2); \; y(0) = \pi$$

12. If $\dfrac{d^2y}{dx^2} = x - 5$ and at $x = 0$, $y' = -2$ and $y = 1$, find a solution of the differential equation.

Part B—Calculators are permitted.

13. Find the average value of $y = \tan x$ from $x = \dfrac{\pi}{4}$ to $x = \dfrac{\pi}{3}$.

14. The acceleration function of a moving particle on a straight line is given by $a(t) = 3e^{2t}$, where t is measured in seconds, and the initial velocity is $\dfrac{1}{2}$. Find the displacement and total distance traveled by the particle in the first 3 seconds.

15. The sales of an item in a company follow an exponential growth/decay model, where t is measured in months. If the sales drop from 5000 units in the first month to 4000 units in the third month, how many units should the company expect to sell during the seventh month?

16. Find an equation of the curve that has a slope of $\dfrac{2y}{x+1}$ at the point (x, y) and passes through the point $(0, 4)$.

17. The population in a city is approximately 750,000 in 1980, and growing at a rate of 3% per year. If the population growth follows an exponential growth model, find the city's population in the year 2002.

18. Find a solution of the differential equation $4e^y = y' - 3xe^y$ and $y(0) = 0$.

19. How much money should a person invest at 6.25% interest compounded continuously so that the person will have $50,000 after 10 years?

20. The velocity function of a moving particle is given as $v(t) = 2 - 6e^{-t}$, $t \geq 0$ and t is measured in seconds. Find the total distance traveled by the particle during the first 10 seconds.

21. Draw a slope field for the differential equation $\dfrac{dy}{dx} = x - y$

22. A rumor spreads through an office of fifty people at a model by $\dfrac{dP}{dt} = .65P\left(1 - \dfrac{P}{50}\right)$. On day zero, one person knows the rumor. Find the model for the population at time t, and use it to predict when more than half the people in the office will have heard the rumor.

23. A college dormitory that houses 200 students experiences an outbreak of influenza. The illness is recognized when two students are diagnosed on the same day. The residents are quarantined to restrict the infection to this one building. On the fifth day of the outbreak, 12 students are ill. Use a logistic model to describe the course of infection and predict the number of infected students on day 10.

24. Use Euler's Method with a step size of $\Delta x = 0.1$ to compute $y(.5)$ if $y(x)$ is the solution of the differential equation $\dfrac{dy}{dx} = x^2 - y^3$ with the condition $y(0) = 1$.

25. Use Euler's Method with a step size of $\Delta x = 0.5$ to compute $y(3)$ if $y(x)$ is the solution of the differential equation $\dfrac{dy}{dx} = y - 2x$ with initial condition $y(0) = 1$.

9.10 CUMULATIVE REVIEW PROBLEMS

("Calculator" indicates that calculators are permitted)

26. If $3e^y = x^2 y$, find $\dfrac{dy}{dx}$

27. Evaluate $\displaystyle\int_0^1 \dfrac{x^2}{x^3 + 1}\, dx$

28. The graph of a continuous function f which consists of three line segments on $[-2, 4]$ is shown in Figure 9.10-1. If $F(x) = \displaystyle\int_{-2}^x f(t)\, dt$ for $-2 \le x \le 4$,

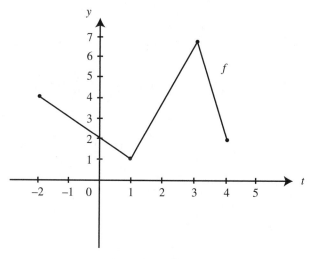

Figure 9.10-1

(a) Find $F(-2)$ and $F(0)$.

(b) Find $F'(0)$ and $F'(2)$.

(c) Find the value of x such that F has a maximum on $[-2, 4]$.

(d) On which interval is the graph of F concave upward?

29. (Calculator) The slope of a function $y = f(x)$ at any point (x, y) is $\dfrac{y}{2x + 1}$ and $f(0) = 2$.

(a) Write an equation of the line tangent to the graph of f at $x = 0$.

(b) Use the tangent in part (a) to find the approximate value of $f(0.1)$.

(c) Find a solution $y = f(x)$ for the differential equation.

(d) Use the result in part (c), find $f(0.1)$.

30. (Calculator) Let R be the region in the first quadrant bounded by $f(x) = e^x - 1$ and $g(x) = 3 \sin x$.

(a) Find the area of region R.

(b) Find the volume of the solid obtained by revolving R about the x-axis.

(c) Find the volume of the solid having R as its base and its cross sections are semicircles perpendicular to the x-axis.

BC 31. An object traveling on a path defined by $\langle x(\theta), y(\theta) \rangle$ has an acceleration vector of $\langle \sin \theta, -\cos \theta \rangle$. If the velocity of the object at time $t = \dfrac{\pi}{3}$ is $\langle -1, 0 \rangle$ and the initial position of the object is the origin, find the position when $\theta = \pi$.

32. $\displaystyle\int x^2 e^{5x-2}\, dx$

33. A projectile follows a path defined by $x = t - 2$, $y = \sin^2 t$ on the interval $0 \le t \le \pi$. Find the point at which the object reaches its maximum y-value.

9.11 SOLUTIONS TO PRACTICE PROBLEMS

Part A—No calculators.

1. $\displaystyle\int_2^4 x^3\, dx = f(c)(4-2)$

$$\int_2^4 x^3\, dx = \frac{x^4}{4}\bigg]_2^4 = \left(\frac{4^4}{4}\right) - \left(\frac{2^4}{4}\right) = 60$$

$$2f(c) = 60 \Rightarrow f(c) = 30$$

$$c^3 = 30 \Rightarrow c(30)^{1/3}.$$

2. Average Value $= \dfrac{1}{8-0}\displaystyle\int_0^1 f(x)\, dx$

$$= \frac{1}{8}\left(\frac{1}{2}(8)(4)\right) = 2.$$

3. Displacement $= s(4) - s(1) = -15 - (-12) = -3$.

Distance Traveled $= \displaystyle\int_1^4 |v(t)|dt$

$$v(t) = s'(t) = 2t - 6$$

Set $2t - 6 = 0 \Rightarrow t = 3$

$$|2t - 6| = \begin{cases} -(2t-6) & \text{if } 0 \le t < 3 \\ 2t - 6 & \text{if } 3 \le t \le 10 \end{cases}$$

$$\int_1^4 |v(t)|\, dt = \int_1^3 -(2t-6)dt + \int_3^4 (2t-6)dt$$

$$= \left[-t^2 + 6t\right]_1^3 + \left[t^2 - 6t\right]_3^4$$

$$= 4 + 1 = 5.$$

4. Position Function $s(t) = \displaystyle\int v(t)dt$

$$= \int (2t + 1)dt$$

$$= t^2 + t + c$$

$$s(1) = -4 \Rightarrow (1)^2 + 1 + c$$

$$= -4 \text{ or } c = -6$$

$$s(t) = t^2 + t - 6.$$

$$s(5) = 5^2 + 5 - 6 = 24.$$

5. Total Loss $= \displaystyle\int_0^5 p(t)dt = \int_0^5 (50t - 600)dt$

$$= 25t^2 - 600t\big]_0^5 = -2375.$$

6. $v(t) = \displaystyle\int a(t)dt = \int -2\, dt = -2t + c$

$$v_0 = 10 \Rightarrow -2(0) + c = 10 \text{ or } c = 10$$

$$v(t) = -2t + 10.$$

Distance Traveled $= \displaystyle\int_0^4 |v(t)|\, dt$.

Set $v(t) = 0 \Rightarrow -2t + 10 = 0$ or $t = 5$.

$$|-2t + 10| = -2t + 10 \text{ if } 0 \le t < 5$$

$$\int_0^4 |v(t)|\, dt = \int_0^4 (-2t + 10)dt$$

$$= -t^2 + 10t\big]_0^4 = 24.$$

7. Total Distance Traveled

$$= \int_0^8 |v(t)|\, dt + \left|\int_8^{12} v(t)\right|$$

$$= \frac{1}{2}(8)(10) + \frac{1}{2}(4)(10)$$

$$= 60 \text{ meters.}$$

8. Total Leakage $= \displaystyle\int_0^3 10e^{0.2t} = 50e^{0.2t}\big]_0^3$

$$= 91.1059 - 50$$

$$= 41.1059 = 41 \text{ gallons}$$

9. Total change in temperature

$$= \int_0^5 -\cos\left(\frac{t}{4}\right) dt$$

$$= -4\sin\left(\frac{t}{4}\right)\bigg]_0^5$$

$$= -3.79594 - 0$$

$$= -3.79594°F.$$

Thus the temperature of coffee after 5 minutes is $(92 - 3.79594) \approx 88.204°F$

10. $y(t) = y_0 e^{kt}$

Half-life = 4500 years $\Rightarrow \dfrac{1}{2} = e^{4500k}$

Take ln of both sides:

$$\ln\left(\frac{1}{2}\right) = \ln e^{4500k}$$

$$\Rightarrow -\ln 2 = 4500k$$

$$\text{or } k = \frac{-\ln 2}{4500}$$

$y(t) = 100e^{\left(\frac{-\ln 2}{4500}\right)(5000)} = 25(2^{2/9}) \approx 46.293.$

There are approximately 46.29 grams left.

11. Step 1. Separate variables: $dy = x\cos(x^2)\,dx$

Step 2. Integrate both sides:

$$\int dy = \int x\cos(x^2)\,dx.$$

$$\int dy = y$$

$\int x\cos(x^2)dx$: Let $u = x^2$;

$$du = 2x\,dx, \frac{du}{2} = x\,dx$$

$$\int x\cos(x^2)dx = \int \cos u \frac{du}{2}$$

$$= \frac{\sin u}{2} + c = \frac{\sin(x^2)}{2} + c.$$

Thus $y = \dfrac{\sin(x^2)}{2} + c$

Step 3. Substitute given values.

$$y(0) = \frac{\sin(0)}{2} + c = \pi \Rightarrow c = \pi$$

$$y = \frac{\sin(x^2)}{2} + \pi.$$

Step 4. Verify result by differentiating

$$\frac{dy}{dx} = \frac{\cos(x^2)(2x)}{2} = x\cos(x^2).$$

12. Step 1. Rewrite $\dfrac{d^2y}{dx^2}$ as $\dfrac{dy'}{dx}$

$$\frac{dy'}{dx} = x - 5$$

Step 2. Separate variables: $dy' = (x - 5)dx$

Step 3. Integrate both sides:

$$\int dy' = \int (x - 5)\,dx$$

$$y' = \frac{x^2}{2} - 5x + c_1$$

Step 4. Substitute given values:

At $x = 0, y' = \dfrac{0}{2} - 5(0) + c_1$

$$= -2 \Rightarrow c_1 = -2$$

$$y' = \frac{x^2}{2} - 5x - 2$$

Step 5. Rewrite: $y' = \dfrac{dy}{dx}; \dfrac{dy}{dx} = \dfrac{x^2}{2} - 5x - 2.$

Step 6. Separate variables:

$$dy = \left(\frac{x^2}{2} - 5x - 2\right)dx.$$

Step 7. Integrate both sides:

$$\int dy = \int \left(\frac{x^2}{2} - 5x - 2\right)dx.$$

$$y = \frac{x^3}{6} - \frac{5x^2}{2} - 2x + c_2.$$

Step 8. Substitute given values:

At $x = 0, y = 0 - 0 - 0 + c_2$

$$= 1 \Rightarrow c_2 = 1$$

$$y = \frac{x^3}{6} - \frac{5x^2}{2} - 2x + 1.$$

Step 9. Verify result by differentiating:

$$\frac{dy}{dx} = \frac{x^2}{2} - 5x - 2$$

$$\frac{d^2y}{dx^2} = x - 5.$$

Part B—Calculators are permitted.

13. Average Value $= \dfrac{1}{\pi/3 - \pi/4} \displaystyle\int_{\pi/4}^{\pi/3} \tan x \, dx$

Enter $= \left(1 \big/ (\pi/3 - \pi/4) \right) \displaystyle\int (\tan x, x, \pi/4, \pi/3)$

and obtain $\dfrac{6 \ln(2)}{\pi} = 1.32381$.

14. $v(t) = \displaystyle\int a(t) dt$

$= \displaystyle\int 3e^{2t} = \dfrac{3}{2} e^{2t} + c$

$v(0) = \dfrac{3}{2} e^0 + c = \dfrac{1}{2} \Rightarrow \dfrac{3}{2} + c = \dfrac{1}{2}$ or $c = -1$

$v(t) = \dfrac{3}{2} e^{2t} - 1$

Displacement $= \displaystyle\int_0^3 \left(\dfrac{3}{2} e^{2t} - 1 \right) dt$

Enter $\displaystyle\int (3/2 * e\char`\^(2x) - 1, x, 0, 3)$

and obtain 298.822.

Distance Traveled $= \displaystyle\int_0^3 |v(t)| \, dt$

since $\dfrac{3}{2} e^{2t} - 1 > 0$ for $t \geq 0$,

$\displaystyle\int_0^3 |v(t)| dt = \displaystyle\int_0^3 \left(\dfrac{3}{2} e^{2t} - 1 \right) dt = 298.822$.

15. Step 1. $y(t) = y_0 e^{kt}$

$y(1) = 5000 \Rightarrow 5000 = y_0 e^k \Rightarrow y_0 = 5000 e^{-k}$

$y(3) = 4000 \Rightarrow 4000 = y_0 e^{3k}$

Substituting
$y(0) = 5000 e^{-k}, 4000 = \left(5000 e^{-k} \right) e^{3k}$

$4000 = 5000 e^{2k}$

$\dfrac{4}{5} = e^{2k}$

$\ln \left(\dfrac{4}{5} \right) = \ln \left(e^{2k} \right) = 2k$

$k = \dfrac{1}{2} \ln \left(\dfrac{4}{5} \right) \approx -0.111572$.

Step 2. $5000 = y_0 e^{-0.111572}$

$y(0) = (5000)/e^{-0.111572} \approx 5590.17$

$y(t) = (5590.17) e^{-0.111572}$

Step 3. $y(7) = (5590.17) e^{-0.111572(7)} \approx 2560$
Thus sales for the 7th month are approximately 2560 units.

16. Step 1. Separate variables:

$\dfrac{dy}{dx} = \dfrac{2y}{x+1}$

$\dfrac{dy}{2y} = \dfrac{dy}{x+1}$

Step 2. Integrate both sides:

$\displaystyle\int \dfrac{dy}{2y} = \displaystyle\int \dfrac{dx}{x+1}$

$\dfrac{1}{2} \ln |y| = \ln |x+1| + c$.

Step 3. Substitute given value $(0, 4)$:

$\dfrac{1}{2} \ln(4) = \ln(1) + c$

$\ln 2 = c$

$\dfrac{1}{2} \ln |y| - \ln |x+1| = \ln 2$

$\ln \left| \dfrac{y^{1/2}}{x+1} \right| = \ln 2$

$e^{\ln \left| \frac{y^{1/2}}{x+1} \right|} = e^{\ln 2}$

$\dfrac{y^{1/2}}{x+1} = 2$

$y^{1/2} = 2(x+1)$

$y = (2)^2 (x+1)^2$

$y = 4(x+1)^2$.

Step 4. Verify result by differentiating:

$\dfrac{dy}{dx} = 4(2)(x+1) = 8(x+1)$

Compare with $\dfrac{dy}{dx} = \dfrac{2y}{x+1}$

$$= \dfrac{2\left(4(x+1)^2\right)}{(x+1)}$$

$$= 8\,(x+1).$$

17. $y\,(t) = y_0 e^{kt}$

$y_0 = 750{,}000$

$y(22) = (750{,}000)\,e^{(0.03)(22)}$

$$\approx \begin{cases} 1.45109E6 \approx 1{,}451{,}090 \text{ using} \\ \qquad\qquad \text{a T1} - 89 \\ 1{,}451{,}094 \text{ using a T1-85.} \end{cases}$$

18. Step 1. Separate variables:

$$4e^y = \dfrac{dy}{dx} - 3xe^y$$

$$4e^y + 3xe^y = \dfrac{dy}{dx}$$

$$e^y(4 + 3x) = \dfrac{dy}{dx}$$

$$(4 + 3x)\,dx = \dfrac{dy}{e^y} = e^{-y}dy$$

Step 2. Integrate both sides:

$$\int (4 + 3x)\,dx = \int e^{-y}dy$$

$$4x + \dfrac{3x^2}{2} = -e^{-y} + c$$

switch sides: $e^{-y} = -\dfrac{3x^2}{2} - 4x + c$

Step 3. Substitute given value: $y\,(0) = 0$
$\Rightarrow e^0 = 0 - 0 + c \Rightarrow c = 1$

Step 4. Take ln of both sides:

$$e^{-y} = -\dfrac{3x^2}{2} - 4x + 1$$

$$\ln(e^{-y}) = \ln\left(-\dfrac{3x^2}{2} - 4x + 1\right)$$

$$y = -\ln\left(1 - 4x - \dfrac{3x^2}{2}\right)$$

Step 5. Verify result by differentiating:
Enter $d(-\ln(1 - 4x - 3(x-^2)/2), x)$
and

obtain $\dfrac{-2(3x+4)}{3x^2 + 8x - 2}$, Which is

equivalent to $e^y(4 + 3x)$.

19. $y\,(t) = y_0 e^{kt}$

$k = 0.0625, y\,(10) = 50{,}000$

$50{,}000 = y_0 e^{10(0.0625)}$

$$y_0 = \dfrac{50{,}000}{e^{0.625}} \begin{cases} \$26763.1 \text{ using a T1-89} \\ \$26763.071426 \approx \$26763.07 \\ \text{using a T1-85.} \end{cases}$$

20. Set $v(t) = 2 - 6e^{-t} = 0$. Using the Zero Function on your calculator, compute $t = 1.09861$.

Distance Traveled $= \displaystyle\int_0^{10} |v\,(t)|\,dt$

$$|2 - 6e^{-t}| = \begin{cases} -\left(2 - 6e^{-t}\right) \text{ if } 0 \le t < 1.09861 \\ 2 - 6e^{-t} \text{ if } t \ge 1.09861 \end{cases}$$

$$\int_0^{10} |2 - 6e^{-t}|dt = \int_0^{1.09861} -\left(2 - 6e^{-t}\right)dt$$

$$+ \int_{1.09861}^{10} \left(2 - 6e^{-t}\right)dt$$

$$= 1.80278 + 15.803 = 17.606$$

Alternatively, use the nInt Function on the calculator.
Enter nInt(abs($2 - 6e^\wedge(-x)$), $x, 0, 10$) and obtain the same result.

21. Build a table of value for $\dfrac{dy}{dx} = x - y$.

	$y = -2$	$y = -1$	$y = 0$	$y = 1$	$y = 2$
$x = -2$	0	−1	−2	−3	−4
$x = -1$	1	0	−1	−2	−3
$x = 0$	2	1	0	−1	−2
$x = 1$	3	2	1	0	−1
$x = 2$	4	3	2	1	0

Draw short lines at each intersection with slopes equal to the value of $\frac{dy}{dx}$ at that point.

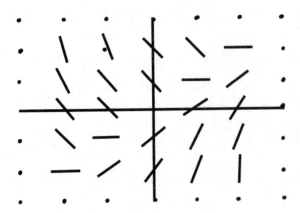

22. $\frac{dP}{dt} = .65P\left(1 - \frac{P}{50}\right)$ can be separated and integrated by partial fractions.

$\int \frac{dP}{P} + \int \frac{dP}{(50 - P)} = \int .65\,dt$ produces $\ln|P| + \ln|50 - P| = .65t + C_1$ and $\frac{P}{50 - P} = C_2 e^{.65t}$, so $P = \frac{50C_2 e^{.65t}}{1 + C_2 e^{.65t}}$. Since one person knows the rumor on day zero, $1 = \frac{50C_2}{1 + C_2}$ and $C_2 = \frac{1}{49}$. The model for the population becomes $P = \frac{50e^{.65t}/49}{1 + \left(e^{.65t}/49\right)} =$

$\frac{50e^{.65t}}{49 + e^{.65t}} = \frac{50}{49e^{-.65t} + 1}$. Half the population of the office would be 25 people, so solve for t in $25 = \frac{50e^{.65t}}{49 + e^{.65t}}$. Since $t \approx 5.987$, half of the office will have heard the rumor by the sixth day.

23. The logistic model becomes $\frac{dP}{dt} = kP\left(1 - \frac{P}{200}\right)$ since the carrying capacity is 200. Separate the variables $\int \frac{200\,dP}{P(200 - P)} = \int k\,dt$ and integrate by partial fractions $\int \frac{dP}{P} + \int \frac{dP}{200 - P} = \int k\,dt$. You find $\ln\left|\frac{P}{200 - P}\right| = kt + C_1$. Exponentiate to get $\frac{P}{200 - P} = e^{kt + C_1} = C_2 e^{kt}$. Solving for P produces $P = \frac{200C_2 e^{kt}}{1 + C_2 e^{kt}}$. On day zero, two students are infected, so $2 = \frac{200C_2}{1 + C_2}$ and $C_2 = \frac{1}{99}$. On day 5, 12 students are infected, so $12 = \frac{200e^{5k}/99}{1 + (e^{5k}/99)}$ and $k \approx .369$. Therefore

$P = \frac{200\left(\frac{1}{99}\right)e^{.369t}}{1 + \left(\frac{1}{99}\right)e^{.369t}} = \frac{200e^{.369t}}{99 + e^{.369t}} =$

$\frac{200}{99e^{-.369t} + 1}$. On day 10,

$P = \frac{200}{99e^{-.369(10)} + 1} = \frac{200}{3.472} \approx 57.600$, so we would predict that approximately 58 students would be infected on the tenth day.

24. Apply $y_n = y_{n-1} + \Delta x \cdot f'(x_{n-1}, y_{n-1})$ with $\Delta x = 0.1$

x	y	dy/dx	Calculate next y-value:
0	1	-1	$1 + 0.1(-1)$
0.1	0.9	-0.719	$0.9 + 0.1(-0.719)$
0.2	0.8281	0.607869	$0.8281 + 0.1(0.607869)$
0.3	0.888886	0.792327	$0.8888860.1(0.792327)$
0.4	0.968119	1.067375	$0.968119 + 0.1(1.067375)$
0.5	1.074857		

$y(5) \approx 1.075$

25. Apply $y_n = y_{n-1} + \Delta x \cdot f'(x_{n-1}, y_{n-1})$ with $\Delta x = 0.5$:

x	y	dy/dx	Calculate next y-value
0	1	1	$1 + 0.5(1)$
0.5	1.5	.5	$1.5 + 0.5(0.5)$
1	1.75	-0.25	$1.75 + 0.5(-0.25)$
1.5	1.625	-1.375	$1.625 + 0.5(-1.375)$
2	0.9375	-3.0625	$0.9375 + 0.5(-3.0625)$
2.5	-0.59375	-5.59375	$-0.59375 + 0.5(-5.59375)$
3	-3.390625		

$y(3) \approx -3.391$

9.12 SOLUTIONS TO CUMULATIVE REVIEW PROBLEMS

26. $3e^y = x^2 y$

$$3e^y \frac{dy}{dx} = 2xy + \frac{dy}{dx}\left(x^2\right)$$

$$3e^y \frac{dy}{dx} - \frac{dy}{dx} x^2 = 2xy$$

$$\frac{dy}{dx}\left(3e^y - x^2\right) = 2xy$$

$$\frac{dy}{dx} = \frac{2xy}{3e^y - x^2}$$

27. Let $u = x^3 + 1$; $du = 3x^2 dx$ or $\dfrac{du}{3} = x^2 dx$

$$\int \frac{x^2}{x^3 + 1} dx = \int \frac{1}{u} \frac{du}{3}$$

$$= \frac{1}{3} \ln|u| + c = \frac{1}{3} \ln|x^3 + 1| + c$$

$$\int_0^3 \frac{x^2}{x^3 + 1} dx = \frac{1}{3} \ln|x^3 + 1|\Big]_0^1$$

$$= \frac{1}{3}(\ln 2 - \ln 1) = \frac{\ln 2}{3}.$$

28. (a) $F(-2) = \displaystyle\int_{-2}^{-2} f(t)\, dt = 0$

$$F(0) = \int_{-2}^0 f(t) dt = \frac{1}{2}(4+2)\,2 = 6$$

(b) $F'(x) = f(x)$; $F'(0) = 2$ and $F'(2) = 4$

(c) Since $f > 0$ on $[-2, 4]$, F has a maximum value at $x = 4$.

(d) The function f is increasing on $(1, 3)$ which implies that $f' > 0$ on $(1, 3)$.

Thus, F is concave upward on $(1, 3)$. (Note: f' is equivalent to the 2nd derivative of F.)

29. (a) $\dfrac{dy}{dx} = \dfrac{y}{2x+1}$; $f(0) = 2$

$$\frac{dy}{dx}\bigg|_{x=0} = \frac{2}{2(0)+1} = 2 \Rightarrow m = 2 \text{ at } x = 0$$

$$y - y_1 = m(x - x_1)$$

$$y - 2 = 2(x - 0) \Rightarrow y = 2x + 2$$

The equation of the tangent to f at $x = 0$ is $y = 2x + 2$.

(b) $f(0.1) = 2(0.1) + 2 = 2.2$

(c) Solve the differential equation:
$$\frac{dy}{dx} = \frac{y}{2x+1}$$

Step 1. Separate variables

$$\frac{dy}{y} = \frac{dx}{2x+1}$$

Step 2. Integrate both sides

$$\int \frac{dy}{y} = \int \frac{dx}{2x+1}$$

$$\ln|y| = \frac{1}{2}\ln|2x+1| + c$$

Step 3. Substitute given values $(0, 2)$

$$\ln 2 = \frac{1}{2}\ln 1 + c \Rightarrow c = \ln 2$$

$$\ln|y| = \frac{1}{2}|2x+1| + \ln 2$$

$$\ln|y| - \frac{1}{2}|2x+1| = \ln 2$$

$$\ln\left|\frac{y}{(2x+1)^{1/2}}\right| = \ln 2$$

$$e^{\ln\left|\frac{y}{(2x+1)^{1/2}}\right|} = e^{\ln 2}$$

$$\frac{y}{(2x+1)^{1/2}} = 2$$

$$y = 2(2x+1)^{1/2}.$$

Step 4. Verify result by differentiating

$$y = 2(2x+1)^{1/2}$$

$$\frac{dy}{dx} = 2\left(\frac{1}{2}\right)(2x+1)^{-1/2}(2)$$

$$= \frac{2}{\sqrt{2x+1}}.$$

Compare this with:

$$\frac{dy}{dx} = \frac{y}{2x+1} = \frac{2(2x+1)^{1/2}}{2x+1}$$

$$= \frac{2}{\sqrt{2x+1}}.$$

Thus the function is
$y = f(x) = 2(2x+1)^{1/2}$.

(d) $f(x) = 2(2x+1)^{1/2}$

$f(0.1) = 2(2(0.1)+1)^{1/2} = 2(1.2)^{1/2}$
≈ 2.191

30. See Figure 9.12-1.

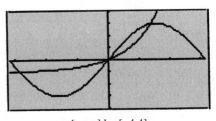

$[-\pi,\pi]$ by $[-4,4]$

Figure 9.12-1

(a) Intersection points: Using the Intersection Function on the calculator, you have $x = 0$ and $x = 1.37131$.

$$\text{Area of } R = \int_0^{1.37131} \left[3\sin x - (e^x - 1)\right]dx$$

Enter $\int (3\sin(x)) - (e^\wedge(x) - 1), x, 0, 1.37131)$ and obtain 0.836303.

The area of region R is approximately 0.836.

(b) Using the Washer Method, volume of

$$R = \pi \int_0^{1.37131} \left[(3\sin x)^2 - (e^x - 1)^2\right]dx$$

Enter $\pi \int ((3\sin(x))^\wedge 2 - (e^\wedge(x) - 1)^\wedge 2, x, 0, 1.37131)$ and obtain 2.54273π or 7.98824.

The volume of the solid is 7.988.

(c) Volume of Solid $= \pi \int_0^{1.37131}$ (Area of Cross Section)dx

Area of Cross Section $= \frac{1}{2}\pi r^2$

$$= \frac{1}{2}\pi\left(\frac{1}{2}(3\sin x - (e^x - 1))\right)^2$$

Enter $\left(\frac{\pi}{2}\right)\frac{1}{4} * \int ((3\sin(x) - (e^\wedge(x) - 1))^\wedge 2, x, 0, 1.37131)$ and obtain $0.077184\,\pi$ or 0.24248.

The volume of the solid is approximately $0.077184\,\pi$ or 0.242.

 31. The acceleration vector $\langle\sin\theta, -\cos\theta\rangle$ is equal to $(\sin\theta)i + (-\cos\theta)j$. Integrating, we find that the velocity is $(-\cos\theta)i + (\sin\theta)j + C_1$. If the velocity of the object at $\theta = \frac{\pi}{3}$ is given as $\langle-1, 0\rangle$, so $\left(-\cos\frac{\pi}{3}\right)i + \left(\sin\frac{\pi}{3}\right)j + C_1 = -i$

which implies $\left(-\dfrac{1}{2}\right)i + \left(\dfrac{\sqrt{3}}{2}\right)j + C_1 = -i$ and

$C_1 = -\dfrac{1}{2}i - \dfrac{\sqrt{3}}{2}j$. The velocity vector therefore

is $\left(-\dfrac{1}{2} - \cos\theta\right)i + \left(\sin\theta - \dfrac{\sqrt{3}}{2}\right)j$. Integrate

again to obtain the position function.

$s = \left(-\dfrac{1}{2}\theta - \sin\theta\right)i + \left(-\cos\theta - \dfrac{\sqrt{3}}{2}\theta\right)j + C_2$.

Since the initial position is the origin,
$0i - 1j + C_2 = 0i + 0j$ and $C_2 = j$. Therefore,
the position function

$s = \left(-\dfrac{1}{2}\theta - \sin\theta\right)i + \left(-\cos\theta - \dfrac{\sqrt{3}}{2}\theta + 1\right)j$.

At $\theta = \pi$,

$s = \left(-\dfrac{1}{2}\pi - \sin\pi\right)i$

$\quad + \left(-\cos\pi - \dfrac{\sqrt{3}}{2}\pi + 1\right)j$

$\quad = \left(-\dfrac{1}{2}\pi\right)i + \left(\dfrac{4 - \sqrt{3}}{2}\pi\right)j$.

32. Integrate $\int x^2 e^{5x-2}\,dx$ by parts. Let $u = x^2$,

$du = 2x\,dx$, $dv = e^{5x-2}\,dx$, and $v = \dfrac{e^{5x-2}}{5}$.

Then $\int x^2 e^{5x-2}\,dx = \dfrac{x^2 e^{5x-2}}{5} - \dfrac{2}{5}\int x e^{5x-2}\,dx$.

The remaining integral can be evaluated by
parts, using $u = x$, $du = dx$, $dv = e^{5x-2}\,dx$, and

$v = \dfrac{e^{5x-2}}{5} \cdot \int x^2 e^{5x-2}\,dx$

$= \dfrac{x^2 e^{5x-2}}{5} - \dfrac{2}{5}\left[\dfrac{x e^{5x-2}}{5} - \dfrac{1}{5}\int e^{5x-2}\,dx\right]$

$= \dfrac{x^2 e^{5x-2}}{5} - \dfrac{2}{5}\left[\dfrac{x e^{5x-2}}{5} - \dfrac{1}{5} \cdot \dfrac{e^{5x-2}}{5}\right]$

$= \dfrac{x^2 e^{5x-2}}{5} - \dfrac{2x e^{5x-2}}{25} + \dfrac{2 e^{5x-2}}{125}$.

33. The path is defined by $x = t - 2$, $y = \sin^2 t$.

Since $\dfrac{dx}{dt} = 1$ and $\dfrac{dy}{dt} = 2\sin t \cos t$,

$\dfrac{dy}{dx} = \dfrac{2\sin t \cos t}{1}$. This is the slope of a tangent
line to the curve, and when the slope is zero,
the curve will reach either a maximum or a
minimum. The slope $2\sin t \cos t = 0$ when
$\sin t = 0$ or when $\cos t = 0$. The first equation

gives us $t = 0$ or $t = \pi$ and the second, $t = \dfrac{\pi}{2}$.

The second derivative,

$\dfrac{d^2 y}{dx^2} = -2\sin^2 t + 2\cos^2 t = 2\cos 2t$. Evaluating

at each of the possible values of t, we find
$2\cos 2t|_{t=0} = 2$, $2\cos 2t|_{t=\pi} = 2$, and

$2\cos 2t|_{t=\frac{\pi}{2}} = -2$. The maximum value will
occur when the second derivative is negative, so

the maximum y-value is achieved when $t = \dfrac{\pi}{2}$,

$x = \dfrac{\pi}{2} - 2 = \dfrac{\pi - 4}{2}$, and $y = \sin^2 \dfrac{\pi}{2} = 1$.

Chapter 10

Series

10.1 SEQUENCES AND SERIES

Main Concepts: *Sequences, Series, Convergence*

Sequences and Series

A sequence is a function whose domain is the non-negative integers. It can be expressed as a list of terms $\{a_n\} = \{a_1, a_2, a_3, \ldots, a_n, \ldots\}$ or by a formula that defines the nth term of the sequence for any value of n. A series $\sum a_n = \sum\limits_{n=1}^{\infty} a_n = a_1 + a_2 + a_3 + \cdots + a_n + \cdots$ is the sum of the terms of a sequence $\{a_n\}$. Associated with each series is a sequence of partial sums, $\{s_n\}$, where $s_1 = a_1$, $s_2 = a_1 + a_2$, $s_3 = a_1 + a_2 + a_3$, and in general, $s_n = a_1 + a_2 + a_3 + \cdots + a_n$.

Example 1

Find the first three partial sums of the series $\sum\limits_{n=1}^{\infty} \dfrac{(-2)^n}{n^3}$

Step 1: Generate the first three terms of the sequence $\left\{ \dfrac{(-2)^n}{n^3} \right\}$. $a_1 = \dfrac{(-2)^1}{1^3} = -2$,

$$a_2 = \dfrac{(-2)^2}{2^3} = \dfrac{4}{8} = \dfrac{1}{2}, \; a_3 = \dfrac{(-2)^3}{3^3} = \dfrac{-8}{27}$$

Step 2: Find the partial sums. $s_1 = a_1 = -2$, $s_2 = a_1 + a_2 = -2 + \dfrac{1}{2} = \dfrac{-3}{2}$,

$$s_3 = a_1 + a_2 + a_3 = -2 + \dfrac{1}{2} + \dfrac{-8}{27} = \dfrac{-97}{54} \approx -1.796$$

Example 2

Find the fifth partial sum of the series $\sum_{n=1}^{\infty} \dfrac{5+n^2}{n+3}$

Step 1: Generate the first five terms of the sequence $\left\{ \dfrac{5+n^2}{n+3} \right\}$

$$a_1 = \frac{5+1^2}{1+3} = \frac{6}{4} = \frac{3}{2}, \; a_2 = \frac{5+2^2}{2+3} = \frac{9}{5}, \; a_3 = \frac{5+3^2}{3+3} = \frac{14}{6} = \frac{7}{3},$$

$$a_4 = \frac{5+4^2}{4+3} = \frac{21}{7} = 3, \; a_5 = \frac{5+5^2}{5+3} = \frac{30}{8} = \frac{15}{4}$$

Step 2: The fifth partial sum is $a_1 + a_2 + a_3 + a_4 + a_5 = \dfrac{3}{2} + \dfrac{9}{5} + \dfrac{7}{3} + 3 + \dfrac{15}{4} = \dfrac{743}{60}$

Convergence

The series $\sum a_n$ converges if the sequence of associated partial sums, $\{s_n\}$, converges. The limit $\lim\limits_{n \to \infty} s_n = S$, where S is a real number, is the sum of series, $\sum_{n=1}^{\infty} a_n = S$. If $\sum_{n=1}^{\infty} a_n$ and $\sum_{n=1}^{\infty} b_n$ are convergent, then $\sum_{n=1}^{\infty} ca_n = c \sum_{n=1}^{\infty} a_n$ and $\sum_{n=1}^{\infty} (a_n \pm b_n) = \sum_{n=1}^{\infty} a_n \pm \sum_{n=1}^{\infty} b_n$.

Example 1

Determine whether the series $\sum_{n=1}^{\infty} \dfrac{1}{5^n}$ converges or diverges. If it converges, find its sum.

Step 1: Find the first few partial sums. $s_1 = \dfrac{1}{5} = 0.2, \; s_2 = \dfrac{1}{5} + \dfrac{1}{25} = \dfrac{6}{25} = 0.24,$

$$s_3 = \frac{1}{5} + \frac{1}{25} + \frac{1}{125} = \frac{31}{125} = 0.248, \; s_4 = \frac{1}{5} + \frac{1}{25} + \frac{1}{125} + \frac{1}{625} =$$

$$\frac{156}{625} = 0.2496$$

Step 2: The sequence of partial sums $\{0.2, 0.24, 0.248, 0.2496, \ldots\}$ converges to 0.25, so the series converges, and its sum $\sum_{n=1}^{\infty} \dfrac{1}{5^n} = 0.25$

Example 2

Find the sum of the series $\sum_{n=1}^{\infty} (5a_n - 3b_n)$, given that $\sum_{n=1}^{\infty} a_n = 4$ and $\sum_{n=1}^{\infty} a_n = 8$.

Step 1: $\sum_{n=1}^{\infty} (5a_n - 3b_n) = \sum_{n=1}^{\infty} 5a_n - \sum_{n=1}^{\infty} 3b_n = 5 \sum_{n=1}^{\infty} a_n - 3 \sum_{n=1}^{\infty} b_n$

Step 2: $5 \sum_{n=1}^{\infty} a_n - 3 \sum_{n=2}^{\infty} b_n = 5(4) - 3(8) = 20 - 24 = -4$

10.2   TYPES OF SERIES

Main Concepts: p-Series, Harmonic Series, Geometric Series, Decimal Expansion

p-Series

The *p*-series is a series of the form $1 + \dfrac{1}{2^p} + \dfrac{1}{3^p} + \dfrac{1}{4^p} + \cdots + \dfrac{1}{n^p} + \cdots = \displaystyle\sum_{n=1}^{\infty} \dfrac{1}{n^p}$. The *p*-series converges when $p > 1$ and diverges when $0 < p \le 1$.

Harmonic Series

The harmonic series $1 + \dfrac{1}{2} + \dfrac{1}{3} + \dfrac{1}{4} + \cdots + \dfrac{1}{n} + \cdots = \displaystyle\sum_{n=1}^{\infty} \dfrac{1}{n}$ is a *p*-series with $p = 1$. The harmonic series diverges.

Geometric Series

A geometric series is a series of the form $\displaystyle\sum_{n=1}^{\infty} ar^{n-1}$ where $a \ne 0$. A geometric series converges when $|r| < 1$. The sum of the first n terms of a geometric series is $s_n = \dfrac{a(1 - r^n)}{1 - r}$. The sum of the series $\displaystyle\sum_{n=1}^{\infty} ar^{n-1} = \lim_{n \to \infty} s_n = \lim_{n \to \infty} \dfrac{a(1 - r^n)}{1 - r} = \dfrac{a}{1 - r}$.

Example 1

Determine whether the series $1 + \dfrac{3}{2} + \dfrac{9}{4} + \dfrac{27}{8} + \cdots$ converges. $1 + \dfrac{3}{2} + \dfrac{9}{4} + \dfrac{27}{8} + \cdots$ is a geometric series with $a = 1$ and $r = \dfrac{3}{2}$. Since $r > 1$, the series diverges.

Example 2

Find the tenth partial sum of the series $12 + 9 + \dfrac{27}{4} + \dfrac{81}{16} + \cdots$.

While it is possible to extend the terms of the series and directly compute the tenth partial sum, it is quicker to recognize that this is a geometric series. The ratio of any two subsequent terms is $r = \dfrac{3}{4}$ and the first term is $a = 12$.

$$s_{10} = \dfrac{12\left(1 - \left(\dfrac{3}{4}\right)^{10}\right)}{1 - \dfrac{3}{4}} \approx 45.297$$

Example 3

Find the sum of the series $12 + 9 + \dfrac{27}{4} + \dfrac{81}{16} + \cdots$. Since $12 + 9 + \dfrac{27}{4} + \dfrac{81}{16} + \cdots$ is a geometric series with $a = 12$ and $r = \dfrac{3}{4}$, so $S = \dfrac{a}{1 - r} = \dfrac{12}{1 - \dfrac{3}{4}} = 48$.

Decimal Expansion

The rational number equal to the repeating decimal is the sum of the geometric series that represents the repeating decimal.

Example:

Find the rational number equivalent to $3.8\overline{76}$

Step 1: $3.8\overline{76} = 3.8 + .076 + .00076 + \cdots = \dfrac{38}{10} + \dfrac{76}{10^3} + \dfrac{76}{10^5} + \dfrac{76}{10^7} + \cdots =$

$\dfrac{38}{10} + 76 \displaystyle\sum_{n=1}^{\infty} \dfrac{1}{10^{2n+1}}$

Step 2: $\displaystyle\sum_{n=1}^{\infty} \dfrac{1}{10^{2n+1}}$ is a geometric series with $a = \dfrac{1}{10^3}$ and $r = \dfrac{1}{10^2}$. The sum of the

series is $\dfrac{a}{1-r} = \dfrac{\dfrac{1}{1000}}{1 - \dfrac{1}{100}} = \dfrac{1}{990}$.

Step 3: $3.8\overline{76} = \dfrac{38}{10} + 76 \displaystyle\sum_{n=1}^{\infty} \dfrac{1}{10^{2n+1}} = \dfrac{38}{10} + \dfrac{76}{990} = \dfrac{3838}{990} = \dfrac{1919}{495}$

10.3 CONVERGENCE TESTS

Main Concepts: *Integral Test, Ratio Test, Comparison Test, Limit Comparison Test*

Integral Test

If $a_n = f(n)$ where f is a continuous, positive, decreasing function on $[c, \infty)$, then the series $\displaystyle\sum_{n=1}^{\infty} a_n$ is convergent if and only if the improper integral $\displaystyle\int_{c}^{\infty} f(x)\,dx$ exists.

Example 1

Determine whether the series $\displaystyle\sum_{n=1}^{\infty} \dfrac{1}{n^2} \sin \dfrac{\pi}{n}$ converges or diverges.

Step 1: $f(x) = \dfrac{1}{x^2} \sin \dfrac{\pi}{x}$ is continuous, positive and decreasing on the interval $[2, \infty)$.

Step 2: $\displaystyle\int_{2}^{\infty} \dfrac{1}{x^2} \sin \dfrac{\pi}{x}\,dx = \dfrac{1}{\pi} \lim_{u \to \infty} \cos\left(\dfrac{\pi}{x}\right)\Big|_{2}^{u} = \dfrac{1}{\pi} \lim_{u \to \infty} \left[\cos\left(\dfrac{\pi}{u}\right) - \cos\left(\dfrac{\pi}{2}\right)\right] =$

$\dfrac{1}{\pi} \lim_{u \to \infty} \left[\cos\left(\dfrac{\pi}{u}\right)\right] = \dfrac{1}{\pi}.$ The improper integral exists so $\displaystyle\sum_{n=1}^{\infty} \dfrac{1}{n^2} \sin \dfrac{\pi}{n}$ converges.

Example 2

Determine whether the series $1 + \dfrac{1}{4} + \dfrac{1}{9} + \dfrac{1}{16} + \cdots + \dfrac{1}{n^2} + \cdots = \displaystyle\sum_{n=1}^{\infty} \dfrac{1}{n^2}$ converges or diverges.

Step 1: $f(x) = \dfrac{1}{x^2}$ is continuous, positive and decreasing on the interval $[1, \infty)$.

Step 2: $\displaystyle\int_{1}^{\infty} \dfrac{1}{x^2}\,dx = \lim_{u \to \infty} \int_{1}^{u} \dfrac{1}{x^2}\,dx = \lim_{u \to \infty}(-x^{-1})\Big|_{1}^{u} = \lim_{u \to \infty}\left[\dfrac{-1}{u} - \dfrac{-1}{1}\right] =$

$\lim_{u \to \infty}\left[1 - \dfrac{1}{u}\right] = 1.$

Since the improper integral exists, the series $\displaystyle\sum_{n=1}^{\infty} \dfrac{1}{n^2}$ converges.

Example 3

Determine whether the series $1 + \dfrac{1}{2} + \dfrac{1}{3} + \dfrac{1}{4} + \cdots + \dfrac{1}{n} + \cdots = \displaystyle\sum_{n=1}^{\infty} \dfrac{1}{n}$ converges or diverges.

Step 1: $f(x) = \dfrac{1}{x}$ is continuous, positive and decreasing on $[1, \infty)$

Step 2: $\displaystyle\int_{1}^{\infty} \dfrac{1}{x} dx = \lim_{u \to \infty} \int_{1}^{u} \dfrac{1}{x} dx = \lim_{u \to \infty} (\ln x)\big|_{1}^{u} = \lim_{u \to \infty} [\ln u - \ln 1] = \infty$. Since the improper integral does not converge, the series $\displaystyle\sum_{n=1}^{\infty} \dfrac{1}{n}$ diverges.

Example 4

Determine whether the series $1 + \dfrac{1}{\sqrt{2}} + \dfrac{1}{\sqrt{3}} + \dfrac{1}{\sqrt{4}} + \cdots + \dfrac{1}{\sqrt{n}} + \cdots = \displaystyle\sum_{n=1}^{\infty} \dfrac{1}{\sqrt{n}}$ converges or diverges.

Step 1: $f(x) = \dfrac{1}{\sqrt{x}}$ is continuous, positive and decreasing on $[1, \infty)$

Step 2: $\displaystyle\int_{1}^{\infty} \dfrac{1}{\sqrt{x}} dx = \lim_{u \to \infty} \int_{1}^{u} \dfrac{1}{\sqrt{x}} dx = \lim_{u \to \infty} (2\sqrt{x})\big|_{1}^{u} = \lim_{u \to \infty} [2\sqrt{u} - 2] = \infty$

Since the improper integral does not converge, the series $\displaystyle\sum_{n=1}^{\infty} \dfrac{1}{\sqrt{n}}$ diverges.

Ratio Test

If $\displaystyle\sum_{n=1}^{\infty} a_n$ is a series with positive terms and $\displaystyle\lim_{n \to \infty} \dfrac{a_{n+1}}{a_n} < 1$, then the series converges. If the limit is greater than 1 or ∞, the series diverges. If the limit is 1, another test must be used.

Example 1

Determine whether the series $\displaystyle\sum_{n=1}^{\infty} \dfrac{(n+1) \cdot 2^n}{n!}$ converges or diverges.

Step 1: For all $n \geq 1$, $\dfrac{(n+1) \cdot 2^n}{n!}$ is positive.

Step 2: $\displaystyle\lim_{n \to \infty} \dfrac{a_{n+1}}{a_n} = \lim_{n \to \infty} \left(\dfrac{(n+2) \cdot 2^{n+1}}{(n+1)!} \cdot \dfrac{n!}{(n+1) \cdot 2^n} \right) = \lim_{n \to \infty} \left(\dfrac{(n+2) \cdot 2}{(n+1)^2} \right) = 0$.
Since this limit is less than 1, the series converges.

Example 2

Determine whether the series $\displaystyle\sum_{n=1}^{\infty} \dfrac{n^3}{(\ln 2)^n}$ converges or diverges.

Step 1: For all $n \geq 1$, $\dfrac{n^3}{(\ln 2)^n}$ is positive.

Step 2: $\displaystyle\lim_{n \to \infty} \dfrac{a_{n+1}}{a_n} = \lim_{n \to \infty} \left(\dfrac{(n+1)^3}{(\ln 2)^{n+1}} \cdot \dfrac{(\ln 2)^n}{n^3} \right) = \dfrac{1}{\ln 2} \lim_{n \to \infty} \left(\dfrac{n+1}{n} \right)^3 = \dfrac{1}{\ln 2} \approx 1.443$.
Since this limit is greater than one, the series diverges.

Comparison Test

Suppose $\sum\limits_{n=1}^{\infty} a_n$ and $\sum\limits_{n=1}^{\infty} b_n$ are series with non-negative terms, and $\sum\limits_{n=1}^{\infty} b_n$ is known to converge. If a term-by-term comparison shows that for all n, $a_n \leq b_n$, then $\sum\limits_{n=1}^{\infty} a_n$ converges. If $\sum\limits_{n=1}^{\infty} b_n$ diverges, and if for all n, $a_n \geq b_n$, then $\sum\limits_{n=1}^{\infty} a_n$ diverges. Common series that may be used for comparison include the geometric series, which converges for $r < 1$ and diverges for $r \geq 1$, and the p-series, which converges for $p > 1$ and diverges for $p \leq 1$.

Example 1

Determine whether the series $\sum\limits_{n=1}^{\infty} \dfrac{1}{n^2 + 5}$ converges or diverges.

Step 1: Choose a series for comparison. The series $\sum\limits_{n=1}^{\infty} \dfrac{1}{n^2 + 5}$ can be compared to $\sum\limits_{n=1}^{\infty} \dfrac{1}{n^2}$, the p-series with $p = 2$. Both series have non-negative terms.

Step 2: A term-by-term comparison shows that $\dfrac{1}{n^2 + 5} < \dfrac{1}{n^2}$ for all values of n.

Step 3: $\sum\limits_{n=1}^{\infty} \dfrac{1}{n^2}$ converges, so $\sum\limits_{n=1}^{\infty} \dfrac{1}{n^2 + 5}$ converges.

Example 2

Determine whether the series $2 + \dfrac{3}{5} + \dfrac{4}{10} + \dfrac{5}{17} + \cdots$ converges or diverges.

Step 1: The series $2 + \dfrac{3}{5} + \dfrac{4}{10} + \dfrac{5}{17} + \cdots = \sum\limits_{n=1}^{\infty} \dfrac{n+1}{n^2 + 1}$ can be compared to $\sum\limits_{n=1}^{\infty} \dfrac{1}{n}$.

Step 2: $\dfrac{n+1}{n^2 + 1} = \dfrac{n^2 + n}{n^3 + n} \geq \dfrac{n^2 + 1}{n^3 + n} = \dfrac{1}{n}$ so $\dfrac{n+1}{n^2 + 1} \geq \dfrac{1}{n}$ for all $n \geq 1$.

Step 3: Since $\sum\limits_{n=1}^{\infty} \dfrac{1}{n}$ diverges, $2 + \dfrac{3}{5} + \dfrac{4}{10} + \dfrac{5}{17} + \cdots$ also diverges.

Limit Comparison Test

If $\sum\limits_{n=1}^{\infty} a_n$ and $\sum\limits_{n=1}^{\infty} b_n$ are series with positive terms, and if $\lim\limits_{n\to\infty} \dfrac{a_n}{b_n} = L$ where $0 < L < \infty$, then either both series converge or both diverge. By choosing, for one of these, a series that is known to converge, or known to diverge, you can determine whether the other series converges or diverges. Choose a series of a similar form so that the limit expression can be simplified.

Example 1

Determine whether the series $1 + \dfrac{1}{5} + \dfrac{1}{9} + \dfrac{1}{13} + \cdots$ converges or diverges.

The given series $1 + \dfrac{1}{5} + \dfrac{1}{9} + \dfrac{1}{13} + \cdots = \sum\limits_{n=1}^{\infty} \dfrac{1}{4n - 3}$. Choose $\sum\limits_{n=1}^{\infty} \dfrac{1}{n}$ for comparison, since it has a similar structure, and we know it is a p-series with $p = 1$, so it diverges.

The limit $\lim\limits_{n\to\infty}\dfrac{1/4n-3}{1/n}=\lim\limits_{n\to\infty}\dfrac{n}{4n-3}=\dfrac{1}{4}$. The limit exists and is greater than zero; therefore, since $\sum\limits_{n=1}^{\infty}\dfrac{1}{n}$ diverges, $1+\dfrac{1}{5}+\dfrac{1}{9}+\dfrac{1}{13}+\cdots=\sum\limits_{n=1}^{\infty}\dfrac{1}{4n-3}$ also diverges.

Example 2

Determine whether the series $\sum\limits_{n=1}^{\infty}\dfrac{n-2}{n^3}$ converges or diverges.

Compare to the known convergent p-series $\sum\limits_{n=1}^{\infty}\dfrac{1}{n^2}$. The limit $\lim\limits_{n\to\infty}\dfrac{(n-2)/n^3}{1/n^2}=$ $\lim\limits_{n\to\infty}\dfrac{(n-2)(n^2)}{(n^3)}=\lim\limits_{n\to\infty}\dfrac{n-2}{n}=1$. Since $0<\lim\limits_{n\to\infty}\dfrac{n-2/n^3}{1/n^2}<\infty$ and $\sum\limits_{n=1}^{\infty}\dfrac{1}{n^2}$ converges, $\sum\limits_{n=1}^{\infty}\dfrac{n-2}{n^3}$ also converges.

10.4   ALTERNATING SERIES

Main Concepts: *Alternating Series, Error Bound, Absolute Convergence*

Alternating Series

A series whose terms alternate between positive and negative are called alternating series. Alternating series have one of two forms $\sum\limits_{n=1}^{\infty}(-1)^n a_n$ or $\sum\limits_{n=1}^{\infty}(-1)^{n+1}a_n$. An alternating series converges if $a_1>a_2>a_3>\cdots>a_n>\cdots$ and $\lim\limits_{n\to\infty}a_n=0$.

Example 1

Determine whether the series $\dfrac{1}{e}-\dfrac{2}{e^2}+\dfrac{3}{e^3}-\dfrac{4}{e^4}+\cdots$ converges or diverges.

Step 1: $\dfrac{1}{e}-\dfrac{2}{e^2}+\dfrac{3}{e^3}-\dfrac{4}{e^4}+\cdots=\sum\limits_{n=1}^{\infty}(-1)^{n+1}\dfrac{n}{e^n}$.

Step 2: Note that $\dfrac{1}{e}>\dfrac{2}{e^2}>\dfrac{3}{e^3}>\dfrac{4}{e^4}$ and in general, $\dfrac{n}{e^n}>\dfrac{n+1}{e^{n+1}}$ since multiplying by e^{n+1} gives $en>n+1$.

Step 3: $\left\{\dfrac{1}{e},\dfrac{2}{e^2},\dfrac{3}{e^3},\dfrac{4}{e^4},\ldots\right\}\approx\{.36788,.27067,.14936,.07326,\ldots\}$ so $\lim\limits_{n\to\infty}\dfrac{n}{e^n}=0$.
Therefore the series converges.

Example 2

Determine whether the series $4-1+\dfrac{1}{4}-\dfrac{1}{16}+\cdots$ converges or diverges. If it converges, find its sum.

Step 1: $4-1+\dfrac{1}{4}-\dfrac{1}{16}+\cdots$ is a geometric series with $a=4$ and $r=\dfrac{-1}{4}$. Since $r<1$, the series converges.

Step 2: $S=\dfrac{a}{1-r}=\dfrac{4}{1-\dfrac{-1}{4}}=\dfrac{4}{\dfrac{5}{4}}=\dfrac{16}{5}=3.2$

Error Bound

If an alternating series converges to the sum S, then S lies between two consecutive partial sums of the series. If S is approximated by a partial sum s_n, the absolute error $|S - s_n|$ is less than the next term of the series a_{n+1} and the sign of $S - s_n$ is the same as the coefficient of a_{n+1}.

Example:

$4 - 1 + \dfrac{1}{4} - \dfrac{1}{16} + \cdots$ converges to 3.2. This value is greater than s_n for n odd, and less than s_n for n even. If S is approximated by the third partial sum, $s_3 = 3.25$, the absolute error $|S - s_3| = |3.2 - 3.25| = |-0.05| = 0.05$, which is clearly less than $a_4 = \dfrac{1}{16} = 0.0625$. The coefficient of a_4 is negative, as is $S - s_3$.

Absolute Convergence

If the series $\displaystyle\sum_{n=1}^{\infty} |a_n|$ converges, then $\displaystyle\sum_{n=1}^{\infty} a_n$ converges. A series that converges absolutely converges.

Example:

Determine whether the series $-1 + \dfrac{1}{3} - \dfrac{1}{9} + \dfrac{1}{27} \cdots = \displaystyle\sum_{n=1}^{\infty} (-1)^n \dfrac{1}{3^{n-1}}$ converges.

Step 1: Consider the series $\displaystyle\sum_{n=1}^{\infty} \left| \dfrac{(-1)^n}{3^{n-1}} \right| = 1 + \dfrac{1}{3} + \dfrac{1}{9} + \dfrac{1}{27} + \cdots$. For this series, $s_1 = 1$, $s_2 = 1 + \dfrac{1}{3} = \dfrac{4}{3} = 1.\bar{3}$, $s_3 = 1 + \dfrac{1}{3} + \dfrac{1}{9} = \dfrac{13}{9} = 1.\bar{4}$, $s_4 = 1 + \dfrac{1}{3} + \dfrac{1}{9} + \dfrac{1}{27} = \dfrac{40}{27} = 1.\overline{481}$. The sequence of partial sums, $\{1, 1.\bar{3}, 1.\bar{4}, 1.\overline{481}, \ldots\}$, converges to 1.5. Or, note that this is a geometric series with $a = 1$, $r = \frac{1}{3}$; thus it converges to $\frac{3}{2}$.

Step 2: Since $\displaystyle\sum_{n=1}^{\infty} \left| (-1)^n \dfrac{1}{3^{n-1}} \right|$ converges, $\displaystyle\sum_{n=1}^{\infty} (-1)^n \dfrac{1}{3^{n-1}}$ converges.

10.5 POWER SERIES

Main Concepts: *Power Series, Radius and Interval of Convergence*

Power Series

A power series is a series of the form $\displaystyle\sum_{n=1}^{\infty} c_n x^n$, where $c_1, c_2, c_3, \ldots$ are constants, and x is a variable.

Radius and Interval of Convergence

A power series centered at $x = a$ converges only for $x = a$, for all real values of x, or for all x in some open interval $(a - R, a + R)$, called the interval of convergence. The radius of convergence is R. If the series converges on $(a - R, a + R)$, then it diverges if $x < a - R$ or $x > a + R$, but convergence or divergence much be investigated individually at $x = a - R$ and at $x = a + R$.

Example 1

Find the values of x for which the series $1 + x + \dfrac{x^2}{2!} + \dfrac{x^3}{3!} + \cdots + \dfrac{x^n}{n!} + \cdots$ converges.

Step 1: Use the ratio test. $\lim\limits_{n \to \infty} \left| \dfrac{a_{n+1}}{a_n} \right| = \lim\limits_{n \to \infty} \left| \dfrac{x^{n+1}}{(n+1)!} \cdot \dfrac{n!}{x^n} \right| = \lim\limits_{n \to \infty} \left| \dfrac{x}{n+1} \right| = 0.$

Step 2: The series converges absolutely, so $\displaystyle\sum_{n=1}^{\infty} \dfrac{x^n}{n!}$ converges for all real x.

Example 2

Find the interval of convergence for the series $\displaystyle\sum_{n=1}^{\infty} \dfrac{(x-2)^n}{n}$

Step 1: Use the ratio test. $\lim\limits_{n \to \infty} \left| \dfrac{a_{n+1}}{a_n} \right| = \lim\limits_{n \to \infty} \left| \dfrac{(x-2)^{n+1}}{n+1} \cdot \dfrac{n}{(x-2)^n} \right| = \lim\limits_{n \to \infty} \left| \dfrac{n(x-2)}{n+1} \right|$
$= |x - 2|.$

Step 2: The series converges absolutely when $|x - 2| < 1$, or $-1 < x - 2 < 1$, or $1 < x < 3$.

Step 3: When $x = 1$, the series becomes $\displaystyle\sum_{n=1}^{\infty} \dfrac{(-1)^n}{n} = -1 + \dfrac{1}{2} - \dfrac{1}{3} + \dfrac{1}{4} - \dfrac{1}{5} \cdots.$
Since $1 > \dfrac{1}{2} > \dfrac{1}{3} > \dfrac{1}{4} > \dfrac{1}{5} > \cdots$ and $\lim\limits_{n \to \infty} \dfrac{1}{n} = 0$, this alternating series converges.
When $x = 3$, the series becomes $\displaystyle\sum_{n=1}^{\infty} \dfrac{1}{n} = 1 + \dfrac{1}{2} + \dfrac{1}{3} + \dfrac{1}{4} + \dfrac{1}{5} \cdots$ which is a
p-series with $p = 1$, and therefore diverges.
Therefore, the interval of convergence is $[1, 3)$.

10.6 TAYLOR SERIES

Main Concepts: *Taylor Series, MacLaurin Series, Common MacLaurin Series*

Taylor Series and MacLaurin Series

A Taylor polynomial approximates the value of a function $f(x)$ at the point $x = a$. If the function and all its derivatives exist at $x = a$, then on the interval of convergence, the Taylor Series $\displaystyle\sum_{n=0}^{\infty} \dfrac{f^{(n)}(a)}{n!}(x - a)^n$ converges to $f(x)$. The MacLaurin Series is the name given to a Taylor Series centered at $x = 0$.

Example 1

Find the Taylor polynomial of degree 3 for $f(x) = \dfrac{1}{x + 2}$ about the point $x = 3$.

Step 1: Differentiate $f'(x) = \dfrac{-1}{(x+2)^2}$, $f''(x) = \dfrac{2}{(x+2)^3}$, $f'''(x) = \dfrac{-6}{(x+2)^4}$

Step 2: Evaluate $f(3) = \dfrac{1}{5}$, $f'(3) = \dfrac{-1}{25}$, $f''(3) = \dfrac{2}{125}$, $f'''(x) = \dfrac{-6}{625}$

Step 3: $\dfrac{f(3)}{0!} = \dfrac{1/5}{1} = \dfrac{1}{5}$ $\dfrac{f'(3)}{1!} = \dfrac{-1/25}{1} = \dfrac{-1}{25}$ $\dfrac{f''(3)}{2!} = \dfrac{2/125}{2} = \dfrac{1}{125}$ $\dfrac{f'''(3)}{3!} =$

$\dfrac{6/625}{6} = \dfrac{1}{125}$

Step 4: $\displaystyle\sum_{n=0}^{3} \dfrac{f^{(n)}(a)}{n!}(x-a)^n = \dfrac{1}{5} - \dfrac{(x-3)}{25} + \dfrac{(x-3)^2}{125} - \dfrac{(x-3)^3}{625}$

Example 2

A function $f(x)$ is approximated by the third order Taylor Series $1 + 2(x-1) - (x-1)^2 + (x-1)^3$ centered at $x = 1$. Find $f'(1)$ and $f'''(1)$.

Step 1: Compare $\displaystyle\sum_{n=0}^{\infty} \dfrac{f^{(n)}(a)}{n!}(x-a)^n$ to the given polynomial. $\dfrac{f(1)}{0!} = 1$, $\dfrac{f'(1)}{1!} = 2$, $\dfrac{f''(1)}{2!} = -1$, and $\dfrac{f'''(1)}{3!} = 1$.

Step 2: $f'(1) = 2 \cdot 1! = 2$ and $f'''(1) = 1 \cdot 3! = 6$

Example 3

Find the MacLaurin polynomial of degree 4 that approximates $f(x) = \ln(1+x)$

Step 1: Differentiate $f'(x) = \dfrac{1}{1+x}$, $f''(x) = \dfrac{-1}{(1+x)^2}$, $f'''(x) = \dfrac{2}{(1+x)^3}$, $f^{(4)}(x) = \dfrac{-6}{(1+x)^4}$

Step 2: Evaluate $f(0) = 0$, $f'(0) = 1$, $f''(0) = -1$, $f'''(0) = 2$, $f^{(4)}(0) = -6$

Step 3: $\dfrac{f(0)}{0!}x^0 + \dfrac{f'(0)}{1!}x^1 + \dfrac{f''(0)}{2!}x^2 + \dfrac{f'''(0)}{3!}x^3 + \dfrac{f^{(4)}(0)}{4!}x^4 = \dfrac{0}{1}x^0 + \dfrac{1}{1}x^1 + \dfrac{-1}{2}x^2 +$

$\dfrac{2}{6}x^3 + \dfrac{-6}{24}x^4 = x - \dfrac{1}{2}x^2 + \dfrac{1}{3}x^3 - \dfrac{1}{4}x^4$

Example 4

Find the Taylor series for the function $f(x) = e^{-x}$ about the point $x = \ln 2$.

Step 1: $f^{(n)}(x) = e^{-x}$ when n is even and $f^{(n)}(x) = -e^{-x}$ when n is odd.

Step 2: Evaluate $f^{(n)}(\ln 2) = e^{-\ln 2} = \dfrac{1}{2}$ when n is even and $f^{(n)}(\ln 2) = \dfrac{-1}{2}$ when n is odd.

Step 3: $f(x) = e^{-x} = \dfrac{1/2}{0!}(x - \ln 2)^0 + \dfrac{-1/2}{1!}(x - \ln 2)^1 + \dfrac{1/2}{2!}(x - \ln 2)^2 + \cdots = \displaystyle\sum_{n=0}^{\infty} \dfrac{(-1)^n}{2 \cdot n!}(x - \ln 2)^n$

Example 5

Find the MacLaurin series for the function $f(x) = xe^x$

Step 1: Investigating the first few derivatives of $f(x) = xe^x$ shows that $f^{(n)}(x) = xe^x + ne^x$.

Step 2: Evaluating $f^{(n)}(x) = xe^x + ne^x$ at $x = 0$ gives $f^{(n)}(0) = n$

Step 3: $f(x) = \sum_{n=0}^{\infty} \frac{f^{(n)}(0)}{n!} x^n = \sum_{n=0}^{\infty} \frac{n}{n!} x^n = \sum_{n=1}^{\infty} \frac{x^n}{(n-1)!}$

MacLaurin Series For The Functions e^x, $\sin x$, $\cos x$, and $\dfrac{1}{1-x}$.

Familiarity with these common MacLaurin series will simplify many problems.

$$f(x) = e^x = \sum_{n=0}^{\infty} \frac{x^n}{n!} = 1 + x + \frac{x^2}{2} + \frac{x^3}{6} + \cdots$$

$$f(x) = \sin x = \sum_{n=0}^{\infty} \frac{(-1)^n x^{2n+1}}{(2n+1)!} = x - \frac{x^3}{3!} + \frac{x^5}{5!} - \frac{x^7}{7!} + \cdots$$

$$f(x) = \cos x = \sum_{n=0}^{\infty} \frac{(-1)^{2n} x^{2n}}{(2n)!} = 1 - \frac{x^2}{2} + \frac{x^4}{24} - \frac{x^6}{6!} + \cdots$$

$$f(x) = \frac{1}{1-x} = \sum_{n=0}^{\infty} x^n = 1 + x + x^2 + x^3 + \cdots$$

10.7 OPERATIONS ON SERIES

Main Concepts: *Substitution, Differentiation, Integration, Error Bounds*

Substitution

New series can be generated by making an appropriate substitution in a known series.

Example 1

Find the MacLaurin series for $f(x) = \dfrac{1}{1+x^2}$.

Step 1: Begin with the known series $f(x) = \dfrac{1}{1-x} = \sum_{n=0}^{\infty} x^n$

Step 2: Substitute $-x^2$ for x.
$$\frac{1}{1+x^2} = \sum_{n=0}^{\infty} (-x^2)^n = \sum_{n=0}^{\infty} (-1)^n x^{2n} = 1 - x^2 + x^4 - x^6 + \cdots$$

Example 2

Find the first four nonzero terms of the MacLaurin series for $f(x) = \cos(2x)$.

Step 1: Begin with the known series $\cos x = 1 - \dfrac{x^2}{2!} + \dfrac{x^4}{4!} - \dfrac{x^6}{6!} + \cdots$

Step 2: Substitute $2x$ for x. $\cos(2x) = 1 - \dfrac{(2x)^2}{2!} + \dfrac{(2x)^4}{4!} - \dfrac{(2x)^6}{6!} + \cdots = 1 - \dfrac{4x^2}{2} +$
$\dfrac{16x^4}{24} - \dfrac{64x^6}{720} + \cdots = 1 - 2x^2 + \dfrac{2}{3}x^4 - \dfrac{4}{45}x^6$

Differentiation and Integration

If a function $f(x)$ is represented by a Taylor series with a non-zero radius of convergence, the derivative $f'(x)$ can be found by differentiating the series term by term. If the series is integrated term-by-term, the resulting series converges to $\int f(x)dx$. In either case, the radius of convergence is identical to that of the original series.

Example 1

Differentiate the MacLaurin series for $f(x) = \ln(x + 1)$ to find the Taylor series expansion for $f(x) = \dfrac{1}{x + 1}$.

Step 1: $f(x) = \ln(x + 1) = x - \dfrac{1}{2}x^2 + \dfrac{1}{3}x^3 - \dfrac{1}{4}x^4 + \cdots$

Step 2: $f'(x) = \dfrac{1}{x + 1} = 1 - x + x^2 - x^3 + \cdots = \displaystyle\sum_{n=0}^{\infty} (-1)^n x^n$

Example 2

Find the MacLaurin series for $f(x) = \dfrac{1}{(x + 1)^2}$

Step 1: We know that $\dfrac{1}{x + 1} = 1 - x + x^2 - x^3 + \cdots = \displaystyle\sum_{n=0}^{\infty} (-1)^n x^n$

Step 2: Differentiate $\dfrac{-1}{(x + 1)^2} = -1 + 2x - 3x^2 + 4x^3 - \ldots = \displaystyle\sum_{n=0}^{\infty} (-1)^{n+1}(n + 1)x^n$

Step 3: Multiply by -1. $\dfrac{1}{(x + 1)^2} = 1 - 2x + 3x^2 - 4x^3 + \cdots = \displaystyle\sum_{n=0}^{\infty} (-1)^n(n + 1)x^n$

Example 3

Use MacLaurin series to approximate the integral $\displaystyle\int_0^1 \sin(x^2)\,dx$ to three decimal place accuracy.

Step 1: Substitute x^2 for x in the MacLaurin series representing $\sin x$.

$$\sin(x^2) = (x^2) - \frac{(x^2)^3}{3!} + \frac{(x^2)^5}{5!} - \frac{(x^2)^7}{7!} \cdots = x^2 - \frac{x^6}{3!} + \frac{x^{10}}{5!} - \frac{x^{14}}{7!} + \cdots$$

Step 2: $\displaystyle\int_0^1 \sin(x^2)dx = \int_0^1 \left(x^2 - \frac{x^6}{3!} + \frac{x^{10}}{5!} - \frac{x^{14}}{7!} \cdots \right) dx = \frac{x^3}{3} - \frac{x^7}{7 \cdot 3!} + \frac{x^{11}}{11 \cdot 5!} -$

$\dfrac{x^{15}}{15 \cdot 7!} + \cdots \Big|_0^1 = \dfrac{1}{3} - \dfrac{1}{7 \cdot 3!} + \dfrac{1}{11 \cdot 5!} - \dfrac{1}{15 \cdot 7!} + \cdots = \displaystyle\sum_{n=0}^{\infty} \dfrac{1}{(4n + 3)(2n + 1)!}$

Step 3: For this alternating series, the absolute error for the nth partial sum is less than the $n + 1$ term so $|S - s_n| < \dfrac{1}{(4n + 4)(2n + 2)!}$. We want three decimal place accuracy, so we need $|S - s_n| < \dfrac{1}{(4n + 4)(2n + 2)!} \leq 0.0005$ or $2000 \leq (4n + 4)(2n + 2)!$ This occurs for $n \geq 2$.

Step 4: Taking the sum $a_0 + a_1 + a_2 = \dfrac{1}{3} - \dfrac{1}{7.3!} + \dfrac{1}{11 \cdot 5!} = 0.3103$ so $\displaystyle\int_0^1 \sin(x^2)dx \approx 0.3103$

Error Bounds

The remainder, $R_n(x)$, for a Taylor series is the difference between the actual value of the function $f(x)$ and the nth partial sum that approximates the function. If the function $f(x)$ can be differentiated $n + 1$ times on an interval containing x_0, and if $|f^{(n+1)}(x)| \le M$ for all x in that interval, then $|R_n(x)| \le \dfrac{M}{(n+1)!}|x - x_0|^{n+1}$ for all x in the interval.

Example 1

Approximate $\sqrt{e}$ accurate to three decimal places.

Step 1: Substitute $\dfrac{1}{2}$ for x in the MacLaurin series representation for e^x.

$$e^{1/2} = \sum_{n=0}^{\infty} \frac{(1/2)^n}{n!} = 1 + \frac{1}{2} + \frac{(1/2)^2}{2} + \frac{(1/2)^3}{6} + \cdots$$

$$= 1 + \frac{1}{2} + \frac{1}{8} + \frac{1}{48} + \cdots = \sum_{n=0}^{\infty} \frac{1}{2^n \cdot n!}$$

Step 2: For three decimal place accuracy, we want to find the value of n for which the remainder is less than or equal to 0.0005. Choose $x_0 = 1$ in the interval $(0, 1)$. All derivatives of e^x are equal to e^x, and therefore $|f^{(n+1)}(x)| \le e$, so $M = e \cdot \left|R_n\left(\dfrac{1}{2}\right)\right| \le \dfrac{e}{(n+1)!}\left|\dfrac{1}{2} - 1\right|^{n+1}$ and $\dfrac{e}{(n+1)!}\left|\dfrac{1}{2} - 1\right|^{n+1} = \dfrac{e}{2^{n+1} \cdot (n+1)!} \le 0.0005$ when $n \ge 4$.

Step 3: $\sqrt{e} = \sum_{n=0}^{4} \dfrac{1}{2^n \cdot n!} = 1 + \dfrac{1}{2} + \dfrac{1}{2^2 \cdot 2!} + \dfrac{1}{2^3 \cdot 3!} + \dfrac{1}{2^4 \cdot 4!} = 1.6484$

Example 2

Estimate $\sin 4°$ accurate to five decimal places.

Step 1: $4° = \dfrac{\pi}{45}$ radians. Substitute $\dfrac{\pi}{45}$ for x in the MacLaurin series that represents $\sin x \cdot \sin\dfrac{\pi}{45} = \dfrac{\pi}{45} - \dfrac{(\pi/45)^3}{3!} + \dfrac{(\pi/45)^5}{5!} - \dfrac{(\pi/45)^7}{7!} + \cdots$

Step 2: For five decimal place accuracy, we must find the value of n for which the absolute error is less than or equal to 5×10^{-6}. For all x, $|f^{(n+1)}(x)| \le 1$. Choose $x_0 = 0$. $\left|R_n\left(\dfrac{\pi}{45}\right)\right| \le \dfrac{1}{(n+1)!}\left|\dfrac{\pi}{45} - 0\right|^{n+1} = \dfrac{(\pi/45)^{n+1}}{(n+1)!}$. The absolute error is less than or equal to 5×10^{-6} for $n \ge 3$.

Step 3: $\sin 4° = \sin\dfrac{\pi}{45} = \dfrac{\pi}{45} - \dfrac{(\pi/45)^3}{3!} + \dfrac{(\pi/45)^5}{5!} - \dfrac{(\pi/45)^7}{7!} = 0.069756$

10.8 RAPID REVIEW

1. Find the sum of the series $81 + 27 + 9 + 3 + 1 + \dfrac{1}{3} + \cdots$.

   Answer: This is a geometric series with first term 81 and a ratio of $\dfrac{1}{3}$, so $S = \dfrac{81}{1 - 1/3} = \dfrac{243}{2}$

2. Determine whether the series $\sum \dfrac{5^n}{n!}$ converges or diverges.

Answer: $\lim\limits_{n\to\infty} \dfrac{5^{n+1}}{(n+1)!} \cdot \dfrac{n!}{5^n} = \lim\limits_{n\to\infty} \dfrac{5}{n+1} = 0$ so $\sum \dfrac{5^n}{n!}$ converges by ratio test.

3. Determine whether the series $\sum \dfrac{\ln n}{\sqrt{n}}$ converges or diverges.

Answer: $\dfrac{\ln n}{\sqrt{n}} > \dfrac{1}{\sqrt{n}}$ for $n > e$. $\sum \dfrac{1}{\sqrt{n}}$ is a p-series with $p < 1$, so it diverges.

$\displaystyle\sum_{n=1}^{\infty} \dfrac{\ln n}{\sqrt{n}} = \dfrac{\ln 2}{\sqrt{2}} + \sum_{n=3}^{\infty} \dfrac{\ln n}{\sqrt{n}}$ and $\displaystyle\sum_{n=3}^{\infty} \dfrac{\ln n}{\sqrt{n}}$ diverges by comparison to $\sum \dfrac{1}{\sqrt{n}}$,

so the series diverges.

4. Determine whether the series $\sum \dfrac{n}{n^2+1}$ converges or diverges.

Answer: $\lim\limits_{n\to\infty} \dfrac{\frac{n}{n^2+1}}{\frac{1}{n}} = \lim\limits_{n\to\infty} \dfrac{n^2}{n^2+1} = 1$ and $\sum \dfrac{1}{n}$ diverges, so $\sum \dfrac{n}{n^2+1}$

diverges by limit comparison with $\sum \dfrac{1}{n}$.

5. Determine whether the series $\sum \dfrac{50}{n(n+1)}$ converges or diverges.

Answer: Since $\displaystyle\int_1^{\infty} \dfrac{50}{x(x+1)}dx = 50 \lim\limits_{k\to\infty} \int_1^k \left(\dfrac{1}{x} + \dfrac{-1}{x+1}\right)dx$

$= 50 \lim\limits_{k\to\infty}[\ln x - \ln(x+1)]_1^k = 50 \lim\limits_{k\to\infty}[\ln k - \ln(k+1) + \ln 2]$

$= 50 \lim\limits_{k\to\infty}\left[\ln \dfrac{k}{k+1} + \ln 2\right] = 50\ln 2, \sum \dfrac{50}{n(n+1)}$ converges by integral test.

6. Find the interval to convergence for the series $\displaystyle\sum_{n=1}^{\infty} \dfrac{(x+1)^n}{\sqrt{n}}$.

Answer: $\lim\limits_{n\to\infty}\left|\dfrac{(x+1)^{n+1}}{\sqrt{n+1}} \cdot \dfrac{\sqrt{n}}{(x+1)^n}\right| = \lim\limits_{n\to\infty}\left|\dfrac{\sqrt{n}(x+1)}{\sqrt{n+1}}\right| = |x+1|$. Since

$|x+1| < 1$ when $-1 < x+1 < 1$ or $-2 \le x < 0$, the series converges on $(-2, 0)$.

When $x = -2$, the series becomes $\displaystyle\sum_{n=1}^{\infty} \dfrac{(-1)^n}{\sqrt{n}}$. Since $1 > \dfrac{1}{\sqrt{2}} > \dfrac{1}{\sqrt{3}} > \dfrac{1}{2} > \cdots$

and $\lim\limits_{n\to\infty} \dfrac{1}{\sqrt{n}} = 0$, this alternating series converges. When $x = 0$, the series

becomes $\displaystyle\sum_{n=1}^{\infty} \dfrac{1}{\sqrt{n}}$, which is a p-series with $p = \dfrac{1}{2}$ and therefore diverges.

Thus, the interval of convergence is $[-2, 0)$.

7. Approximate the function $f(x) = \dfrac{1}{x+2}$ with a fourth degree Taylor polynomial centered at $x = 3$.

Answer: $f(3) = \dfrac{1}{5}, f'(x) = \dfrac{-1}{(x+2)^2} \Rightarrow f'(3) = \dfrac{-1}{25}$,

$f''(x) = \dfrac{2}{(x+2)^3} \Rightarrow f''(3) = \dfrac{2}{125}, f'''(x) = \dfrac{-6}{(x+2)^4} \Rightarrow f'''(3) = \dfrac{-6}{625}$,

$$f^{(4)}(x) = \frac{24}{(x+2)^5} \Rightarrow f^{(4)}(3) = \frac{24}{3125} \text{ so}$$

$$P(x) = \frac{1/5}{0!}(x-3)^0 + \frac{-1/25}{1!}(x-3)^1 + \frac{2/125}{2!}(x-3)^2$$

$$+ \frac{-6/625}{3!}(x-3)^3 + \frac{24/3125}{4!}(x-3)^4$$

$$= \frac{1}{5} - \frac{x-3}{25} + \frac{(x-3)^2}{125} - \frac{(x-3)^3}{625} + \frac{(x-3)^4}{3125}$$

8. Find the MacLaurin series for the function $f(x) = e^{-x}$ and determine its interval of convergence.

Answer: Since $e^x = \sum \frac{x^n}{n!}$, substitute $-x$ to find

$$e^{-x} = \sum \frac{(-x)^n}{n!} = 1 - x + \frac{x^2}{2} - \frac{x^3}{6} + \cdots. \text{ The ratio } \lim_{n\to\infty} \left| \frac{(-x)^{n+1}}{(n+1)!} \frac{n!}{(-x)^n} \right|$$

$$= \lim_{n\to\infty} \left| \frac{-x}{n+1} \right| = 0, \text{ so the series converges on the interval } (-\infty, \infty).$$

10.9 PRACTICE PROBLEMS

Determine whether each series converges or diverges.

1. $\sum 5^{-n}$

2. $\sum \frac{1}{n \cdot 2^n}$

3. $\sum \frac{n}{e^n}$

4. $\sum \frac{n+1}{n(n+2)}$

5. $\sum \frac{(-1)^{n-1}}{n!}$

6. $\sum \frac{n}{(n+1)^n}$

7. $\sum (-1)^{n-1} \frac{n+1}{n}$

8. $\sum \frac{1}{n^{5n+1}}$

9. Find the sum of the geometric series $\sum 4\left(\frac{1}{3}\right)^n$.

10. If the sum of the alternating series $\sum \frac{(-1)^{n-1}}{2n-1}$ is approximated by s_{50}, find the maximum absolute error.

Find the interval of convergence for each series.

11. $\sum \frac{x^n}{1+n^2}$

12. $\sum \frac{3^n}{n^2}x^n$

13. The Taylor Series representation of $\ln x$, centered at $x = a$.
Approximate each function with a fourth degree Taylor polynomial centered at the given value of x.

14. $f(x) = e^{x^2}$ at $x = 1$

15. $f(x) = \cos \pi x$ at $x = \frac{1}{2}$

16. $f(x) = \ln x$ at $x = e$

Find the MacLaurin series for each function and determine its interval of convergence.

17. $f(x) = \dfrac{1}{1-x}$

18. $f(x) = \dfrac{1}{1+x^2}$

19. Estimate $\sin 9°$ accurate to three decimal places.

20. Find the rational number equivalent to $1.\overline{83}$.

 ## 10.10 CUMULATIVE REVIEW PROBLEMS

21. The movement of an object in the plane is defined by $x(t) = \ln t$, $y(t) = t^2$. Find the speed of the object at the moment when the acceleration is $a(t) = \langle -1, 2 \rangle$

22. Find the slope of the tangent line to the curve $r = 5\cos 3\theta$ when $\theta = \dfrac{2\pi}{3}$.

23. $\displaystyle\int_1^e x^3 \ln x \, dx$

24. $\displaystyle\int_0^1 \dfrac{5}{x^2 - x - 6} dx$

25. $\displaystyle\lim_{x \to 1} \dfrac{\ln x}{x^2 - 1}$

10.11 SOLUTIONS TO PRACTICE PROBLEMS

1. $\sum 5^{-n} = \sum \dfrac{1}{5^n} = 1 + \dfrac{1}{5} + \dfrac{1}{25} + \dfrac{1}{125} + \cdots$ is a geometric series with an initial term of one and a ratio of $\dfrac{1}{5}$. Since the ratio is less than one, the series converges, and $\sum 5^{-n} = \dfrac{1}{1 - 1/5} = \dfrac{5}{4}$.

2. By the ratio test, $\displaystyle\lim_{n \to \infty} \dfrac{1}{(n+1) \cdot 2^{n+1}} \cdot \dfrac{n \cdot 2^n}{1} =$
$\displaystyle\lim_{n \to \infty} \dfrac{n}{2(n+1)} = \dfrac{1}{2}$, therefore $\sum \dfrac{1}{n \cdot 2^n}$ converges.

3. Consider the integral $\displaystyle\int_0^\infty \dfrac{x}{e^x} dx = \int_0^\infty xe^{-x}dx$. Integrate by parts, with $u = x$, $du = dx$, $dv = e^{-x}dx$, and $v = -e^{-x}$. $\displaystyle\int xe^{-x}dx =$
$-xe^{-x} + \displaystyle\int e^{-x}dx = -xe^{-x} - e^{-x} + C$.
Therefore, $\displaystyle\int_0^\infty xe^{-x}dx = \lim_{k \to \infty} \int_0^k xe^{-x}dx =$
$\displaystyle\lim_{k \to \infty} \left[-xe^{-x} - e^{-x}\right]_0^k = \lim_{k \to \infty} \left[-ke^{-k} - e^{-k} + 1\right]_0^k = 1$.
Since the improper integral converges, $\sum \dfrac{n}{e^n}$ converges.

4. Use the limit comparison test, comparing to the series $\displaystyle\sum_{n=1}^\infty \dfrac{1}{n}$, which is known to diverge. Divide $\dfrac{n+1}{n(n+2)} \div \dfrac{1}{n} = \dfrac{(n+1)/n}{n/(n+2)} = \dfrac{n+1}{n+2}$. The limit

$\displaystyle\lim_{n \to \infty} \dfrac{n+1}{n+2} = 1$, and $\displaystyle\sum_{n=1}^\infty \dfrac{1}{n}$ diverges, so the series $\displaystyle\sum \dfrac{n+1}{n(n+2)}$ diverges.

5. $\displaystyle\lim_{n \to \infty} \left| \dfrac{(-1)^n}{(n+1)!} \cdot \dfrac{n!}{(-1)^{n-1}} \right| = \lim_{n \to \infty} \left| \dfrac{-1}{n+1} \right| =$
$\displaystyle\lim_{n \to \infty} \dfrac{1}{n+1} = 0$ so the series $\displaystyle\sum \dfrac{(-1)^{n-1}}{n!}$ converges absolutely.

6. Use the ratio test. $\displaystyle\lim_{n \to \infty} \left[\dfrac{n+1}{(n+2)^{n+1}} \cdot \dfrac{(n+1)^n}{n} \right] =$
$\displaystyle\lim_{n \to \infty} \left[\dfrac{(n+1)^{n+1}}{(n+2)^{n+1}} \cdot \dfrac{1}{n} \right] = 0$, therefore $\displaystyle\sum \dfrac{n}{(n+1)^n}$ converges.

7. $\displaystyle\sum (-1)^{n-1} \dfrac{n+1}{n} = \sum (-1)^{n-1} \left(1 + \dfrac{1}{n} \right) =$
$\displaystyle\sum \left((-1)^{n-1} + \dfrac{(-1)^{n-1}}{n} \right) = \sum (-1)^{n-1} +$
$\displaystyle\sum \dfrac{(-1)^{n-1}}{n}$. Both series diverge, so
$\displaystyle\sum (-1)^{n-1} \dfrac{n+1}{n}$ diverges.

8. Compare to the p-series with $p = 2$,
$1 + \dfrac{1}{2^2} + \dfrac{1}{3^2} + \dfrac{1}{4^2} + \cdots$. This series,
$\displaystyle\sum \dfrac{1}{n^{5n+1}} = 1 + \dfrac{1}{2^{11}} + \dfrac{1}{3^{16}} + \dfrac{1}{4^{21}} + \cdots$, is

term-by-term less than or equal to the p-series with $p = 2$. Since that p-series converges, $\sum \dfrac{1}{n^{5n+1}}$ converges.

9. The sum of the geometric series $\sum 4\left(\dfrac{1}{3}\right)^n$ is
$$S = \frac{a_n}{1-r} = \frac{4}{1-1/3} = 6.$$

10. For the alternating series $\sum \dfrac{(-1)^{n-1}}{2n-1}$ approximated by s_{50}, the maximum absolute error $|R_n| < a_{n+1}$ so $|R_{50}| < a_{51} = \dfrac{(-1)^{50}}{101} = \dfrac{1}{101} \approx 0.0099$.

11. Examine the ratio of successive terms.
$$\left|\frac{x^{n+1}}{1+(n+1)^2} \cdot \frac{1+n^2}{x^n}\right| = \left|\frac{x(1+n^2)}{n^2+2n+2}\right| \text{ Since}$$
$$\lim_{n\to\infty}\left|\frac{x(1+n^2)}{n^2+2n+2}\right| = |x|, \text{ the series will converge}$$
when $|x| < 1$ or $-1 < x < 1$. When $x = 1$, the series becomes $\sum \dfrac{1}{1+n^2}$. This series is term-by-term smaller than the p-series with $p = 2$; therefore the series converges. When $x = -1$, the series becomes $\sum \dfrac{(-1)^n}{1+n^2}$, which also converges. Therefore, the interval of convergence is $[-1, 1]$.

12. The ratio is $\left|\dfrac{(3x)^{n+1}}{(n+1)^2} \cdot \dfrac{n^2}{(3x)^n}\right| = \left|\dfrac{3xn^2}{(n+1)^2}\right|$ and
$$\lim_{n\to\infty}\left|\frac{3xn^2}{(n+1)^2}\right| = |3x| \text{ so the series will converge}$$
when $|3x| < 1$. This tells you $-1 < 3x < 1$ and $\dfrac{-1}{3} < x < \dfrac{1}{3}$. When $x = \dfrac{1}{3}$, the series becomes $\sum \dfrac{3^n}{n^2}\left(\dfrac{1}{3}\right)^n = \sum \dfrac{1}{n^2}$, which is a convergent p-series. When $x = -\dfrac{1}{3}$, the series becomes $\sum \dfrac{3^n}{n^2}\left(-\dfrac{1}{3}\right)^n = \sum \dfrac{(-1)^n}{n^2}$, a convergent alternating series. Therefore the interval of convergence is $\left[-\dfrac{1}{3}, \dfrac{1}{3}\right]$.

13. Represent $\ln x$ by a Taylor Series. Investigate the first few terms by finding and evaluating the derivatives and generating the first few terms.
$f(a) = \ln a$, $f'(a) = \dfrac{1}{a}$, $f''(a) = \dfrac{-1}{a^2}$, $f'''(a) = \dfrac{2}{a^3}$

so $\ln x$ can be represented by the series
$$= \frac{\ln a}{0!}(x-a)^0 + \frac{1/a}{1!}(x-a)^1 + \frac{-1/a^2}{2!}(x-a)^2 +$$
$$\frac{2/a^3}{3!}(x-a)^3 + \cdots = \ln a + \frac{(x-a)}{a} - \frac{(x-a)^2}{2a^2} +$$
$$\frac{(x-a)^3}{3a^3} + \cdots + \frac{(-1)^{n-1}(x-a)^n}{na^n} + \cdots$$
Using the ratio test,
$$\lim_{n\to\infty}\left|\frac{(-1)^n(x-a)^{n+1}}{(n+1)a^{n+1}} \cdot \frac{na^n}{(-1)^{n-1}(x-a)^n}\right| =$$
$$\lim_{n\to\infty}\left|\frac{n}{(n+1)} \cdot \frac{(x-a)}{a}\right| = \left|\frac{x-a}{a}\right|. \text{ The series}$$
converges when $\left|\dfrac{x-a}{a}\right| < 1$, that is,
$-1 < \dfrac{x-a}{a} < 1$. Solving the inequality, you find $-a < x - a < a$ or $0 < x < 2a$. When $x = 0$, the series becomes $\sum \dfrac{(-1)^{n-1}(-a)^n}{na^n} =$
$$\sum \frac{(-1)^{2n-1}}{n} = \sum \frac{-1}{n} = -\sum \frac{1}{n}. \text{ Since } \sum \frac{1}{n}$$
diverges, this series diverges as well. When $x = 2a$, the series becomes
$$\sum \frac{(-1)^{n-1}(2a-a)^n}{na^n} = \sum \frac{(-1)^{n-1}}{n}. \text{ This}$$
alternating series converges; therefore, the interval of convergence is $(0, 2a]$.

14. Calculate the derivatives and evaluate at $x = 1$. $f(x) = e^{x^2}$ and $f(1) = e$. $f'(x) = 2xe^{x^2}$ and $f'(1) = 2e$. $f''(x) = 4x^2e^{x^2} + 2e^{x^2}$ and $f''(1) = 6e$. $f'''(x) = 8x^3e^{x^2} + 12xe^{x^2}$ and $f'''(1) = 20e$. $f^{(4)}(x) = 16x^4e^{x^2} + 48x^2e^{x^2} + 12e^{x^2}$ and $f^{(4)}(1) = 76e$. Then the function $f(x) = e^{x^2}$ can be approximated by
$$\frac{e}{0!}(x-1)^0 + \frac{2e}{1!}(x-1)^1 + \frac{6e}{2!}(x-1)^2 +$$
$$\frac{20e}{3!}(x-1)^3 + \frac{76e}{4!}(x-1)^4. \text{ Simplifying}$$
$$f(x) = e^{x^2} \approx e + 2e(x-1) + 3e(x-1)^2 + \frac{10e}{3}(x-1)^3 + \frac{19e}{6}(x-1)^4.$$

15. $f\left(\dfrac{1}{2}\right) = \cos\left(\dfrac{\pi}{2}\right) = 0$. Find the derivatives and evaluate at $x = \dfrac{1}{2}$.
$$f'\left(\frac{1}{2}\right) = -\pi\sin\pi x|_{x=1/2} = -\pi,$$
$$f''\left(\frac{1}{2}\right) = -\pi^2\cos\pi x|_{x=1/2} = 0,$$
$$f'''\left(\frac{1}{2}\right) = \pi^3\sin\pi x|_{x=1/2} = \pi^3, \text{ and}$$
$$f^{(4)}\left(\frac{1}{2}\right) = \pi^4\cos\pi x|_{x=1/2} = 0. \text{ Then}$$

$f(x) = \cos \pi x$ around $x = \dfrac{1}{2}$ can be approximated by

$$\dfrac{0}{0!}\left(x - \dfrac{1}{2}\right)^0 + \dfrac{-\pi}{1!}\left(x - \dfrac{1}{2}\right)^1 + \dfrac{0}{2!}\left(x - \dfrac{1}{2}\right)^2 +$$

$$\dfrac{\pi^3}{3!}\left(x - \dfrac{1}{2}\right)^3 + \dfrac{0}{4!}\left(x - \dfrac{1}{2}\right)^4 \text{ or}$$

$$-\pi\left(x - \dfrac{1}{2}\right) + \dfrac{\pi^3}{6}\left(x - \dfrac{1}{2}\right)^3$$

16. At $x = e$, $f(e) = \ln e = 1$, $f'(e) = \dfrac{1}{x}\Big|_{x=e} = \dfrac{1}{e}$,

$f''(e) = \dfrac{-1}{x^2}\Big|_{x=e} = \dfrac{-1}{e^2}$, $f'''(e) = \dfrac{2}{x^3}\Big|_{x=e} = \dfrac{2}{e^3}$,

and $f^{(4)}(e) = \dfrac{-6}{x^4}\Big|_{x=e} = \dfrac{-6}{e^4}$. $f(x) = \ln x$ can be

approximated by $\dfrac{1}{0!}(x - e)^0 + \dfrac{1/e}{1!}(x - e)^1 +$

$\dfrac{-1/e^2}{2!}(x - e)^2 + \dfrac{2/e^3}{3!}(x - e)^3 + \dfrac{-6/e^4}{4!}(x - e)^4 =$

$1 + \dfrac{x - e}{e} - \dfrac{(x - e)^2}{2e^2} + \dfrac{(x - e)^3}{3e^3} - \dfrac{(x - e)^4}{4e^4}$

17. Calculate the derivatives and evaluate at $x = 0$.

$f(x) = \dfrac{1}{1 - x}$, $f(0) = 1$, $f'(x) = \dfrac{1}{(1 - x)^2}$,

$f'(0) = 1$, $f''(x) = \dfrac{2}{(1 - x)^3}$, $f''(x) = 2$,

$f'''(x) = \dfrac{6}{(1 - x)^4}$, $f'''(x) = 6$, $f^{(4)}(x) = \dfrac{24}{(1 - x)^5}$,

$f^{(4)}(x) = 24$. In general, $f^{(n)}(0) = n!$, so the MacLaurin series

$$f(x) = \dfrac{1}{1 - x} = \sum \dfrac{f^{(n)}(x)}{n!}x^n = \sum \dfrac{n!}{n!}x^n = \sum x^n.$$

The series converges to $f(x) = \dfrac{1}{1 - x}$ when

$\lim\limits_{n\to\infty}\left|\dfrac{x^{n+1}}{x^n}\right| = \lim\limits_{n\to\infty}|x| < 1$. The series converges

on $(-1, 1)$. When $x = 1$, the series becomes $\sum 1^n$ which diverges. When $x = -1$, the series becomes $\sum(-1)^n$ which diverges. Therefore the interval of convergence is $(-1, 1)$.

18. Begin with the known series

$$f(x) = \dfrac{1}{1 - x} = \sum_{n=0}^{\infty} x^n = 1 + x + x^2 + x^3 + \cdots$$

and replace x with $-x^2$. Then $\dfrac{1}{1 + x^2} =$

$\dfrac{1}{1 - (-x^2)} = 1 + (-x^2) + (-x^2)^2 + (-x^2)^3 + \cdots =$

$1 - x^2 + x^4 - x^6 + \cdots = \sum_{n=0}^{\infty}(-x^2)^n = \sum_{n=0}^{\infty}(-1)^n x^{2n}$.

The series converges to $f(x) = \dfrac{1}{1 + x^2}$ when

$\lim\limits_{n\to\infty}\left|\dfrac{x^{2n+1}}{x^{2n}}\right| = \lim\limits_{n\to\infty}|x| < 1$. The series converges

on $(-1, 1)$. When $x = 1$, the series becomes $\sum_{n=0}^{\infty}(-1)^n$, which diverges. When $x = -1$, the

series becomes $\sum_{n=0}^{\infty}(-1)^{3n}$, which diverges.

Therefore the interval of convergence is $(-1, 1)$.

19. Use the MacLaurin series $f(x) = \sin x =$

$\sum_{n=0}^{\infty}\dfrac{(-1)^n x^{2n+1}}{(2n + 1)!} = x - \dfrac{x^3}{3!} + \dfrac{x^5}{5!} - \dfrac{x^7}{7!} + \cdots$

with $9° = \dfrac{\pi}{20}$. Then $\sin 9° = \sin\dfrac{\pi}{20} =$

$\sum_{n=0}^{\infty}\dfrac{(-1)^n(\pi/20)^{2n+1}}{(2n + 1)!} = \dfrac{\pi}{20} - \dfrac{(\pi/20)^3}{3!} +$

$\dfrac{(\pi/20)^5}{5!} - \dfrac{(\pi/20)^7}{7!} + \cdots$

20. $1.\overline{83} = 1 + \dfrac{83}{10^2} + \dfrac{83}{10^4} + \dfrac{83}{10^6} + \cdots =$

$1 + \sum_{n=1}^{\infty}\dfrac{83}{10^{2m}} = 1 + 83\sum_{n=1}^{\infty}\dfrac{1}{10^{2m}}$. The

geometric series $\sum_{n=1}^{\infty}\dfrac{1}{10^{2m}}$ converges to

$s = \dfrac{a}{1 - r} = \dfrac{1/100}{1 - 1/100} = \dfrac{1}{99}$, Therefore

$1.\overline{83} = 1 + 83\sum_{n=1}^{\infty}\dfrac{1}{10^{2m}} = 1 + 83\left(\dfrac{1}{99}\right) = \dfrac{182}{99}$.

10.12 SOLUTIONS TO CUMULATIVE REVIEW PROBLEMS

21. $x'(t) = \dfrac{1}{t}$ $x''(t) = \dfrac{-1}{t^2}$ $y'(t) = 2t$ $y''(t) = 2$

The acceleration $\left\langle\dfrac{-1}{t^2}, 2\right\rangle = \langle-1, 2\rangle$ when $t = 1$.

The speed of the object at time $t = 1$ is

$\sqrt{\left(\dfrac{1}{t}\right)^2 + (2t)^2} = \sqrt{5}$.

22. $x = 5\cos 3\theta \cos\theta$ and $y = 5\cos 3\theta \sin\theta$.

$\dfrac{dx}{d\theta} = -5\sin\theta\cos 3\theta - 15\cos\theta\sin 3\theta$ and

$\dfrac{dy}{d\theta} = 5\cos\theta\cos 3\theta - 15\sin\theta\sin 3\theta$.

$$\frac{dy}{dx} = \frac{5\cos\theta\cos 3\theta - 15\sin\theta\sin 3\theta}{-5\sin\theta\cos 3\theta - 15\cos\theta\sin 3\theta} = \frac{3\sin\theta\sin 3\theta - \cos\theta\cos 3\theta}{3\cos\theta\sin 3\theta + \sin\theta\cos 3\theta}.$$

Evaluated at $\theta = \dfrac{2\pi}{3}$, $\dfrac{dy}{dx} =$

$$\frac{3\sin\left(\dfrac{2\pi}{3}\right)\sin 3\left(\dfrac{2\pi}{3}\right) - \cos\left(\dfrac{2\pi}{3}\right)\cos 3\left(\dfrac{2\pi}{3}\right)}{3\cos\left(\dfrac{2\pi}{3}\right)\sin 3\left(\dfrac{2\pi}{3}\right) + \sin\left(\dfrac{2\pi}{3}\right)\cos 3\left(\dfrac{2\pi}{3}\right)}$$

$= \dfrac{1}{\sqrt{3}} = \dfrac{\sqrt{3}}{3}$. The slope of the tangent is $\dfrac{\sqrt{3}}{3}$.

23. Integrate by parts, using $u = \ln x$, $dv = x^3 dx$,

$du = \dfrac{1}{x}dx$, $v = \dfrac{x^4}{4}$. Then $\displaystyle\int x^3 \ln x\, dx =$

$\dfrac{x^4}{4}\ln x - \displaystyle\int \dfrac{x^4}{4}\cdot\dfrac{1}{x}dx = \dfrac{x^4}{4}\ln x - \dfrac{1}{4}\displaystyle\int x^3 dx =$

$\dfrac{x^4}{4}\ln x - \dfrac{x^4}{16}$. Consider the limits of integration,

$$\int_1^e x^3 \ln x\, dx = \frac{x^4}{4}\ln x - \frac{x^4}{16}\Big|_1^e =$$

$$\left(\frac{e^4}{4}\ln e - \frac{e^4}{16}\right) - \left(\frac{1^4}{4}\ln 1 - \frac{1^4}{16}\right) =$$

$$\frac{3e^4 + 1}{16} \approx 10.300.$$

24. Use partial fraction decomposition.
$$\frac{5}{x^2 - x - 6} = \frac{5}{(x-3)(x+2)} = \frac{A}{x-3} + \frac{B}{x+2}.$$
Solving the system
$$\begin{cases} A + B = 0 \\ 2A - 3B = 5 \end{cases} \quad \text{gives } A = 1, B = -1, \text{ and}$$
$$\int_0^1 \frac{5}{x^2 - x - 6}\, dx = \int_0^1 \frac{1}{x-3}dx + \int_0^1 \frac{-1}{x+2}\, dx.$$
Then $\ln|x-3|\,\big|_0^1 - \ln|x+2|\,\big|_0^1 =$
$2\ln 2 - 2\ln 3 \approx -0.811.$

25. $\displaystyle\lim_{x\to 1}\frac{\ln x}{x^2 - 1} = \lim_{x\to 1}\frac{1/x}{2x} = \lim_{x\to 1}\frac{1}{2x^2} = \frac{1}{2}.$

PART IV

PRACTICE MAKES PERFECT

AB PRACTICE EXAM 1

Answer Sheet for AB Practice Exam 1—Section I

Part A

1. _____

2. _____

3. _____

4. _____

5. _____

6. _____

7. _____

8. _____

9. _____

10. _____

11. _____

12. _____

13. _____

14. _____

15. _____

16. _____

17. _____

18. _____

19. _____

20. _____

21. _____

22. _____

23. _____

24. _____

25. _____

26. _____

27. _____

28. _____

Part B

76. _____

77. _____

78. _____

79. _____

80. _____

81. _____

82. _____

83. _____

84. _____

85. _____

86. _____

87. _____

88. _____

89. _____

90. _____

91. _____

92. _____

Section I—Part A

Number of Questions	Time	Use of Calculator
28	55 Minutes	No

Directions:

Use the answer sheet provided in the previous page. All questions are given equal weight. There is no penalty for unanswered questions. However, 1/4 of the number of the incorrect answers will be subtracted from the number of correct answers. Unless otherwise indicated, the domain of a function f is the set of all real numbers. The use of a calculator is *not* permitted in this part of the exam.

1. The $\displaystyle\lim_{x \to -\infty} \frac{2x-1}{1+2x}$ is

   (A) -1 (B) 0 (C) 1
   (D) 2 (E) nonexistent

2. $\displaystyle\int_{\pi/2}^{x} \cos t \, dt$

   (A) $\cos x$ (B) $-\sin x$ (C) $\sin x - 1$
   (D) $\sin x + 1$ (E) $-\sin x + 1$

3. The radius of a sphere is increasing at a constant of 2 cm/sec. At the instant when the volume of the sphere is increasing at 32π cm^3/sec, the surface area of the sphere is

   (A) 8π (B) $\dfrac{32\pi}{3}$ (C) 16π

   (D) 64π (E) $\dfrac{256\pi}{3}$

4. Given the equation $A = \dfrac{\sqrt{3}}{4}(5s-1)^2$, what is the instantaneous rate of change of A with respect to s at $s = 1$?

   (A) $2\sqrt{3}+5$ (B) $2\sqrt{3}$ (C) $\dfrac{5}{2}\sqrt{3}$
   (D) $4\sqrt{3}$ (E) $10\sqrt{3}$

5. What is the $\displaystyle\lim_{x \to \ln 2} g(x)$, if

   $$g(x) = \begin{cases} e^x & \text{if } x > \ln 2 \\ 4 - e^x & \text{if } x \le \ln 2 \end{cases}?$$

   (A) -2 (B) $\ln 2$ (C) e^2
   (D) 2 (E) nonexistent

6. The graph of f' is shown in Figure 1T-1.

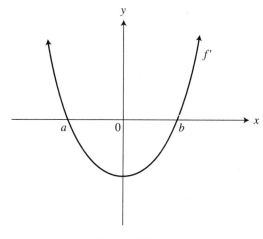

Figure 1T-1

A possible graph of f is (see Figure 1T-2):

7. If $g(x) = -2|x+3|$, what is the $\displaystyle\lim_{x \to -3^-} g(x)$?

   (A) -6 (B) -2 (C) 2
   (D) 6 (E) nonexistent

8. What is $\displaystyle\lim_{\Delta x \to 0} \frac{\sin\left(\dfrac{\pi}{3}+\Delta x\right) - \sin\left(\dfrac{\pi}{3}\right)}{\Delta x}$?

   (A) $-\dfrac{1}{2}$ (B) 0 (C) $\dfrac{1}{2}$

   (D) $\dfrac{\sqrt{3}}{2}$ (E) nonexistent

(A)

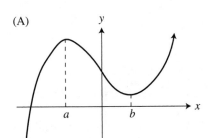

(B)

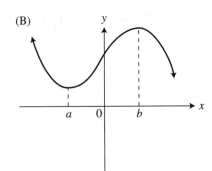

(C)

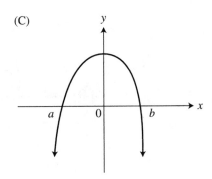

(D)

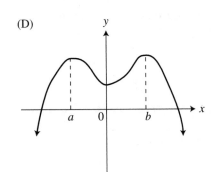

(E)

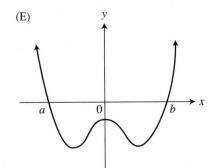

Figure 1T-2

9. If $f(x)$ is an antiderivative of xe^{-x^2} and $f(0) = 1$, then $f(1) =$

(A) $\dfrac{1}{e}$ (B) $\dfrac{1}{2e} - \dfrac{3}{2}$ (C) $\dfrac{1}{2e} - \dfrac{1}{2}$

(D) $-\dfrac{1}{2e} + \dfrac{3}{2}$ (E) $-\dfrac{1}{2e} + \dfrac{1}{2}$

10. If $g(x) = 3\tan^2(2x)$, then $g'\left(\dfrac{\pi}{8}\right)$ is

(A) 6   (B) $6\sqrt{2}$ (C) 12
(D) $12\sqrt{2}$ (E) 24

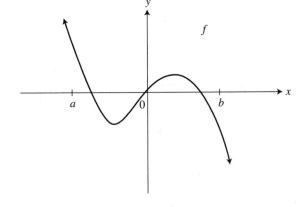

Figure 1T-3

11. The graph of the function f is shown in Figure 1T-3. Which of the following statements is/are true?

   I. $f'(0) = 0$
  II. f has an absolute maximum value on $[a, b]$
 III. $f'' < 0$ on $(0, b)$

    (A) III only
    (B) I and II only
    (C) II and III only
    (D) I and III only
    (E) I, II, and III

12. $\displaystyle\int \dfrac{1+x}{\sqrt{x}}\, dx =$

(A) $2\sqrt{x} + \dfrac{x^2}{2} + c$

(B) $\dfrac{\sqrt{x}}{2} + \dfrac{3}{2}x^{3/2} + c$

(C) $2\sqrt{x} + \dfrac{2}{3}x^{3/2} + c$

(D) $x + \dfrac{2}{3}x^{3/2} + c$

(E) 0

13. The graph of f is shown in Figure 1T-4 and f is twice differentiable. Which of the following has the smallest value?

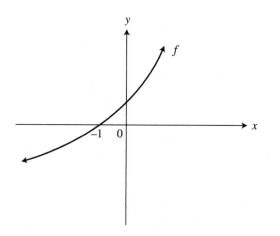

Figure 1T-4

   I. $f(-1)$
  II. $f'(-1)$
 III. $f''(-1)$

(A) I   (B) II   (C) III
(D) I and II   (E) II and III

14. If $\dfrac{dy}{dx} = 3e^{2x}$, and at $x = 0$, $y = \dfrac{5}{2}$, a solution to the differential equation is

(A) $3e^{2x} - \dfrac{1}{2}$ (B) $3e^{2x} + \dfrac{1}{2}$ (C) $\dfrac{3}{2}e^{2x} + 1$

(D) $\dfrac{3}{2}e^{2x} + 2$ (E) $\dfrac{3}{2}e^{2x} + 5$

15. The graph of the velocity function of a moving particle is shown in Figure 1T-5. What is the total displacement of the particle during $0 \le t \le 20$?

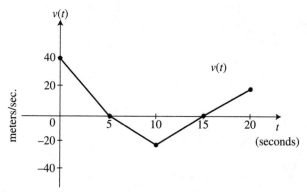

Figure 1T-5

(A) 20 m   (B) 50 m   (C) 100 m
(D) 250 m   (E) 500 m

16. The position function of a moving particle is $s(t) = \dfrac{t^3}{6} - \dfrac{t^2}{2} + t - 3$ for $0 \le t \le 4$. What is the maximum velocity of the particle on the interval $0 \le t \le 4$?

(A) $\dfrac{1}{2}$ (B) 1   (C) $\dfrac{14}{16}$ (D) 4   (E) 5

17. If $\displaystyle\int_{-k}^{k} |2x|\,dx = 18$ and $k > 0$, the value(s) of k are

(A) -3 (B) $-3\sqrt{2}$ (C) 3
(D) $3\sqrt{2}$ (E) 9

18. A function f is continuous on $[-1, 1]$ and some of the values of f are shown below:

x	-1	0	1
$f(x)$	2	b	-2

If $f(x) = 0$ has only one solution, r, and $r < 0$, then a possible value of b is

(A) 3   (B) 2   (C) 1   (D) 0   (E) -1

19. $\displaystyle\int_{0}^{\ln 2} e^{2x}\,dx =$

(A) $\dfrac{3}{2}$ (B) 3   (C) 4

(D) $e^2 - \dfrac{1}{2}$ (E) $2e^2 - 1$

20. The area of the region enclosed by the graph of $y = \sqrt{9 - x^2}$ and the x-axis is

(A) 36   (B) $\dfrac{9\pi}{2}$ (C) 9π
(D) 18π (E) 36π

21. If a function f is continuous for all values of x, and $a > 0$ and $b > 0$, which of the following integrals always have the same value?

   I. $\displaystyle\int_{0}^{a} f(x)\,dx$

  II. $\displaystyle\int_{b}^{a+b} f(x - b)\,dx$

 III. $\displaystyle\int_{b}^{a+b} f(x + b)\,dx$

(A) I and II only
(B) I and III only

(C) II and III only

(D) I, II, and III

(E) None

22. What is the average value of the function $y = 2\sin(2x)$ on the interval $\left[0, \frac{\pi}{6}\right]$?

(A) $-\dfrac{3}{\pi}$ (B) $\dfrac{1}{2}$ (C) $\dfrac{3}{\pi}$

(D) $\dfrac{3}{2\pi}$ (E) 6π

23. Given the equation $y = 3\sin^2\left(\dfrac{x}{2}\right)$, what is an equation of the tangent line to the graph at $x = \pi$?

(A) $y = 3$

(B) $y = \pi$

(C) $y = \pi + 3$

(D) $y = x - \pi + 3$

(E) $y = 3(x - \pi) + 3$

24. The position function of a moving particle on the x-axis is given as $s(t) = t^3 + t^2 - 8t$ for $0 \le t \le 10$. For what values of t is the particle moving to the right?

(A) $t < -2$ (B) $t > 0$ (C) $t < \dfrac{4}{3}$

(D) $0 < t < \dfrac{4}{3}$ (E) $t > \dfrac{4}{3}$

25. See Figure 1T-6.

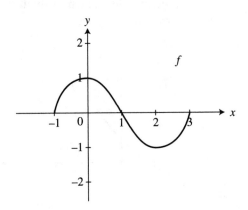

Figure 1T-6

The graph of f consist of two semicircles, for $-1 \le x \le 3$ as shown in Figure 1T-6. What is the value of $\displaystyle\int_{-1}^{3} f(x)\,dx$?

(A) 0   (B) π (C) 2π

(D) 4π (E) 8π

26. If $f(x) = \displaystyle\int_{1}^{x} t(t^3 + 1)^{3/2}\,dt$, then $f'(2)$ is

(A) $2^{3/2}$ (B) $54 - 2^{3/2}$ (C) 54

(D) $135 - \dfrac{13\sqrt{2}}{2}$ (E) 135

(A)

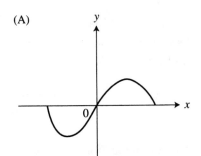

(B)

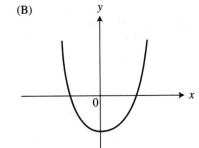

(C)

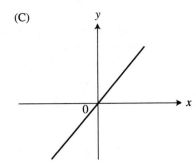

(D)

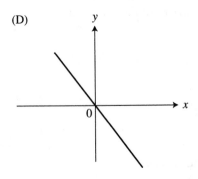

(E)

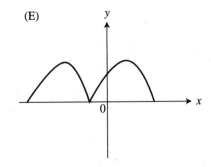

Figure 1T-7

27. If $\displaystyle\int_{-k}^{k} f(x)dx = 2\int_{-k}^{0} f(x)dx$ for all positive values of k, then which of the following could be the graph of f? (See Figure 1T-7.)

28. If $h'(x) = k(x)$ and k is a continuous function for all real values of x, then $\displaystyle\int_{-1}^{1} k(5x)dx$ is

(A) $h(5) - h(-5)$

(B) $5h(5) + 5h(-5)$

(C) $5h(5) - 5h(-5)$

(D) $\dfrac{1}{5}h(5) + \dfrac{1}{5}h(-5)$

(E) $\dfrac{1}{5}h(5) - \dfrac{1}{5}h(-5)$

Section I—Part B

Number of Questions	Time	Use of Calculator
17	50 Minutes	Yes

Directions:

Use the same answer sheet for Part A. *Please note that the questions begin with number 76. This is not an error. It is done to be consistent with the numbering system of the actual AP Calculus AB Exam. All questions are given equal weight. There is no penalty for unanswered questions. However, 1/4 of the number of incorrect answers will be subtracted from the number of correct answers. Unless otherwise indicated, the domain of a function f is the set of all real numbers. If the exact numerical value does not appear among the given choices, select the best approximate value. The use of a calculator is permitted in this part of the exam.*

76. If $f(x) = \int_0^x -\cos t\, dt$ on $[0, 2\pi]$, then f has a local maximum at $x =$

    (A) 0   (B) $\dfrac{\pi}{2}$ (C) π

    (D) $\dfrac{3\pi}{2}$ (E) 2π

77. The equation of the normal line to the graph $y = e^{2x}$ at the point where $\dfrac{dy}{dx} = 2$ is

    (A) $y = -\dfrac{1}{2}x - 1$

    (B) $y = -\dfrac{1}{2}x + 1$

    (C) $y = 2x + 1$

    (D) $y = -\dfrac{1}{2}\left(x - \dfrac{\ln 2}{2}\right) + 2$

    (E) $y = 2\left(x - \dfrac{\ln 2}{2}\right) + 2$

78. The graph of f', the derivative of f, is shown in Figure 1T-8. At which value of x does the graph of f have a point of inflection?

    (A) 0   (B) x_1 (C) x_2
    (D) x_3 (E) x_4

79. The temperature of a metal is dropping at the rate of $g(t) = 10e^{-0.1t}$ for $0 \le t \le 10$ where g is measured in degrees in Fahrenheit and t in minutes. If the metal is initally 100°F, what is

the temperature to the nearest degree Fahrenheit after 6 minutes?

(A) 37   (B) 45   (C) 55
(D) 63   (E) 82

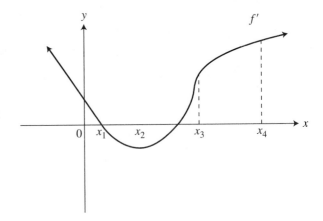

Figure 1T-8

80. What is the approximate volume of the solid obtained by revolving about the x-axis the region in the first quadrant enclosed by the curves $y = x^3$ and $y = \sin x$?

    (A) 0.061π (B) 0.139π (C) 0.215π
    (D) 0.225π (E) 0.278π

81. Let f be a differentiable function on (a, b). If f has a point of inflection on (a, b), which of the following could be the graph of f'' on (a, b)? See Figure 1T-9.

    (A) A   (B) B   (C) C   (D) D   (E) None

(A)

(B)

(C)

(D)

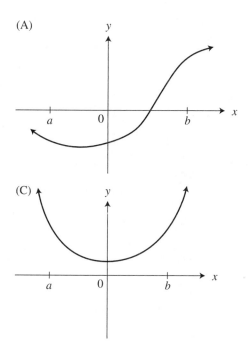

Figure 1T-9

82. The base of a solid is a region bounded by the lines $y = x$, $y = -x$ and $x = 4$ as shown in Figure 1T-10. What is the volume of the solid if the cross sections perpendicular to the x-axis are equilateral triangles?

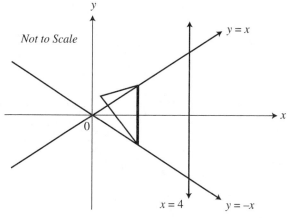

Figure 1T-10

(A) $\dfrac{16\sqrt{3}}{3}$ (B) $\dfrac{32\sqrt{3}}{3}$ (C) $\dfrac{64\sqrt{3}}{3}$

(D) $\dfrac{256\pi}{3}$ (E) $\dfrac{3072\pi}{5}$

83. Let f be a continuous function on $[0, 6]$ and have selected values as shown below.

x	0	2	4	6
$f(x)$	0	1	2.25	6.25

If you use the subintervals $[0, 2]$, $[2, 4]$, and $[4, 6]$, what is the trapezoidal approximation of $\displaystyle\int_0^6 f(x)dx$?

(A) 9.5 (B) 12.75 (C) 19
(D) 25.5 (E) 38.25

84. The amount of a certain bacteria y in a petri dish grows according to the equation $\dfrac{dy}{dt} = ky$, where k is a constant and t is measured in hours.

If the amount of bacteria triples in 10 hours, then $k \approx$

(A) -1.204 (B) -0.110 (C) 0.110
(D) 1.204 (E) 0.3

85. The volume of the solid generated by revolving about the y-axis the region bounded by the graphs of $y = \sqrt{x}$ and $y = x$ is

(A) $\dfrac{2\pi}{15}$ (B) $\dfrac{\pi}{6}$ (C) $\dfrac{2\pi}{3}$

(D) $\dfrac{16\pi}{15}$ (E) $\dfrac{56\pi}{15}$

86. How many points of inflection does the graph of $y = \dfrac{\sin x}{x}$ have on the interval $(-\pi, \pi)$?

(A) 0   (B) 1   (C) 2   (D) 3   (E) 4

87. Given $f(x) = x^2 e^x$, what is an approximate value of $f(1.1)$, if you use a tangent line to the graph of f at $x = 1$.

(A) 3.534   (B) 3.635   (C) 7.055
(D) 8.155   (E) 10.244

88. The area under the curve $y = \sin x$ from $x = b$ to $x = \pi$ is 0.2. If $0 \le b < \pi$, then $b =$

(A) -0.927 (B) -0.201 (C) 0.644
(D) 1.369   (E) 2.498

89. At what value(s) of x do the graphs of $y = x^2$ and $y = -\sqrt{x}$ have perpendicular tangent lines?

(A) -1 (B) 0   (C) $\dfrac{1}{4}$
(D) 1   (E) none

90. What is the approximate slope of the tangent to the curve $x^3 + y^3 = xy$ at $x = 1$?

(A) -2.420 (B) -1.325 (C) -1.014
(D) -0.698 (E) 0.267

91. The graph of f is shown in Figure 1T-11, and $g(x) = \displaystyle\int_{a}^{x} f(t)\,dt$, $x > a$. Which of the following is a possible graph of g? See Figure 1T-12.

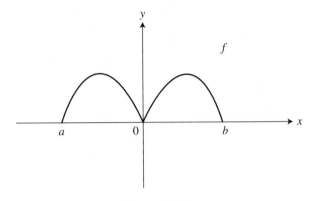

Figure 1T-11

92. If $g(x) = |xe^x|$, which of the following statements about g are true?

 I. g has a relative minimum at $x = 0$.
 II. g changes concavity at $x = 0$.
 III. g is differentiable at $x = 0$.

(A) I only
(B) II only
(C) III only
(D) I and II only
(E) I and III only

(A)
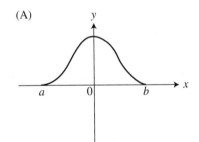

(B)
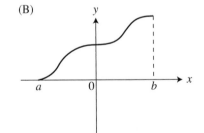

(C)
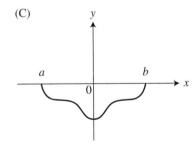

(D)
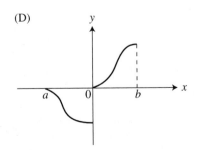

(E)
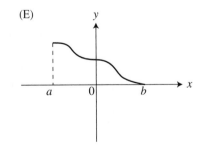

Figure 1T-12

Section II—Part A

Number of Questions	Time	Use of Calculator
3	45 Minutes	Yes

Directions:

Show all work. You may *not* receive any credit for correct answers without supporting work. You may use an approved calculator to help solve a problem. However, you must clearly indicate the setup of your solution using mathematical notations and *not* calculator syntax. Calculators may be used to find the derivative of a function at a point, compute the numerical value of a definite integral, or solve an equation. Unless otherwise indicated, you may assume the following: (a) the numeric or algebraic answers need not be simplified, (b) your answer, if expressed in approximation, should be rounded to 3 places after the decimal point, and (c) the domain of a function f is the set of all real numbers.

1. The slope of a function at any point (x, y) is $\dfrac{e^x}{e^x + 1}$. The point $(0, 2\ln 2)$ is on the graph of f.

    (A) Write an equation of the tangent line to the graph of f at $x = 0$.
    (B) Use the tangent line in part (A) to approximate $f(0.1)$ to the nearest thousandth.
    (C) Solve the differential equation $\dfrac{dy}{dx} = \dfrac{e^x}{e^x + 1}$ with the initial condition $f(0) = 2\ln 2$.
    (D) Use the solution in part (C) and find $f(0.1)$ to the nearest thousandth.

2. The temperature in a greenhouse from 7:00 p.m. to 7:00 a.m. is given by $f(t) = 96 - 20\sin\left(\dfrac{t}{4}\right)$, where $f(t)$ is measured in Fahrenheit and t is measured in hours.

    (A) What is the temperature of the greenhouse at 1:00 a.m. to the nearest degree Fahrenheit?
    (B) Find the average temperature between 7:00 p.m. and 7:00 a.m. to the nearest tenth of a degree Fahrenheit.
    (C) When the temperature of the greenhouse drops below 80°F, a heating system will automatically be turned on to maintain the temperature at a minimum of 80°F. At what value of t to the nearest tenth is the heating system turned on?
    (D) The cost of heating the greenhouse is $0.25 per hour for each degree. What is the total cost to the nearest dollar to heat the greenhouse from 7:00 p.m. and 7:00 a.m.?

3. A particle is moving on a straight line. The velocity of the particle for $0 \le t \le 30$ is shown in the table below for selected values of t.

t (sec)	0	3	6	9	12	15	18	21	24	27	30
$v(t)$ (m/sec)	0	7.5	10.1	12	13	13.5	14.1	14	13.9	13	12.2

    (A) Using the midpoints of five subintervals of equal length, find the approximate value of $\displaystyle\int_0^{30} v(t)\,dt$.
    (B) Using the result in part (A), find the average velocity over the interval $0 \le t \le 30$.
    (C) Find the average acceleration over the interval $0 \le t \le 30$.
    (D) Find the approximate acceleration at $t = 6$.
    (E) During what intervals of time is the acceleration negative?

Section II—Part B

Number of Questions	Time	Use of Calculator
3	45 Minutes	No

Directions:

The use of a calculator is not permitted in this part of the exam. When you have finished this part of the exam, you may return to the problems in Part A of Section II and continue to work on them. However, you may *not* use a calculator. You should *show all work*. You may *not* receive any credit for correct answers without supporting work. Unless otherwise indicated, the numeric or algebraic answers need not be simplified, and the domain of a function f is the set of all real numbers.

4. See Figure 1T-13.

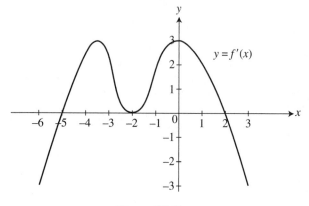

Figure 1T-13

The graph of f', the derivative of a function f, for $-6 \le x \le 3$ is shown in Figure 1T-13.

(A) At what value(s) of x does f have a relative maximum value? Justify your answer.

(B) At what value(s) of x does f have a relative minimum value? Justify your answer.

(C) At what value(s) of x does the function have a point of inflection? Justify your answer.

(D) If $f(-5) = 2$, draw a possible sketch of f on $-6 < x < 3$.

5. Given the equation $y^2 - x + 2y - 3 = 0$:

(A) Find $\dfrac{dy}{dx}$.

(B) Write an equation of the line tangent to the graph of the equation at the point $(0, -3)$.

(C) Write an equation of the line normal to the graph of the equation at the point $(0, -3)$.

(D) The line $y = \dfrac{1}{4}x + 3$ is tangent to the graph at point P. Find the coordinates of point P.

6. Let R be the region enclosed by the graph of $y = x^2$ and the line $y = 4$.

(A) Find the area of region R.

(B) If the line $x = a$ divides region R into two regions of equal area, find a.

(C) If the line $y = b$ divides the region R into two regions of equal area, find b.

(D) If region R is revolved about the x-axis, find the volume of the resulting solid.

Answers to AB Practice Exam 1—Section I

Part A			
1. C	12. C	24. E	81. A
2. C	13. A	25. A	82. C
3. C	14. C	26. C	83. B
4. E	15. B	27. B	84. C
5. D	16. E	28. E	85. A
6. A	17. C		86. C
7. C	18. E	**Part B**	87. A
8. C	19. A	76. D	88. E
9. D	20. B	77. B	89. D
10. E	21. A	78. C	90. C
11. C	22. C	79. C	91. B
	23. A	80. B	92. D

Answers to AB Practice Exam 1—Section II

Part A

1. (A) $y = \dfrac{1}{2}x + 2\ln 2$ (3 pts.)
   (B) 1.436   (1 pt.)
   (C) $y = \ln(e^x + 1) + \ln 2$ (4 pts.)
   (D) 1.438   (1 pt.)

2. (A) $76°$ (2 pts.)
   (B) $82.7°$ (2 pts.)
   (C) $3.7 \leq t \leq 8.9$ (2 pts.)
   (D) $3   (3 pts.)

3. (A) 360   (3 pts.)
   (B) 12 m/sec   (1 pt.)
   (C) 0.407 m/sec^2 (2 pts.)
   (D) 0.75 m/sec^2 (1 pt.)
   (E) $18 < t < 30$ (2 pts.)

Part B

4. (A) $x = 2$ (2 pts.)
   (B) $x = -5$ (2 pts.)
   (C) $x = -4, x = -2$ and $x = 0$ (2 pts.)
   (D) See solution.   (3 pts.)

5. (A) $\dfrac{dy}{dx} = \dfrac{1}{2y + 2}$ (3 pts.)
   (B) $y = -\dfrac{1}{4}x - 3$ (2 pts.)
   (C) $y = 4x - 3$ (2 pts.)
   (D) $(0, 1)$ (2 pts.)

6. (A) $\dfrac{32}{3}$ (3 pts.)
   (B) $a = 0$ (1 pt.)
   (C) $b = 4^{2/3}$ (2 pts.)
   (D) $\dfrac{256\pi}{5}$ (3 pts.)

Solutions to AB Practice Exam 1—Section I

Part A—No calculators.

1. The correct answer is (C).

$$\lim_{x \to -\infty} \frac{2x - 1}{1 + 2x} = \lim_{x \to -\infty} \frac{2 - (1/x)}{(1/x) + 2} = 1.$$

2. The correct answer is (C).

$$\int_{\pi/2}^{x} \cos t \, dt = \sin t]_{\pi/2}^{x} = \sin x - (\sin \pi/2)$$

$$= \sin x - 1.$$

3. The correct answer is (C).

$$V = \frac{4}{3}\pi r^3 \text{ and } \frac{dV}{dt} = 4\pi r^2 (2) = 8\pi r^2$$

$$\frac{dV}{dt} = 32\pi \text{ cm}^3/\text{sec}; \ 8\pi r^2 = 32\pi \Rightarrow r = 2$$

Surface Area $= 4\pi r^2 = 4\pi(2)^2 = 16\pi.$

4. The correct answer is (E).

$$A = \frac{\sqrt{3}}{4}(5s - 1)2, \frac{dA}{ds} = (2)\left(\frac{\sqrt{3}}{4}\right)(5s - 1)(5)$$

$$= \frac{5\sqrt{3}}{2}(5s - 1)$$

$$\left.\frac{dA}{ds}\right|_{s=1} = \frac{5\sqrt{3}}{2}(4) = 10\sqrt{3}.$$

5. The correct answer is (D).

$$\lim_{x \to (\ln 2)^+} (e^x) = e^{\ln 2} = 2 \text{ and } \lim_{x \to (\ln 2)^-} (4 - e^x)$$

$$= 4 - e^{\ln 2} = 4 - 2 = 2$$

Since the two, one-sided limits are the same,
$\lim_{x \to (\ln 2)} g(x) = 2.$

6. The correct answer is (A).

See Figure 1TS-1.
The only graph that satisfies the behavior of f is (A).

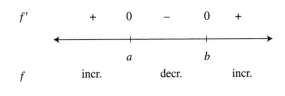

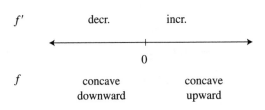

Figure 1TS-1

7. The correct answer is (C).

$$g(x) = \begin{cases} -2(x + 3) & \text{if } x \geq -3 \\ (-2)[-(x + 3)] & \text{if } x < -3 \end{cases}$$

$$= \begin{cases} -2x - 6 & \text{if } x \geq -3 \\ 2x + 6 & \text{if } x < -3 \end{cases}$$

$$g'(x) = \begin{cases} -2 & \text{if } x > -3 \\ 2 & \text{if } x < -3 \end{cases}$$

Thus $\lim_{x \to -3^-} g'(x) = 2.$

8. The correct answer is (C).

The definition of $f'(x)$ is $f'(x)$

$$= \lim_{\Delta x \to 0} \frac{f(x + \Delta x) - f(x)}{\Delta x}.$$

Thus $\lim_{\Delta x \to 0} \frac{\sin((\pi/3) + \Delta x) - \sin(\pi/3)}{\Delta x}$

$$= \left.\frac{d(\sin x)}{dx}\right|_{x=\pi/3}$$

$$= \cos\left(\frac{\pi}{3}\right) = \frac{1}{2}.$$

9. The correct answer is (D).

Since $f(x) = \int xe^{-x^2} dx$, let $u = -x^2$,

$$du = -2x \, dx \text{ or } \frac{-du}{2} = x \, dx.$$

Thus $f(x) = \int e^u \left(-\dfrac{du}{2}\right) = -\dfrac{1}{2}e^u + c$

$= -\dfrac{1}{2}e^{-x^2} + c$

And $f(0) = 1 \Rightarrow -\dfrac{1}{2} = (e^0) + c = 1 \Rightarrow -\dfrac{1}{2} + c$

$= 1 \Rightarrow c = \dfrac{3}{2}$

Therefore, $f(x) = -\dfrac{1}{2}e^{-x^2} + \dfrac{3}{2}$ and

$$f(1) = -\dfrac{1}{2}e^{-1} + \dfrac{3}{2} = -\dfrac{1}{2e} + \dfrac{3}{2}.$$

10. The correct answer is (E).

$g(x) = 3\,[\tan(2x)]^2\,;$

$g'(x) = 6\,[\tan(2x)]\sec^2(2x)2$

$= 12\tan(2x)\sec^2(2x)$

$g'(\pi/8) = 12\tan(\pi/4)\sec^2(\pi/4)$

$= 12(1)^2\left(\sqrt{2}\right)^2 = 24.$

11. The correct answer is (C).

I. $f'(0) \neq 0$ since the tangent to $f(x)$ at $x = 0$ is not parallel to the x-axis.

II. f has an absolute maximum at $x = a$.

III. f'' is less than 0 on $(0, b)$ since f is concave downward.

Thus only statements II and III are true.

12. The correct answer is (C).

$\displaystyle\int \dfrac{1+x}{\sqrt{x}}dx = \int \left(\dfrac{1}{\sqrt{x}} + \dfrac{x}{\sqrt{x}}\right)dx$

$= \displaystyle\int \left(x^{-1/2} + x^{1/2}\right)dx$

$= \dfrac{x^{1/2}}{1/2} + \dfrac{x^{3/2}}{3/2} + c$

$= 2x^{1/2} + 2/3x^{3/2} + c$

$= 2\sqrt{x} + 2/3x^{3/2} + c.$

13. The correct answer is (A).

I. $f(-1) = 0;$

II. Since f is increasing, $f'(-1) > 0$

III. Since f is concave upward, $f''(-1) > 0$

Thus $f(-1)$ has the smallest value.

14. The correct answer is (C).

Since $dy = 3e^{2x}dx \Rightarrow \int 1dy \Rightarrow \int 3e^{2x}dx \Rightarrow$

$y = \dfrac{3e^{2x}}{2} + c.$

At $x = 0,\quad \dfrac{5}{2} = \dfrac{3(e^0)}{2} + c \Rightarrow \dfrac{5}{2} = \dfrac{3}{2} + c$

$\Rightarrow c = 1$

Therefore, $y = \dfrac{3e^{2x}}{2} + 1.$

15. The correct answer is (B).

$\displaystyle\int_0^{20} v(t)dt = \dfrac{1}{2}(40)(5) + \dfrac{1}{2}(10)(-20)$

$+ \dfrac{1}{2}(5)(20) = 50.$

16. The correct answer is (E).

$v(t) = s'(t) = \dfrac{t^2}{2} - t + 1$ and $a(t) = t - 1$ and $a'(t) = 1.$

Set $a(t) = 0 \Rightarrow t = 1.$ Thus, $v(t)$ has a relative minimum at $t = 1$ and $v(1) = \dfrac{1}{2}.$ Since it is the only relative extremum, it is an absolute minimum. And since $v(t)$ is continuous on the closed interval $[0, 4]$, thus $v(t)$ has an absolute maximum at the endpoints.

$v(0) = 1$ and $v(4) = 8 - 4 + 1 = 5.$

Therefore, the maximum velocity of the particle on $[1, 4]$ is 5.

17. The correct answer is (C).

Since $y = |2x|$ is symmetrical with respect to the y-axis,

$\displaystyle\int_{-k}^{k} |2x|dx = 2\int_0^k 2x\,dx$

$= 2\left[x^2\right]_0^k = 2k^2$

Set $2k^2 = 18 \Rightarrow k^2 = 9 \Rightarrow k = \pm 3.$ Since $k > 0, k = 3.$

18. The correct answer is (E).

See Figure 1TS-2.
If $b = 0$, then 0 is a root and thus $r = 0$.
If $b = 1, 2,$ or 3, then the graph of f must cross the x-axis which implies there is another root. Thus, $b = -1$.

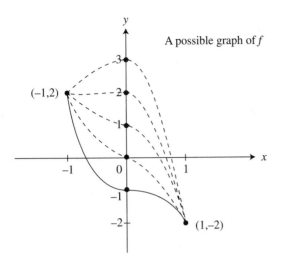

A possible graph of f

Figure 1TS-2

19. The correct answer is (A).

$$\int_0^{\ln 2} e^{2x}dx = \frac{e^{2x}}{2}\bigg]_0^{\ln 2} = \frac{e^{2(\ln 2)}}{2} - \frac{e^{2(0)}}{2}$$

$$= \frac{(e^{\ln 2})^2}{2} - \frac{e^0}{2} = \frac{(2)^2}{2} - \frac{1}{2} = \frac{3}{2}.$$

20. The correct answer is (B).

The graph of $y = \sqrt{9 - x^2}$ is a semicircle above the x-axis and whose endpoints are $(-3, 0)$ and $(3, 0)$. Thus the radius of the circle is $r = 3$.
Area $= \frac{1}{2}\pi r^2 = \frac{9\pi}{2}$.

21. The correct answer is (A).

See Figure 1TS-3.
The graphs $f(x - b)$ and $f(x + b)$ are the same as the graph of $f(x)$ shifted b units to the right and left, respectively. Looking at Figure 1TS-3, only I and II have the same value.

22. The correct answer is (C).

$$\text{Average value} = \frac{1}{(\pi/6) - 0}\int_0^{\pi/6} 2\sin(2x)dx$$

$$= \frac{6}{\pi}\big[-\cos(2x)\big]_0^{\pi/6}$$

$$= \frac{6}{\pi}\left[-\cos\left(\frac{\pi}{3}\right) - (-\cos 0)\right]$$

$$= \frac{6}{\pi}\left[-\frac{1}{2} + 1\right] = \frac{3}{\pi}.$$

23. The correct answer is (A).

$$y = 3\sin^2\left(\frac{x}{2}\right); \quad \frac{dy}{dx} = 6\sin\left(\frac{x}{2}\right)\left[\cos\left(\frac{x}{2}\right)\right]\frac{1}{2}$$

$$= 3\sin\left(\frac{x}{2}\right)\cos\left(\frac{x}{2}\right)$$

$$\frac{dy}{dx}\bigg|_{x=\pi} = 3\sin\left(\frac{\pi}{2}\right)\cos\left(\frac{\pi}{2}\right) = 3(1)(0) = 0.$$

At $x = \pi$, $y = 3\sin^2\left(\frac{\pi}{2}\right) = 3(1)^2 = 3; (\pi, 3)$
Equation of tangent at $x = \pi; y = 3$.

24. The correct answer is (E).

$$s(t) = t^3 + t^2 - 8t; \quad v(t) = 3t^2 + 2t - 8$$

Set $v(t) = 0 \Rightarrow 3t^2 + 2t - 8$

$$= 0 \Rightarrow (3t - 4)(t + 2)$$

$$= 0 \text{ or } t = \frac{4}{3} \text{ or } t = -2$$

Since $0 \le t \le 10$, thus $t = -2$ is not in the domain.
If $t > \frac{4}{3}, v(t) > 0 \Rightarrow$ the particle is moving to the right.

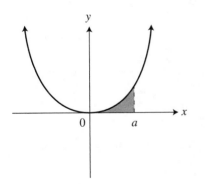

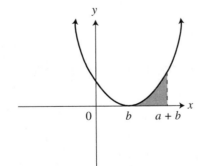

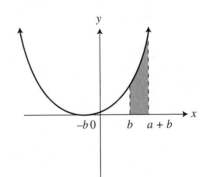

Figure 1TS-3

25. The correct answer is (A).

$$\int_{-1}^{3} f(x)dx = \int_{-1}^{1} f(x)dx + \int_{1}^{3} f(x)dx$$

$$= \frac{1}{2}\pi(1)^2 - \frac{1}{2}\pi(1)^2 = 0.$$

26. The correct answer is (C).
$$f'(x) = x(x^3 + 1)^{3/2}; \ f'(2) = 2(2^3 + 1)^{3/2}$$

$$= 2(9)^{3/2} = 54.$$

27. The correct answer is (B).

$$\int_{-k}^{k} f(x)dx = 2\int_{-k}^{0} f(x)dx \Rightarrow f(x) \text{ is an even}$$

function, i.e., $f(x) = f(-x)$.
The graph in (B) is the only even function.

28. The correct answer is (E).

Let $u = 5x; \ du = 5dx$ or $\dfrac{du}{5} = dx$

$$\int k(5x)dx = \frac{1}{5}\int k(u)du = \frac{1}{5}h(u) + c$$

$$= \frac{1}{5}h(5x) + c$$

$$\int_{-1}^{1} k(5x)dx = \frac{1}{5}h(5x)\Big]_{-1}^{1} = \frac{1}{5}h(5)$$

$$- \frac{1}{5}h(-5).$$

Part B—Calculators are permitted.

76. The correct answer is (D).

$$f(x) = \int_{0}^{x} -\cos t \, dt; \ f'(x) = -\cos x$$

Let $f'(x) = 0 \Rightarrow -\cos x = 0 \Rightarrow x = \pi/2$ or $3\pi/2$
See Figure 1TS-4.

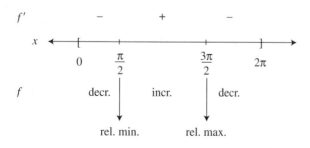

Figure 1TS-4

Thus f has a local maximum at $x = \dfrac{3\pi}{4}$.

77. The correct answer is (B).

$$y = e^{2x}; \quad \frac{dy}{dx} = (e^{2x})2 = 2e^{2x}$$

Set $\dfrac{dy}{dx} = 2 \Rightarrow 2e^{2x} = 2 \Rightarrow e^{2x} = 1 \Rightarrow \ln(e^{2x})$

$$= \ln 1 \Rightarrow 2x = 0 \text{ or } x = 0.$$

At $x = 0, \ y = e^{2x} = e^{2(0)}$

$$= 1; \ (0, 1) \text{ or } y = -\frac{1}{2}x + 1.$$

78. The correct answer is (C).

See Figure 1TS-5.

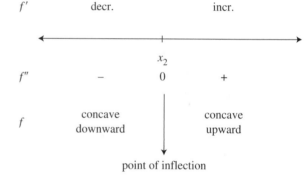

Figure 1TS-5

The graph of f has a point of inflection at $x = x_2$.

79. The correct answer is (C).

Temperature of metal $= 100 - \displaystyle\int_{0}^{6} 10e^{-0.1t}dt$
Using your calculator, you obtain:
Temperature of metal $= 100 - 45.1188$
$$= 54.8812 \approx 55°F.$$

80. The correct answer is (B).

See Figure 1TS-6.

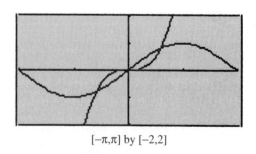

$[-\pi,\pi]$ by $[-2,2]$

Figure 1TS-6

Using the Intersection function on your calculator, you obtain the points of intersection: (0, 0) and (0.929, 0.801).

$$v = \pi \int_0^{0.929} ((\sin x)^2 - (x^3)^2)dx = 0.139\pi.$$

81. The correct answer is (A).

A point of inflection $\Rightarrow$ the graph of f changes its concavity $\Rightarrow f''$ changes signs. Thus, the graph in (A) is the only one that goes from below the x-axis (negative) to above the x-axis (positive).

82. The correct answer is (C).

Area of a cross section $= \dfrac{\sqrt{3}}{4}(2x)^2 = \sqrt{3}x^2$

Using your calculator, you have:

Volume of solid $= \displaystyle\int_0^4 \sqrt{3}(x^2)dx = \dfrac{64\sqrt{3}}{3}.$

83. The correct answer is (B).

$$\int_0^6 f(x)dx \approx \frac{6-0}{2(3)}[0 + 2(1) + 2(2.25) + 6.25]$$

$$\approx 12.75.$$

84. The correct answer is (C).

$$\frac{dy}{dx} = ky \Rightarrow y = y_0 e^{kt}$$

Triple in 10 hours $\Rightarrow y = 3y_0$ at $t = 10$.

$$3y_0 = y_0 e^{10k} \Rightarrow 3 = e^{10k} \Rightarrow \ln 3 = \ln(e^{10k})$$

$$\Rightarrow \ln 3 = 10k \text{ or } k = \frac{\ln 3}{10}$$

$$\approx 0.109861 \approx 0.110.$$

85. The correct answer is (A).

See Figure 1TS-7.

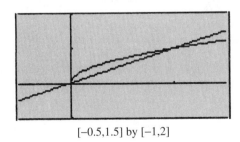

[−0.5,1.5] by [−1,2]

Figure 1TS-7

Points of intersection: (0, 0) and (1, 1)

Volume of solid $= \pi \displaystyle\int_0^1 (y^2 - (y^2)^2)dy$

Using your calculator, you obtain:

Volume of solid $= \dfrac{2\pi}{15}.$

86. The correct answer is (C).

See Figure 1TS-8.

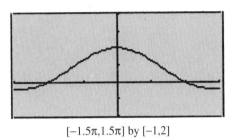

[−1.5π,1.5π] by [−1,2]

Figure 1TS-8

Using the Inflection function on your calculator, you obtain $x = -2.08$ and $x = 2.08$. Thus, there are two points of inflection on $(-\pi, \pi)$.

87. The correct answer is (A).

$$f(x) = x^2 e^x$$

Using your calculator, you obtain $f(1) \approx 2.7183$ and $f'(1) \approx 8.15485$. Equation of tangent line at $x = 1$:

$$y - 2.7183 = 8.15485(x - 1)$$

$$y = 8.15485(x - 1) + 2.7183$$

$$f(0.01) \approx 8.15485(1.1 - 1) + 2.7183$$

$$\approx 3.534.$$

88. The correct answer is (E).

See Figure 1TS-9.

$$\text{Area} = \int_b^\pi \sin x \, dx = -\cos x\Big]_b^\pi$$

$$= -\cos \pi - (-\cos b)$$

$$= -(-1) + \cos b = 1 + \cos b$$

Set $1 + \cos b = 0.2 \Rightarrow \cos b = -0.8 \Rightarrow$
$b = \cos^{-1}(-0.8) \Rightarrow b \approx 2.498.$

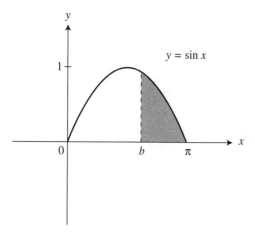

Figure 1TS-9

89. The correct answer is (D).

$$y = x^2; \quad \frac{dy}{dx} = 2x$$

$$y = -\sqrt{x} = -x^{1/2}; \quad \frac{dy}{dx} = -\frac{1}{2}x^{-1/2} = -\frac{1}{2\sqrt{x}}$$

Perpendicular tangent lines ⇒ slopes are negative reciprocals.

Thus $(2x)\left(-\dfrac{1}{2\sqrt{x}}\right) = -1$

$-\sqrt{x} = -1 \Rightarrow \sqrt{x} = 1$ or $x = 1$.

90. The correct answer is (C).

$$x^3 + y^3 = xy$$

$$3x^2 + 3y^2\frac{dy}{dx} = (1)y + x\frac{dy}{dx}$$

$$3y^2\frac{dy}{dx} - x\frac{dy}{dx} = y - 3x^2$$

$$\frac{dy}{dx} = \frac{y - 3x^2}{3y^2 - x}$$

At $x = 1$, $x^3 + y^3 = xy$ becomes $1 + y^3 = y$

$\Rightarrow y^3 - y + 1 = 0$

Using your calculator, you obtain: $y \approx -1.325$

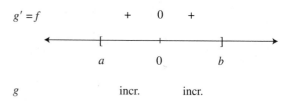

91. The correct answer is (B).

$$g(x) = \int_a^x f(t)dt \Rightarrow g'(x) = f(x)$$

See Figure 1TS-10.

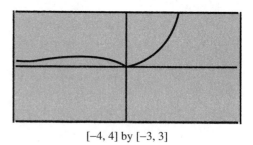

Figure 1TS-10

The graph in (B) is the only one that satisfies the behavior of g.

92. The correct answer is (D).

See Figure 1TS-11.

[−4, 4] by [−3, 3]

Figure 1TS-11

At $x = 0$, the graph of $g(x)$ shows: (1) a relative minimum; (2) a change of concavity; (3) a cusp (i.e., not differentiable at $x = 0$). Thus, only statements I and II are true.

Solutions—Section II

Part A—Calculators are permitted.

1. (A) $\dfrac{dy}{dx} = \dfrac{e^x}{e^x + 1}$

At $x = 0$, $\dfrac{dy}{dx} = \dfrac{e^0}{e^0 + 1} = \dfrac{1}{2}$

Equation of tangent line at $x = 0$:

$y - 2\ln 2 = \dfrac{1}{2}(x - 0)$

$y - 2\ln 2 = \dfrac{1}{2}x$ or $y = \dfrac{1}{2}x + 2\ln 2.$

(B) $f(0.1) \approx \dfrac{1}{2}(0.1) + 2\ln 2 \approx 1.43629 \approx 1.436.$

(C) $\dfrac{dy}{dx} = \dfrac{e^x}{e^x + 1}$

$dy = \dfrac{e^x}{e^x + 1}dx$

$\displaystyle\int dy = \int \dfrac{e^x}{e^x + 1}dx$

Let $u = e^x + 1$, $du = e^x\, dx$

$\displaystyle\int \dfrac{e^x}{e^x + 1}dx = \int \dfrac{1}{u}du = \ln|u| + c$

$\qquad\qquad = \ln|e^x + 1| + c$

$\qquad y = \ln(e^x + 1) + c$

The point $(0, 2\ln 2)$ is on the graph of f.

$2\ln 2 = \ln(e^0 + 1) + c$

$2\ln 2 = \ln 2 + c \Rightarrow c = \ln 2$

$\qquad y = \ln(e^x + 1) + \ln 2.$

(D) $f(0.1) = \ln(e^{0.1} + 1) + \ln 2$

$\qquad \approx 1.43754 \approx 1.438$

2. (A) At 1:00 am, $t = 6$.

$f(6) = 96 - 20\sin\left(\dfrac{6}{4}\right)$

$\qquad = 76.05° \approx 76°$ Fahrenheit.

(B) Average temperature

$\qquad = \dfrac{1}{12}\displaystyle\int_0^{12}\left[96 - 20\sin\left(\dfrac{t}{4}\right)\right]dt$

Using your calculator, you have:

Average temperature $= \dfrac{1}{12}(992.80)$

$\qquad\qquad\qquad\quad = 82.73 \approx 82.7.$

(C) Let $y_1 = f(x) = 96 - 20\sin\left(\dfrac{x}{4}\right)$ and
$y_2 = 80$

Using the Intersection function of your calculator, you obtain

$x = 3.70 \approx 3.7$ or $x = 8.85 \approx 8.9$

Thus, heating system is turned on when $3.7 \le t \le 8.9$.
(See Figure 1TS-12.)

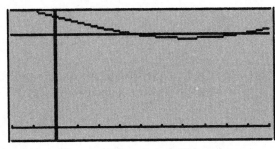

[−2,10] by [−10,100]

Figure 1TS-12

(D) Total cost

$\qquad = (0.25)\displaystyle\int_{3.7}^{8.9}(80 - f(t))dt$

$\qquad = (0.25)\displaystyle\int_{3.7}^{8.9}\left[80 - \left(96 - 20\sin\left(\dfrac{t}{4}\right)\right)\right]dt$

$\qquad = (0.25)\displaystyle\int_{3.7}^{8.9}\left(-16 + 20\sin\left(\dfrac{t}{4}\right)\right)dt.$

Using your calculator, you have:

$\qquad = (0.25)(13.629) = 3.407$

$\qquad \approx 3$ dollars.

3. (A) Midpoints of 5 subintervals of equal length are $t = 3, 9, 15, 21,$ and 27.
The length of each subinterval is
$\dfrac{30 - 0}{5} = 6.$

Thus, $\int_0^{30} v(t)dt = 6[v(3) + v(9) + v(15)$

$$+ v(21) + v(27)]$$

$$= 6[7.5 + 12 + 13.5$$

$$+ 14 + 13]$$

$$= 6[60] = 360.$$

(B) Average velocity $= \dfrac{1}{30 - 0} \int_0^{30} v(t)dt$

$$\approx \dfrac{1}{30}(360)$$

$$\approx 12 \text{ m/sec.}$$

(C) Average acceleration $= \dfrac{12.2 - 0}{30 - 0} \text{ m/sec}^2$

$$= 0.407 \text{ m/sec}^2.$$

(D) Approximate acceleration at $t = 6$

$$= \dfrac{v(9) - v(3)}{9 - 3} = \dfrac{12 - 7.5}{6} = 0.75 \text{ m/sec}^2.$$

(E) Looking at the velocity in the table, you see that the velocity decreases from $t = 18$ to $t = 30$. Thus the acceleration is negative for $18 < t < 30$.

Part B—No calculators.

4. (A) See Figure 1TS-13.

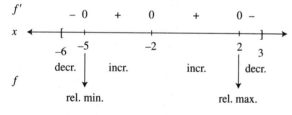

Figure 1TS-13

Since f increases on $(-5, 2)$ and decreases on $(2, 3)$, it has a relative maximum at $x = 2$.

(B) Since f decreases on $(-6, -5)$ and increases on $(-5, 2)$, it has a relative minimum at $x = -5$.

(C) See Figure 1TS-14.
A change of concavity occurs at $x = -4, -2,$ and 0, and since f' exists at these x-values, f has a point of inflection at $x = -4$, $x = -2$, and $x = 0$.

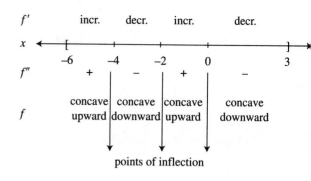

Figure 1TS-14

(D) See Figure 1TS-15.

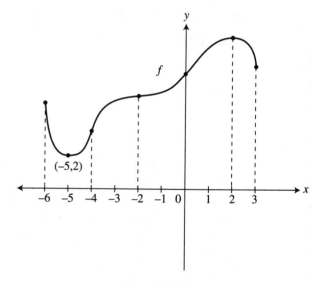

Figure 1TS-15

5. (A) Differentiating: $2y\dfrac{dy}{dx} - 1 + 2\dfrac{dy}{dx} = 0$

$$\dfrac{dy}{dx}(2y + 2) = 1 \Rightarrow \dfrac{dy}{dx} = \dfrac{1}{2y + 2}.$$

(B) At $(0, -3)\dfrac{dy}{dx} = \dfrac{1}{2(-3) + 2} = -\dfrac{1}{4}$

$$y - y_1 = m(x - x_1)$$

$$y - (-3) = -\tfrac{1}{4}(x - 0)$$

$$y + 3 = -\dfrac{1}{4}x \text{ or } y = -\dfrac{1}{4}x - 3.$$

(C) $m_{\text{normal}} = \dfrac{-1}{m_{\text{tangent}}}$

At $(0, -3), m_{\text{normal}} = \dfrac{-1}{-1/4} = 4.$

$$y - (-3) = 4(x - 0)$$

$$y + 3 = 4x \text{ or } y = 4x - 3.$$

(D) $y = \frac{1}{4}x + 3 \Rightarrow m = \frac{1}{4}$

Set $\frac{dy}{dx} = \frac{1}{2y+2} = \frac{1}{4} \Rightarrow y = 1$

$y^2 - x + 2y - 3 = 0$.

At $y = 1, 1^2 - x + 2(1) - 3 = 0 \Rightarrow x = 0$.
Thus, point P is $(0, 1)$.

6. See Figure 1TS-16.

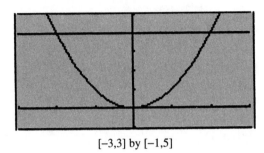

[−3,3] by [−1,5]

Figure 1TS-16

(A) Set $x^2 = 4 \Rightarrow x = \pm 2$

$$\text{Area of } R = \int_{-2}^{2}(4 - x^2)dx = 4x - \frac{x^3}{3}\Bigg]_{-2}^{2}$$

$$= \left(4(2) - \frac{2^3}{3}\right) - \left(4(-2) - \frac{(-2)^3}{3}\right)$$

$$= \frac{16}{3} - \left(-\frac{16}{3}\right) = \frac{32}{3}.$$

(B) Since $y = x^2$ is an even function, $x = 0$ divides R into two regions of equal area. Thus $a = 0$.

(C) See Figure 1TS-17.

$$\text{Area } R_1 = \text{Area } R_2 = \frac{16}{3}$$

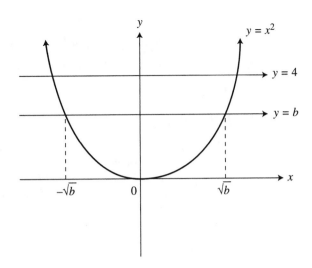

Figure 1TS-17

$$\text{Area } R_2 = \int_{-\sqrt{b}}^{\sqrt{b}}(b - x^2)dx$$

$$= 2\int_{0}^{\sqrt{b}}(b - x^2)dx$$

$$= 2\left[bx - \frac{x^3}{3}\right]_{0}^{\sqrt{b}} = 2\left[b\left(\sqrt{b}\right) - \frac{\left(\sqrt{b}^3\right)}{3}\right]$$

$$= 2\left(b^{3/2} - \frac{b^{3/2}}{3}\right) = 2\left(\frac{2b^{3/2}}{3}\right) = \frac{4b^{3/2}}{3}$$

Set $\frac{4b^{3/2}}{3} = \frac{16}{3} \Rightarrow b^{3/2} = 4$ or $b = 4^{2/3}$.

(D) Washer Method

$$V = \pi\int_{-2}^{2}(4^2 - (x^2)^2)dx = \pi\int_{-2}^{2}(16 - x^4)dx$$

$$= \pi\left[16x - \frac{x^5}{5}\right]_{-2}^{2} = \frac{256\pi}{5}.$$

Scoring Sheet for AB Practice Exam 1

Section I—Part A

$$\underline{\hspace{3cm}} \times 1.2 = \underline{\hspace{3cm}}$$
No. Correct $\qquad\qquad$ Subtotal A

$$\underline{\hspace{3cm}} \times (0.25) \times 1.2 = \underline{\hspace{3cm}}$$
No. Incorrect $\qquad\qquad$ Subtotal B

$$\text{Part A (Subtotal A} - \text{Subtotal B)} = \underline{\hspace{3cm}}$$
$\qquad\qquad\qquad\qquad\qquad\qquad\qquad\qquad$ Subtotal C

Section I—Part B

$$\underline{\hspace{3cm}} \times 1.2 = \underline{\hspace{3cm}}$$
No. Correct $\qquad\qquad$ Subtotal D

$$\underline{\hspace{3cm}} \times (0.25) \times 1.2 = \underline{\hspace{3cm}}$$
No. Incorrect $\qquad\qquad$ Subtotal E

$$\text{Part B (Subtotal D} - \text{Subtotal E)} = \underline{\hspace{3cm}}$$
$\qquad\qquad\qquad\qquad\qquad\qquad\qquad\qquad$ Subtotal F

Section II—Part A (Each question is worth 9 points.)

$$\underline{\hspace{2.5cm}} + \underline{\hspace{2.5cm}} + \underline{\hspace{2.5cm}} = \underline{\hspace{2.5cm}}$$
Q#1 $\qquad\qquad$ Q#2 $\qquad\qquad$ Q#3 $\qquad$ Subtotal G

Section II—Part B (Each question is worth 9 points)

$$\underline{\hspace{2.5cm}} + \underline{\hspace{2.5cm}} + \underline{\hspace{2.5cm}} = \underline{\hspace{2.5cm}}$$
Q#1 $\qquad\qquad$ Q#2 $\qquad\qquad$ Q#3 $\qquad$ Subtotal H

$$\text{Total Raw Score (Subtotals C} + \text{F} + \text{G} + \text{H)} = \boxed{}$$

Approximate Conversion Scale:	
Total Raw Score	Approximate AP Grade
75–108	5
60–74	4
45–59	3
31–44	2
0–30	1

AB PRACTICE EXAM 2

Answer Sheet for AB Practice Exam 2—Section I

Part A

1. _____

2. _____

3. _____

4. _____

5. _____

6. _____

7. _____

8. _____

9. _____

10. _____

11. _____

12. _____

13. _____

14. _____

15. _____

16. _____

17. _____

18. _____

19. _____

20. _____

21. _____

22. _____

23. _____

24. _____

25. _____

26. _____

27. _____

28. _____

Part B

76. _____

77. _____

78. _____

79. _____

80. _____

81. _____

82. _____

83. _____

84. _____

85. _____

86. _____

87. _____

88. _____

89. _____

90. _____

91. _____

92. _____

Section I—Part A

Number of Questions	Time	Use of Calculator
28	55 Minutes	No

Directions:

Use the answer sheet provided in the previous page. All questions are given equal weight. There is no penalty for unanswered questions. However, 1/4 of the number of the incorrect answers will be subtracted from the number of correct answers. Unless otherwise indicated, the domain of a function f is the set of all real numbers. The use of a calculator is *not* permitted in this part of the exam.

1. $\int_0^8 x^{2/3}\,dx$

   (A) $\dfrac{1}{3}$ (B) $\dfrac{96}{5}$ (C) $\dfrac{4}{3}$

   (D) $-\dfrac{1}{3}$ (E) $-\dfrac{96}{5}$

2. The $\lim\limits_{x\to-\infty} \dfrac{x^2+4x-5}{x^3-1}$ is

   (A) 0   (B) $\dfrac{1}{3}$ (C) 5

   (D) $-\infty$ (E) ∞

3. What is the $\lim\limits_{x\to-2} f(x)$, if

   $$f(x) = \begin{cases} |x-1| & \text{if } x > -2 \\ 2x+7 & \text{if } x \le -2 \end{cases}?$$

   (A) -3 (B) 1   (C) 3
   (D) 11   (E) nonexistent

4. The graph of f' is shown in Figure 2T-1.

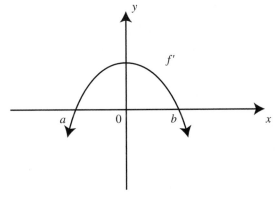

Figure 2T-1

Which of the graphs in Figure 2T-2 on page 352 is a possible graph of f?

5. $\int_{\pi/2}^x 2\cos t\,dt =$

   (A) $2\cos x$ (B) $-2\cos x$ (C) $2\sin x$
   (D) $-2\sin x + 2$ (E) $2\sin x - 2$

6. Given the equation $y = 3e^{-2x}$, what is an equation of the normal line to the graph at $x = \ln 2$?

   (A) $y = \dfrac{2}{3}(x - \ln 2) + \dfrac{3}{4}$

   (B) $y = \dfrac{2}{3}(x + \ln 2) - \dfrac{3}{4}$

   (C) $y = -\dfrac{3}{2}(x - \ln 2) + \dfrac{3}{4}$

   (D) $y = -\dfrac{3}{2}(x - \ln 2) - \dfrac{3}{4}$

   (E) $y = 24(x - \ln 2) + 12$

7. What is the $\lim\limits_{h\to 0} \dfrac{\csc(\pi/4 + h) - \csc(\pi/4)}{h}$?

   (A) $\sqrt{2}$ (B) $-\sqrt{2}$ (C) 0

   (D) $-\dfrac{\sqrt{2}}{2}$ (E) undefined

8. If $f(x)$ is an antiderivative of $x^2\sqrt{x^3+1}$ and $f(2) = 0$, then $f(0) =$

   (A) -6 (B) 6   (C) $\dfrac{2}{9}$

   (D) $\dfrac{-52}{9}$ (E) $\dfrac{56}{9}$

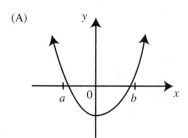
(A)

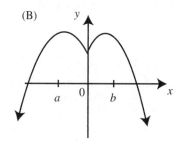
(B)

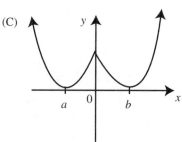
(C)

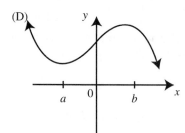
(D)

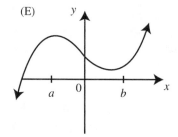
(E)

Figure 2T-2

9. If a function f is continuous for all values of x, which of the following statements is/are always true?

I. $2\displaystyle\int_a^b f(x)\,dx = \int_{2a}^{2b} f(x)\,dx$

II. $\displaystyle\int_a^b f(x)\,dx = \int_b^a -f(x)\,dx$

III. $\left|\displaystyle\int_a^b f(x)\,dx\right| = \int_a^b |f(x)|\,dx$

    (A) I only
    (B) I and II only
    (C) II only
    (D) II and III only
    (E) I, II, and III

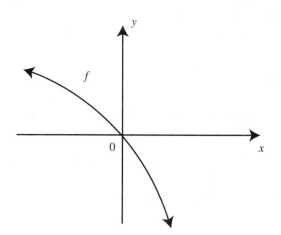

Figure 2T-3

10. The graph of f is shown in Figure 2T-3 and f is twice differentiable. Which of the following has the largest value: $f(0), f'(0), f''(0)$?

    (A) $f(0)$
    (B) $f'(0)$
    (C) $f''(0)$
    (D) $f(0)$ and $f'(0)$
    (E) $f'(0)$ and $f''(0)$

11. $\displaystyle\int \frac{x^4 - 1}{x^2}\,dx =$

    (A) $\dfrac{x^3}{3} + x + c$

    (B) $\dfrac{x^3}{3} - x + c$

    (C) $\dfrac{x^3}{3} + \dfrac{3}{x^3} + c$

    (D) $\dfrac{x^3}{3} + \dfrac{1}{x} + c$

    (E) $\dfrac{x^3}{3} - \dfrac{1}{x} + c$

12. If $p'(x) = q(x)$ and q is a continuous function for all values of x, then $\int_{-1}^{0} q(4x)dx$ is

   (A) $p(0) - p(-4)$

   (B) $4p(0) - 4p(-4)$

   (C) $\frac{1}{4}p(0) - \frac{1}{4}p(-4)$

   (D) $\frac{1}{4}p(0) + \frac{1}{4}p(-4)$

   (E) $p(0) + P(-4)$

13. Water is leaking from a tank at a rate represented by $f(t)$ whose graph is shown in Figure 2T-4. Which of the following is the best approximation of the total amount of water leaked from the tank for $1 \le t \le 3$?

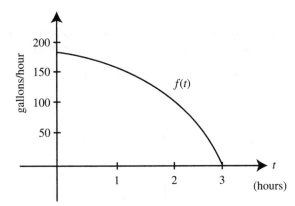

Figure 2T-4

(A) $\frac{9}{2}$ gallons

(B) 5 gallons

(C) 175 gallons

(D) 350 gallons

(E) 450 gallons

14. If $f(x) = 5\cos^2(\pi - x)$, then $f'\left(\frac{\pi}{2}\right)$ is

   (A) 0   (B) -5 (C) 5

   (D) -10 (E) 10

15. $g(x) = \int_{1}^{x} \frac{3t}{t^3 + 1}dt$, then $g'(2)$ is

   (A) 0   (B) $-\frac{2}{3}$ (C) $\frac{2}{3}$

   (D) $\frac{-5}{6}$ (E) $\frac{5}{6}$

16. If $\int_{k}^{2} (2x - 2)dx = -3$, a possible value of k is

   (A) -2 (B) 0   (C) 1

   (D) 2   (E) 3

17. If $\int_{0}^{a} f(x)dx = -\int_{-a}^{0} f(x)dx$ for all positive values of a, then which of the following could be the graph of f? See Figure 2T-5.

(A)

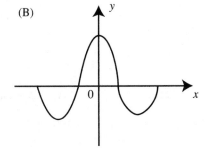

(B)

(C)

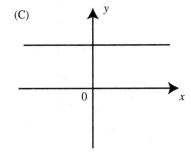

(D)

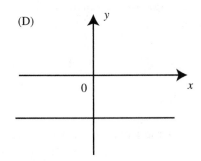

(E)

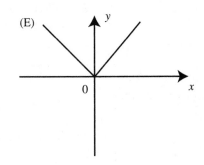

Figure 2T-5

18. A function f is continuous on $[1, 5]$ and some of the values of f are shown below:

x	1	3	5
$f(x)$	-2	b	-1

If f has only one root, r, on the closed interval $[1, 5]$, and $r \neq 3$, then a possible value of b is

(A) -1 (B) 0 (C) 1
(D) 3 (E) 5

19. Given the equation $V = \dfrac{1}{3}\pi r^2 (5 - r)$, what is the instantaneous rate of change of V with respect to r at $r = 5$?

(A) $-\dfrac{25\pi}{3}$ (B) $\dfrac{25\pi}{3}$ (C) $\dfrac{50\pi}{3}$

(D) 25π (E) $\dfrac{125\pi}{3}$

20. What is the slope of the tangent to the curve $x^3 - y^2 = 1$ at $x = 1$?

(A) $-\dfrac{3}{2}$ (B) 0 (C) $\dfrac{3}{2\sqrt{2}}$

(D) $\dfrac{3}{2}$ (E) undefined

21. The graph of function f is shown in Figure 2T-6. Which of the following is true for f on the interval (a, b)?

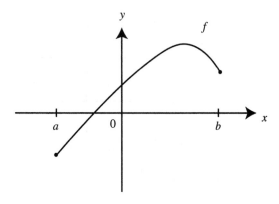

Figure 2T-6

   I. The function f is differentiable on (a, b).
  II. There exists a number k on (a, b) such that $f'(k) = 0$.
 III. $f'' > 0$ on (a, b)

(A) I only
(B) II only

(C) I and II only
(D) II and III only
(E) I, II and III

22. The velocity function of a moving particle on the x-axis is given as $v(t) = t^2 - 3t - 10$. For what positive values of t is the particle's speed increasing?

(A) $0 < t < \dfrac{3}{2}$ only

(B) $t > \dfrac{3}{2}$ only

(C) $t > 5$ only

(D) $0 < t < \dfrac{3}{2}$ and $t > 5$ only

(E) $\dfrac{3}{2} < t < 5$ only

23. The graph of f consists of two line segments and a semicircle for $-2 \leq x \leq 2$ as shown in Figure 2T-7. What is the value of $\displaystyle\int_{-2}^{2} f(x)dx$?

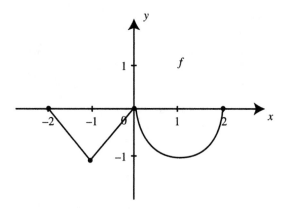

Figure 2T-7

(A) $-2 - 2\pi$ (B) $-2 - \pi$ (C) $-1 - \dfrac{\pi}{2}$

(D) $1 + \dfrac{\pi}{2}$ (E) $-1 - \pi$

24. What is the average value of the function $y = 3\cos(2x)$ on the interval $\left[-\dfrac{\pi}{2}, \dfrac{\pi}{2}\right]$?

(A) -2 (B) $-\dfrac{2}{\pi}$ (C) 0

(D) $\dfrac{1}{\pi}$ (E) $\dfrac{3}{2\pi}$

25. If $f(x) = |x^3|$, what is the value of $\displaystyle\lim_{x \to -1} f'(x)$?

(A) -3 (B) 0 (C) 1
(D) 3 (E) undefined

26. A spherical balloon is being inflated. At the instant when the rate of increase of the volume of the sphere is four times the rate of increase of the radius, the radius of the sphere is

(A) $\dfrac{1}{4\sqrt{\pi}}$ (B) $\dfrac{1}{\sqrt{\pi}}$ (C) $\dfrac{1}{\pi}$

(D) $\dfrac{1}{16\pi}$ (E) π

27. If $\dfrac{dy}{dx} = \dfrac{x^2}{y}$ and at $x = 0, y = 4$, a solution to the differential equation is

(A) $y = \dfrac{x^3}{3}$

(B) $y = \dfrac{x^3}{3} + 4$

(C) $\dfrac{y^2}{2} = \dfrac{x^3}{3}$

(D) $\dfrac{y^2}{2} = \dfrac{x^3}{3} + 4$

(E) $\dfrac{y^2}{2} = \dfrac{x^3}{3} + 8$

28. The area of the region enclosed by the graph of $x = y^2 - 1$ and the y-axis is

(A) $-\dfrac{4}{3}$ (B) 0 (C) $\dfrac{2}{3}$

(D) $\dfrac{4}{3}$ (E) $\dfrac{8}{3}$

Section I—Part B

Number of Questions	Time	Use of Calculator
17	50 Minutes	Yes

Directions:

Use the same answer sheet from Part A. *Please note that the questions begin with number 76. This is not an error.* It is done to be consistent with the numbering system of the actual AP Calculus AB Exam. All questions are given equal weight. There is no penalty for unanswered questions. However, 1/4 of the number of incorrect answers will be subtracted from the number of correct answers. Unless otherwise indicated, the domain of a function f is the set of all real numbers. If the exact numerical value does not appear among the given choices, select the best approximate value. The use of a calculator is *permitted* in this part of the exam.

76. The graph of f', the derivative of f, is shown in Figure 2T-8. At which value of x does the graph f have a horizontal tangent?

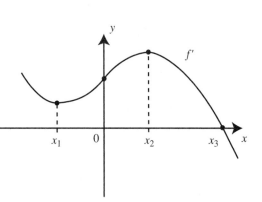

Figure 2T-8

(A) x_1 (B) 0   (C) x_2
(D) x_1 and x_2 (E) x_3

77. The position function of a moving particle is $s(t) = 5 + 4t - t^2$ for $0 \leq t \leq 10$ where s is in meters and t is measured in seconds. What is the maximum speed in m/sec of the particle on the interval $0 \leq t \leq 10$?

(A) -16 (B) 0   (C) 2
(D) 4   (E) 16

78. How many points of inflection does the graph of $y = \cos(x^2)$ have on the interval $(0, \pi)$?

(A) 0   (B) 1   (C) 2
(D) 3   (E) 4

79. Let f be a continuous function on [4, 10] and have selected values as shown below:

x	4	6	8	10
$f(x)$	2	2.4	2.8	3.2

Using three right endpoint rectangles of equal length, what is the approximate value of $\int_4^{10} f(x)dx$?

(A) 8.4   (B) 9.6   (C) 14.4
(D) 16.8   (E) 20.8

80. Given a differentiable function f with $f(-1) = 2$ and $f'(-1) = \dfrac{1}{2}$. Using a tangent line to the graph of f at $x = -1$, find an approximate value of $f(-1.1)$?

(A) -3.05 (B) -1.95 (C) 0.95
(D) 1.95   (E) 3.05

81. The area under the curve of $y = \dfrac{\ln x}{x}$ from $x = 1$ to $x = b$, where $b > 1$ is 0.66. Then the value of b is approximately,

(A) 1.93   (B) 2.25   (C) 3.15
(D) 3.74   (E) 5.71

82. The base of a solid is a region enclosed by the circle $x^2 + y^2 = 4$. What is the approximate volume of the solid if the cross sections of the solid perpendicular to the x-axis are semicircles?

(A) 8π (B) $\dfrac{16\pi}{3}$ (C) $\dfrac{32\pi}{3}$

(D) $\dfrac{64\pi}{3}$ (E) $\dfrac{512\pi}{15}$

83. The temperature of a cup of coffee is dropping at the rate of $f(t) = 4\sin\left(\dfrac{t}{4}\right)$ degrees for $0 \le t \le 5$, where f is measured in Fahrenheit and t in minutes. If initially, the coffee is 95°F, find its temperature to the nearest degree Fahrenheit 5 minutes later.

(A) 84   (B) 85   (C) 91
(D) 92   (E) 94

84. The graphs of f', g', p', and q' are shown in Figure 2T-9. Which of the functions f, g, p, or q have a relative minimum on (a, b)?

(A) f only   (B) g only
(C) p only   (D) q only
(E) q and p only

85. What is the volume of the solid obtained by revolving about the y-axis the region enclosed by the graphs of $x = y^2$ and $x = 9$?

(A) 36π (B) $\dfrac{81\pi}{2}$ (C) $\dfrac{486\pi}{2}$

(D) $\dfrac{1994}{5}$ (E) $\dfrac{1944\pi}{5}$

86. At what value(s) of x do the graphs of $y = e^x$ and $y = x^2 + 5x$ have parallel tangent lines?

(A) -2.5 (B) 0
(C) 0 and 5   (D) -5 and 0.24
(E) -2.45 and 2.25

87. Let y represent the population in a town. If y decreases according to the equation $\dfrac{dy}{dt} = ky$, t in years, and the population decreases by 25% in 6 years, then $k =$

(A) -8.318 (B) -1.726 (C) -0.231
(D) -0.120 (E) -0.048

88. If $h(x) = \displaystyle\int_4^x (t-5)^3\,dt$ on [4, 8], then h has a local minimum at $x =$

(A) 4   (B) 5   (C) 6
(D) 7   (E) 8

89. If $p(x) = \displaystyle\int_a^x q(t)\,dt$ $a < x < b$ and the graph of q is shown in Figure 2T-10, which of the graphs shown in Figure 2T-11 is a possible graph of p?

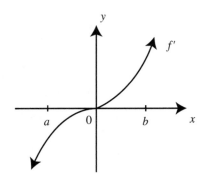

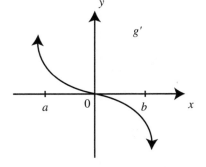

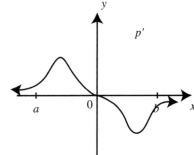

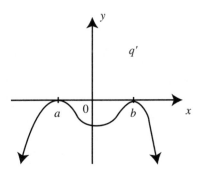

Figure 2T-9

90. The volume of the solid generated by revolving about the y-axis the region bounded by the graph of $y = x^3$, the line $y = 1$ and the y-axis is

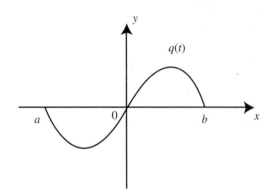

Figure 2T-10

(A) $\dfrac{\pi}{4}$ (B) $\dfrac{2\pi}{5}$ (C) $\dfrac{3\pi}{5}$

(D) $\dfrac{2\pi}{3}$ (E) $\dfrac{3\pi}{4}$

91. If $f(x) = -|x - 3|$, which of the following statement about f is true?

I. f is differentiable at $x = 3$.

II. f has an absolute minimum at $x = 3$.

III. f has a point of inflection at $x = 3$.

(A) II only
(B) III only
(C) II and III only
(D) I, II, and III
(E) none

92. The equation of the tangent line to the graph of $y = \sin x$ for $0 \le x \le \pi$ at the point where $\dfrac{dy}{dx} = \dfrac{1}{2}$ is

(A) $y = \dfrac{1}{2}\left(x - \dfrac{\pi}{3}\right) - \dfrac{\sqrt{3}}{2}$

(B) $y = \dfrac{1}{2}\left(x - \dfrac{\pi}{3}\right) + \dfrac{\sqrt{3}}{2}$

(C) $y = \dfrac{1}{2}\left(x - \dfrac{1}{2}\right) + \dfrac{\pi}{3}$

(D) $y = \dfrac{1}{2}\left(x - \dfrac{1}{2}\right) - \dfrac{\pi}{3}$

(E) $y = \dfrac{1}{2}\left(x + \dfrac{1}{2}\right) - \dfrac{\pi}{3}$

(A)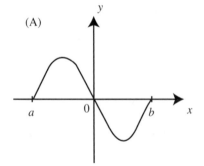

(B)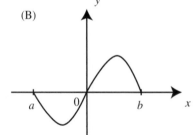

(C)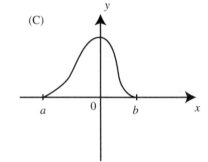

(D)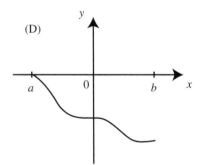

(E)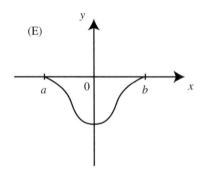

Figure 2T-11

Section II—Part A

Number of Questions	Time	Use of Calculator
3	45 Minutes	Yes

Directions:

Show all work. You may *not* receive any credit for correct answers without supporting work. You may use an approved calculator to help solve a problem. However, you must clearly indicate the setup of your solution using mathematical notations and *not* calculator syntax. Calculators may be used to find the derivative of a function at a point, compute the numerical value of a definite integral, or solve an equation. Unless otherwise indicated, you may assume the following: (a) the numeric or algebraic answers need not be simplified, (b) your answer, if expressed in approximation, should be correct to 3 places after the decimal point, and (c) the domain of a function f is the set of all real numbers.

1. A particle is moving on a coordinate line. The graph of its velocity function $v(t)$ for $0 \leq t \leq 24$ seconds is shown in Figure 2T-12.

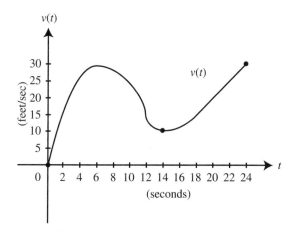

Figure 2T-12

(A) Using midpoints of the three subintervals of equal length, find the approximate value of
$$\int_0^{24} v(t)dt.$$

(B) Using the result in part (A), find the average velocity over the interval $0 \leq t \leq 24$ seconds.

(C) Find the average acceleration over the interval $0 \leq t \leq 24$ seconds.

(D) When is the acceleration of the particle equal to zero?

(E) Find the approximate acceleration at $t = 20$ seconds.

2. Let R be the region in the first quadrant enclosed by the graph of $y = 2\cos x$, the x-axis and the y-axis.

(A) Find the area of the region R.

(B) If the line $x = a$ divides the region R into two regions of equal area, find a.

(C) Find the volume of the solid obtained by revolving region R about the x-axis.

(D) If R is the base of a solid whose cross sections perpendicular to the x-axis are semi-circles, find the volume of the solid.

3. The temperature of a liquid at a chemical plant during a 20-minute period is given as
$$g(t) = 90 - 4\tan\left(\frac{t}{20}\right),$$ where $g(t)$ is measured in degrees Fahrenheit and $0 \leq t \leq 20$, t is measured in minutes.

(A) Sketch the graph of g on the provided grid. What is the temperature of the liquid to the nearest hundredth of a degree Fahrenheit when $t = 10$? See Figure 2T-13.

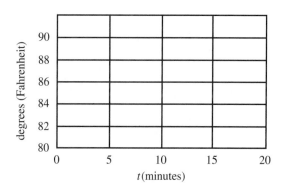

Figure 2T-13

(B) What is the average temperature of the liquid to the nearest hundredth of a degree Fahrenheit during the 20-minute period.

(C) At what values of t is the temperature of the liquid below 86°F?

(D) During the time within the 20-minute period when the temperature is below 86°F, what is the average temperature to the nearest hundredth of a degree Fahrenheit?

Section II—Part B

Number of Questions	Time	Use of Calculator
3	45 Minutes	No

Directions:

The use of a calculator is *not* permitted in this part of the exam. When you have finished this part of the exam, you may return to the problems in Part A of Section II and continue to work on them. However, you may not use a calculator. You should *show all work*. You may *not* receive any credit for correct answers without supporting work. Unless otherwise indicated, the numeric or algebraic answers need not be simplified, and the domain of a function f is the set of all real numbers.

4. Given the function $f(x) = 3e^{-2x^2}$,

   (A) At what value(s) of x, if any, is $f'(x) = 0$?
   (B) At what value(s) of x, if any, is $f''(x) = 0$?
   (C) Find $\lim_{x \to \infty} f(x)$ and $\lim_{x \to -\infty} f(x)$.
   (D) Find the absolute maximum value of f and justify your answer.
   (E) Show that if $f(x) = ae^{-bx^2}$ where $a > 0$ and $b > 0$, the absolute maximum value of f is a.

5. The function f is defined as $f(x) = \int_0^x g(t)\,dt$ where the graph of g consists of five line segments as shown in Figure 2T-14.

   (A) Find $f(-3)$ and $f(3)$.
   (B) Find all values of x on $(-3, 3)$ such that f has a relative maximum or minimum. Justify your answer.
   (C) Find all values of x on $(-3, 3)$ such that the graph f has a point of inflection. Justify your answer.
   (D) Write an equation of the line tangent to the graph to f at $x = 1$.

6. The slope of a function f at any point (x, y) is $\dfrac{y}{2x^2}$. The point $(2, 1)$ is on the graph of f.

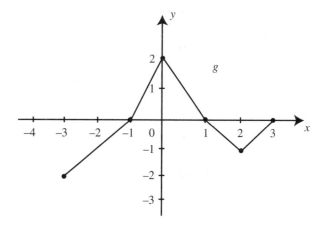

Figure 2T-14

   (A) Write an equation of the tangent line to the graph of f at $x = 2$.
   (B) Use the tangent line in part (A) to approximate $f(2.5)$.
   (C) Solve the differential equation $\dfrac{dy}{dx} = \dfrac{y}{2x^2}$ with the initial condition $f(2) = 1$.
   (D) Use the solution in part (C) and find $f(2.5)$.

Answers to AB Practice Exam 2—Section I

Part A	12. C	24. C	81. C
1. B	13. C	25. A	82. B
2. A	14. A	26. B	83. A
3. C	15. C	27. E	84. A
4. D	16. E	28. D	85. E
5. E	17. A		86. E
6. A	18. A	Part B	87. E
7. B	19. A	76. E	88. B
8. D	20. E	77. E	89. E
9. C	21. C	78. D	90. C
10. A	22. D	79. D	91. E
11. D	23. C	80. D	92. B

Answers to AB Practice Exam 2—Section II

Part A

Part B

1. (A) 480 (3 pts.)
   (B) 20 ft/s (2 pts.)
   (C) 1.25 ft/s^2 (1 pt.)
   (D) $t = 6$ and $t = 14$ (2 pts.)
   (E) 2.5 ft/s^2 (1 pt.)

2. (A) 2 (2 pts.)
   (B) $a = \dfrac{\pi}{6}$ (3 pts.)
   (C) π^2 (2 pts.)
   (D) $\dfrac{\pi^2}{8}$ (2 pts.)

3. (A) See graph, and $g(10) = 87.82°$ (3 pts.)
   (B) 87.54° (2 pts.)
   (C) $15.708 < t \le 20$ (2 pts.)
   (D) 84.99° (2 pts.)

4. (A) $x = 0$ (1 pt.)
   (B) $x = \pm\dfrac{1}{2}$ (2 pts.)
   (C) $\lim\limits_{x \to \infty} f(x) = 0$ and $\lim\limits_{x \to -\infty} f(x) = 0$ (2 pts.)
   (D) 3 (2 pts.)
   (E) See solution. (2 pts.)

5. (A) $f(-3) = 1$ and $f(3) = 0$ (2 pts.)
   (B) $x = -1, 1$ (3 pts.)
   (C) $x = 0$ and $x = 2$ (2 pts.)
   (D) $y = 1$ (2 pts.)

6. (A) $y = \dfrac{1}{8}(x - 2) + 1$ (3 pts.)
   (B) 1.063 (1 pt.)
   (C) $y = e^{(-1/2x)+(1/4)}$ (4 pts.)
   (D) $e^{1/20}$ (or 1.051) (1 pt.)

Solutions to AB Practice Exam 2—Section I

Part A—No calculators.

1. The correct answer is (B).

$$\int_0^8 x^{2/3}\,dx = \frac{x^{5/3}}{5/3}\bigg]_0^8 = \frac{3x^{5/3}}{5}\bigg]_0^8$$

$$= \frac{3(8)^{5/3}}{5} - 0 = \frac{3(32)}{5} = \frac{96}{5}.$$

2. The correct answer is (A).

$$\lim_{x\to-\infty}\frac{x^2+4x-5}{x^3-1} = \lim_{x\to-\infty}\frac{\dfrac{x^2}{x^3}-\dfrac{4x}{x^3}-\dfrac{5}{x^3}}{\dfrac{x^3}{x^3}-\dfrac{1}{x^3}}$$

$$= \lim_{x\to-\infty}\frac{\dfrac{1}{x}-\dfrac{4}{x^2}-\dfrac{5}{x^3}}{1-\dfrac{1}{x^3}} = 0.$$

3. The correct answer is (C).

$$\lim_{x\to-2^+}|x-1| = |-2-1| = 3$$

$$\lim_{x\to-2^-}(2x+7) = 2(-2)+7 = 3$$

Thus $\lim_{x\to-2} f(x) = 3$.

4. The correct answer is (D).

See Figure 2TS-1.

5. The correct answer is (E).

$$\int_{\pi/2}^x 2\cos t\,dt = 2\sin t]_{\pi/2}^x = 2\sin x - 2(1)$$

$$= 2\sin x - 2.$$

6. The correct answer is (A).

$$y = 3e^{-2x};\quad \frac{dy}{dx} = 3e^{-2x}(-2) = -6e^{-2x}$$

$$\frac{dy}{dx}\bigg|_{x=\ln 2} = -6e^{-2\ln 2} = -6\left(e^{\ln 2}\right)^{-2} = -6(2)^{-2}$$

$$= -6\left(\frac{1}{4}\right) = -\frac{3}{2}$$

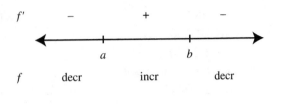

$$f' \qquad\qquad -\qquad\qquad +\qquad\qquad -$$

$$a \qquad\qquad b$$

$$f \qquad \text{decr} \qquad \text{incr} \qquad \text{decr}$$

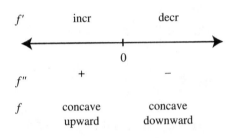

$$f' \qquad \text{incr} \qquad \text{decr}$$

$$0$$

$$f'' \qquad + \qquad -$$

$$f \qquad \text{concave} \qquad \text{concave}$$
$$\text{upward} \qquad \text{downward}$$

Figure 2TS-1

Slope of normal line at $x = \ln 2$ is $\dfrac{2}{3}$.

At $x = \ln 2$, $y = 3e^{-2\ln 2} = \dfrac{3}{4}$; point $\left(\ln 2, \dfrac{3}{4}\right)$

Equation of normal line:

$$y - \frac{3}{4} = \frac{2}{3}(x-\ln 2) \text{ or } y = \frac{2}{3}(x-\ln 2) + \frac{3}{4}.$$

7. The correct answer is (B).

$$f'(x) = \lim_{h\to0}\frac{f(x+h)-f(x)}{h}$$

Thus $\lim_{h\to0}\dfrac{\csc\left(\dfrac{\pi}{4}+h\right)-\csc\left(\dfrac{\pi}{4}\right)}{h}$

$$= \frac{d(\csc x)}{dx}\bigg|_{x=\frac{\pi}{4}} = -\csc\left(\frac{\pi}{4}\right)\cot\left(\frac{\pi}{4}\right)$$

$$= -\sqrt{2}(1) = -\sqrt{2}.$$

8. The correct answer is (D).

Let $u = x^3+1$, $du = 3x^2\,dx$ or $\dfrac{du}{3} = x^2\,dx$

$$f(x) = \int x^2\sqrt{x^3+1}\,dx = \int \sqrt{u}\frac{du}{3}$$

$$= \frac{1}{3}\int u^{1/2}\,du$$

$$= \frac{1}{3} \frac{u^{3/2}}{3/2} + c$$

$$= \frac{2}{9}(x^3 + 1)^{3/2} + c$$

$$f(2) = 0 \Rightarrow \frac{2}{9}(2^3 + 1)^{3/2} + c = 0$$

$$\Rightarrow \frac{2}{9}(9)^{3/2} + c = 0$$

$$\Rightarrow 6 + c = 0 \text{ or } c = -6$$

$$f(x) = \frac{2}{9}(x^3 + 1)^{3/2} - 6$$

$$f(0) = \frac{2}{9} - 6 = \frac{-52}{9}.$$

9. The correct answer is (C).

Statement I is *not true*, e.g.,
$$2\int_0^4 x \, dx \neq 2\int_0^8 x \, dx.$$
Statement II is always *true* since
$$\int_a^b f(x)dx = -\int_b^a f(x)dx \text{ by properties of}$$
definite integrals.
Statement III is *not true*, e.g.,
$$\left| \int_{-2}^2 x \, dx \right| \neq \int_{-2}^2 |x| \, dx.$$

10. The correct answer is (A).

$f(0) = 0$; $f'(0) \leq 0$ since f is concave download.
Thus $f(0)$ has the largest value.

11. The correct answer is (D).

$$\int \frac{x^4 - 1}{x^2} dx = \int \left(x^2 - \frac{1}{x^2} \right) dx$$

$$= \int (x^2 - x^{-2})dx$$

$$= \frac{x^3}{3} - \frac{x^{-1}}{-1} + c = \frac{x^3}{3} + \frac{1}{x} + c$$

12. The correct answer is (C).

Let $u = 4x$; $du = 4dx$ or $\frac{du}{4} = dx$

$$\int q(4x)dx = \int q(u)\frac{du}{4}$$

$$= \frac{1}{4} p(u) + c = \frac{1}{4}p(4x) + c$$

Thus $\int_{-1}^0 q(4x)dx = \frac{1}{4}p(4x) \Big]_{-1}^0$

$$= \frac{1}{4}p(0) - \frac{1}{4}p(-4).$$

13. The correct answer is (C).

The total amount of water leaked from the tank for

$$1 \leq t \leq 3 = \int_1^3 f(t)dt$$

$$\approx 100 + 25 + 50 \approx 175 \text{ gallons.}$$

14. The correct answer is (A).

$$f'(x) = 10[\cos(\pi - x)][-\sin(\pi - x)](-1)$$

$$= 10\cos(\pi - x)\sin(\pi - x)$$

$$f'\left(\frac{\pi}{2}\right) = 10\cos\left(\pi - \frac{\pi}{2}\right)\sin\left(\pi - \frac{\pi}{2}\right)$$

$$= 10\cos\left(\frac{\pi}{2}\right)\sin\left(\frac{\pi}{2}\right) = 10(0)(1) = 0.$$

15. The correct answer is (C).

$$g'(x) = \frac{3x}{x^3 + 1}; \ g'(2) = \frac{3(2)}{2^3 + 1} = \frac{6}{9} = \frac{2}{3}.$$

16. The correct answer is (E).

$$\int_k^2 (2x - 2)dx = x^2 - 2x\Big]_k^2 = \left(2^2 - 2(2)\right) - \left(k^2 - 2k\right)$$

$$= 0 - \left(k^2 - 2k\right) = -k^2 + 2k$$

Set $-k^2 + 2k = -3 \Rightarrow 0 = k^2 - 2k - 3$

$$0 = (k - 3)(k + 1) \Rightarrow k = 3 \text{ or } k = -1$$

17. The correct answer is (A).

$$\int_0^a f(x)dx = -\int_{-a}^0 f(x)dx \Rightarrow f(x) \text{ is an}$$
odd function. The function whose graph is
shown in (A) is the only odd function.

18. The correct answer is (A).

See Figure 2TS-2.
If $b = 0$, then $r = 3$, but r cannot be 3.
If $b = 1, 3,$ or 5, f would have more than one
root. Thus, of all the choices, the only possible
value for b is -1.

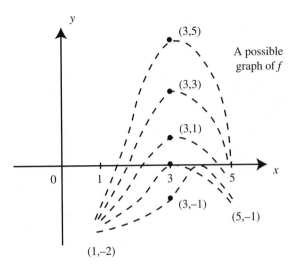

Figure 2TS-2

19. The Correct answer is (A).

$$V = \frac{1}{3}\pi r^2 (5) - \frac{1}{3}\pi r^2 (r)$$

$$= \frac{5}{3}\pi r^2 - \frac{1}{3}\pi r^3$$

$$\frac{dV}{dr} = \frac{10}{3}\pi r - \pi r^2$$

$$\frac{dV}{dr}\bigg|_{r=5} = \frac{10}{3}\pi (5) - \pi (25) = \frac{-25\pi}{3}.$$

20. The correct answer is (E).

$$x^3 - y^2 = 1; 3x^2 - 2y\frac{dy}{dx} = 0 \Rightarrow \frac{dy}{dx} = \frac{3x^2}{2y}$$

At $x = 1, 1^3 - y^2 = 1 \Rightarrow y = 0 \Rightarrow (1, 0)$

$$\frac{dy}{dx}\bigg|_{x=1} = \frac{3(1^2)}{2(0)} \text{ undefined.}$$

21. The correct answer is (C).

   I. f is differentiable on (a, b) since the graph is a smooth curve.

   II. There exists a horizontal tangent to the graph on (a, b); thus $f'(k) = 0$ for some k on (a, b); thus $f'(k) = 0$ for some k on (a, b).

   III. The graph is concave downward; thus $f'' < 0$.

22. The correct answer is (D).

$$v(t) = t^2 - 3t - 10; \text{ set } v(t) = 0 \Rightarrow (t - 5)(t + 2)$$

$$= 0 \Rightarrow t = 5 \text{ or } t = -2$$

$$a(t) = 2t - 3; \text{ set } a(t) = 0 \Rightarrow 2t - 3 = 0$$

$$\text{or } t = \frac{3}{2}.$$

See Figure 2TS-3.

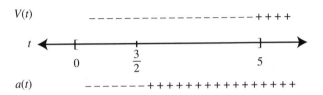

Figure 2TS-3

Since $v(t)$ and $a(t)$ are both negative on $(0, 3/2)$ and are both positive on $(5, \infty)$, the particle's speed is increasing on these intervals.

23. The correct answer is (C).

$$\int_{-2}^{2} f(x)\,dx = \int_{-2}^{0} f(x)\,dx + \int_{0}^{2} f(x)\,dx$$

$$= \frac{1}{2}(2)(-1) + \left(-\left(\frac{1}{2}\right)\pi(1)^2\right) = -1 - \frac{\pi}{2}.$$

24. The correct answer is (C).

$$\text{Average value} = \frac{1}{\frac{\pi}{2} - \left(-\frac{\pi}{2}\right)}\int_{-\pi/2}^{\pi/2} 3\cos(2x)\,dx$$

$$= \frac{1}{\pi}\left[\frac{3\sin(2x)}{2}\right]_{-\pi/2}^{\pi/2}$$

$$= \frac{3}{2\pi}[\sin\pi - (\sin[-\pi])] = 0.$$

25. The correct answer is (A).

$$f(x) = |x^3| = \begin{cases} x^3 & \text{if } x \geq 0 \\ -x^3 & \text{if } x < 0 \end{cases}$$

$$f'(x) = \begin{cases} 3x^2 & \text{if } x \geq 0 \\ -3x^2 & \text{if } x < 0 \end{cases}$$

$$\lim_{x \to -1} f'(x) = \lim_{x \to -1}\left(-3x^2\right) = -3.$$

26. The correct answer is (B).

$$V = \frac{4}{3}\pi r^3; \quad \frac{dV}{dt} = 4\pi r^2 \frac{dr}{dt}$$

Since $\frac{dV}{dt} = 4\frac{dr}{dt} \Rightarrow 4 = 4\pi r^2$ or

$$r^2 = \frac{1}{\pi} \text{ or } r = \frac{1}{\sqrt{\pi}}.$$

27. The correct answer is (E).

$$\frac{dy}{dx} = \frac{x^2}{y}; \quad y\,dy = x^2 dx$$

$$\int y\,dy = \int x^2 dx$$

$$\frac{y^2}{2} = \frac{x^3}{3} + c. \text{ Substituting } (0,4)$$

$$\frac{4^2}{2} = 0 + c \Rightarrow c = 8$$

Thus a solution is $\frac{y^2}{2} = \frac{x^3}{3} + 8$.

28. The correct answer is (D).

See Figure 2TS-4.

$$A = \left| \int_{-1}^{1} (y^2 - 1)dy \right| = \left| \left[\frac{y^3}{3} - y \right]_{-1}^{1} \right|$$

$$= \left| \left(\frac{1}{3} - 1\right) - \left(-\frac{1}{3} - (-1)\right) \right| = \frac{4}{3}.$$

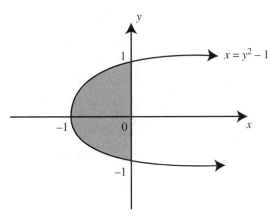

Figure 2TS-4

Part B—Calculators are permitted.

76. The correct answer is (E).

At $x = x_3$, $f' = 0$. Thus the tangent to the graph of f at $x = x_3$ is horizontal.

77. The correct answer is (E).

$$s(t) = 5 + 4t - f^2; \quad v(t) = s'(t) = 4 - 2t.$$

See Figure 2TS-5.

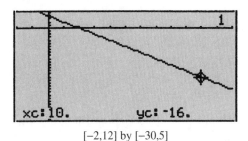

[−2,12] by [−30,5]

Figure 2TS-5

Since $v(t)$ is a straight line with a negative slope, the maximum speed for $0 \leq t \leq 10$ occurs at $t = 10$ where $v(t) = 4 - 2(10) = -16$. Thus maximum speed $= 16$.

78. The correct answer is (D).

See Figure 2TS-6.

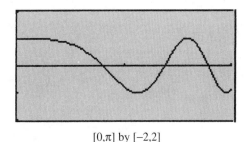

[0,π] by [−2,2]

Figure 2TS-6

Using the Inflection function of your calculator, you will find three points of inflection. They occur at $x = 1.335$, 2.195, and 2.8.

79. The correct answer is (D).

$$\int_{4}^{10} f(x)\,dx \approx 2\big(f(6) + f(8) + f(10)\big)$$

$$\approx 2(2.4 + 2.8 + 3.2) \approx 16.8.$$

80. The correct answer is (D).

$$f(-1) = 2 \Rightarrow \text{ a point } (-1, 2)$$

$f'(-1) = \dfrac{1}{2} \Rightarrow$ the slope at $x = -1$ is $\dfrac{1}{2}$.

Equation of tangent at $x = -1$ is

$y - 2 = \dfrac{1}{2}(x+1)$ or $y = \dfrac{1}{2}(x+1) + 2$.

Thus $f(-1.1) \approx \dfrac{1}{2}(-1.1+1) + 2 \approx 1.95$.

81. The correct answer is (C).

Area $= \displaystyle\int_1^b \dfrac{\ln x}{x}\,dx = 0.66$

Let $u = \ln x;\quad du = \dfrac{1}{x}\,dx$

$\displaystyle\int \dfrac{\ln x}{x}\,dx = \int u\,du = \dfrac{u^2}{2} + c = \dfrac{(\ln x)^2}{2} + c$

$\displaystyle\int_1^b \dfrac{\ln x}{x}\,dx = \dfrac{(\ln x)^2}{2}\bigg]_1^b = \dfrac{(\ln b)^2}{2} - \dfrac{(\ln 1)^2}{2}$

$= \dfrac{(\ln b)^2}{2}$

Let $\dfrac{(\ln b)^2}{2} = 0.66\,(\ln b)^2 = 1.32$

$\ln b = \sqrt{1.32}$

$e^{\ln b} = e^{\sqrt{1.32}}$

$b \approx 3.15$.

82. The correct answer is (B).

See Figure 2TS-7.

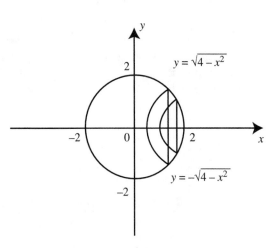

Figure 2TS-7

Area of a cross section $= \dfrac{1}{2}\pi\left(\sqrt{4-x^2}\right)^2$

$= \dfrac{1}{2}\pi(4-x^2)$

Volume of the solid $= \displaystyle\int_{-2}^2 \dfrac{1}{2}\pi(4-x^2)\,dx$

Using your calculator, you obtain $V = \dfrac{16\pi}{3}$.

83. The correct answer is (A).

Temperature of coffee $= 95 - \displaystyle\int_0^5 4\sin\left(\dfrac{t}{4}\right)dt$

$\approx 95 - 10.9548 \approx 84$.

84. The correct answer is (A).

See Figure 2TS-8.

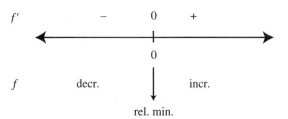

Figure 2TS-8

Only f has a relative minimum on (a, b).

85. The correct answer is (E).

See Figure 2TS-9.

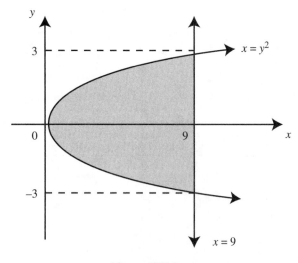

Figure 2TS-9

$$\text{Volume} = \pi \int_{-3}^{3} \left(9^2 - (y^2)^2\right) dy$$

$$= \frac{1944\pi}{5}.$$

86. The correct answer is (E).

$$y = e^x; \ \frac{dy}{dx} = e^x$$

$$y = x^2 + 5x; \ \frac{dy}{dx} = 2x + 5$$

If the graphs have parallel tangents at a point, then the slopes of the tangents are equal. Enter $y_1 = e^x$ and $y_2 = 2x + 5$. Using the Intersection function on your calculator, you obtain $x = -2.45$ and $x = 2.25$. (See Figure 2TS-10.)

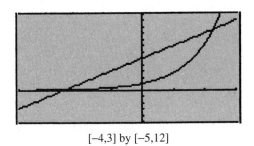

[−4,3] by [−5,12]

Figure 2TS-10

87. The correct answer is (E).

Since $\frac{dy}{dx} = ky \Rightarrow y = y_0 e^{kt}$

$$\frac{3}{4}y_0 = y_0 e^{k(6)} \Rightarrow \frac{3}{4} = e^{6k} \Rightarrow \ln\left(\frac{3}{4}\right) = \ln\left(e^{6k}\right)$$

$$\Rightarrow \ln\frac{3}{4} = 6k \text{ or } k = \frac{\ln\left(\frac{3}{4}\right)}{6} = -0.048.$$

88. The correct answer is (B).

$$h'(x) = (x - 5)^3$$

See Figure 2TS-11.

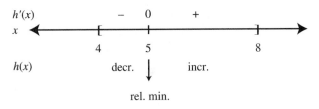

Figure 2TS-11

Thus h has a relative minimum at $x = 5$.

89. The correct answer is (E).

See Figure 2TS-12.

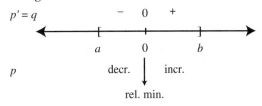

Figure 2TS-12

The graph in choice (E) is the one that satisfies the behavior of p.

90. The correct answer is (C).

See Figure 2TS-13.

$$\text{Volume} = \pi \int_{0}^{1} \left(y^{1/3}\right)^2 dy$$

Using your calculator, you obtain $V = 3\pi/5$.

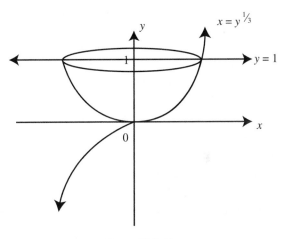

Figure 2TS-13

91. The correct answer is (E).

See Figure 2TS-14.

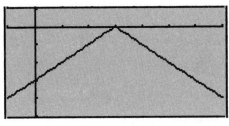

[−1,7] by [−5,1]

Figure 2TS-14

The function f is not differentiable at $x = 3$; has a relative maximum at $x = 3$ and has no point of inflection. Thus all three statements are not true.

92. The correct answer is (B).

$$y = \sin x; \frac{dy}{dx} = \cos x$$

Set $\dfrac{dy}{dx} = \dfrac{1}{2} \Rightarrow \cos x = \dfrac{1}{2}$ or $x = \dfrac{\pi}{3}$.

At $x = \dfrac{\pi}{3}, y = \sin\left(\dfrac{\pi}{3}\right) = \dfrac{\sqrt{3}}{2}; \left(\dfrac{\pi}{3}, \dfrac{\sqrt{3}}{2}\right)$

Equation of tangent line at $x = \dfrac{\pi}{3}$:

$$y - \frac{\sqrt{3}}{2} = \frac{1}{2}\left(x - \frac{\pi}{3}\right) \text{ or }$$

$$y = \frac{1}{2}\left(x - \frac{\pi}{3}\right) + \frac{\sqrt{3}}{2}.$$

Solutions—Section II

Part A—Calculators are permitted.

1. **(A)** The midpoints of 3 subintervals of equal length are:

$t = 4, 12,$ and 20.

The length of each interval is $\dfrac{24 - 0}{3} = 8.$

Thus $\displaystyle\int_0^{24} v(t)dt \approx 8\left[v(4) + v(12) + v(20)\right]$

$\approx 8[25 + 15 + 20]$

$= 8(60) = 480.$

(B) Average velocity $= \dfrac{1}{24}\displaystyle\int_0^{24} v(t)dt$

$\approx \dfrac{1}{24}(480) = 20$ ft/s.

(C) Average acceleration $= \dfrac{v(24) - v(0)}{24 - 0} = \dfrac{30}{24}$

$= 1.25$ ft/s^2

(D) $a(t) = 0$ at $t = 6$ and $t = 14$, since the slopes of tangents at $t = 6$ and $t = 14$ are 0.

(E) $a(20) \approx \dfrac{v(22) - v(18)}{22 - 18} \approx \dfrac{25 - 15}{4} \approx \dfrac{10}{4}$

≈ 2.5 ft/s^2.

2. See Figure 2TS-15.

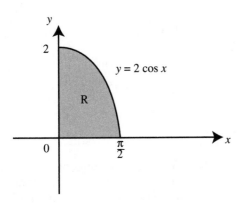

Figure 2TS-15

(A) Area of $R = \displaystyle\int_0^{\pi/2} 2\cos x\,dx = 2\sin x]_0^{\pi/2}$

$= 2\sin\left(\dfrac{\pi}{2}\right) - 2\sin(0) = 2.$

(B) $\displaystyle\int_0^{a} \cos x\,dx = 1$

$2\sin x]_0^a = 2\sin a - 2\sin(0) = 2\sin a$

$2\sin a = 1 \Rightarrow \sin a = \dfrac{1}{2}$

$a = \sin^{-1}\left(\dfrac{1}{2}\right) = \dfrac{\pi}{6}.$

(C) Volume $= \pi \displaystyle\int_0^{\pi/2} (2\cos x)^2 dx$

$= \pi \displaystyle\int_0^{\pi/2} 4\cos^2 x\,dx$

$= 4\pi \displaystyle\int_0^{\pi/2} \cos^2 dx$

$= 4\pi \displaystyle\int_0^{\pi/2} \dfrac{1 + \cos(2x)}{2}dx$

$= 2\pi \displaystyle\int_0^{\pi/2} \left[1 + \cos(2x)\right]dx$

$= 2\pi\left[x + \dfrac{\sin(2x)}{2}\right]_0^{\pi/2}$

$= 2\pi\left[\left(\dfrac{\pi}{2} + \dfrac{\sin \pi}{2}\right) - 0\right] = \pi^2.$

(D) Area of cross section $= \dfrac{1}{2}\pi\left(\dfrac{2\cos x}{2}\right)^2$

$= \dfrac{1}{2}\pi \cos^2 x$

$V = \displaystyle\int_0^{\pi/2} \dfrac{1}{2}\pi \cos^2 x\,dx$

$= \dfrac{1}{2}\pi \displaystyle\int_0^{\pi/2} \cos^2 x\,dx$

$= \dfrac{1}{2}\pi \displaystyle\int_0^{\pi/2} \dfrac{1 + \cos(2x)}{2}dx$

$= \dfrac{\pi}{4}\displaystyle\int_0^{\pi/2} (1 + \cos(2x)dx)$

$$= \frac{\pi}{4}\left[x + \frac{\sin(2x)}{2}\right]_0^{\pi/2}$$

$$= \frac{\pi}{4}\left[\left(\frac{\pi}{2} + \frac{\sin \pi}{2}\right) - 0\right] = \frac{\pi^2}{8}.$$

3. (A) See Figure 2TS-16.

$$g(10) = 90 - 4\tan\left(\frac{10}{20}\right) = 90 - 4\tan\left(\frac{1}{2}\right)$$

$$\approx 90 - 4(0.5463) \approx 90 - 2.1852$$

$$\approx 87.815 \approx 87.82° \text{ F.}$$

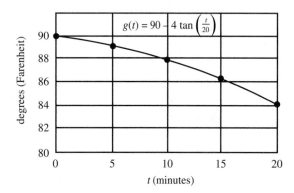

Figure 2TS-16

(B) Average Temperature

$$= \frac{1}{20}\int_0^{20}\left[90 - 4\tan\left(\frac{t}{20}\right)\right] dt$$

Using your calculator, you obtain:

$$= \frac{1}{20}(1750.75) = 87.5375$$

$$\approx 87.54°\text{F.}$$

(C) Set the temperature of the liquid equal to 86°F. Using your calculator, let

$$y_1 = 90 - 4\tan\left(\frac{x}{20}\right); \text{ and } y_2 = 86.$$

To find the intersection point of y_1 and y_2, let $y_3 = y_1 - y_2$ and find the zeros of y_5.
Using the zero function of your calculator, you obtain $x = 15.708$.

Since $y_1 < y_2$ on the interval $15.708 < x \leq 20$, the temperature of the liquid is below 86°F when $15.708 < t \leq 20$. (Note that the intersection function of a calculator capabilities allowed without supporting explanation.)

(D) Average temperature below 86°

$$= \frac{1}{20 - 15.708}\int_{15.708}^{20}\left(90 - 4\tan\left(\frac{x}{20}\right)\right) dx.$$

Using your calculator, you obtain:

$$\text{Average temperature} = \frac{1}{4.292}(364.756)$$

$$\approx 84.9851$$

$$\approx 84.99° \text{ F.}$$

Part B—No calculators.

4. (A) $f'(x) = 3(e^{-2x^2})(-4x) = -12xe^{-2x^2}$
Setting $f'(x) = 0$, $-12xe^{-2x^2} = 0 \Rightarrow x = 0$.

(B) $f''(x) = (-12)\left(e^{-2x^2}\right)$

$$+ (-12x)\left(e^{-2x^2}\right)(-4x)$$

$$= -12e^{-2x^2} + 48x^2e^{-2x^2}$$

Setting $f''(x) = 0$, $12e^{-2x^2} + 48x^2e^{-2x^2} = 0$

$$\Rightarrow 48x^2e^{-2x^2} = 12e^{-2x^2}$$

$$\Rightarrow 48x^2 = 12 \Rightarrow x^2 = \frac{1}{4} \text{ or } x = \pm\frac{1}{2}.$$

(C) $\lim\limits_{x\to\infty} 3e^{-2x^2} = \lim\limits_{x\to\infty}\dfrac{3}{e^{2x^2}} = 0$

$$\lim\limits_{x\to-\infty} 3e^{-2x^2} = \lim\limits_{x\to-\infty}\dfrac{3}{e^{2x^2}} = 0.$$

(D) See Figure 2TS-17.

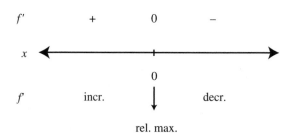

Figure 2TS-17

$$f'(x) = -12xe^{-2x^2} = \frac{-12x}{e^{2x^2}}$$

$f(0) = 3$, since f has only one critical point (at $x = 0$), thus at $x = 0$, f has an absolute maximum. The absolute maximum value is 3.

(E) $f(x) = ae^{-bx^2}$, $a > 0$, $b > 0$

$f'(x) = ae^{-bx^2}(-2bx) = -2abxe^{-bx^2}$

Setting $f'(x) = 0$, $-2abxe^{-bx^2} = 0 \Rightarrow x = 0$

$f'(x) = \dfrac{-2abx}{e^{bx^2}}$

$f'(x) > 0$ if $x < 0$ and $f'(x)$ if $x > 0$. Thus f has a relative maximum at $x = 0$ and since it is the only critical point, f has an absolute maximum at $x = 0$. Since $f(0) = a$, the absolute maximum for f is a.

5. (A) $f(-3) = \displaystyle\int_0^{-3} g(t)dt = -\int_{-3}^0 g(t)dt$

$= -\displaystyle\int_{-3}^{-1} g(t)dt - \int_{-1}^0 g(t)dt$

$= -\left(-\dfrac{1}{2}(2)(2)\right) - \left(\dfrac{1}{2}(1)(2)\right)$

$= 2 - 1 = 1.$

$f(3) = \displaystyle\int_0^3 g(t)dt$

$= \displaystyle\int_0^1 g(t)dt + \int_1^3 g(t)dt$

$= \dfrac{1}{2}(1)(2) + \left(-\dfrac{1}{2}(1)(2)\right)$

$= 1 - 1 = 0$

(B) The function f increases on $(0, 1)$ and decreases on $(1, 3)$. Thus f has a relative maximum at $x = 1$. And f decreases on $(-3, -1)$ and increases on $(-1, 0)$. Thus f has a relative minimum at $x = -1$.

(C) $f'(x) = g(x)$ and $f''(x) = g'(x)$.
See Figure 2TS-18.

Figure 2TS-18

The function f has a change of concavity at $x = 0$ and $x = 2$. Thus f has a point of inflection at $x = 0$ and $x = 2$.

(D) $f(1) = \displaystyle\int_0^1 g(t)dt = \dfrac{1}{2}(1)(2) = 1$

$f'(1) = g(1) = 0$

Thus, $m = 0$, point $(1, 1)$; the equation of the tangent line to $f(x)$ at $x = 1$ is $y = 1$.

6. (A) $\dfrac{dy}{dx} = \dfrac{y}{2x^2}$; $(2,1)$

$\dfrac{dy}{dx}\bigg|_{x=2,y=1} = \dfrac{1}{2(2)^2} = \dfrac{1}{8}$

Equation of tangent:

$y - 1 = \dfrac{1}{8}(x - 2)$ or $y = \dfrac{1}{8}(x - 2) + 1$.

(B) $f(2.5) \approx \dfrac{1}{8}(2.5 - 2) + 1 = 1.0625 \approx 1.063$

(C) $\dfrac{dy}{dx} = \dfrac{y}{2x^2}$

$\dfrac{dy}{y} = \dfrac{dx}{2x^2}$ and $\displaystyle\int \dfrac{dy}{y} = \int \dfrac{dx}{2x^2}$

$\ln|y| = \displaystyle\int \dfrac{1}{2}x^{-2}dx = \dfrac{1}{2}\dfrac{(x^{-1})}{-1} + c = -\dfrac{1}{2x} + c$

$e^{\ln|y|} = e^{\left(-\frac{1}{2x} + c\right)}$

$y = e^{-\frac{1}{2x}+c}$; $f(2) = 1$

$1 = e^{-\frac{1}{2(2)}+c} \Rightarrow 1 = e^{-\frac{1}{4}+c}$

Since $e^0 = 1$, $-\dfrac{1}{4} + c = 0 \Rightarrow c = \dfrac{1}{4}$.

Thus, $y = e^{-\frac{1}{2x} + \frac{1}{4}}$.

(D) $f(2.5) = e^{\left(-\frac{1}{2(2.5)} + \frac{1}{4}\right)} = e^{\left(-\frac{1}{5} + \frac{1}{4}\right)} = e^{\frac{1}{20}}$
$(or \approx 1.05127 \approx 1.051)$.

Scoring Sheet for AB Practice Exam 2

Section I—Part A

$$\underline{\hspace{3cm}} \times 1.2 \quad = \quad \underline{\hspace{3cm}}$$

No. Correct   Subtotal A

$$\underline{\hspace{3cm}} \times (0.25) \times 1.2 \quad = \quad \underline{\hspace{3cm}}$$

No. Incorrect   Subtotal B

Part A (Subtotal A − Subtotal B) = $\underline{\hspace{3cm}}$

                                                 Subtotal C

Section I—Part B

$$\underline{\hspace{3cm}} \times 1.2 \quad = \quad \underline{\hspace{3cm}}$$

No. Correct   Subtotal D

$$\underline{\hspace{3cm}} \times (0.25) \times 1.2 \quad = \quad \underline{\hspace{3cm}}$$

No. Incorrect   Subtotal E

Part B (Subtotal D − Subtotal E) = $\underline{\hspace{3cm}}$

                                                 Subtotal F

Section II—Part A (Each question is worth 9 points.)

$$\underline{\hspace{2.5cm}} + \underline{\hspace{2.5cm}} + \underline{\hspace{2.5cm}} = \underline{\hspace{2.5cm}}$$

Q#1   Q#2   Q#3   Subtotal G

Section II—Part B (Each question is worth 9 points.)

$$\underline{\hspace{2.5cm}} + \underline{\hspace{2.5cm}} + \underline{\hspace{2.5cm}} = \underline{\hspace{2.5cm}}$$

Q#1   Q#2   Q#3   Subtotal H

Total Raw Score (Subtotals C + F + G + H) = $\boxed{}$

Approximate Conversion Scale:	
Total Raw Score	Approximate AP Grade
75–108	5
60–74	4
45–59	3
31–44	2
0–30	1

BC PRACTICE EXAM 1

Answer Sheet for BC Practice Exam 1—Section I

Part A

1. _____

2. _____

3. _____

4. _____

5. _____

6. _____

7. _____

8. _____

9. _____

10. _____

11. _____

12. _____

13. _____

14. _____

15. _____

16. _____

17. _____

18. _____

19. _____

20. _____

21. _____

22. _____

23. _____

24. _____

25. _____

26. _____

27. _____

28. _____

Part B

76. _____

77. _____

78. _____

79. _____

80. _____

81. _____

82. _____

83. _____

84. _____

85. _____

86. _____

87. _____

88. _____

89. _____

90. _____

91. _____

92. _____

Section I—Part A

Number of Questions	Time	Use of Calculator
28	55 Minutes	No

Directions:

Use the answer sheet provided on the previous page. All questions are given equal weight. There is no penalty for unanswered questions. However, 1/4 of the number of the incorrect answers will be subtracted from the number of correct answers. Unless otherwise indicated, the domain of a function f is the set of all real numbers. The use of a calculator is *not* permitted in this part of the exam

1. $\int_{-2}^{2} 3e^{-x}\,dx =$

   (A) $-3e^{-2}$
   (B) $-3e^{2}$
   (C) $6(1 - e^{-2})$
   (D) $3(e^{2} - e^{-2})$
   (E) $3(e^{-2} - e^{2})$

2. If $f(x) = x^3 + 3x^2 + cx + 4$ has a horizontal tangent and a point of inflection at the same value of x, what is the value of c?

   (A) 0
   (B) 1
   (C) −1
   (D) −3
   (E) 3

3. Find $\dfrac{dy}{dx}$ if $\tan y = (x - y)^2$

   (A) $\dfrac{dy}{dx} = \dfrac{2(x - y)}{\sec^2 y + 2(x - y)}$
   (B) $\dfrac{dy}{dx} = \dfrac{2(x - y)}{\sec^2 y}$
   (C) $\dfrac{dy}{dx} = \dfrac{\sec^2 y - 2(x - y)}{-2(x - y)}$
   (D) $\dfrac{dy}{dx} = \dfrac{1}{1 + \sec^2 y}$
   (E) $\dfrac{dy}{dx} = 1 + \sec^2 y$

4. Find $\dfrac{dy}{dx}$ if $y = 3^{(4 - x^2)}$

   (A) $\dfrac{dy}{dx} = (\ln 3)3^{(4 - x^2)}$
   (B) $\dfrac{dy}{dx} = -2x(\ln 3)3^{(4 - x^2)}$
   (C) $\dfrac{dy}{dx} = -2x(4 - x^2)(\ln 3)$
   (D) $\dfrac{dy}{dx} = (-2x)3^{(4 - x^2)}$
   (E) $\dfrac{dy}{dx} = (4 - x^2)3^{(3 - x^2)}$

5. If $x = \cos t$ and $y = \sin^2 t$, then $\dfrac{d^2y}{dx^2}$ at $t = \dfrac{\pi}{4}$

   (A) $\sqrt{2}$
   (B) $-\sqrt{2}$
   (C) 1
   (D) 0
   (E) −1

6. If $g(x)$ is continuous for all real values of x, then $\displaystyle\int_{a/3}^{b/3} g(3x)\,dx =$

   (A) $\dfrac{1}{3}\displaystyle\int_{a}^{b} g(x)\,dx$
   (B) $3\displaystyle\int_{a}^{b} g(x)\,dx$
   (C) $\dfrac{1}{3}\displaystyle\int_{3a}^{3b} g(x)\,dx$
   (D) $\displaystyle\int_{a}^{b} g(x)\,dx$
   (E) $3\displaystyle\int_{3a}^{3b} g(x)\,dx$

7. The area enclosed by the parabola $y = x - x^2$, the line $x = 1$, and the x-axis is revolved about the x-axis. The volume of the resulting solid is

   (A) $\dfrac{1}{30}$
   (B) $\dfrac{1}{6}$

(C) $\dfrac{\pi}{30}$

(D) $\dfrac{\pi}{15}$

(E) $\dfrac{\pi}{6}$

8. $\displaystyle\sum_{n=1}^{\infty} \dfrac{1}{(2n-1)(2n+1)} =$

(A) $\dfrac{1}{2}$

(B) 1

(C) 0

(D) 4

(E) $\dfrac{1}{4}$

9. The function $f(x)$ is defined on the interval $(-2, 2)$ such that for all x, $-2 < x < 2$, $f'(x) > 0$ and $f''(x) > 0$. Which of the following could be the graph of $f(x)$ on $(-2, 2)$?

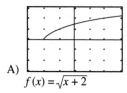

A) $f(x) = \sqrt{x+2}$

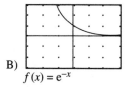

B) $f(x) = e^{-x}$

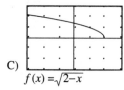

C) $f(x) = \sqrt{2-x}$

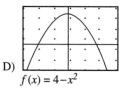

D) $f(x) = 4 - x^2$

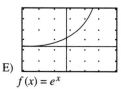

E) $f(x) = e^x$

10. Which of the following series are convergent?

I. $12 - 8 + \dfrac{16}{3} - \dfrac{32}{9} + \cdots$

II. $5 + \dfrac{5\sqrt{2}}{2} + \dfrac{5\sqrt{3}}{3} + \dfrac{5}{2} + \sqrt{5} + \dfrac{5\sqrt{6}}{6} + \cdots$

III. $8 + 20 + 50 + 125 + \cdots$

(A) I only
(B) II only
(C) III only
(D) I and II
(E) II and III

11. Find the values of a and b that assure that
$$f(x) = \begin{cases} \ln(3 - x) & \text{if } x < 2 \\ a - bx & \text{if } x \geq 2 \end{cases}$$ is differentiable at $x = 2$.

(A) $a = 3, b = 1$
(B) $a = 1, b = 2$
(C) $a = 2, b = 1$
(D) $a = -2, b = -1$
(E) $a = 1, b = 3$

12. A particle moves in the xy-plane so that its velocity vector at time t is $v(t) = \langle 2 - 3t^2, \pi \sin(\pi t) \rangle$ and the particle's position vector at time $t = 2$ is $\langle 4, 3 \rangle$. What is the position vector of the particle when $t = 3$?

(A) $\langle -25, 0 \rangle$
(B) $\langle -21, 1 \rangle$
(C) $\langle -10, 0 \rangle$
(D) $\langle -13, 5 \rangle$
(E) $\langle 4, 3 \rangle$

13. The $\displaystyle\lim_{h \to 0} \dfrac{\ln(x - 3 + h) - \ln(x - 3)}{h}$ is

(A) $\ln(x + 3)$
(B) $\ln(x - 3)$
(C) $\dfrac{1}{\ln(x - 3)}$
(D) $\dfrac{1}{x + 3}$
(E) $\dfrac{1}{x - 3}$

14. The area of the region enclosed by the polar curve $r = 3 - \sin\theta$ is

(A) 19π
(B) $\dfrac{19\pi}{2}$
(C) $\dfrac{19\pi}{4}$
(D) $\dfrac{19\pi - 12}{4}$
(E) 6π

15. The slope of the normal line to the graph of $y = 5 + 5\sin\theta$ at $\theta = \dfrac{\pi}{3}$ is

(A) 1
(B) -1
(C) 5
(D) $\dfrac{5\sqrt{3}}{2}$
(E) $\dfrac{-1}{5}$

16. Use the trapezoidal method with 4 divisions to approximate the area of the region bounded by the graph of $y = \dfrac{1}{2x}$, the lines $x = 1$ and $x = 3$, and the x-axis.

(A) $\dfrac{67}{60}$
(B) $\dfrac{67}{120}$
(C) $\dfrac{91}{240}$
(D) $\dfrac{91}{120}$
(E) $\dfrac{67}{30}$

17. The shortest distance from the origin to the graph of $y = \dfrac{-4}{x}$ is

(A) 2
(B) -2
(C) $2\sqrt{2}$
(D) $-2\sqrt{2}$
(E) 4

18. $\displaystyle\int_2^4 \dfrac{3x}{3x^2 - 4}\,dx =$

(A) $\dfrac{1}{2}\left[\ln 36\right]$
(B) $\left[\ln 18\right]$
(C) $\left[\ln 22 - \ln 4\right]$
(D) $\dfrac{1}{2}\left[\ln 44 - \ln 8\right]$
(E) $\dfrac{1}{2}\left[\ln 4 - \ln 2\right]$

19. The graph of $y = e^{\sin x}$ has a relative minimum at

(A) $x = \dfrac{\pi}{2}$
(B) $x = \pi$

(C) $x = \dfrac{2\pi}{3}$
(D) $x = \dfrac{3\pi}{2}$
(E) $x = 2\pi$

20. Which of the following is an equation of the line tangent to the curve with parametric equations $x = 3t^2 - 2$, $y = 2t^3 + 2$ at the point when $t = 1$?

(A) $y = 3x^2 + 7x$
(B) $y = 9$
(C) $y = 6x - 2$
(D) $y = x$
(E) $y = x + 3$

21. The series expansion for $\displaystyle\int_0^x \cos\sqrt{t}\,dt$ is

(A) $x - \dfrac{x^2}{2\cdot 2!} + \dfrac{x^3}{3\cdot 4!} - \dfrac{x^4}{4\cdot 6!} + \cdots$
$\quad + \dfrac{(-1)^{2n}x^{n+1}}{(n+1)(2n)!} + \cdots$

(B) $x - \dfrac{x^2}{2!} + \dfrac{x^3}{4!} - \dfrac{x^4}{6!} + \cdots + \dfrac{(-1)^{2n}x^{n+1}}{(2n)!} + \cdots$

(C) $x + \dfrac{x^2}{2\cdot 2!} + \dfrac{x^3}{3\cdot 4!} + \dfrac{x^3}{4\cdot 6!} \cdots$
$\quad + \dfrac{x^{n+1}}{(n+1)(2n)!} + \cdots$

(D) $1 - \dfrac{x}{2\cdot 2!} + \dfrac{x^2}{3\cdot 4!} - \dfrac{x^3}{4\cdot 6!} \cdots$
$\quad + \dfrac{(-1)^{2n}x^n}{(n+1)(2n)!} + \cdots$

(E) $1 - \dfrac{x^2}{2\cdot 2!} + \dfrac{x^4}{3\cdot 4!} - \dfrac{x^6}{4\cdot 6!} \cdots$
$\quad + \dfrac{(-1)^{2n}x^{2n}}{(n+1)(2n)!} + \cdots$

22. $\displaystyle\int \dfrac{dx}{2x^2 + 9x - 5} =$

(A) $\ln\left|\dfrac{2x - 1}{x + 5}\right|$
(B) $\ln\left|2x^2 + 9x - 5\right|$
(C) $\dfrac{1}{11}\ln\left|\dfrac{2x - 1}{x + 5}\right|$
(D) $\dfrac{1}{11}\ln\left|2x^2 + 9x - 5\right|$
(E) $\dfrac{1}{11}\ln\dfrac{(2x - 1)^2}{|x + 5|}$

23. A solid has a circular base of radius 4. If every plane cross section perpendicular to the x-axis is a square, then the volume of the solid is

    (A) 16π

    (B) 32π

    (C) $\dfrac{64}{3}$

    (D) $\dfrac{256}{3}$

    (E) $\dfrac{1024}{3}$

24. If a particle moves in the xy-plane on a path defined by $x = \sin^2 t$ and $y = \cos(2t)$ for $0 \le t \le \dfrac{\pi}{2}$, then the length of the arc the particle traces out is

    (A) $\sqrt{2}$
    (B) 2
    (C) $\sqrt{5}$
    (D) 5
    (E) $\sqrt{10}$

25. Find the interval of convergence for $\sum \dfrac{(-1)^n x^n}{e^n}$

    (A) $(0, 1)$
    (B) $(-1, 1)$
    (C) $[-e, e)$
    (D) $(-e, e)$
    (E) $(-e, e]$

26. If n is a positive integer, then
    $$\lim_{n \to \infty} \frac{1}{n}\left[\left(\frac{1}{n}\right)^2 + \left(\frac{2}{n}\right)^2 + \cdots \left(\frac{n-1}{n}\right)^2\right]$$

    (A) $\displaystyle\int_0^1 \frac{1}{x^2}\,dx$

    (B) $\displaystyle\int_0^1 x^2\,dx$

    (C) $\displaystyle\int_0^1 \frac{2}{x^2}\,dx$

    (D) $\displaystyle\int_0^1 \frac{1}{x}\,dx$

    (E) $\displaystyle\int_0^2 x^2\,dx$

27. $\displaystyle\int_{-3}^{-2} \frac{5x}{(x+2)(x-3)}\,dx =$

    (A) $\displaystyle\lim_{n \to 0}\int_{-3}^{n} \frac{5x}{(x+2)(x-3)}\,dx$

    (B) $\displaystyle\lim_{n \to -3^+}\int_{n}^{-2} \frac{5x}{(x+2)(x-3)}\,dx$

    (C) $\displaystyle\lim_{n \to -2^-}\int_{-3}^{n} \frac{5x}{(x+2)(x-3)}\,dx$

    (D) $\displaystyle\lim_{n \to -3}\int_{-3}^{n} \frac{5x}{(x+2)(x-3)}\,dx$

    (E) $\displaystyle\lim_{n \to 0}\int_{n}^{-2} \frac{5x}{(x+2)(x-3)}\,dx$

28. $\displaystyle\int_1^2 e^{4-3\ln x}\,dx =$

    (A) $\dfrac{-1}{2}$

    (B) $\dfrac{e^4}{4}$

    (C) $\dfrac{1}{4}$

    (D) $\dfrac{3}{8}$

    (E) $\dfrac{3e^4}{8}$

Section I—Part B

Number of Questions	Time	Use of Calculator
17	50 Minutes	Yes

Directions:

Use the same answer sheet from Part A. *Please note that the questions begin with number 76.* This is not an error. It is done to be consistent with the numbering system of the actual AP Calculus BC Exam. All questions are given equal weight. There is no penalty for unanswered questions. However, 1/4 of the number of incorrect answers will be subtracted from the number of correct answers. Unless otherwise indicated, the domain of a function f is the set of all real numbers. If the exact numerical value does not appear among the given choices, select the best approximate value. The use of a calculator is *permitted* in this part of the exam.

76. If $f(x) = \sqrt[3]{x^3 - x}$ then $f'(2)$ is approximately

    (A) 1.110
    (B) 2.245
    (C) 0.101
    (D) 12.107
    (E) 18.161

77. $\displaystyle\int_1^{e^\pi} \frac{\cos(\ln x)}{x}\,dx =$

    (A) −0.913
    (B) −0.043
    (C) −1.754
    (D) 0
    (E) 72.699

78. A rumor spreads through a community of 200 people at a rate modeled by
    $\dfrac{dy}{dt} = 0.2y\left(1 - \dfrac{y}{200}\right)$. If the rumor began with two people, find the number of people who have heard the rumor after thirty days.

    (A) 5
    (B) 32
    (C) 161
    (D) 199
    (E) 200

79. Which best approximates
    $\displaystyle\lim_{h\to 0} \frac{\cos 2(2+h) - \cos 4}{h}$?

    (A) −0.757
    (B) 0.757
    (C) −0.654
    (D) 0.654
    (E) 1.514

80. The area under the curve $y = 3x^2 - kx + 1$ bounded by the lines $x = 1$ and $x = 2$ is approximately −5.5. Find the value of k.

    (A) 9
    (B) 11
    (C) 5.5
    (D) 16.5
    (E) 1

81. The rate of growth of a population is proportional to the population and increases by 23% at the end of the first 12 years. What is the constant of proportionality, correct to three decimal places?

    (A) 0.230
    (B) 0.023
    (C) 0.017
    (D) 0.019
    (E) 2.760

82. $\displaystyle\int_{\frac{1}{2}}^1 \csc 3x\,dx$

    (A) −0.285
    (B) 0.04704
    (C) 0.906
    (D) 1.193
    (E) ∞

83. Let $F(x) = \displaystyle\int_0^x \sqrt{\sin t}\,dt$. Which of the following is the best approximation for $F'(0.2)$?

    (A) 0.040
    (B) 0.060
    (C) 0.137
    (D) 0.446
    (E) 2.199

84. Use a series to estimate $\int_0^1 e^{-x}dx$ accurate to three decimal places.

   (A) 0.632
   (B) 1.718
   (C) 0.001
   (D) −1.000
   (E) −0.368

85. If $y = \sec^2(3x)$, then $\dfrac{dy}{dx}$ at $\dfrac{\pi}{9}$ is approximately

   (A) $8\sqrt{3}$
   (B) $8\sqrt{3}/3$
   (C) $12\sqrt{3}$
   (D) $24\sqrt{3}$
   (E) $24\sqrt{3}/3$

86. Find the approximate value of y when $x = 3.1$ if $2x\dfrac{dy}{dx} - 7 = 1$ and $y = 5$ when $x = 3$?

   (A) 1.290
   (B) −9.104
   (C) 4.632
   (D) −2.666
   (E) 4.525

87. The slope of the normal line to $y = e^{-2x}$ when $x = 1.158$ is approximately

   (A) 5.068
   (B) 0.864
   (C) −0.197
   (D) 0.099
   (E) 10.135

88. The volume of the solid generated by revolving about the x-axis the region bounded by $y = \sin x + \cos x$ and the x-axis between $x = 0$ and $x = \dfrac{\pi}{2}$ is approximately

   (A) 1
   (B) 1.071
   (C) 2.071

   (D) 8.076
   (E) 16.153

89. The absolute minimum of $f(x) = \ln(3x) + \cos x$ on the interval $\left[\dfrac{\pi}{2}, \pi\right]$ is approximately

   (A) 1.186 when $x = 2.773$
   (B) 2.773 when $x = 1.186$
   (C) 1.550 when $x = \dfrac{\pi}{2}$
   (D) 1.243 when $x = \pi$
   (E) −1.186 when $x = 2.773$

90. A particle moves along the y-axis so that its position at time t is $y(t) = 5t^3 - 9t^2 + 2t - 1$. At the moment when the particle first changes direction, its position is

   (A) (0, 0.124)
   (B) (0.124, −0.881)
   (C) (0.124, 0)
   (D) (0, −0.881)
   (E) (−0.881, 0)

91. The area of the region enclosed by the graphs of $y = \cos x + 1$ and $y = 2 + 2x - x^2$ is approximately

   (A) 3.002
   (B) 2.424
   (C) 2.705
   (D) 0.094
   (E) 0.009

92. The interval of convergence of the series $\sum_{n=1}^{\infty} \dfrac{(x-3)^n}{n^2}$ is

   (A) (−4, 4)
   (B) (2, 4)
   (C) (2, 4]
   (D) [2, 4)
   (E) [2, 4]

Section II—Part A

Directions:

Show all work. You may not receive any credit for correct answers without supporting work. You may use an approved calculator to help solve a problem. However, you must clearly indicate the setup of your solution using mathematical notations and *not* calculator syntax. Calculators may be used to find the derivative of a function at a point, compute the numerical value of a definite integral, or solve an equation. Unless otherwise indicated, you may assume the following: (a) the numeric or algebraic answers need not be simplified, (b) your answer, if expressed in approximation, should be correct to 3 places after the decimal point, and (c) the domain of a function f is the set of all real numbers.

1. The slope of a function f at any point (x, y) is $\dfrac{4x + 1}{2y}$. The point $(2, 4)$ is on the graph of f

   (A) Write an equation of the line tangent to the graph of f at $x = 2$.
   (B) Use the tangent line in part (A) to approximate $f(2.1)$.
   (C) Solve the differential equation $\dfrac{dy}{dx} = \dfrac{4x + 1}{2y}$ with the initial condition $f(2) = 4$.
   (D) Use the solution in part (C) and find $f(2.1)$.

2. Let f be a function that has derivatives of all orders for all real numbers. Assume $f(0) = 1$, $f'(0) = 6$, $f''(0) = -4$, and $f'''(0) = 30$.

   (A) Write the third-degree Taylor polynomial for f about $x = 0$ and use it to approximate $f(0.1)$.
   (B) Write the sixth-degree Taylor polynomial for g, where $g(x) = f(x^2)$, about $x = 0$.
   (C) Write the seventh-degree Taylor polynomial for h, where $h(x) = \displaystyle\int_0^x g(t)\,dt$, about $x = 0$.

3. Consider the differential equation given by $\dfrac{dy}{dx} = \dfrac{2xy}{3}$.

   (A) On the axes provided, sketch a slope field for the given differential equation at the points indicated.

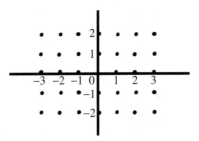

   (B) Let $y = f(x)$ be the particular solution to the given differential equation with the initial condition $f(0) = 2$. Use Euler's method, starting at $x = 0$, with a step size of 0.1, to approximate $f(0.3)$. Show the work that leads to your answer.
   (C) Find the particular solution $y = f(x)$ to the given differential equation with the initial condition $f(0) = 2$. Use your solution to find $f(0.3)$.

Section II—Part B

Directions:

The use of a calculator is *not* permitted in this part of the exam. When you have finished this part of the test, you may return to the problems in Part A of Section II and continue to work on them. However, you may not use a calculator. You should *show all work*. You may *not* receive any credit for correct answers without supporting work. Unless otherwise indicated, the numeric or algebraic answers need not be simplified, and the domain of a function f is the set of all real numbers.

4. Given the parametric equations $x = 2(\theta - \sin\theta)$ and $y = 2(1 - \cos\theta)$.

   (A) Find $\dfrac{dy}{dx}$ in terms of θ.
   (B) Find an equation of the line tangent to the graph at $\theta = \pi$.
   (C) Find an equation of the line tangent to the graph at $\theta = 2\pi$.
   (D) Set up but do not evaluate an integral representing the length of the curve over the interval $0 \le \theta \le 2\pi$.

5. Let R be the region enclosed by the graph of $y = x^3$, the x-axis and the line $x = 2$.

   (A) Find the area of region R.
   (B) Find the volume of the solid obtained by revolving region R about the x-axis.

   (C) The line $x = a$ divides region R into two regions such that when the regions are revolved about x-axis, the resulting solids have equal volume. Find a.
   (D) If region R is the base of a solid whose cross sections perpendicular to the x-axis are squares, find the volume of the solid.

6. Given the function $f(x) = xe^{2x}$.

   (A) At what value(s) of x, if any, is $f'(x) = 0$?
   (B) At what value(s) of x, if any, is $f''(x) = 0$?
   (C) Find $\lim\limits_{x \to \infty} f(x)$ and $\lim\limits_{x \to -\infty} f(x)$.
   (D) Find the absolute extrema of f and justify your answer.
   (E) Show that if $f(x) = xe^{ax}$ where $a > 0$, the absolute minimum value of f is $\dfrac{-1}{ae}$.

Answers to BC Practice Exam 1—Section I

Part A	13. E	26. B	85. D
1. D	14. B	27. C	86. C
2. E	15. A	28. E	87. A
3. A	16. B	**Part B**	88. D
4. B	17. C	76. A	89. B
5. A	18. D	77. D	90. D
6. A	19. D	78. C	91. A
7. C	20. E	79. E	92. E
8. A	21. A	80. A	
9. E	22. C	81. C	
10. A	23. E	82. C	
11. C	24. C	83. D	
12. D	25. D	84. A	

Answers to BC Practice Exam 1—Section II

Part A

1. (A) $y = \dfrac{9}{8}(x-2)+4$

   (B) 4.113
   (C) $y = \sqrt{2x^2+x+6}$
   (D) 4.113

2. (A) $f(x) = 1+6x-2x^2+5x^3$; $f(0.1) \approx 1.585$
   (B) $g(x) = 1+6x^2-2x^4+5x^6$
   (C) $h(x) = x+2x^3-\dfrac{2}{5}x^5+\dfrac{5}{7}x^7$

3. (A)

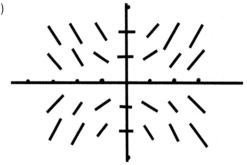

   (B) 2.03984
   (C) $y = 2e^{x^{2/3}}$; 2.06091

Part B

4. (A) $\dfrac{dy}{dx} = \dfrac{\sin\theta}{1-\cos\theta}$

   (B) $y = 4$

   (C) $x = 4\pi$

   (D) $L = \displaystyle\int_0^{2\pi} \sqrt{[2(1-\cos\theta)]^2 + [2\sin\theta]^2}\,d\theta$

   $\qquad = 2\sqrt{2}\displaystyle\int_0^{2\pi} \sqrt{1-\cos\theta}\,d\theta$

5. (A) 4

   (B) $\dfrac{128\pi}{7}$

   (C) $2^{6/7}$

   (D) $\dfrac{128}{7}$

6. (A) $f'(x) = e^{2x}(1+2x)$, $x = -0.5$

   (B) $x = -1$

   (C) $\displaystyle\lim_{x\to\infty} f(x) = \infty$ and $\displaystyle\lim_{x\to-\infty} f(x) = 0$

   (D) $-\dfrac{1}{2e}$

   (E) See solution.

Solutions to BC Practice Exam 1—Section I

Section I—Part A

1. The correct answer is (D).

$$\int_{-2}^{2} 3e^{-x}dx = -3e^{-x}\big|_{-2}^{2}$$

$$= -3e^{-2} + 3e^{2} = 3(e^2 - e^{-2})$$

2. The correct answer is (E).

$f(x) = x^3 + 3x^2 + cx + 4 \Rightarrow f'(x) = 3x^2 + 6x + c \Rightarrow f''(x) = 6x + 6$. Set $6x + 6 = 0$ so $x = -1$. $f'' > 0$ if $x > -1$ and $f'' < 0$ if $x < -1$. $f'(-1) = 3(-1)^2 + 6(-1) + c = 0 \Rightarrow 3 - 6 + c = 0 \Rightarrow -3 + c = 0 \Rightarrow c = 3$.

3. The correct answer is (A).

$$\sec^2 y\frac{dy}{dx} = 2(x - y)\left(1 - \frac{dy}{dx}\right)$$

$$\sec^2 y\frac{dy}{dx} = 2x - 2y - 2x\frac{dy}{dx} + 2y\frac{dy}{dx}$$

$$\sec^2 y\frac{dy}{dx} + 2x\frac{dy}{dx} - 2y\frac{dy}{dx} = 2x - 2y$$

$$\frac{dy}{dx}\left(\sec^2 y + 2x - 2y\right) = 2x - 2y$$

$$\frac{dy}{dx} = \frac{2x - 2y}{\sec^2 y + 2x - 2y}$$

4. The correct answer is (B).

$$y = 3^{(4-x^2)}$$

$$\ln y = (4 - x^2)\ln 3$$

$$\frac{1}{y}\frac{dy}{dx} = \ln 3\,(-2x)$$

$$\frac{dy}{dx} = y\ln 3\,(-2x)$$

$$\frac{dy}{dx} = -2x(\ln 3)3^{(4-x^2)}$$

5. The correct answer is (A).

$$x = \cos t \Rightarrow \frac{dx}{dt} = -\sin t \text{ and}$$

$$y = \sin^2 t \Rightarrow \frac{dy}{dt} = 2\sin t\cos t.$$

Then $\dfrac{dy}{dx} = \dfrac{dy}{dt}\cdot\dfrac{dt}{dx} = \dfrac{2\sin t\cos t}{-\sin t} = -2\cos t.$

Then $\dfrac{d^2y}{dx^2} = 2\sin t$. Evaluate at $t = \dfrac{\pi}{4}$ for

$$\frac{d^2y}{dx^2}\bigg|_{t=\frac{\pi}{4}} = 2\sin\left(\frac{\pi}{4}\right) = 2\left(\frac{\sqrt{2}}{2}\right) = \sqrt{2}.$$

6. The correct answer is (A).

Let $u = 3x$, $du = 3dx \Rightarrow dx = \dfrac{1}{3}du$,

$x = \dfrac{a}{3} \Rightarrow u = a$, and $x = \dfrac{b}{3} \Rightarrow u = b$. Then

$$\int_{a/3}^{b/3} g(3x)dx = \int_{a}^{b}\frac{1}{3}g(u)du$$

$$= \frac{1}{3}\int_{a}^{b} g(u)du = \frac{1}{3}\left[G(b) - G(a)\right]$$

$$= \frac{1}{3}\int_{a}^{b} g(x)dx$$

7. The correct answer is (C).

$$V = \pi\int_{0}^{1} (x - x^2)^2 dx$$

$$= \pi\int_{0}^{1} (x^4 - 2x^3 + x^2)dx$$

$$= \pi\left[\frac{x^5}{5} - \frac{x^4}{2} + \frac{x^3}{3}\right]_{0}^{1}$$

$$= \frac{\pi}{30}$$

8. The correct answer is (A)

$$\sum_{n=1}^{\infty}\frac{1}{(2n-1)(2n+1)} = \sum_{n=1}^{\infty}\frac{1}{4n^2 - 1}.$$

The sequence of partial sums

$$\left\{\frac{1}{3}, \frac{2}{5}, \frac{3}{7}, \frac{4}{9}, \cdots, \frac{n}{2n+1}, \cdots\right\} \text{ and}$$

$$= \lim_{n\to\infty}\left[\frac{n}{2n+1}\right] = \frac{1}{2}$$

9. The correct answer is (E)

The graph must be increasing and concave up. $f(x) = e^{-x}$ and $f(x) = \sqrt{2-x}$ are decreasing. $f(x) = 4 - x^2$ is increasing on $(-2, 0)$ but decreasing on $(0, 2)$. $f(x) = \sqrt{x+2}$ is increasing, but concave down. Only $f(x) = e^x$ is increasing and concave up.

10. The correct answer is (A).

Which of the following series are convergent?

I. $12 - 8 + \dfrac{16}{3} - \dfrac{32}{9} + \cdots = \sum 12 \left(\dfrac{-2}{3}\right)^n$ is a geometric series with $r = \dfrac{-2}{3}$. Since $|r| < 1$, the series converges.

II. $5 + \dfrac{5\sqrt{2}}{2} + \dfrac{5\sqrt{3}}{3} + \dfrac{5}{2} + \sqrt{5} + \dfrac{5\sqrt{5}}{6} + \cdots$

$= 5 + \dfrac{5}{\sqrt{2}} + \dfrac{5}{\sqrt{3}} + \dfrac{5}{\sqrt{4}} + \dfrac{5}{\sqrt{5}} + \dfrac{5}{\sqrt{6}} + \cdots$

$= \sum \dfrac{5}{\sqrt{n}}$ is a p-series, with $p = \dfrac{1}{2}$. Since $p < 1$, the series diverges.

III. $8 + 20 + 50 + 125 + \cdots = \sum 8 \left(\dfrac{5}{2}\right)^n$ is also a geometric series but since $r = \dfrac{5}{2} > 1$, the series diverges.
Therefore, only series I converges.

11. The correct answer is (C).

To assure that $f(x) = \begin{cases} \ln(3-x) & \text{if } x < 2 \\ a - bx & \text{if } x \geq 2 \end{cases}$ is differentiable at $x = 2$, we must first be certain that the function is continuous. As $x \to 2$, $\ln(3-x) \to 0$, so we want $a - 2b = 0 \Rightarrow a = 2b$. Continuity does not guarantee differentiability, however; we must assure that $\lim\limits_{h \to 0} \dfrac{f(2+h) - f(2)}{h}$ exists. We must be certain that $\lim\limits_{h \to 0^-} \dfrac{\ln(3 - (2+h)) - \ln(3-2)}{h}$ is equal to $\lim\limits_{h \to 0^+} \dfrac{(a - b(x+h)) - (a - bx)}{h}$.

$\lim\limits_{h \to 0^-} \dfrac{\ln(3 - (2+h)) - \ln(3-2)}{h} =$

$\lim\limits_{h \to 0^-} \dfrac{\ln(1-h)}{h} = \dfrac{0}{0}$. Thus, $\lim\limits_{h \to 0^-} \left(\dfrac{1}{1-h}\right)(-1)$

$= -1$. $\lim\limits_{h \to 0^+} \dfrac{(a - b(2+h)) - (a - 2bh)}{h} = -b$

$\Rightarrow -b = -1 \Rightarrow b = 1 \Rightarrow a = 2$

12. The correct answer is (D).

The velocity vector $v(t) = \langle 2 - 3t^2, \pi \sin(\pi t)\rangle = (2 - 3t^2)i + (\pi \sin(\pi t))j$. Integrate to find the position. $s(t) = (2t - t^3)i + \left(\dfrac{-\pi}{\pi} \cos(\pi t)\right)j + c = 4i + 3j$. Evaluate at $t = 2$ to find the constant.

$s(2) = (4-8)i + (-1\cos(2\pi))j + c = 4i + 3j$

$s(2) = (-4)i - j + c = 4i + 3j$

$c = 8i + 4j$

Therefore $s(t) = (8 + 2t - t^3)i + (4 - \cos(\pi t))j = \langle 8 + 2t - t^3, 4 - \cos(\pi t)\rangle$

Evaluate at $t = 3$.

$s(3) = (8 + 6 - 27)i + (4 - \cos(3\pi))j$

$s(3) = -13i + 5j$

The position vector is $\langle -13, 5\rangle$.

13. The correct answer is (E).

The $\lim\limits_{h \to 0} \dfrac{\ln(x - 3 + h) - \ln(x - 3)}{h}$ is the definition of the derivative for the function $y = \ln(x - 3)$, therefore the limit is equal to $y' = \dfrac{1}{x - 3}$.

14. The correct answer is (B).

To enclose the area, θ must sweep through the interval from 0 to 2π. The area of the region enclosed by $r = 3 - \sin\theta$ is

$A = \dfrac{1}{2} \int_0^{2\pi} r^2\, d\theta = \dfrac{1}{2} \int_0^{2\pi} (3 - \sin\theta)^2\, d\theta$

$= \dfrac{1}{2} \int_0^{2\pi} (9 - 6\sin\theta + \sin^2\theta) d\theta$

$= \dfrac{1}{2} \int_0^{2\pi} \left(9 - 6\sin\theta + \dfrac{1 - \cos 2\theta}{2}\right) d\theta$

$= \dfrac{1}{2} \int_0^{2\pi} \left(\dfrac{19}{2} - 6\sin\theta - \dfrac{1}{2}\cos 2\theta\right) d\theta$

$= \dfrac{1}{2} \left[\dfrac{19}{2}\theta + 6\cos\theta - \dfrac{1}{4}\sin 2\theta\right]_0^{2\pi}$

$= \dfrac{1}{2} [(19\pi + 6) - (6)] = \dfrac{19\pi}{2}$

15. The correct answer is (A).

Differentiate $r = 5 + 5\sin\theta$ to get $\dfrac{dr}{d\theta} = 5\cos\theta$. Use a parametric representation of the curve.

$$x = r\cos\theta \Rightarrow \frac{dx}{d\theta} = -r\sin\theta + \cos\theta\frac{dr}{d\theta}$$

$$= -(5 + 5\sin\theta)\sin\theta + \cos\theta(5\cos\theta)$$

$$= -5\sin\theta - 5\sin^2\theta + 5\cos^2\theta$$

$$= -5\sin\theta + 5\cos(2\theta) \text{ and}$$

$$y = r\sin\theta \Rightarrow \frac{dy}{d\theta} = r\cos\theta + \sin\theta\frac{dr}{d\theta}$$

$$= (5 + 5\sin\theta)\cos\theta + \sin\theta(5\cos\theta)$$

$$= 5\cos\theta + 5\sin(2\theta). \text{ Then}$$

$$\frac{dy}{dx} = \frac{5\cos\theta + 5\sin(2\theta)}{-5\sin\theta + 5\cos\theta} = \frac{\cos\theta + \sin(2\theta)}{-\sin\theta + \cos(2\theta)}.$$

At $\theta = \frac{\pi}{3}$, $\dfrac{dy}{dx} = \dfrac{\frac{1}{2} + \frac{\sqrt{3}}{2}}{\frac{\sqrt{3}}{2} - \frac{1}{2}} = \dfrac{1 + \sqrt{3}}{-\sqrt{3} + 1} =$

$\dfrac{1 + \sqrt{3}}{-(\sqrt{3} + 1)} = -1$. The slope of the tangent line
is equal to 1, so the slope of the normal line is
the negative reciprocal; thus, the slope of the
normal line is 1.

16. The correct answer is (B).

Use the trapezoidal method with 4 divisions:
$x = 1, y(1) = \frac{1}{2}, x = 1.5, y(1.5) = \frac{1}{3}, x = 2,$
$y(2) = \frac{1}{4}, x = 2.5, y(2.5) = \frac{1}{5}$, and
$x = 3, y(3) = \frac{1}{6}$. The area is approximated by
the sum of the areas of the four trapezoids.

$$A = \frac{1}{2}\cdot\frac{1}{2}\left(\frac{1}{2} + \frac{1}{3}\right) + \frac{1}{2}\cdot\frac{1}{2}\left(\frac{1}{3} + \frac{1}{4}\right)$$

$$+ \frac{1}{2}\cdot\frac{1}{2}\left(\frac{1}{4} + \frac{1}{5}\right) + \frac{1}{2}\cdot\frac{1}{2}\left(\frac{1}{5} + \frac{1}{6}\right)$$

$$= \frac{1}{2}\cdot\frac{1}{2}\left(\frac{1}{2} + \frac{1}{3} + \frac{1}{3} + \frac{1}{4} + \frac{1}{4} + \frac{1}{5} + \frac{1}{5} + \frac{1}{6}\right)$$

$$= \frac{1}{2}\cdot\frac{1}{2}\left(\frac{1}{2} + \frac{2}{3} + \frac{2}{4} + \frac{2}{5} + \frac{1}{6}\right)$$

$$= \frac{1}{4}\left(\frac{30}{60} + \frac{40}{60} + \frac{30}{60} + \frac{24}{60} + \frac{10}{60}\right)$$

$$= \frac{1}{4}\left(\frac{134}{60}\right) = \frac{1}{2}\left(\frac{67}{60}\right) = \frac{67}{120}$$

17. The correct answer is (C).

Let $\left(x, \dfrac{-4}{x}\right)$ be a point on $y = \dfrac{-4}{x}$. Then the
distance from the origin to the point $\left(x, \dfrac{-4}{x}\right)$

is $d = \sqrt{x^2 + \left(\dfrac{-4}{x}\right)^2} = \sqrt{x^2 + \dfrac{16}{x^2}}$

$$= \sqrt{\frac{x^4 + 16}{x^2}} = \frac{\sqrt{x^4 + 16}}{x}$$

to find the minimum distance, differentiate

$$d' = \frac{x\left(\frac{1}{2}\right)(x^4 + 16)^{-1/2}(4x^3) - \sqrt{x^4 + 16}}{x^2}$$

$$= \frac{(x^2 - 4)(x^2 + 4)}{x^2\sqrt{x^4 + 16}}.$$

Set the derivative equal to zero and solve for x.

$$\frac{(x^2 - 4)(x^2 + 4)}{x^2\sqrt{x^4 + 16}} = 0 \Rightarrow (x^2 + 4)(x^2 - 4) = 0.$$

The first factor gives $x^2 = -4$ which has no real
solution. The second gives $x^2 = 4 \Rightarrow x = \pm 2$.
There are two points at minimum distance from
the origin. $x = 2 \Rightarrow y = -2 \Rightarrow (2, -2)$ and
$x = -2, y = 2 \Rightarrow (-2, 2)$. Calculate the distance
from the origin to one of those points.

$$d = \sqrt{2^2 + (-2)^2} = \sqrt{4 + 4} = \sqrt{8} = 2\sqrt{2}.$$

18. The correct answer is (D).

$$\int_2^4 \frac{3x}{3x^2 - 4}dx = \frac{1}{2}\int_2^4 \frac{6x}{3x^2 - 4}dx.$$

Let $u = 3x^2 - 4$, $du = 6x\,dx$, $x = 2 \Rightarrow u = 8$,
and $x = 4 \Rightarrow u = 44$. Then the integral

$$= \frac{1}{2}\int_8^{44}\frac{du}{u} = \frac{1}{2}\ln|u|\Big]_8^{44} = \frac{1}{2}\big[\ln 44 - \ln 8\big].$$

19. The correct answer is (D).

The graph of $y = e^{\sin x}$ has a relative extremum
when $\dfrac{dy}{dx} = (e^{\sin x})(\cos x) = 0 \Rightarrow e^{\sin x} = 0$ or
$\cos x = 0$. Since we know $e^{\sin x} > 0$, it must be
the case that $\cos x = 0 \Rightarrow x = \dfrac{\pi}{2}, \dfrac{3\pi}{2}$. The graph
of $y = e^{\sin x}$ has a relative extremum at $x = \dfrac{\pi}{2}$
and $x = \dfrac{3\pi}{2}$. Find the second derivative

$$\frac{d^2y}{dx^2} = -\sin x e^{\sin x} + \cos^2 x\, e^{\sin x}$$

and evaluate at each critical number.

$$\frac{d^2y}{dx^2}\bigg|_{\pi/2} = -e \Rightarrow \text{max but } \frac{d^2y}{dx^2}\bigg|_{3\pi/2} = \frac{1}{e} \Rightarrow \text{min}.$$

Therefore, the graph of $y = e^{\sin x}$ has a relative
minimum when $x = \dfrac{3\pi}{2}$.

20. The correct answer is (E).

$$x = 3t^2 - 2 \Rightarrow \frac{dx}{dt} = 6t \text{ and}$$

$$y = 2t^3 + 2 \Rightarrow \frac{dy}{dt} = 6t^2 \text{ so}$$

$$\left.\frac{dy}{dx}\right|_{t=1} = \left.\frac{6t^2}{6t}\right|_{t=1} = 1 \text{ is the slope of the tangent}$$

line. At $t = 1$, $x = 1$, $y = 4$ so the point of tangency is (1, 4). Equation of tangent: $y - 4 = 1(x - 1) \Rightarrow y = x + 3$.

21. The correct answer is (A).

We know $f(x) = \cos x = \sum_{n=0}^{\infty} \frac{(-1)^{2n} x^{2n}}{(2n)!}$

$$= 1 - \frac{x^2}{2!} + \frac{x^4}{4!} - \frac{x^6}{6!} + \cdots .$$

Substitute $\sqrt{t}$ for x, and

$$\cos\sqrt{t} = 1 - \frac{t}{2!} + \frac{t^2}{4!} - \frac{t^3}{6!} + \cdots + \frac{(-1)^{2n} t^n}{(2n)!} + \cdots .$$

The required integral, $\int_0^x \cos\sqrt{t}\, dt$, will be equal to

$$\int_0^x \left(1 - \frac{t}{2!} + \frac{t^2}{4!} - \frac{t^3}{6!} + \cdots + \frac{(-1)^{2n} t^n}{(2n)!} + \cdots\right) dt$$

which, when integrated term by term, is

$$t - \frac{t^2}{2 \cdot 2!} + \frac{t^3}{3 \cdot 4!} - \frac{t^4}{4 \cdot 6!} + \cdots$$

$$\left. + \frac{(-1)^{2n} t^{n+1}}{(n+1)(2n)!} + \cdots \right|_0^x$$

$$= x - \frac{x^2}{2 \cdot 2!} + \frac{x^3}{3 \cdot 4!} - \frac{x^4}{4 \cdot 6!} + \cdots$$

$$+ \frac{(-1)^{2n} x^{n+1}}{(n+1)(2n)!} + \cdots .$$

The series expansion for $\int_0^x \cos\sqrt{t}\, dt$ is

$$x - \frac{x^2}{2 \cdot 2!} + \frac{x^3}{3 \cdot 4!} - \frac{x^4}{4 \cdot 6!} + \cdots$$

$$+ \frac{(-1)^{2n} x^{n+1}}{(n+1)(2n)!} + \cdots$$

22. The correct answer is (C).

$$\int \frac{dx}{2x^2 + 9x - 5} = \int \frac{dx}{(2x - 1)(x + 5)}. \text{ Use a}$$

partial fractions decomposition with

$$\int \frac{A}{2x - 1}\, dx + \int \frac{B}{x + 5}\, dx.$$

Then $A(x + 5) + B(2x - 1) = 1 \Rightarrow$

$Ax + 2Bx = 0 \Rightarrow A = -2B$.
Substituting and solving,

$$5A - B = 1 \Rightarrow 5(-2B) - B = 1 \text{ so } B = -\frac{1}{11}$$

and $A = \frac{2}{11}$. Then

$$\int \frac{dx}{(2x - 1)(x + 5)} = \frac{1}{11}\int \frac{2\, dx}{(2x - 1)} -$$

$$\frac{1}{11}\int \frac{dx}{(x + 5)} = \frac{1}{11}\ln|2x - 1| - \frac{1}{11}\ln|x + 5| =$$

$$\frac{1}{11}\ln\left|\frac{2x - 1}{x + 5}\right| + c.$$

23. The correct answer is (E).

Assume that the base of the solid is the circle $x^2 + y^2 = 16 \Rightarrow y = \sqrt{16 - x^2}$. Then the area of each cross section is $s^2 = \left(2\sqrt{16 - x^2}\right)^2$. Then the volume is

$$V = \int_{-4}^4 \left(2\sqrt{16 - x^2}\right)^2 dx = 4\int_{-4}^4 \left(16 - x^2\right) dx$$

$$= 4\left[16x - \frac{x^3}{3}\right]_{-4}^4$$

$$= 4\left[\left(64 - \frac{64}{3}\right) - \left(-64 - \frac{-64}{3}\right)\right]$$

$$= 4\left[128 - \frac{128}{3}\right] = 4\left[\frac{3(128)}{3} - \frac{128}{3}\right]$$

$$= 4\left[\frac{2(128)}{3}\right] = \frac{8(128)}{3} = \frac{1024}{3}$$

24. The correct answer is (C).

$$x = \sin^2 t \Rightarrow \frac{dx}{dt} = 2\sin t \cos t = \sin(2t) \text{ and}$$

$$y = \cos(2t) \Rightarrow \frac{dy}{dt} = -2\sin(2t). \text{ Then}$$

$$\left(\frac{dx}{dt}\right)^2 = \sin^2(2t) \text{ and}$$

$$\left(\frac{dy}{dt}\right)^2 = (-2\sin(2t))^2 = 4\sin^2(2t).$$

For $\frac{\pi}{2} \leq t \leq \pi$, the length of the arc the particle traces out is

$$L = \int_{\pi/2}^{\pi} \sqrt{\sin^2(2t) + 4\sin^2(2t)}\, dt$$

$$= \int_{\pi/2}^{\pi} \sqrt{5\sin^2(2t)}\, dt = \sqrt{5}\int_{\pi/2}^{\pi} \sin(2t)\, dt$$

$$= \frac{\sqrt{5}}{2}\cos(2t)\Big]_{\pi/2}^{\pi} = \frac{\sqrt{5}}{2}(\cos(2\pi) - \cos(\pi))$$

$$= \frac{\sqrt{5}}{2}(1 - (-1)) = \sqrt{5}.$$

25. The correct answer is (C).

The series $\sum \frac{(-1)^n x^n}{e^n}$ will converge when the

ratio $\left| \frac{(-1)^{n+1} x^{n+1}}{e^{n+1}} \cdot \frac{e^n}{(-1)^n x^n} \right| < 1.$

Simplify, and the series converges when

$$\left| \frac{xe^n}{e^{n+1}} \right| = \left| \frac{x}{e} \right| < 1. \ \left| \frac{x}{e} \right| < 1 \Rightarrow -1 < \frac{x}{e} < 1 \Rightarrow$$

$-e < x < -e.$ when $x = e$, the series becomes $\sum (-1)^n$, which diverges. When $x = -e$, the series becomes $\sum (-1)^{2n} = \sum 1^n$, which diverges. So the interval of convergence is $(-e, e)$.

26. The correct answer is (B).

$$\frac{1}{n}\left[\left(\frac{1}{n}\right)^2 + \left(\frac{2}{n}\right)^2 + \cdots \left(\frac{n-1}{n}\right)^2 \right]$$

represents the sum of the areas of n rectangles each of width $\frac{1}{n}$. The heights of the rectangles are the squares of the division points, $\frac{1}{n}, \frac{2}{n}, \frac{3}{n}, \ldots, \frac{n-1}{n}$, all of which are between 0 and 1. Thus,

$$\lim_{n \to \infty} \frac{1}{n}\left[\left(\frac{1}{n}\right)^2 + \left(\frac{2}{n}\right)^2 + \cdots \left(\frac{n-1}{n}\right)^2 \right]$$

represents the area under the $y = x^2$ from 0 to 1, or $\int_0^1 x^2 dx$

27. The correct answer is (C).

$\int_{-3}^{-2} \frac{5x}{(x+2)(x-3)} dx$ is an improper integral

since $f(x) = \frac{5x}{(x+2)(x-3)}$ has an infinite

discontinuity at $x = -2$, one of the limits of integration. Therefore

$\int_{-3}^{-2} \frac{5x}{(x+2)(x-3)} dx$ is equal to

$$\lim_{n \to -2^-} \int_{-3}^{n} \frac{5x}{(x+2)(x-3)} dx$$

28. The correct answer is (E).

$$\int_1^2 e^{4-3\ln x} dx = \int_1^2 e^4 \cdot e^{-3\ln x} dx$$

$$= \int_1^2 e^4 \left(e^{\ln x}\right)^{-3} dx$$

$$= \int_1^2 e^4 x^{-3} dx = \frac{e^4}{-2x^2}\Big|_1^2$$

$$= \frac{e^4}{-2}\left[\frac{1}{4} - \frac{1}{1}\right] = \frac{3e^4}{8}$$

Section I—Part B

76. The correct answer is (A).

If $f(x) = \sqrt[3]{x^3 - x}$ then
$f'(x) = \frac{1}{3}\left(x^3 - x\right)^{-2/3}\left(3x^2 - 1\right)$
Evaluating at $x = 2$,
$f'(2) = \frac{1}{3}(8 - 2)^{-2/3}(12 - 1)$

$$= \frac{1}{3}(6)^{-2/3}(11) \approx 1.11046$$

77. The correct answer is (D).

Let $u = \ln x$, $du = \frac{dx}{x}$, $x = 1 \Rightarrow u = 0$, and
$x = e^{\pi} \Rightarrow u = \pi$. Then

$$\int_1^{e^{\pi}} \frac{\cos(\ln x)}{x} dx = \int_0^{\pi} \cos u \, du = \sin u\Big|_0^{\pi}$$

$$= \sin \pi - \sin 0 = 0.$$

78. The correct answer is (C).

$\frac{dy}{dx} = 0.2y\left(1 - \frac{y}{200}\right)$ is separable and can be integrated by partial fractions.

$$\frac{200}{y(200 - y)} dy = 0.2 \, dt$$

$$\Rightarrow \int \frac{1}{y} dy + \int \frac{1}{200 - y} dy$$

$$= \int 0.2 \, dt$$

$$\ln |y| - \ln |200 - y|$$

$$= 0.2t + c_1 \Rightarrow \ln \left| \frac{y}{200 - y} \right|$$

$$= 0.2t + c_1$$

Exponentiate and solve for y:

$$\frac{y}{200 - y} = c_2 e^{0.2t} \Rightarrow y = \frac{200 c_2 e^{0.2t}}{1 + c_2 e^{0.2t}}$$

or $y = \dfrac{200}{(1/c_2)\, e^{-0.2t} + 1}$.

Since the rumor begins with two people,

$$2 = \frac{200 c_2}{1 + c_2} \Rightarrow c_2 = \frac{1}{99} \Rightarrow \frac{1}{c_2} = 99,$$

so $y = \dfrac{200}{99 e^{-0.2t} + 1}$. Evaluate at $t = 30$,

$$y(30) = \frac{200}{99 e^{(-0.2)(30)} + 1} \approx 160.591.$$

The total number of people who have heard the rumor after thirty days is about 161.

79. The correct answer is (E).

$$\lim_{h \to 0} \frac{\cos 2(2 + h) - \cos 4}{h} = \frac{d}{dx}\big(\cos(2x)\big)$$

$$= -2\sin(2x)\big|_{x=2} = 1.514$$

80. The correct answer is (A).

The area under the curve $y = 3x^2 - kx + 1$ bounded by the lines $x = 1$ and $x = 2$ is

$$A = \int_1^2 \big(3x^2 - kx + 1\big)\, dx = x^3 - \frac{k}{2}x^2 + x \bigg|_1^2$$

$$= \left(2^3 - \frac{k}{2} 2^2 + 2\right) - \left(1^3 - \frac{k}{2} 1^2 + 1\right)$$

$$= (10 - 2k) - \left(2 - \frac{k}{2}\right).$$

Since the area is known to be -5.5

set $A = 8 - \dfrac{3}{2}k = -5.5$ and solve:

$$-\frac{3}{2}k = -5.5 - 8 = -13.5 \Rightarrow$$

$$-\frac{3}{2}k = -13.5 \Rightarrow k = 9.$$

81. The correct answer is (C).

$\dfrac{dP}{dt} = kP$ is separable, so $\dfrac{1}{P} dP = k\, dt$ can be integrated to $\ln |P| = kt + C$. Exponentiate for $P = ce^{kt}$. Since the population increases 23% in 12 years, $1.23 = 1e^{k(12)}$

$$\Rightarrow \ln 1.23 = \ln 1 + 12k \Rightarrow k = \frac{\ln 1.23}{12}$$

$$\approx 0.0172511808 \approx 0.017$$

82. The correct answer is (C).

$$\int_{1/2}^1 \csc 3x\, dx = \int_{1/2}^1 \frac{1}{\sin 3x}\, dx.$$

Use the integral function on your calculator to find the integral is ≈ 0.90571039.

83. The correct answer is (D).

If $F(x) = \displaystyle\int_0^x \sqrt{\sin t}\, dt$. Then $F'(x) = \sqrt{\sin x}$,

so $F'(0.2) = \sqrt{\sin(0.2)} \approx 0.4457$

84. The correct answer is (A).

Begin with the known series

$$f(x) = e^x = \sum_{n=0}^{\infty} \frac{x^n}{n!} = 1 + x + \frac{x^2}{2} + \frac{x^3}{6} + \cdots.$$

Substitute $-x$ for x and

$$e^{-x} = 1 - x + \frac{x^2}{2} - \frac{x^3}{6} + \cdots + \frac{(-1)^n x^n}{n!} + \cdots.$$

Integrate term by term:

$$\int_0^1 e^{-x} dx = \int_0^1 \left(1 - x + \frac{x^2}{2} - \frac{x^3}{6} + \cdots \right.$$

$$\left. + \frac{(-1)^n x^n}{n!} + \cdots \right) dx$$

$$= \left[x - \frac{x^2}{2 \cdot 1!} + \frac{x^3}{3 \cdot 2!} - \frac{x^4}{4 \cdot 3!} + \cdots \right.$$

$$\left. + \frac{(-1)^n x^{n+1}}{(n+1)\, n!} + \cdots \right]_0^1$$

$$= \sum \frac{(-1)^n}{(n+1)!}$$

The function $f(x) = e^{-x}$ can be differentiated repeatedly on the interval $(0, 2)$, with $f^{(n)}(x) = (-1)^n e^{-x}$ and $\big|f^{(n+1)}(x)\big| \le 1$ for all x in that interval. The remainder

$$|R_n(x)| \le \frac{1}{(n+1)!}$$ and for three digit accuracy,

we want $\dfrac{1}{(n+1)!} \le 0.0005$. With the help of

a calculator, we find that $\dfrac{1}{(n+1)!} \approx 0.0002$ for $n = 6$. Therefore we estimate

$$\int_0^1 e^{-x} dx = \sum_{n=0}^6 \frac{(-1)^n}{(n+1)!} \approx 0.632143.$$

85. The correct answer is (D).

Rewrite $y = \sec^2(3x)$ as $y = [\sec(3x)]^2$

$$\frac{dy}{dx} = 2\left[\sec(3x)\right]\left[\sec(3x)\tan(3x)\right](3)$$

$$= 6\sec^2(3x)\tan(3x)$$

$$\left.\frac{dy}{dx}\right|_{x=\pi/9} = 6(2^2)(\sqrt{3}) = 24\sqrt{3}$$

Then $\dfrac{dy}{dx}$ at $x = 0.5$ is $\dfrac{dy}{dx} = \left(0.5^{3(0.5)-2}\right)$

$$\left[3\,(0.5) - 1 + 3\,(0.5)\ln(0.5)\right]$$

$$\Rightarrow \frac{dy}{dx} = \sqrt{2}\left[0.5 - 1.5\ln 2\right].$$

86. The correct answer is (C).

If $2x\dfrac{dy}{dx} - 7 = 1 \Rightarrow \dfrac{dy}{dx} = \dfrac{8}{2x} = \dfrac{4}{x}.$

Integrate $dy = \dfrac{4\,dx}{x}$ to get $\Rightarrow y = 4\ln|x| + c.$
Since $y = 4.5$ when $x = 3$, $4.5 = 4\ln 3 + c \Rightarrow$
$0.106 = c$. Thus $y = 4\ln|x| + 0.106$. At
$x = 3.1$, $y = 4.632$

87. The correct answer is (A).

The slope of the *tangent* line to $y = e^{-2x}$

is $\dfrac{dy}{dx} = -2e^{-2x}$. The slope of the *normal*

is the negative reciprocal, $m = \dfrac{1}{2e^{-2x}} = \dfrac{e^{2x}}{2}.$

When $x = 1.158$, $m = \dfrac{e^{2(1.158)}}{2} \approx 5.068.$

The slope of the normal line is approximately 5.068.

88. The correct answer is (D).

The volume of the solid generated by revolving about the x-axis is

$$V = 2\pi \int_0^{\pi/2} (\sin x + \cos x)^2 dx.$$

Since $(\sin x + \cos x)^2 = \sin^2 x + 2\sin x \cos x$

$$+\cos^2 x = 1 + 2\sin x \cos x$$

$$= 1 + \sin 2x,$$

$V = \pi \displaystyle\int_0^{\pi/2} (1 + \sin 2x)\,dx.$ This integral is

$$= \pi\left[x - \frac{1}{2}\cos 2x\right]_0^{\pi/2}$$

$$= \pi\left[\left(\frac{\pi}{2} - \frac{1}{2}\cos \pi\right) - \left(0 - \frac{1}{2}\cos 0\right)\right]$$

$$= \pi\left[\frac{\pi}{2} + \frac{1}{2} + \frac{1}{2}\right] \approx 8.076$$

89. The correct answer is (B).

The absolute minimum of $f(x) = \ln(3x) + \cos x$ on the closed interval $\left[\dfrac{\pi}{2}, \pi\right]$. First find the relative extrema of
$f(x) = \ln(3x) + \cos x$ on $\left[\dfrac{\pi}{2}, \pi\right].$

Set the derivative
$f'(x) = \dfrac{1}{x} - \sin x = 0$ and

$\dfrac{1}{x} = \sin x \Rightarrow x = 1.1141571 < \dfrac{\pi}{2}$ (not in the interval) or $x = 2.7726047.$

$f''(x) = \dfrac{-1}{x^2} - \cos x \Rightarrow f''$ is positive at
$x = 2.7726047$, indicating a relative minimum.
$f(2.7726047) = 1.1857067$, so the relative minimum is approximately $(2.773, 1.186)$.
Checking the endpoints of the interval,

$f\left(\dfrac{\pi}{2}\right) = 1.550$ and $f(\pi) = 1.243$. Therefore the

absolute minimum is $(2.773, 1.186)$.

90. The correct answer is (D).

The position of the particle is
$y(t) = 5t^3 - 9t^2 + 2t - 1$, and the velocity is
$v(t) = y'(t) = 15t^2 - 18t + 2$. At the moment
the particle changed direction, its velocity was
zero, so $15t^2 - 18t + 2 = 0$. Solving tells us that
the particle changes direction twice, first at
$t \approx 0.124$ and later at $t = 1.076$. Taking the
first of these, and evaluating the position
function, $y \approx -0.881$. At the moment when the
particle first changes direction, its position is
$(0, -0.881)$.

91. The correct answer is (A).

Use the intersect function to find that the points of intersection of $y = \cos x + 1$ and $y = 2 + 2x - x^2$ are $(0, 2)$ and $(2.705, 0.094)$. The area enclosed by the curves is

$$\int_0^{2.705} [2 + 2x - x^2 - (\cos x + 1)]dx$$

$$= x + x^2 - \frac{x^3}{3} - \sin x \Big|_0^{2.705} \approx 3.002.$$

92. The correct answer is (E).

Use the ratio test for absolute convergence.

$$\lim_{n \to \infty} \left| \frac{(x-3)^{n+1}}{(n+1)^2} \cdot \frac{n^2}{(x-3)^n} \right|$$

$$= \lim_{n \to \infty} \left| \frac{(x-3)n^2}{(n+1)^2} \right|$$

$$= |x-3| \lim_{n \to \infty} \left(\frac{n}{n+1} \right)^2$$

$$= |x-3|$$

Set $|x-3| < 1 \Rightarrow -1 < (x-3) < 1 \Rightarrow 2 < x < 4$. At $x = 4$, series becomes $\sum_{n=1}^{\infty} \frac{1}{n^2}$ which is a p-series with $p = 2$. Series converges. At $x = 2$, series becomes $\sum_{n=1}^{\infty} \frac{(-1)^n}{n^2}$ which converges absolutely. Thus, the interval of convergence is $[2,4]$.

Section II—Part A

1. (A) $\dfrac{dy}{dx} = \dfrac{4x+1}{2y}; (2,4)\ \dfrac{dy}{dx}\Big|_{(2,4)} = \dfrac{4(2)+1}{2(4)} = \dfrac{9}{8}$

Equation of tangent line: $y - 4 = \dfrac{9}{8}(x-2)$

or $y = \dfrac{9}{8}(x-2) + 4$

(B) $f(2.1) = \dfrac{9}{8}(2.1-2) + 4 \approx \dfrac{0.9}{8} + 4$

$\approx 4.1125 \approx 4.113$

(C) $2y\,dy = (4x+1)\,dx$

$\displaystyle\int 2y\,dy = \int (4x+1)\,dx$

$y^2 = 2x^2 + x + c;\ f(2) = 4$

$4^2 = 2(2)^2 + 2 + c \Rightarrow c = 6$

Thus, $y^2 = 2x^2 + x + 6$ or $y = \pm\sqrt{2x^2 + x + 6}$. Since the point $(2, 4)$ is on the graph of f, $y = \sqrt{2x^2 + x + 6}$

(D) $y = \sqrt{2x^2 + x + 6}$

$f(2.1) = \sqrt{2(2.1)^2 + 2.1 + 6} = \sqrt{16.92}$

$\approx 4.11339 \approx 4.113$

2. Given $f(0) = 1$, $f'(0) = 6$, $f''(0) = -4$, and $f'''(0) = 30$

(A) The third-degree Taylor polynomial for f about $x = 0$ is

$$f(x) \approx \frac{f(0)}{0!}x^0 + \frac{f'(0)}{1!}x^1 + \frac{f''(0)}{2!}x^2$$

$$+ \frac{f'''(0)}{3!}x^3$$

$$\approx 1 + 6x + \frac{-4}{2}x^2 + \frac{30}{6}x^3$$

$$\approx 1 + 6x - 2x^2 + 5x^3$$

To approximate $f(0.1)$:

$$f(0.1) \approx 1 + 6(0.1) - 2(0.1)^2 + 5(0.1)^3$$

$$\approx 1 + 0.6 - 2(0.01) + 5(0.001)$$

$$\approx 1 + 0.6 - 0.02 + 0.005$$

$$\approx 1.585$$

(B) The sixth degree Taylor polynomial for $g(x) = f(x^2)$, about $x = 0$, is

$$g(x) = f(x^2) = 1 + 6(x^2) - 2(x^2)^2 + 5(x^2)^3$$

$$g(x) = 1 + 6x^2 - 2x^4 + 5x^6$$

(C) The seventh degree Taylor polynomial for

$$h(x) = \int_0^x g(t)\,dt,\ \text{about } x = 0,\ \text{is}$$

$$h(x) = \int_0^x \left(1 + 6t^2 - 2t^4 + 5t^6\right)dt$$

$$h(x) = \left[t + \frac{6}{3}t^3 - \frac{2}{5}t^5 + \frac{5}{7}t^7\right]_0^x$$

$$h(x) = x + 2x^3 - \frac{2}{5}x^5 + \frac{5}{7}x^7$$

3. Given the differential equation $\dfrac{dy}{dx} = \dfrac{2xy}{3}$.

(A) Calculate slopes

	$y = -2$	$y = -1$	$y = 0$	$y = 1$	$y = 2$
$x = -3$	4	2	0	-2	-4
$x = -2$	$\dfrac{8}{3}$	$\dfrac{4}{3}$	0	$-\dfrac{4}{3}$	$-\dfrac{8}{3}$
$x = -1$	$\dfrac{4}{3}$	$\dfrac{2}{3}$	0	$-\dfrac{2}{3}$	$-\dfrac{4}{3}$
$x = 0$	0	0	0	0	0
$x = 1$	$-\dfrac{4}{3}$	$-\dfrac{2}{3}$	0	$\dfrac{2}{3}$	$\dfrac{4}{3}$
$x = 2$	$-\dfrac{8}{3}$	$-\dfrac{4}{3}$	0	$\dfrac{4}{3}$	$\dfrac{8}{3}$
$x = 3$	-4	-2	0	2	4

Sketch the slope field.

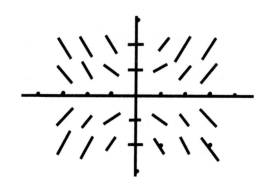

(B) $f(0.1) = f(0) + 0.1 \left. \dfrac{2xy}{3} \right|_{x=0,\,y=2}$

$\qquad = 2 + 0.1\,(0) = 2.$

$f(0.2) = f(0.1) + 0.1 \left. \dfrac{dy}{dx} \right|_{x=0.1,\,y=2}$

$\qquad = 2 + 0.1\left(\dfrac{0.4}{3}\right) = 2 + \dfrac{0.04}{3} = 2.01\overline{3}$

$f(0.3) = f(0.2) + 0.1 \left. \dfrac{dy}{dx} \right|_{x=0.2,\,y=2.01\overline{3}}$

$\qquad = 2.01\overline{3} + 0.02684 = 2.03984$

(C) $\dfrac{dy}{dx} = \dfrac{2xy}{3}$

$\dfrac{1}{y}\,dy = \dfrac{2}{3}x\,dx$

$\ln |y| = \dfrac{1}{3}x^2 + c_1$

$y = c_2 e^{x^2/3}$

According to the initial condition, $2 = c_2 e^{0/3} \Rightarrow c_2 = 2$ so the particular solution is $y = 2e^{x^2/3}$. Evaluate at $x = 0.3$ and $y(0.3) = 2e^{0.09/3} = 2e^{0.03} \approx 2.06091$.

Section II—Part B

4. Given $x = 2(\theta - \sin\theta)$ and $y = 2(1 - \cos\theta)$.

(A) $\dfrac{dx}{d\theta} = 2(1 - \cos\theta)$ and $\dfrac{dy}{d\theta} = 2\sin\theta$. Divide

to find $\dfrac{dy}{dx} = \dfrac{2\sin\theta}{2(1-\cos\theta)} = \dfrac{\sin\theta}{1-\cos\theta}$

(B) At $\theta = \pi$, $x = 2(\pi - \sin\pi) = 2\pi$,

$y = 2(1 - \cos\pi) = 4$, and

$\left. \dfrac{dy}{dx} \right|_{\theta=\pi} = \dfrac{\sin\pi}{1-\cos\pi} = 0.$

The tangent line at $(2\pi,\,4)$ is horizontal, so the equation of the tangent is $y = 4$.

(C) At $\theta = 2\pi$, $x = 2(2\pi - \sin 2\pi) = 4\pi$,

$y = 2(1 - \cos 2\pi) = 0$, and

$\left. \dfrac{dy}{dx} \right|_{\theta=2\pi} = \dfrac{\sin 2\pi}{1-\cos 2\pi} = \dfrac{0}{0}$. Since the derivative is undefined, the tangent line at

$(4\pi,\,0)$ is vertical, so the equation of the tangent is $x = 4\pi$.

(D) $L = \displaystyle\int_0^{2\pi} \sqrt{[2(1-\cos\theta)]^2 + [2\sin\theta]^2}\,d\theta$

$= \displaystyle\int_0^{2\pi} \sqrt{4\left(1 - 2\cos\theta + \cos^2\theta\right) + 4\sin^2\theta}\,d\theta$

$= \displaystyle\int_0^{2\pi} \sqrt{4 - 8\cos\theta + 4\cos^2\theta + 4\sin^2\theta}\,d\theta$

$= \displaystyle\int_0^{2\pi} \sqrt{4 - 8\cos\theta + 4}\,d\theta$

$$= \int_0^{2\pi} \sqrt{8 - 8\cos\theta}\, d\theta$$

$$= 2\sqrt{2} \int_0^{2\pi} \sqrt{1 - \cos\theta}\, d\theta$$

5. See Figure DS-18

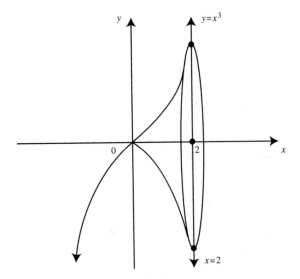

Figure DS-18

(A) Area of R $= \displaystyle\int_0^2 x^3\, dx = \dfrac{x^4}{4}\Big]_0^2$

$$= \dfrac{2^4}{4} - 0 = 4$$

(B) Volume of solid $= \pi \displaystyle\int_0^2 \left(x^3\right)^2 dx$

$$= \pi \left[\dfrac{x^7}{7}\right]_0^2 = \dfrac{2^7(\pi)}{7}$$

$$= \dfrac{128\pi}{7}$$

(C) $\pi \displaystyle\int_0^a \left(x^3\right)^2 dx = \dfrac{1}{2}\left(\dfrac{128\pi}{7}\right)$

$$\pi \left[\dfrac{x^7}{7}\right]_0^a = \dfrac{64\pi}{7};\ \dfrac{\pi a^7}{7} = \dfrac{64\pi}{7};$$

$$a^7 = 64 = 2^6;\quad a = 2^{6/7}$$

(D) Area of cross section $= (x^3)^2 = x^6$

Volume of solid

$$= \int_0^2 x^6\, dx = \dfrac{x^7}{7}\Big]_0^2 = \dfrac{128}{7}.$$

6. (A) $f(x) = xe^{2x}$

$$f'(x) = e^{2x} + x(e^{2x})(2) = e^{2x} + 2xe^{2x}$$

$$= e^{2x}(1 + 2x)$$

Set $f'(x) = 0 \Rightarrow e^{2x}(1 + 2x) = 0$. Since $e^{2x} > 0$, $1 + 2x = 0 \Rightarrow x = -0.5$.

(B) $f'(x) = e^{2x} + 2xe^{2x}$

$$f''(x) = e^{2x}(2) + 2e^{2x} + 2xe^{2x}(2)$$

$$= 2e^{2x} + 2e^{2x} + 4xe^{2x} = 4e^{2x} + 4xe^{2x}$$

$$= 4e^{2x}(1 + x)$$

Set $f''(x) = 0 \Rightarrow 4e^{2x}(1 + x) = 0$. Since $e^{2x} > 0$, thus $1 + x = 0$ or $x = -1$

(C) $\displaystyle\lim_{x\to\infty} xe^{2x} = \infty$, since xe^{2x} increases without bound as x approaches $+\infty$.

$\displaystyle\lim_{x\to-\infty} xe^{2x} = \lim_{x\to-\infty} \dfrac{x}{e^{-2x}}$. As $x \to -\infty$, the numerator $\to -\infty$. As $x \to -\infty$, the denominator $e^{-2x} \to \infty$. However, the denominator increases at a much greater rate and thus $\displaystyle\lim_{x\to-\infty} xe^{2x} = 0$

(D) Since as $x \to \infty$, xe^{2x} increases without bound, f has no absolute maximum value. From part (A), $f(x)$ has one critical point at $x = -0.5$. Since $f'(x) = e^{2x}(1 + 2x)$, $f'(x) < 0$ for $x < -0.5$ and $f'(x) > 0$ for $x > -0.5$, thus f has a relative minimum at $x = -0.5$, and it is the absolute minimum because $x = -0.5$ is the only critical point on an open interval. The absolute minimum value is $-0.5e^{2(-0.5)} = -\dfrac{1}{2e}$

(E) $f(x) = xe^{ax}$, $a > 0$

$$f'(x) = e^{ax} + x(e^{ax})(a) = e^{ax} + axe^{ax}$$

$$= e^{ax}(1 + ax)$$

Set $f'(x) = 0 \Rightarrow e^{ax}(1 + ax) = 0$ or $x = -\dfrac{1}{a}$.

If $x < -\dfrac{1}{a}$, $f'(x) < 0$ and if $x > -\dfrac{1}{a}$, $f'(x) > 0$.

Thus $x = -\dfrac{1}{a}$ is the only critical point, and f has an absolute minimum at $x = -\dfrac{1}{a}$.

$$f\left(-\dfrac{1}{a}\right) = \left(-\dfrac{1}{a}\right)e^{a\left(-\frac{1}{a}\right)} = -\dfrac{1}{a}e^{-1} = -\dfrac{1}{ae}.$$

The absolute minimum value of f is $-\dfrac{1}{ae}$ for all $a > 0$.

Scoring Sheet for BC Practice Exam 1

Section I—Part A

$$\underline{\hspace{3cm}} \times 1.2 = \underline{\hspace{3cm}}$$
No. Correct $\quad\quad\quad\quad\quad$ Subtotal A

$$\underline{\hspace{3cm}} \times (0.25) \cdot 1.2 = \underline{\hspace{3cm}}$$
No. Incorrect $\quad\quad\quad\quad$ Subtotal B

Part A (Subtotal A – Subtotal B) = $\underline{\hspace{3cm}}$
$\quad\quad\quad\quad\quad\quad\quad\quad\quad\quad\quad\quad\quad\quad$ Subtotal C

Section I—Part B

$$\underline{\hspace{3cm}} \times 1.2 = \underline{\hspace{3cm}}$$
No. Correct $\quad\quad\quad\quad\quad$ Subtotal D

$$\underline{\hspace{3cm}} \times (0.25) \cdot 1.2 = \underline{\hspace{3cm}}$$
No. Incorrect $\quad\quad\quad\quad$ Subtotal E

Part B (Subtotal D – Subtotal E) = $\underline{\hspace{3cm}}$
$\quad\quad\quad\quad\quad\quad\quad\quad\quad\quad\quad\quad\quad\quad$ Subtotal F

Section II—Part A (Each question is worth 9 points.)

$$\underline{\hspace{2.5cm}} + \underline{\hspace{2.5cm}} + \underline{\hspace{2.5cm}} = \underline{\hspace{2.5cm}}$$
Q#1 $\quad\quad\quad\quad$ Q#2 $\quad\quad\quad\quad$ Q#3 $\quad\quad\quad$ Subtotal G

Section II—Part B (Each question is worth 9 points.)

$$\underline{\hspace{2.5cm}} + \underline{\hspace{2.5cm}} + \underline{\hspace{2.5cm}} = \underline{\hspace{2.5cm}}$$
Q#1 $\quad\quad\quad\quad$ Q#2 $\quad\quad\quad\quad$ Q#3 $\quad\quad\quad$ Subtotal H

Total Raw Score (Subtotals C + F + G + H) = $\boxed{}$

Approximate Conversion Scale:	
Total Raw Score	Approximate AP Grade
75–108	5
60–74	4
45–59	3
31–44	2
0–30	1

BC PRACTICE EXAM 2

Answer Sheet for BC Practice Exam 2—Section I

Part A

1. _____

2. _____

3. _____

4. _____

5. _____

6. _____

7. _____

8. _____

9. _____

10. _____

11. _____

12. _____

13. _____

14. _____

15. _____

16. _____

17. _____

18. _____

19. _____

20. _____

21. _____

22. _____

23. _____

24. _____

25. _____

26. _____

27. _____

28. _____

Part B

76. _____

77. _____

78. _____

79. _____

80. _____

81. _____

82. _____

83. _____

84. _____

85. _____

86. _____

87. _____

88. _____

89. _____

90. _____

91. _____

92. _____

Section I—Part A

Number of Questions	Time	Use of Calculator
28	55 Minutes	No

Directions:

Use the answer sheet provided in the previous page. All questions are given equal weight. There is no penalty for unanswered questions. However, 1/4 of the number of the incorrect answers will be subtracted from the number of correct answers. Unless otherwise indicated, the domain of a function f is the set of all real numbers. The use of a calculator is *not* permitted in this part of the exam.

1. If $f(3) = -2$ and $f'(3) = 5$, then the equation of the tangent to the curve $y = f(x)$ at $x = 3$ is

   (A) $y = 5x - 2$
   (B) $y = 5x - 17$
   (C) $y = -2x + 5$
   (D) $y = -2x - 17$
   (E) $y = 5x + 3$

2. On the interval $-3 < x < -1$, the curve $y = \dfrac{2x + 1}{5x + 2}$ is

   (A) Increasing and concave up
   (B) Increasing and concave down
   (C) Decreasing and concave up
   (D) Decreasing and concave down
   (E) Horizontal

3. The relative extremum of the function $y = \sqrt[3]{x^2 + 4ax + 12a^2}$, $a > 0$ is

   (A) Relative maximum at $\left(-2a, 2a^{2/3}\right)$
   (B) Relative maximum at $(-2a, 0)$
   (C) Relative minimum at $\left(-2a, 2a^{2/3}\right)$
   (D) Relative minimum at $(0, -2a)$
   (E) No relative extrema

4. A function is defined for all real numbers and has the property $f(x + h) - f(x) = 4x^2 h + 2xh - 6x^3 h^2$. Find $f'(3)$

   (A) -18
   (B) -3
   (C) 0
   (D) 3
   (E) 42

5. For what positive value of k is the line $y = -9x + k$ tangent to the curve $y = x^3 - 6x^2$?

   (A) -27
   (B) 0

   (C) 3
   (D) 5
   (E) 4

6. The tangent to the parabola $y = ax^2$ at $x = p$ intersects the y-axis at

   (A) $(-ap^2, 0)$
   (B) $(-2a, 0)$
   (C) $(0, -ap^2)$
   (D) $(0, ap^2)$
   (E) $(0, -ap)$

The table show some of the values of differentiable functions f and g and their derivatives. Use this information for question 10 and 11.

$x =$	1	2	3
$f(x)$	0	4	2
$f'(x)$	-2	5	-1
$g(x)$	-1	3	3
$g'(x)$	5	1	0

7. If $h(x) = f(x)g(x)$, then $h'(1) =$

   (A) 1
   (B) 2
   (C) 3
   (D) 4
   (E) 5

8. If $h(x) = f(g(x))$, then $h'(2) =$

   (A) -2
   (B) -1
   (C) 0

(D) 1

(E) 2

9. If $f(x)$ is a continuous function and $f(5) = 2$ and $f'(5) = 3$, then $f(5.01)$ is approximately

(A) 0.02

(B) 0.03

(C) 2.01

(D) 2.03

(E) 3.02

10. Evaluate $\int_1^2 \frac{x^3 - 1}{x^2} dx$

(A) $\ln 2$

(B) $-\ln 2$

(C) 0

(D) 1

(E) Does not exist

11. A cube is expanding so that its surface area is increasing at a constant rate of 4 square inches per second. How fast is the volume increasing at the instant when the surface area is 24 square inches?

(A) 2 cubic inches per second

(B) 4 cubic inches per second

(C) 8 cubic inches per second

(D) 16 cubic inches per second

(E) 24 cubic inches per second

12. If $\int_0^4 f(x)\,dx = 10$, $\int_0^5 f(x)\,dx = 9$, and $\int_4^7 f(x)\,dx = 1$, then $\int_5^7 f(x)\,dx =$

(A) 2

(B) 8

(C) 9

(D) 10

(E) 11

13. An antiderivative for $\dfrac{1}{x^2 - 3x + 2}$ is

(A) $\ln\left|\dfrac{x-1}{x-2}\right| + C$

(B) $\dfrac{x-2}{x-1} + C$

(C) $\dfrac{\ln|x-2|}{\ln|x-1|} + C$

(D) $\dfrac{x-1}{x-2} + C$

(E) $\ln\left|\dfrac{x-2}{x-1}\right| + C$

14. $\displaystyle\lim_{\theta \to 0} \frac{1 - \cos\theta}{2\sin^2\theta}$ is

(A) indeterminate

(B) 1

(C) 0

(D) $\dfrac{1}{4}$

(E) $\dfrac{1}{2}$

15. $\displaystyle\int xf(x)\,dx =$

(A) $xf(x) - \displaystyle\int xf'(x)\,dx$

(B) $\dfrac{x^2}{2}f(x) - \displaystyle\int \dfrac{x^2}{2}f'(x)\,dx$

(C) $xf(x) - \dfrac{x^2}{2}f(x) + C$

(D) $xf(x) - \displaystyle\int f'(x)\,dx$

(E) $\dfrac{x^2}{2}\displaystyle\int f(x)\,dx$

16. If $x^3 + 3xy + 2y^3 = 17$, then in terms of x and y, $\dfrac{dy}{dx} =$

(A) $-\dfrac{x^2 + y}{x + 2y^2}$

(B) $-\dfrac{x^2 + y}{x + y^2}$

(C) $-\dfrac{x^2 + y}{x + 2y}$

(D) $-\dfrac{x^2 + y}{2y^2}$

(E) $\dfrac{-x^2}{1 + 2y^2}$

17. If $f(x) = 3\tan^5(2x)$, then $f'(x)$ is:

(A) $3\sec^{10}(2x)$

(B) $30\sec^{10}(2x)$

(C) $15\tan^4(2x)$

(D) $30\tan^4(2x)$

(E) $30\tan^4(2x)\sec^2(2x)$

18. If $f'(x) = \cos x \sin^2 x$ and $f(0) = 2$, then $f\left(\dfrac{3\pi}{2}\right)$ is:

(A) 0

(B) -1

(C) $\dfrac{1}{2}$

(D) $\dfrac{5}{3}$

(E) 2

19. The particular solution for the differential equation $\sqrt{1-x^2}\dfrac{dy}{dx} = \sqrt{1-y^2}$ satisfying the initial condition $y(0) = 1$ is

 (A) $y = \sqrt{1-x^2}$
 (B) $y = \sqrt{1+x^2}$
 (C) $y = 1 + x^2$
 (D) $y = 1 - x^2$
 (E) $y = 1 + x$

20. Let $f(x)$ be a differentiable function on the closed interval $[1, 3]$. The average value of $f'(x)$ on $[1, 3]$ is

 (A) $f'(3) - f'(1)$
 (B) $2(f'(3) - f'(1))$
 (C) $\dfrac{1}{2}(f'(3) - f'(1))$
 (D) $f(3) - f(1)$
 (E) $\dfrac{1}{2}(f(3) - f(1))$

21. Which of the following improper integrals converge?

 I. $\displaystyle\int_0^\infty e^{-x}dx$

 II. $\displaystyle\int_0^1 \dfrac{1}{x^2}dx$

 III. $\displaystyle\int_0^1 \dfrac{1}{\sqrt{x}}dx$

 (A) I only
 (B) III only
 (C) I and II
 (D) II and III
 (E) I and III

22. A particle moves in the xy-plane so that its coordinates at time t are given by $x = t^2$ and $y = 4 - t^3$. At $t = 1$, its acceleration vector is

 (A) $\langle 2, -3\rangle$
 (B) $\langle 2, -6\rangle$
 (C) $\langle 1, -6\rangle$
 (D) $\langle 2, 6\rangle$
 (E) $\langle 1, -2\rangle$

23. If $\displaystyle\int x\sec^2 x\, dx = f(x) + \ln|\cos x| + C$, then $f(x) =$

 (A) $\tan x$
 (B) $\dfrac{1}{2}x^2$
 (C) $x\tan x$
 (D) $x^2\tan x$
 (E) $\tan^2 x$

24. $\displaystyle\int_0^\pi \left(x^2 - \dfrac{x^4}{3!} + \dfrac{x^6}{5!} - \dfrac{x^8}{7!} + \cdots\right)dx =$

 (A) $-\pi$
 (B) π
 (C) π^2
 (D) $\dfrac{\pi}{2}$
 (E) $\pi + 1$

25. Which of the following equations has a graph that is a circle with radius 2 and center at the origin?

 I. $y = \sqrt{4-x^2};\ -2 \le x \le 2$

 II. $x = \cos t,\ y = \sin t;\ 0 \le t \le 2\pi$

 III. $r = 4\cos\theta;\ 0 \le \theta \le \pi$

 (A) I
 (B) II
 (C) III
 (D) All of these
 (E) None of these

26. If $r = f(\theta)$ is continuous and non-negative for $\alpha \le x \le \beta$, then the area enclosed by the polar curve $r = f(\theta)$ and the lines $\theta = \alpha$ and $\theta = \beta$ is given by

 (A) $\dfrac{1}{2}\displaystyle\int_\alpha^\beta f(\theta^2)d\theta$

 (B) $\dfrac{1}{2}\displaystyle\int_\alpha^\beta f(\theta)d\theta$

 (C) $\dfrac{1}{2}\displaystyle\int_\alpha^\beta \theta f(\theta^2)d\theta$

 (D) $\dfrac{1}{2}\displaystyle\int_\alpha^\beta \theta f(\theta)d\theta$

 (E) $\dfrac{1}{2}\displaystyle\int_\alpha^\beta [f(\theta)]^2 d\theta$

27. The first five nonzero terms in the series expansion of e^{-4x} about $x = 0$ are

(A) $1 + 4x + 8x^2 + \dfrac{32x^3}{3} + \dfrac{32x^4}{3}$

(B) $1 - 4x^2 + 8x^4 - \dfrac{32x^6}{3} + \dfrac{32x^8}{3}$

(C) $1 - 4x + 8x^2 - \dfrac{32x^3}{3} + \dfrac{32x^4}{3}$

(D) $1 - 2x + 4x^2 - \dfrac{16x^3}{3} + \dfrac{32x^4}{3}$

(E) $1 + 2x + 4x^2 + \dfrac{8x^3}{3} + \dfrac{16x^4}{3}$

28. The general solution of $\dfrac{dy}{dx} = \dfrac{4x}{y}$ is a family of

(A) Lines
(B) Circles
(C) Ellipses
(D) Hyperbolas
(E) Parabolas

Section I—Part B

Number of Questions	Time	Use of Calculator
17	50 Minutes	Yes

Directions:

Use the same answer sheet for Part A. *Please note that the questions begin with number 76. This is not an error. It is done to be consistent with the numbering system of the actual AP Calculus BC Exam.* All questions are given equal weight. There is no penalty for unanswered questions. However, $\frac{1}{4}$ of the number of incorrect answers will be subtracted from the number of correct answers. Unless otherwise indicated, the domain of a function f is the set of all real numbers. If the exact numerical value does not appear among the given choices, select the best approximate value. The use of a calculator is *permitted* in this part of the exam.

76. If $y = 5^{(7-2x^2)}$ then $\dfrac{dy}{dx}$ at $x = 1$ is approximately

   (A) 5
   (B) 25
   (C) 3125
   (D) 5029.494
   (E) −20117.974

77. Given $f(x) = 2^x$ and $2^{3.03} \approx 8.168$. Using differentials, determine which of the following is closest to $f'(3)$?

   (A) 3.2
   (B) 4.6
   (C) 5.6
   (D) 8.2
   (E) 9.1

78. The rate of growth of a population is proportional to the population. If the population in 2000 was 3 million and in 2010, the population was 3.21 million, what is the predicted population for the year 2020?

   (A) 3.315 million
   (B) 3.435 million
   (C) 6.420 million
   (D) 7.913 million
   (E) 11.609 million

79. The function $f(x) = 5x^3 - 7x^2 - 5$ has which of these relative extrema?

   (A) Relative maximum at $(0.933, -7.033)$
   (B) Relative maximum at $(-5, 0)$
   (C) Relative minimum at $(0, -5)$
   (D) Relative minimum at $(0.933, -7.033)$
   (E) No relative maxima or minima

80. The slope of the tangent line to the curve $r = 4 - 3\cos\theta$ at $\theta = \dfrac{\pi}{6}$ is approximately

   (A) 3.284
   (B) 0.524
   (C) 1.001
   (D) 0.027
   (E) −4.837

81. A particle moves in the xy-plane in such a way that its path is defined by $x = e^t \cos t$ and $y = e^t \sin 2t$. Find the speed of the particle when $t = \dfrac{\pi}{2}$.

   (A) 10.757
   (B) −4.810
   (C) −9.621
   (D) 14.431
   (E) 4.810

82. At what value of x does the graph of $y = \dfrac{1}{x^2} - \dfrac{1}{x^3}$ have a point of inflection?

   (A) $x = 1$
   (B) $x = 2$
   (C) $x = 3$
   (D) $x = 4$
   (E) $x = 5$

83. $\displaystyle\sum_{n=1}^{\infty} \dfrac{n! + 2^n}{2^n \cdot n!} =$

   (A) e
   (B) $1 + e$
   (C) $2 + e$

(D) $\dfrac{2}{3} + e$

(E) $\dfrac{2}{3} + 2e$

84. The base of a solid is the region enclosed by the ellipse $4x^2 + y^2 = 1$. If all plane cross sections perpendicular to the x-axis are semicircles, then its volume is

(A) $\dfrac{\pi}{3}$

(B) $\dfrac{\pi}{6}$

(C) $\dfrac{\pi}{2}$

(D) $\dfrac{2\pi}{3}$

(E) $\dfrac{\pi}{4}$

85. The area under the curve $y = \sqrt{4-x}$ bounded by the line $x = 0$ and $x = 3$ is approximately

(A) 0.633
(B) 1.937
(C) 3.874
(D) 4.667
(E) 5.333

86. Which of the following is the interval of convergence for the power series given by $\sum_{n=0}^{\infty}(x-2)^n 3^n$

(A) $(-2, 2)$
(B) $(-1, 5)$
(C) $(-1, 1)$
(D) $(-4, 0)$
(E) $(-\infty, \infty)$

87. Find the coordinates of the point where the line tangent to $y = 4 - 3x - x^2$ at $x = 1$ intersects the axis of symmetry of the parabola.

(A) $(5, 0)$

(B) $\left(-\dfrac{3}{2}, 5\right)$

(C) $(0, 5)$

(D) $\left(-\dfrac{3}{2}, \dfrac{25}{2}\right)$

(E) $\left(-\dfrac{3}{2}, 0\right)$

88. A rectangular field is fenced off along the bank of a river, and no fence is required along the river. If the material for the fence costs

$5.00 per foot for the two ends and $7.50 per foot for the side parallel to the river, find the largest possible area that can be enclosed with $9,000 worth of fence.

(A) 270,000 square feet
(B) 27,000 square feet
(C) 600 square feet
(D) 450 square feet
(E) 270 square feet

89. Let R be the region in the first quadrant enclosed by the graph of $y = \sqrt[3]{x+1}$, the line $x = 7$, the x-axis and the y-axis. The volume of the solid generated when R is revolved about the x-axis is

(A) $\dfrac{5\pi}{2}$

(B) $\dfrac{15\pi}{2}$

(C) $\dfrac{45\pi}{4}$

(D) $\dfrac{45\pi}{2}$

(E) $\dfrac{93\pi}{5}X$

90. If $\dfrac{dy}{dx} = 2y^2$ and if $y = -1$ when $x = 1$, then when $x = 2$, $y =$

(A) $-\dfrac{1}{2}$

(B) $-\dfrac{1}{3}$

(C) 0

(D) $\dfrac{1}{3}$

(E) $\dfrac{1}{2}$

91. For all x, if $f(x) = \sum_{n=0}^{\infty} \dfrac{(-1)^{n+1}x^{2n+1}}{(2n+1)!}$, then $f'(x) =$

(A) $\sum_{n=0}^{\infty} \dfrac{(-1)^{n+1}x^{2n}}{(2n+1)!}$

(B) $\sum_{n=0}^{\infty} \dfrac{(-1)^n x^{2n}}{(2n)!}$

(C) $\sum_{n=0}^{\infty} \dfrac{(-1)^{n+1}x^{2n}}{(2n+2)!}$

(D) $\displaystyle\sum_{n=0}^{\infty} \frac{(-1)^{n+1}x^{2n}}{(2n)!}$

(E) $\displaystyle\sum_{n=0}^{\infty} \frac{(-1)^{n}x^{2n}}{(2n+1)!}$

92. Find the equation of the tangent line to $y = x^3 + 2x$ at its point of inflection.

(A) $y = 2x$

(B) $y = 3x + 2$

(C) $y = 6x$

(D) $y = 2x - 3$

(E) $y = -2x + 5$

Section II—Part A

Directions:

Show all work. You may *not* receive any credit for correct answers without supporting work. You may use an approved calculator to help solve a problem. However, you must clearly indicate the setup of your solution using mathematical notations and *not* calculator syntax. Calculators may be used to find the derivative of a function at a point, compute the numerical value of a definite integral, or solve an equation. Unless otherwise indicated, you may assume the following: (a) the numeric or algebraic answers need not be simplified, (b) your answer, if expressed in approximation, should be correct to 3 places after the decimal point, and (c) the domain of a function f is the set of all real numbers.

1. Let R represent the region in the first quadrant bounded by the graphs of $y = \cos x$, $y = 1 - \cos x$, and the y-axis.

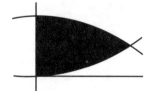

   (A) Find the area of region R.
   (B) Find the volume of the solid created when the region R is revolved about the x-axis.
   (C) Find the volume of the solid that has R as its base if each of the cross section is an isosceles triangle of height x.

2. A drum containing 100 gallons of oil is punctured by a nail and begins to leak at the rate of $10 \sin\left(\dfrac{\pi t}{12}\right)$ gallons/minute where t is measured in minutes and $0 \leq t \leq 10$.

   (A) How much oil to the nearest gallon leaked out after $t = 6$ minutes?
   (B) What is the average amount of oil leaked out per minute from $t = 0$ to $t = 6$ to the nearest gallon?

   (C) Write an expression for $f(t)$ to represent the total amount of oil in the drum at time t, where $0 \leq t \leq 10$.
   (D) At what value if t to the nearest minute will there be 40 gallons of oil remaining in the drum?

3. A population is modeled by a function P that satisfies the logistic differential equation

$$\frac{dP}{dt} = \frac{1}{4}P\left(1 - \frac{P}{20}\right)$$

   (A) If $P(0) = 4$, what is $\lim\limits_{t \to \infty} P(t)$? If $P(0) = 5$, what is $\lim\limits_{t \to \infty} P(t)$?
   (B) If $P(0) = 4$, for what value of P is the population growing fastest?
   (C) A different population is modeled by a function Q that satisfies the separable differential equation $\dfrac{dQ}{dt} = \dfrac{1}{4}Q\left(1 - \dfrac{t}{20}\right)$. Find $Q(t)$ if $Q(0) = 3$.
   (D) For the function found in part (C), what is $\lim\limits_{t \to \infty} Q(t)$?
   (E) If $P(0) = 5$ and $Q(0) = 3$, which population is larger after 20 years, and how much larger?

Section II—Part B

Directions:

The use of a calculator is not permitted in this part of the test. When you have finished this part of the test, you may return to the problems in part A of Section II and continue to work on them. However, you may not use a calculator. You should *show all work*. You may *not* receive any credit for correct answers without supporting work. Unless otherwise indicated, the numeric or algebraic answers need not be simplified, and the domain of a function f is the set of all real numbers.

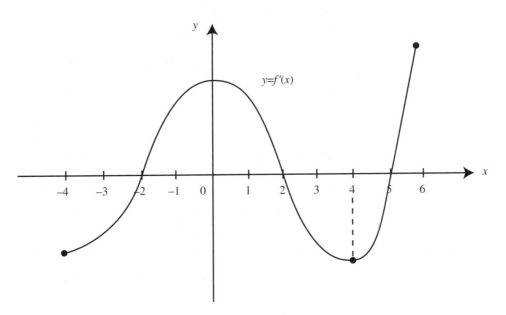

Figure D-11

4. The graph of f', the derivative of the function f, for $-4 \leq x \leq 6$ is shown above.

   (A) At what value(s) of x does f have a relative maximum value? Justify your answer.
   (B) At what value(s) of x does f have a relative minimum value? Justify your answer.
   (C) At what value(s) of x is $f''(x) > 0$? Justify your answer.
   (D) At what value(s) of x, if any, does the graph of f'' have a point of inflection? Justify your answer.
   (E) Draw a possible sketch of $f(x)$, if $f(-2) = 3$.

5. The MacLaurin series for the function f is given by

$$f(x) = \sum_{n=0}^{\infty} \frac{(3x)^{2n}}{(2n)!} = 1 + \frac{(3x)^2}{2!} + \frac{(3x)^4}{4!}$$
$$+ \frac{(3x)^6}{6!} \cdots + \frac{(3x)^{2n}}{(2n)!} + \cdots$$

on its interval of convergence.

   (A) Find the interval of convergence.
   (B) Find the first four terms and the general term for the MacLaurin series for $\int f(x)\,dx$
   (C) Find the first four terms of $f'(x)$ and approximate $f'\left(\dfrac{1}{2}\right)$

6. Given the equation $x^2 y^2 = 4$,

   (A) Find $\dfrac{dy}{dx}$
   (B) Write an equation of the line tangent to the graph of the equation at the point $(1, -2)$.
   (C) Write an equation of the line normal to the curve at point $(1, -2)$.
   (D) The line $y = \dfrac{1}{2}x + 2$ is tangent to the curve P. Find the coordinates of point P.

Answers to BC Practice Exam 2—Section I

Part A	12. A	24. B	81. A
1. B	13. E	25. E	82. B
2. D	14. D	26. E	83. B
3. C	15. B	27. C	84. A
4. E	16. A	28. D	85. D
5. E	17. E		86. B
6. C	18. D	Part B	87. D
7. B	19. A	76. E	88. A
8. B	20. E	77. C	89. E
9. D	21. E	78. B	90. B
10. D	22. B	79. D	91. D
11. A	23. C	80. A	92. A

Answers to BC Practice Exam 2—Section II

Part A

1. (A) 0.685
   (B) 2.152
   (C) 0.133

2. (A) 38 gallons
   (B) 6 gallons
   (C) $f(t) = 100 - \int_0^t 10 \sin(\pi x / 12)\, dx$
   (D) 8 minutes

3. (A) 20; 20
   (B) when $P = 10$
   (C) $Q = 3e^{\frac{t}{4}} - \frac{t^2}{160}$
   (D) 0
   (E) Q is 16.944 larger than P

Part B

4. (A) $x = 2$
   (B) $x = 5$ and $x = -2$
   (C) $(-4, 0)$ and $(4, 6)$
   (D) $x = 0$ and $x = 4$
   (E) See solution

5. (A) $(-\infty, \infty)$
   (B) $x + \dfrac{3^2 x^3}{3!} + \dfrac{3^4 x^5}{5!} + \dfrac{3^6 x^7}{7!} \cdots + \dfrac{3^{2n} x^{2n+1}}{(2n+1)!} + \cdots$
   (C) $\dfrac{3^2 x}{1!} + \dfrac{3^4 x^3}{3!} + \dfrac{3^6 x^5}{5!} + \dfrac{3^8 x^7}{7!}$ and 6.388

6. (A) $\dfrac{dy}{dx} = \dfrac{-y}{x}$
   (B) $y = 2x - 4$
   (C) $y = -\dfrac{1}{2}x - \dfrac{3}{2}$
   (D) $(-2, 1)$

Solutions to BC Practice Exam 2—Section I

Section I—Part A

1. The correct answer is (B).

   At the point $(3, -2)$, $m = f'(3) = 5$, so the equation of the tangent line is $y + 2 = 5(x - 3)$ or $y = 5x - 17$.

2. The correct answer is (D).

   If $y = \dfrac{2x + 1}{5x + 2}$, then

   $$y' = \frac{(5x + 2)(2) - (2x + 1)(5)}{(5x + 2)^2} = \frac{-1}{(5x + 2)^2}.$$

   Since the numerator is always negative and the denominator is always positive, y' is always negative and so the function is monotonically decreasing.

   $$y'' = -1(-2)(5x + 2)^{-3}(5) = \frac{10}{(5x + 2)^3}.$$

   On the interval $-3 < x < -1$, $-13 \le 5x + 2 \le -3 \Rightarrow (5x + 2)^3 < 0 \Rightarrow y'' < 0$. Therefore, the function is concave down on the given interval.

3. The correct answer is (C).

   If $y = \sqrt[3]{x^2 + 4ax + 12a^2}$,
   $y' = \dfrac{1}{3}\left(x^2 + 4ax + 12a^2\right)^{-2/3}(2x + 4a)$. Set the derivative equal to zero.

   $$\frac{2x + 4a}{3\left(x^2 + 4ax + 12a^2\right)^{2/3}} = 0 \Rightarrow 2x + 4a = 0 \Rightarrow$$

   $2x = -4a \Rightarrow x = -2a$. Evaluate

   $$y = \sqrt[3]{(-2a)^2 + 4a(-2a) + 12a^2}$$
   $$= \sqrt[3]{4a^2 - 8a^2 + 12a^2}$$
   $$= \sqrt[3]{8a^2} = 2a^{2/3}.$$

   The second derivative,
   $$y'' = \frac{2}{3}\left(x^2 + 4ax + 12a^2\right)^{-2/3}$$

   $$-\frac{2}{9}(2x + 4a)^2\left(x^2 + 4ax + 12a^2\right)^{-5/3},$$

   evaluated at $\Rightarrow x = -2a$, is

   $$y'' = \frac{-24a^2 + 48a^2 - 8a^2}{9\left(4a^2 - 8a^2 + 12a^2\right)^{5/3}} = \frac{16a^2}{9\left(8a^2\right)^{5/3}} > 0.$$

Therefore, the point $\left(-2a, 2a^{2/3}\right)$ is a relative minimum.

4. The correct answer is (E).
   $$f(x + h) - f(x) = 4x^2h + 2xh - 6x^3h^2$$

   $$\frac{f(x + h) - f(x)}{h} = 4x^2 + 2x - 6x^3h$$

   $$\lim_{h \to 0} \frac{f(x + h) - f(x)}{h} = \lim_{h \to 0}\left(4x^2 + 2x - 6x^3h\right)$$

   $$= 4x^2 + 2x$$

   Find $f'(3) = 4(3)^2 + 2(3) = 42$.

5. The correct answer is (E).

   $y = x^3 - 6x^2 \Rightarrow y' = 3x^2 - 12x$. Slope of the tangent line is -9, so $3x^2 - 12x = -9$, and solving gives $x = 3$ or $x = 1$. For each value, determine the equation of the tangent line. For $x = 3$, $y = 27 - 54 = -27$ and $y + 27 = -9(x - 3)$ which gives $k = 0$. Reject this possibility since k must be positive. For $x = 1$, $y = 1 - 6 = -5$ and the tangent is $y + 5 = -9(x - 1)$, which implies $k = 4$.

6. The correct answer is (C).

   $y = ax^2 \Rightarrow y' = 2ax$. Evaluate when $x = p$ to find the slope of the tangent line is $m = 2ap$. The point of tangency is (p, ap^2), so the equation of the tangent line is $y - ap^2 = 2ap(x - p)$ or $y = 2apx - ap^2$. This tangent intersects the y-axis at $(0, -ap^2)$.

7. The correct answer is (B).

   If $h(x) = f(x)g(x)$, then
   $h'(x) = f(x)g'(x) + g(x)f'(x)$ and
   $h'(1) = f(1)g'(1) + g(1)f'(1)$
   $\quad = (0)(5) + (-1)(-2) = 2$.

8. The correct answer is (B).

   If $h(x) = f(g(x))$, then
   $h'(x) = f'(g(x))g'(x)$ and
   $h'(2) = f'(g(2))g'(2) \Rightarrow$
   $h'(2) = f'(3)g'(2) \Rightarrow$
   $h'(2) = (-1)(1) = -1$.

9. The correct answer is (D).

If $f(x)$ is a continuous function and
$f(5) = 2$ and $f'(5) = 3$,then
$f(5.01) = f(5) + 0.01f'(5) =$
$2 + 0.01(3) = 2.03$

10. The correct answer is (D).

Simplify the expression $\dfrac{x^3 - 1}{x^2}$ as $\left(\dfrac{x^3}{x^2} - \dfrac{1}{x^2}\right)$

which is equivalent to $\left(x - \dfrac{1}{x^2}\right)$

Thus $\displaystyle\int_1^2 \frac{x^3 - 1}{x^2}\,dx = \int_1^2 \left(x - \frac{1}{x^2}\right) dx$

$= \displaystyle\int_1^2 \left(x - x^2\right) dx = \left[\frac{x^2}{2} - \frac{x^{-1}}{-1}\right]_1^2$

$= \left[\dfrac{x^2}{2} + \dfrac{1}{x}\right]_1^2 = \left[\dfrac{(2)^2}{2} + \dfrac{1}{2}\right] - \left[\dfrac{(1)^2}{2} + \dfrac{1}{1}\right]$

$= 2\dfrac{1}{2} - 1\dfrac{1}{2} = 1$

11. The correct answer is (A).

At the moment the surface area is 24 square inches, $A = 6x^2 = 24 \Rightarrow x^2 = 4 \Rightarrow x = 2$, so the edge of the cube is 2 inches. The Surface area is changing so

$\dfrac{dA}{dt} = 12x\dfrac{dx}{dt} = 4 \Rightarrow \dfrac{dx}{dt} = \dfrac{4}{12x} = \dfrac{1}{3x}$,

and when the edge is 2 inches, $\dfrac{dx}{dt} = \dfrac{1}{3(2)} = \dfrac{1}{6}$.

The volume $V = s^3$ is changing at the rate $\dfrac{dV}{dt} = 3x^2\dfrac{dx}{dt}$. When the surface area is 24 square inches and the edge is 2 inches,

$\dfrac{dV}{dt} = 3(2)^2\left(\dfrac{1}{6}\right) = 2$. The volume increasing at 2 cubic inches per second.

12. The correct answer is (A).

$\displaystyle\int_0^7 f(x)\,dx = \int_0^4 f(x)\,dx + \int_4^7 f(x)\,dx$

$= 10 + 1 = 11$

$\displaystyle\int_5^7 f(x)\,dx = \int_0^7 f(x)\,dx - \int_0^5 f(x)\,dx$

$= 11 - 9 = 2$

13. The correct answer is (E).

Integrate $\displaystyle\int \frac{1}{x^2 - 3x + 2}\,dx$

$= \displaystyle\int \frac{1}{(x - 2)(x - 1)}\,dx$ by partial fractions.

$A(x - 1) + B(x - 2) = 1 \Rightarrow Ax + Bx = 0 \Rightarrow$
$A = -B$. Substituting, $-A - 2B = 1 \Rightarrow$
$-A + 2A = 1$ and so $A = 1$, $B = -1$. Then the integral becomes

$= \displaystyle\int \frac{1}{x - 2}\,dx - \int \frac{1}{x - 1}\,dx$ and

$= \ln|x - 2| - \ln|x - 1| + C = \ln\left|\dfrac{x - 2}{x - 1}\right| + C.$

14. The correct answer is (D).

Since $\theta \to 0 \Rightarrow 1 - \cos\theta \to 0$ and $\theta \to 0 \Rightarrow 2\sin^2\theta \to 0$, $\displaystyle\lim_{\theta\to0} \frac{1 - \cos\theta}{2\sin^2\theta} \to \frac{0}{0}$

and is indeterminate.

Differentiate $\dfrac{d}{dx}(1 - \cos\theta) = \sin\theta$

and $\dfrac{d}{dx}(2\sin^2\theta) = 4\sin\theta\cos\theta$.

Then by L'Hôpital's Rule,

$\displaystyle\lim_{\theta\to0} \frac{1 - \cos\theta}{2\sin^2\theta} = \lim_{\theta\to0} \frac{\sin\theta}{4\sin\theta\cos\theta}$

$= \displaystyle\lim_{\theta\to0} \frac{1}{4\cos\theta} = \frac{1}{4}.$

15. The correct answer is (B).

$\displaystyle\int xf(x)\,dx$ requires integration by parts. Let

$u = f(x), du = f'(x)\,dx, dv = x\,dx$, and $v = \dfrac{1}{2}x^2$.

Then $\displaystyle\int xf(x)\,dx = \frac{1}{2}x^2f(x) - \int \frac{1}{2}x^2f'(x)\,dx.$

16. The correct answer is (A).

Differentiate $x^3 + 3xy + 2y^3 = 17$ implicitly.

$3x^2 + 3x\dfrac{dy}{dx} + 3y + 6y^2\dfrac{dy}{dx} = 0$

$3x\dfrac{dy}{dx} + 6y^2\dfrac{dy}{dx} = -3x^2 - 3y$

$\left(3x + 6y^2\right)\dfrac{dy}{dx} = -3x^2 - 3y$

$\dfrac{dy}{dx} = \dfrac{-3x^2 - 3y}{(3x + 6y^2)} = -\dfrac{x^2 + y}{x + 2y^2}.$

17. The correct answer is (E).

Rewrite $3\tan^5(2x)$ as $3[\tan(2x)]^5$.

Thus $f'(x) = 15[\tan(2x)]^4\left[\sec^2(2x)\right](2)$

$$= 30[\tan(2x)]^4\left[\sec^2(2x)\right]$$

18. The correct answer is (D)s.

Letting $u = \sin x$ gives us $du = \cos x\,dx$

$\int \cos x \sin^2 x\,dx = \int u^2 du = \dfrac{u^3}{3} + c = \dfrac{\sin^3 x}{3} + c$

set $2 = \dfrac{\sin^3(0)}{3} + c \Rightarrow c = 2$. Since

$f'(x) = \dfrac{\sin^3}{3} + 2,\ f'\left(\dfrac{3\pi}{2}\right) = \dfrac{-1}{3} + 2 = \dfrac{5}{3}$.

19. The correct answer is (A).

Separate $\sqrt{1-x^2}\dfrac{dy}{dx} = \sqrt{1-y^2}$ as

$\dfrac{1}{\sqrt{1-y^2}}dy = \dfrac{1}{\sqrt{1-x^2}}dx$.

Integrate to produce $\sin^{-1}(y) = \sin^{-1}(x) + C$. With the initial condition $y(0) = 1$, $\sin^{-1}(1) = \sin^{-1}(0) + C,\ C = \dfrac{\pi}{2}$. So

$y = \sin\left(\sin^{-1}(x) + \dfrac{\pi}{2}\right) = \sin(\sin^{-1}(x))\cos\dfrac{\pi}{2} + \cos(\sin^{-1}(x))\sin\dfrac{\pi}{2} = x(0) + \cos(\sin^{-1}(x))$.

Thus $y = \sqrt{1-x^2}$.

20. The correct answer is (E).

The average value of $f'(x)$ on $[1, 3]$ is

$f'_{average} = \dfrac{1}{3-1}\int_1^3 f'(x)\,dx = \dfrac{1}{2}[f(3) - f(1)]$

21. The correct answer is (E).

I. $\displaystyle\int_0^\infty e^{-x}\,dx = \lim_{k\to\infty}\int_0^k e^{-x}dx$

$$= \lim_{k\to\infty}\left[-e^{-x}\right]_0^k = \lim_{k\to\infty}\left[-e^{-k} + e^0\right]$$

$$= \lim_{k\to\infty}\left[-\dfrac{1}{e^k} + 1\right] = 1$$

II. $\displaystyle\int_0^1 \dfrac{1}{x^2}\,dx = \lim_{k\to0^+}\int_k^1 \dfrac{1}{x^2}dx = \lim_{k\to0^+}\dfrac{-1}{x}\Big]_k^1$

$$= \lim_{k\to0^+}\left[-1 + \dfrac{1}{k}\right] \to \infty$$

III. $\displaystyle\int_0^1 \dfrac{1}{\sqrt{x}}\,dx = \lim_{k\to0^+}\int_k^1 \dfrac{1}{\sqrt{x}}dx$

$$= \lim_{k\to0^+}\left[2\sqrt{x}\right]_k^1$$

$$= \lim_{k\to0^+}\left[2 - 2\sqrt{k}\right] = 2$$

22. The correct answer is (B).

$x = t^2 \Rightarrow \dfrac{dx}{dt} = 2t \Rightarrow \dfrac{d^2x}{dt^2} = 2$ and

$y = 4 - t^3 \Rightarrow \dfrac{dy}{dt} = -3t^2 \Rightarrow \dfrac{d^2y}{dt^2} = -6t$.

At $t = 1$, the acceleration vector is $\langle 2, -6t\rangle|_{t=1} = \langle 2, -6\rangle$

23. The correct answer is (C).

Integrate $\displaystyle\int x\sec^2 x\,dx$ by parts. Let $u = x$, $du = dx$, $dv = \sec^2 x\,dx$, and $v = \tan x$. Then $\displaystyle\int x\sec^2 x\,dx = x\tan x - \int \tan x\,dx = x\tan x + \ln|\cos x| + C$. Comparing to the given information, $\displaystyle\int x\sec^2 x\,dx = f(x) + \ln|\cos x| + C$. tells us that $f(x) = x\tan x$.

24. The correct answer is (B).

The integrand of

$$\int_0^\pi \left(x^2 - \dfrac{x^4}{3!} + \dfrac{x^6}{5!} - \dfrac{x^8}{7!} + \cdots\right)dx \text{ can be}$$

factored as $\displaystyle\int_0^\pi x\left(x - \dfrac{x^3}{3!} + \dfrac{x^5}{5!} - \dfrac{x^7}{7!} + \cdots\right)dx$,

which can be recognized as $\displaystyle\int_0^\pi x\sin x\,dx$. Integrate by parts with $u = x$, $du = dx$, $dv = \sin x\,dx$, and $v = -\cos x$. Then

$\displaystyle\int_0^\pi x\sin x\,dx = -x\cos x - \int_0^\pi (-\cos x)\,dx = \left[-x\cos x + \sin x\right]_0^\pi = (-\pi\cos\pi + \sin\pi) - (-0\cos 0 + \sin 0) = \pi$.

25. The correct answer is (E).

Consider each curve.
(I) $y = \sqrt{4 - x^2}$; $-2 \le x \le 2$ defines a semicircle, not a circle. (II) $x = \cos t$, $y = \sin t$; $0 \le t \le 2\pi$ is a circle, but its radius is 1. (III) $r = 4\cos\theta$; $0 \le \theta \le \pi$ is a circle of radius 2, but its center is at $(2, 0)$. Therefore, none of these equations defines a circle of radius 2 with center at the origin.

26. The correct answer is (E).

The area enclosed by a polar curve is

$$\frac{1}{2}\int_\alpha^\beta r^2\, d\theta \text{ so } A = \frac{1}{2}\int_\alpha^\beta [f(\theta)]^2\, d\theta$$

27. The correct answer is (C).

Begin with the known MacLaurin series

$$f(x) = e^x = \sum_{n=0}^\infty \frac{x^n}{n!} = 1 + x + \frac{x^2}{2} + \frac{x^3}{6} + \cdots \text{ and }$$

substitute $-4x$ for x.

$$e^{-4x} = \sum_{n=0}^\infty \frac{(-4x)^n}{n!} = \frac{(-4x)^0}{0!} + \frac{(-4x)^1}{1!}$$

$$+ \frac{(-4x)^2}{2!} + \frac{(-4x)^3}{3!} + \frac{(-4x)^4}{4!}$$

$$= 1 - 4x + \frac{16x^2}{2!} + \frac{(-64x^3)}{3!} + \frac{(256x^4)}{4!}$$

$$= 1 - 4x + 8x^2 - \frac{64x^3}{6} + \frac{256x^4}{24}$$

$$= 1 - 4x + 8x^2 - \frac{32x^3}{3} + \frac{32x^4}{3}$$

28. The correct answer is (D).

$\dfrac{dy}{dx} = \dfrac{4x}{y}$ separates to $y\,dy = 4x\,dx$ which

integrates to $\dfrac{y^2}{2} = 2x^2 + c \Rightarrow y^2 = 4x^2 + 2c$

$\Rightarrow 4x^2 - y^2 = -2c$. This form is the family of hyperbolas.

Section I—Part B

76. The correct answer is (E).

If $y = 5^{(7-2x^2)}$ then $\ln y = (7 - 2x^2)\ln 5$.
Differentiating implicitly,
$\dfrac{1}{y}\dfrac{dy}{dx} = -4(\ln 5)x \Rightarrow \dfrac{dy}{dx} = -4(\ln 5)x(5)^{(7-2x^2)}$.
Evaluating at $x = 1$ produces
$\dfrac{dy}{dx} = -4(\ln 5)(5)^5 \approx -20117.97391$.

77. The correct answer is (C).

If $f(x) = 2^x$, $f(3) = 8$ and $f'(x) = 2^x \ln 2$ so
$f'(3) = 8 \ln 2$. We could estimate $f(3.03)$
$\approx f(3) + 0.03f'(3)$, and since we know
$2^{3.03} \approx 8.168$, $8.168 \approx 8 + 0.03f'(3)$. Solve
$f'(3) \approx \dfrac{0.168}{0.03} \approx 5.6$.

78. The answer is (B).

Separate $\dfrac{dP}{dt} = kP$ as $\dfrac{1}{P}\,dP = k\,dt$. Integrate

$\displaystyle\int \frac{1}{P}\,dP = \int k\,dt$ to produce $\ln|P| = kt + c_1$.

The initial condition tells us $P = c_2 e^{kt} \Rightarrow$
$3 = c_2 e^0 \Rightarrow c_2 = 3$ so $P = 3e^{kt}$. In 2010, the
population was 3.21 million. Thus
$3.21 = 3e^{10k} \Rightarrow 1.07 = e^{10k} \Rightarrow \ln 1.07 = 10k$
$\Rightarrow 0.00677 = k$. Therefore, $P = 3e^{0.00677t}$
and $P = 3e^{0.00677(20)} = 3.435$.

79. The correct answer is (D).

The function $f(x) = 5x^3 - 7x^2 - 5 \Rightarrow f'(x) =$
$15x^2 - 14x$. Set the first derivative equal to
zero. $15x^2 - 14x = 0 \Rightarrow x(15x - 14) = 0 \Rightarrow$
$x = 0 \text{ or } x = \dfrac{14}{15}$. Evaluate the second derivative
$f''(x) = 30x - 14$ at each of the critical points.
$f''(0) = -14 \Rightarrow \max \Rightarrow (0, -5) f''\left(\dfrac{14}{15}\right) =$

$30\left(\dfrac{14}{15}\right) - 14 = 14 \Rightarrow \min \Rightarrow$

$(0.9333, -7.0326)$

80. The correct answer is (A).

$r = 4 - 3\cos\theta \Rightarrow \dfrac{dr}{d\theta} = 3\sin\theta$. Since

$\theta = \dfrac{\pi}{6} \Rightarrow r = 4 - 3\cos\dfrac{\pi}{6} = \left(4 - \dfrac{3\sqrt{3}}{2}\right)$ and

$\dfrac{dr}{d\theta} = 3\sin\dfrac{\pi}{6} = \dfrac{3}{2}$

$\text{slope} = \tan\theta = \dfrac{\left(r + (\tan\theta)\dfrac{dr}{d\theta}\right)}{\left(-r\tan\theta + \dfrac{dr}{d\theta}\right)}$

$\tan\left(\dfrac{\pi}{6}\right) = \dfrac{1}{\sqrt{3}}$

$\text{slope} = \dfrac{\left[4 - \dfrac{3\sqrt{3}}{2} + \left(\dfrac{1}{\sqrt{3}}\right)\left(\dfrac{3}{2}\right)\right]}{\left[-\left(4 - \dfrac{3\sqrt{3}}{2}\right)\left(\dfrac{1}{\sqrt{3}}\right) + \dfrac{3}{2}\right]} \approx 3.284$

81. The correct answer is (A).

A particle $x = e^t \cos t \Rightarrow$
$\dfrac{dx}{dt} = -e^t \sin t + e^t \cos t \Big|_{t=\pi/2} = -e^{\pi/2}$ and
$y = e^t \sin 2t \Rightarrow \dfrac{dy}{dt} = 2e^t \cos 2t + e^t \sin 2t \Big|_{t=\pi/2}$
$= -2e^{\pi/2}$. Then speed $= \sqrt{\left(-e^{\pi/2}\right)^2 + \left(-2e^{\pi/2}\right)^2}$
$= \sqrt{e^\pi + 4e^\pi} = \sqrt{5e^\pi} \approx 10.7566$

82. The correct answer is (B).

$$y = \frac{1}{x^2} - \frac{1}{x^3} = x^{-2} - x^{-3} \Rightarrow$$

$y' = -2x^{-3} + 3x^{-4} \Rightarrow y'' = 6x^{-4} - 12x^{-5}$. Set the second derivative equal to zero and solve.

$$\frac{6}{x^4} - \frac{12}{x^5} = \frac{6x - 12}{x^5} = 0 \Rightarrow$$

$6x - 12 = 0 \Rightarrow x = 2$. Also $y'' < 0$ for $x < 2$ and $y'' > 0$ for $x > 2$.

83. The Correct answer is (B).

$$\sum_{n=1}^{\infty} \frac{n! + 2^n}{2^n \cdot n!} = \sum_{n=1}^{\infty} \frac{n!}{2^n \cdot n!} + \sum_{n=1}^{\infty} \frac{2^n}{2^n \cdot n!}$$

$$= \sum_{n=1}^{\infty} \frac{1}{2^n} + \sum_{n=1}^{\infty} \frac{1}{n!}.$$

The series $\sum_{n=1}^{\infty} \frac{1}{2^n}$ is geometric series, so

$$\sum_{n=1}^{\infty} \frac{1}{2^n} = \frac{1/2}{1 - 1/2} = 1.$$

Compare the series $\sum_{n=1}^{\infty} \frac{1}{n!}$ to the known

MacLaurin series $\sum_{n=0}^{\infty} \frac{x^n}{n!} = e^x$, and

$$\sum_{n=1}^{\infty} \frac{1}{n!} = \sum_{n=1}^{\infty} \frac{1^n}{n!} = e^1.$$

Therefore $\sum_{n=1}^{\infty} \frac{1}{2^n} = \sum_{n=1}^{\infty} \frac{1}{n!} = 1 + e.$

84. The correct answer is (A).

The ellipse $4x^2 + y^2 = 1$ has x-intercepts $x = \pm\frac{1}{2}$. The semicircles will have radius of $r = \sqrt{1 - 4x^2}$ and area of

$$A = \frac{1}{2}\pi\left(\sqrt{1 - 4x^2}\right)^2 = \frac{1}{2}\pi(1 - 4x^2).$$

Then the volume is

$$V = \int_{-1/2}^{1/2} \frac{1}{2}\pi(1 - 4x^2)\,dx = \frac{1}{2}\pi\left[x - \frac{4}{3}x^3\right]_{-1/2}^{1/2}$$

$$V = \frac{1}{2}\pi\left[\left(\frac{1}{2} - \frac{4}{3}\cdot\frac{1}{8}\right) - \left(\frac{-1}{2} - \frac{4}{3}\cdot\frac{-1}{8}\right)\right]$$

$$= \frac{1}{2}\pi \cdot \frac{2}{3} = \frac{\pi}{3}$$

85. The correct answer is (D).

$$A = \int_0^3 \sqrt{4 - x}\,dx = -\frac{2}{3}(4 - x)^{3/2}\Big]_0^3$$

$$= \left(-\frac{2}{3}(4 - 3)^{3/2}\right) - \left(-\frac{2}{3}(4 - 0)^{3/2}\right)\bigg]$$

$$= \left(-\frac{2}{3}\right) - \left(-\frac{2}{3}(8)\right) = -\frac{2}{3} + \frac{16}{3} = \frac{14}{3} = 4.667$$

86. The correct answer is (B).

The series $\sum_{n=0}^{\infty} \frac{(x - 2)^n}{3^n}$ will converge when

$$\left|\frac{(x - 2)^{n+1}}{3^{n+1}} \cdot \frac{3^n}{(x - 2)^n}\right| = \left|\frac{x - 2}{3}\right| < 1. \text{ This}$$

means $-1 < \frac{x - 2}{3} < 1 \Rightarrow -3 < x - 2 < 3 \Rightarrow$

$-1 < x < 5$ when $x = -1$, the series becomes

$\sum_{n=0}^{\infty} \frac{(-3)^n}{3^n} = \sum_{n=0}^{\infty} (-1)^n$, which diverges. When

$x = 5$, the series becomes

$\sum_{n=0}^{\infty} \frac{(3)^n}{3^n} = \sum_{n=0}^{\infty} 1^n$, which diverges. Therefore,

the interval of convergence is $(-1, 5)$.

87. The correct answer is (D).

If $y = 4 - 3x - x^2, \Rightarrow y' = -3 - 2x$. The slope of the tangent line is $y'(1) = -3 - 2(1) = -5$, and the point is $(1, 0)$ since $y(1) = 4 - 3(1) - (1)^2 = 0$. The equation of the tangent line is $y - 0 = -5(x - 1)$ or $y = -5x + 5$. The axis of symmetry of the

parabola is $x = \frac{-(-3)}{2(-1)} = -\frac{3}{2}$. Evaluating the

tangent equation when $x = -\frac{3}{2}$ gives

$y\left(-\frac{3}{2}\right) = -5\left(-\frac{3}{2}\right) + 5 = \frac{25}{2}$. The line

$y = -5x + 5$ intersects the axis of symmetry at

$\left(-\frac{3}{2}, \frac{25}{2}\right)$.

88. The correct answer is (A).

Let l be the length of the section of fence parallel to the river and w be the length of the two remaining sides. The cost of the fence is $C = 5(2w) + 7.50(l) = 9000$. Set cost equal to 9000 and solve for l.

$$5(2w) + 7.50(l) = 9000 \Rightarrow l = \frac{9000 - 10w}{7.50}.$$

The area is

$$A = wl = w\left(\frac{9000 - 10w}{7.50}\right) = \frac{9000 - 10w^2}{7.50}.$$

Differentiate $A' = \dfrac{9000}{7.50} - \dfrac{20w}{7.50}$ and set the

derivative equal to zero. $\dfrac{9000}{7.50} - \dfrac{20w}{7.50} = 0 \Rightarrow$

$9000 - 20w = 0 \Rightarrow w = 450$. If the width is

450, the length $l = \dfrac{9000 - 10(450)}{7.50} = 600$,

$dA'' < 0$; so the maximum area is
$450 \cdot 600 = 270{,}000$.

89. The correct answer is (E).

$V = \pi \int_0^7 \left(\sqrt[3]{x+1}\right)^2 dx = \pi \int_0^7 (x+1)^{2/3}\, dx$.
Using your calculator and obtain $93\pi/5$.

90. The correct answer is (B).

Separate $\dfrac{dy}{dx} = 2y^2$ into $\dfrac{dy}{y^2} = 2\,dx$, and

integrate $\displaystyle\int \dfrac{dy}{y^2} = \int 2\,dx \Rightarrow -\dfrac{1}{y} = 2x + c$.

If $y = -1$ when $x = 1$,

$-\dfrac{1}{(-1)} = 2(1) + c \Rightarrow 1 = 2 + c \Rightarrow c = -1$.

Thus $-\dfrac{1}{y} = 2x - 1 \Rightarrow y = \dfrac{-1}{2x-1}$. If $x = 2$,

$y = \dfrac{-1}{2(2) - 1} = \dfrac{-1}{3}$.

91. The correct answer is (D).

If $f(x) = \displaystyle\sum_{n=0}^{\infty} \dfrac{(-1)^{n+1}\, x^{2n+1}}{(2n+1)!}$, then

$f'(x) = \dfrac{d}{dx}\left(\displaystyle\sum_{n=0}^{\infty} \dfrac{(-1)^{n+1} x^{2n+1}}{(2n+1)!}\right)$

$= \dfrac{d}{dx}\left(\dfrac{-x}{1!} + \dfrac{x^3}{3!} + \dfrac{-x^5}{5!} + \cdots\right)$

$= \dfrac{-1}{1} + \dfrac{3x^2}{3 \cdot 2 \cdot 1} + \dfrac{-5 - x^4}{5 \cdot 4 \cdot 3 \cdot 2 \cdot 1} + \cdots$

$= -1 + \dfrac{x^2}{2!} - \dfrac{x^4}{4!} + \cdots$

$= \displaystyle\sum_{n=0}^{\infty} (2n+1)\dfrac{(-1)^{n+1} x^{2n}}{(2n+1)(2n)!}$

$= \displaystyle\sum_{n=0}^{\infty} \dfrac{(-1)^{n+1} x^{2n}}{(2n)!}$

92. The correct answer is (A).

$y = x^3 + 2x \Rightarrow y' = 3x^2 + 2 \Rightarrow y'' = 6x$.
Set the second derivative equal to zero.
$y'' = 6x = 0 \Rightarrow x = 0$. When $x < 0$, $y'' < 0$ and
when $x > 0$, $y'' > 0$. The point of inflection is $(0, 0)$ and the slope at that point is $y'|_{x=0} = 2$. The
equation of the tangent line to the curve
at its point of inflection is
$y - 0 = 2(x - 0) \Rightarrow y = 2x$.

Section II—Part A

Part A—Calculators are permitted.

1. (A) Determine the point of intersection of the
two curves, either by calculator, or
algebraically. $\cos x = 1 - \cos x \Rightarrow 2\cos x$
$= 1 \Rightarrow \cos x = \dfrac{1}{2}$, so $x = \dfrac{\pi}{3}$ is the
x-coordinate of the point of intersection.
The area of region R is

$A = \displaystyle\int_0^{\pi/3} [\cos x - (1 - \cos x)]\, dx$

$= \displaystyle\int_0^{\pi/3} (2\cos x - 1)\, dx = (2\sin x - x)\,|_0^{\pi/3}$

$= \left(2\sin\dfrac{\pi}{3} - \dfrac{\pi}{3}\right) - (2\sin 0 - 0)$

$= 2\left(\dfrac{\sqrt{3}}{2}\right) - \dfrac{\pi}{3} = \sqrt{3} - \dfrac{\pi}{3} \approx 0.685$

(B) The volume of the solid created when the
region R is revolved about the x-axis can
be found by the method of washers.

$V = \pi \displaystyle\int_0^{\pi/3} \left[\cos^2 x - (1 - \cos x)^2\right] dx$

$= \pi \displaystyle\int_0^{\pi/3} \left[\cos^2 x - (1 - 2\cos x + \cos^2 x)\right] dx$

$= \pi \displaystyle\int_0^{\pi/3} \left[\cos^2 x - 1 + 2\cos x - \cos^2 x\right] dx$

$= \pi \displaystyle\int_0^{\pi/3} (2\cos x - 1)\, dx.$

Integrate and evaluate.

$V = \pi \displaystyle\int_0^{\pi/3} (2\cos x - 1)\, dx$

$= \pi [2\sin x - x]\,|_0^{\pi/3}$

$$= \pi \left[\left(2 \sin \frac{\pi}{3} - \frac{\pi}{3} \right) - (2 \sin 0 - 0) \right]$$

$$= \pi \left[2 \left(\frac{\sqrt{3}}{2} \right) - \frac{\pi}{3} \right] = \pi \left[\sqrt{3} - \frac{\pi}{3} \right]$$

$$\approx 2.152$$

(C) To find the volume of the solid that has R as its base and cross sections that are isosceles triangles of height x, recognize that the area of each cross section is $\frac{1}{2}x[\cos x - (1 - \cos x)] = \frac{1}{2}x(2 \cos x - 1)$, so the volume is $V = \frac{1}{2} \int_0^{\pi/3} x(2 \cos x - 1) \, dx$.

Integrate by parts, with $u = x$, $du = dx$, $dv = (2 \cos x - 1) \, dx$, and $v = 2 \sin x - x$.

$$V = \frac{1}{2} \left[x(2 \sin x - x) - \int (2 \sin x - x) \, dx \right]_0^{\pi/3}$$

$$= \frac{1}{2} \left[x(2 \sin x - x) - \left(-2 \cos x - \frac{1}{2}x^2 \right) \right]_0^{\pi/3}$$

$$= \frac{1}{2} \left[2x \sin x - x^2 + 2 \cos x + \frac{1}{2}x^2 \right]_0^{\pi/3}$$

$$= \frac{1}{2} \left[2x \sin x + 2 \cos x - \frac{1}{2}x^2 \right]_0^{\pi/3}$$

$$= \frac{1}{2} \left[\left(2 \left(\frac{\pi}{3} \right) \sin \frac{\pi}{3} + 2 \cos \frac{\pi}{3} - \frac{1}{2} \left(\frac{\pi}{3} \right)^2 \right) \right.$$

$$\left. - \left(2(0) \sin(0) + 2 \cos(0) - \frac{1}{2}(0)^2 \right) \right]$$

$$= \frac{1}{2} \left[\left(\frac{\pi \sqrt{3}}{3} + 1 - \frac{\pi^2}{18} \right) - 2 \right]$$

$$= \frac{1}{2} \left[\frac{\pi \sqrt{3}}{3} - \frac{\pi^2}{18} - 1 \right]$$

$$\approx \frac{1}{2}[1.8138 - 0.548311 - 1]$$

$$\approx \frac{1}{2}[0.265489] \approx 0.133$$

Or simply use your calculator and obtain the same results

2. (A) The amount of oil leaked out after 6 minutes

$$= \int_0^6 10 \sin \left(\frac{\pi t}{12} \right) dt$$

$$= \left[\frac{-10 \cos \left(\frac{\pi t}{12} \right)}{\pi/12} \right]_0^6$$

$$= \left[\frac{-120}{\pi} \cos \left(\frac{\pi t}{12} \right) \right]_0^6$$

$$= \frac{120}{\pi} \approx 38.1972 \text{ gallons} \approx 38 \text{ gallons}$$

(B) Average amount of oil leaked out per minute from $t = 0$ to $t = 6$:

$$= \frac{1}{6 - 0} \int_0^6 10 \sin \left(\frac{\pi t}{12} \right) dt = \frac{1}{6} \left(\frac{120}{\pi} \right)$$

$$= 6.3662 \approx 6 \text{ gallons}$$

(C) The amount of oil in the drum at t:

$$f(t) = 100 - \int_0^t 10 \sin \left(\frac{\pi x}{12} \right) dx$$

(D) Let a be the value of t:

$$100 - \int_0^a 10 \sin \left(\frac{\pi t}{12} \right) dt = 40$$

$$100 - \left[\left(\frac{-120}{\pi} \right) \cos \left(\frac{\pi t}{12} \right) \right]_0^a = 40$$

$$100 - \left\{ \left(\frac{-120}{\pi} \right) \cos \left(\frac{a\pi}{12} \right) \right.$$

$$\left. - \left[\left(\frac{-120}{\pi} \right) \cos(0) \right] \right\} = 40$$

$$100 + \left(\frac{120}{\pi} \right) \cos \left(\frac{a\pi}{12} \right) - \left(\frac{120}{\pi} \right) = 40$$

$$\left(\frac{120}{\pi} \right) \cos \left(\frac{a\pi}{12} \right) = \left(\frac{120}{\pi} \right) + 40 - 100$$

$$\cos \left(\frac{a\pi}{12} \right) = \left(\frac{120}{\pi} - 60 \right) \left(\frac{\pi}{12} \right)$$

$$\cos \left(\frac{a\pi}{12} \right) = (-\pi + 2)/2 \approx -0.570796$$

$$\frac{a\pi}{12} = \cos^{-1}(-0.570796) \approx 2.17827$$

$$a = (2.17827)\left(\frac{12}{\pi}\right) \approx 8.32038$$

$$a \approx 8 \text{ minutes}$$

3. If $\dfrac{dP}{dt} = \dfrac{1}{4}P\left(1 - \dfrac{P}{20}\right) \Rightarrow \dfrac{1}{P\left(1 - \dfrac{P}{20}\right)}\,dP =$

$$\frac{1}{4}\,dt \Rightarrow \frac{20}{P(20-P)}\,dP = \frac{1}{4}\,dt. \text{ Integrate}$$

$$\int \frac{dP}{P} + \int \frac{dP}{20-P} = \int \frac{1}{4}\,dt \Rightarrow$$

$$\ln|P| - \ln|20-P| = \frac{t}{4} + c_1 \Rightarrow$$

$$\ln\left|\frac{P}{20-P}\right| = \frac{t}{4} + c_1.$$

Exponentiate for $\dfrac{P}{20-P} = c_2 e^{t/4}$ and isolate P.

$$P = \frac{20c_2 e^{t/4}}{1 + c_2 e^{t/4}}.$$

(A) If $P(0) = 4$, $\dfrac{20c_2}{1+c_2} = 4 \Rightarrow 20c_2 =$

$$4(1+c_2) \Rightarrow 16c_2 = 4 \Rightarrow c_2 = \frac{1}{4},$$

so $P = \dfrac{20}{4e^{-t/4}+1}$ and

$$\lim_{t\to\infty} P(t) = \lim_{t\to\infty} \frac{20}{4e^{-t/4}+1} = 20. \text{ If } P(0) = 5,$$

$$\frac{20c_2}{1+c_2} = 5 \Rightarrow 20c_2 = 5(1+c_2) \Rightarrow$$

$$15c_2 = 5 \Rightarrow c_2 = \frac{1}{3} \text{ so } P = \frac{20}{3e^{-t/4}+1}$$

and $\lim_{t\to\infty} P(t) = \lim_{t\to\infty} \dfrac{20}{3e^{-t/4}+1} = 20.$

(B) The rate of growth is given by

$$\frac{dP}{dt} = \frac{1}{4}P\left(1 - \frac{P}{20}\right). \text{ To find the maximum}$$

value of the growth function, take the derivative of the function.

$$\frac{d^2P}{dt^2} = \frac{1}{4}P\left(\frac{-1}{20}\right) + \frac{1}{4}\left(1 - \frac{P}{20}\right)$$

$$= \frac{-P}{80} + \frac{1}{4} - \frac{P}{80} = \frac{1}{4} - \frac{P}{40}.$$

Set $\dfrac{1}{4} - \dfrac{P}{40} = 0 \Rightarrow \dfrac{P}{40} = \dfrac{1}{4} \Rightarrow P = 10.$

When $P = 10$, $\dfrac{dp}{dt} = \dfrac{1}{4}(10)\left(1 - \dfrac{10}{20}\right) = \dfrac{5}{4}.$

The population is growing fastest, at a rate of 125%, when the population is 10.

(C) Seperate $\dfrac{dQ}{dt} = \dfrac{1}{4}Q\left(1 - \dfrac{t}{20}\right)$ to $\dfrac{1}{Q}dQ =$

$$\frac{1}{4}\left(\frac{20-t}{20}\right)dt \text{ and integrate.}$$

$$\int \frac{1}{Q}dQ = \frac{1}{80}\int (20-t)\,dt. \Rightarrow \ln|Q| =$$

$$\frac{1}{80}\left(20t - \frac{t^2}{2}\right) + c_1 \Rightarrow \ln|Q| = \frac{t}{4} - \frac{t^2}{160} +$$

$$c_1 \Rightarrow Q = c_2 e^{\left(\frac{t}{4} - \frac{t^2}{160}\right)}. \text{ If } Q(0) = 3. \ Q(0) =$$

$$c_2 = 3, \text{ so } \Rightarrow Q = 3e^{\left(\frac{t}{4} - \frac{t^2}{160}\right)}.$$

(D) For the function found in part (C),

as t increases, $\dfrac{t}{4} - \dfrac{t^2}{160} = \dfrac{40t - t^2}{160} \to -\infty$

so $e^{\frac{t}{4} - \frac{t^2}{160}} \to 0$; therefore,

$$\lim_{t\to\infty} Q(t) = \lim_{t\to\infty} 3e^{\frac{t}{4} - \frac{t^2}{160}} = 0$$

(E) $P(20) = \dfrac{20}{3e^{-20/4} + 1} \approx 19.6037$

$$Q(20) = 3e^{\frac{20}{4} - \frac{20^2}{160}} \approx 36.5475$$
$$36.5475 - 19.6037 = 16.9437$$
Population Q is 16.944 larger than population P after 20 years.

Section II—Part B

Part B—No calculators

4. (A) See Figure DS-15. Since f increases on $(-2, 2)$ and decreases on $(2, 5)$, f has a relative maximum at $x = 2$.

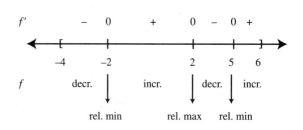

Figure DS-15

(B) Since f decreases on $(-4, -2)$ and increases on $(-2, 2)$, f has a relative minimum at $x = -2$. Since f decreases on $(2, 5)$ and increases on $(5, 6)$, so f has a relative minimum at $x = 5$.

(C) See Figure DS-16.

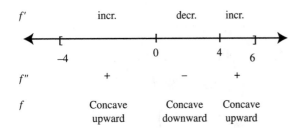

Figure DS-16

Since f' is increasing on the intervals $(-4, 0)$ and $(4, 6)$, $f'' > 0$ on $(-4, 0)$ and $(4, 6)$.

(D) A change of concavity occurs at $x = 0$ and at $x = 4$. (See Figure DS-16). Thus f has a point of inflection at $x = 0$ and at $x = 4$.

(E) See Figure DS-17.

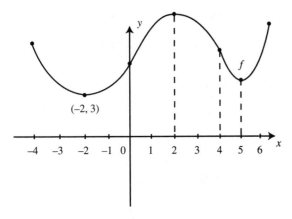

Figure DS-17

5. $f(x) = \displaystyle\sum_{n=0}^{\infty} \frac{(3x)^{2n}}{(2n)!} = 1 + \frac{(3x)^2}{2!} + \frac{(3x)^4}{4!}$

$\qquad + \dfrac{(3x)^6}{6!} \cdots + \dfrac{(3x)^{2n}}{(2n)!} + \cdots$

(A) $\displaystyle\lim_{n\to\infty} \left| \frac{(3x)^{2n+2}}{(2n+2)!} \cdot \frac{(2n)!}{(3x)^{2n}} \right| =$

$\displaystyle\lim_{n\to\infty} \left| \frac{(3x)^2}{(2n+2)(2n+1)} \right| = 0 \Rightarrow (-\infty, \infty)$

(B) The first terms and the general term for the MacLaurin series for

$\displaystyle\int f(x) = \int \left(1 + \frac{(3x)^2}{2!} + \frac{(3x)^4}{4!} + \frac{(3x)^6}{6!} \right.$

$\qquad \left. + \cdots + \frac{(3x)^{2n}}{(2n)!} + \cdots \right) dx$

$= \displaystyle\int \left(1 + \frac{3^2 x^2}{2!} + \frac{3^4 x^4}{4!} + \frac{3^6 x^6}{6!} \right.$

$\qquad \left. + \cdots + \frac{3^{2n} x^{2n}}{(2n)!} + \cdots \right) dx$

$= x + \dfrac{3^2 x^3}{3 \cdot 2!} + \dfrac{3^4 x^5}{5 \cdot 4!} + \dfrac{3^6 x^7}{7 \cdot 6!} \cdots$

$\qquad + \dfrac{3^{2n} x^{2n+1}}{(2n+1)(2n)!} + \cdots$

$= x + \dfrac{3^2 x^3}{3!} + \dfrac{3^4 x^5}{5!} + \dfrac{3^6 x^7}{7!} \cdots$

$\qquad + \dfrac{3^{2n} x^{2n+1}}{(2n+1)!} + \cdots$

(C) The first four terms of $f'(x)$:

$f(x) = 1 + \dfrac{3^2 x^2}{2!} + \dfrac{3^4 x^4}{4!} + \dfrac{3^6 x^6}{6!} + \cdots + \dfrac{3^{2n} x^{2n}}{(2n)!} + \cdots$

$f'(x) \approx \dfrac{3^2 x}{1!} + \dfrac{3^4 x^3}{3!} + \dfrac{3^6 x^5}{5!} + \dfrac{3^8 x^7}{7!}$

The approximate value of $f'\left(\dfrac{1}{2}\right) \approx \dfrac{3^2 (1/2)}{1!}$

$+ \dfrac{3^4 (1/2)^3}{3!} + \dfrac{3^6 (1/2)^5}{5!} + \dfrac{3^8 (1/2)^7}{7!} \approx 6.388.$

6. (A) $x^2 y^2 = 4$. Differentiating using product and chain rules:

$2xy^2 + \left(x^2\right) 2y \dfrac{dy}{dx} = 0$

$2x^2 y \dfrac{dy}{dx} = -2xy^2$

$\dfrac{dy}{dx} = \dfrac{-2xy^2}{2x^2 y} = \dfrac{-y}{x}$

(B) $\left. \dfrac{dy}{dx} \right|_{(1, -2)} = \dfrac{-(-2)}{1} = 2$

Equation of tangent: $y - (2) = 2(x - 1)$

$\Rightarrow y + 2 = 2x - 2$

or $y = 2x - 4$

(C) Slope of normal $= \dfrac{-1}{\text{slope of tangent}} = \dfrac{-1}{2}$

Equation of normal: $y - (-2) = -\dfrac{1}{2}(x - 1)$

$$\Rightarrow y + 2 = -\dfrac{1}{2}x + \dfrac{1}{2}$$

$$\text{or } y = -\dfrac{1}{2}x - \dfrac{3}{2}$$

(D) $y = \dfrac{1}{2}x + 2$; $m = \dfrac{1}{2}$ and $\dfrac{dy}{dx} = \dfrac{-y}{x}$

Set $\dfrac{-y}{x} = \dfrac{1}{2} \Rightarrow -2y = x$

$x^2 y^2 = 4$; substitute $x = -2y$

$(-2y)^2 y^2 = 4$; $4y^2 \cdot y^2 = 4$;

$4y^2 = 4$; $y^4 = 1$

$y = \pm 1$

If $y = 1$, $x^2 y^2 = 4 \Rightarrow x^2 (1)^2 = 4 \Rightarrow$

$$x^2 = 4 \Rightarrow x = \pm 2$$

If $y = -1$, $x^2 y^2 = 4 \Rightarrow x^2 (-1)^2 = 4 \Rightarrow$

$$x^2 = 4 \Rightarrow x = \pm 2$$

Possible points for P are: $(2, 1)$, $(2, -1)$, $(-2, 1)$, and $(-2, -1)$.
The only point on the tangent line is $(-2, 1)$.

Scoring Sheet for BC Practice Exam 2

Section I—Part A

$$\underline{\hspace{3cm}} \times 1.2 = \underline{\hspace{3cm}}$$
No. Correct $\qquad\qquad$ Subtotal A

$$\underline{\hspace{3cm}} \times (0.25) \times 1.2 = \underline{\hspace{3cm}}$$
No. Incorrect $\qquad\qquad$ Subtotal B

$$\text{Part A (Subtotal A} - \text{Subtotal B)} = \underline{\hspace{3cm}}$$
$\qquad\qquad\qquad\qquad\qquad\qquad\qquad$ Subtotal C

Section I—Part B

$$\underline{\hspace{3cm}} \times 1.2 = \underline{\hspace{3cm}}$$
No. Correct $\qquad\qquad$ Subtotal D

$$\underline{\hspace{3cm}} \times (0.25) \times 1.2 = \underline{\hspace{3cm}}$$
No. Incorrect $\qquad\qquad$ Subtotal E

$$\text{Part B (Subtotal D} - \text{Subtotal E)} = \underline{\hspace{3cm}}$$
$\qquad\qquad\qquad\qquad\qquad\qquad\qquad$ Subtotal F

Section II—Part A (Each question is worth 9 points.)

$$\underline{\hspace{2cm}} + \underline{\hspace{2cm}} + \underline{\hspace{2cm}} = \underline{\hspace{2cm}}$$
Q#1 $\qquad$ Q#2 $\qquad$ Q#3 $\qquad$ Subtotal G

Section II—Part B (Each question is worth 9 points.)

$$\underline{\hspace{2cm}} + \underline{\hspace{2cm}} + \underline{\hspace{2cm}} = \underline{\hspace{2cm}}$$
Q#1 $\qquad$ Q#2 $\qquad$ Q#3 $\qquad$ Subtotal H

$$\text{Total Raw Score (Subtotals C} + \text{F} + \text{G} + \text{H)} = \boxed{}$$

Approximate Conversion Scale:	
Total Raw Score	Approximate AP Grade
75–108	5
60–74	4
45–59	3
31–44	2
0–30	1

APPENDIXES

Formulas and Theorems

1. Quadratic Formula:

$$ax^2 + bx + c = 0 \ (a \neq 0)$$

$$x = \frac{-b \pm \sqrt{b^2 - 4ac}}{2a}$$

2. Distance Formula:

$$d = \sqrt{(x_2 - x_1)^2 + (y_2 - y_1)^2}$$

3. Equation of a Circle:

$$x^2 + y^2 = r^2 \text{ center at } (0, 0) \text{ and radius } = r$$

4. Equation of an Ellipse:

$$\frac{x^2}{a^2} + \frac{y^2}{b^2} = 1 \text{ center at } (0, 0)$$

$$\frac{(x - h)^2}{a^2} + \frac{(y - k)^2}{b^2} = 1 \text{ center at } (h, k)$$

5. Area and Volume Formulas:

Figure	Area Formula
Trapezoid	$\frac{1}{2}[\text{base}_1 + \text{base}_2]\,(\text{height})$
Parallelogram	(base)(height)
Equilateral triangle	$\dfrac{s^2\sqrt{3}}{4}$
Circle	πr^2 (circumference $= 2\pi r$)

Solid	Volume	Surface Area
Sphere	$\dfrac{4}{3}\pi r^3$	$4\pi r^2$
Right circular cylinder	$\pi r^2 h$	$2\pi r h$
Right circular cone	$\dfrac{1}{3}\pi r^2 h$	Lateral S.A.: $\pi r\sqrt{r^2 + h^2}$ Total S.A.: $\pi r^2 + \pi r\sqrt{r^2 + h^2}$

6. Special Angles:

Angle Function	0°	$\pi/6$ 30°	$\pi/4$ 45°	$\pi/3$ 60°	$\pi/2$ 90°	π 180°	$3\pi/2$ 270°	2π 360°
Sin	0	$1/2$	$\sqrt{2}/2$	$\sqrt{3}/2$	1	0	-1	0
Cos	1	$\sqrt{3}/2$	$\sqrt{2}/2$	$1/2$	0	-1	0	1
Tan	0	$\sqrt{3}/3$	1	$\sqrt{3}$	Undefined	0	Undefined	0

7. Double Angles:

- $\sin 2\theta = 2\sin\theta\cos\theta$
- $\cos 2\theta = \cos^2\theta - \sin^2\theta$ or
  $1 - 2\sin^2\theta$ or $2\cos^2\theta - 1$

- $\cos^2\theta = \dfrac{1 + \cos 2\theta}{2}$

- $\sin^2\theta = \dfrac{1 - \cos 2\theta}{2}$

8. Pythagorean Identities:

- $\sin^2\theta + \cos^2\theta = 1$
- $1 + \tan^2\theta = \sec^2\theta$
- $1 + \cot^2\theta = \csc^2\theta$

9. Limits:

$$\lim_{x\to\infty}\frac{1}{x} = 0 \qquad \lim_{x\to 0}\frac{\cos x - 1}{x} = 0$$

$$\lim_{x\to 0}\frac{\sin x}{x} = 1 \qquad \lim_{x\to\infty}\left(1 + \frac{1}{h}\right)^h = e$$

$$\lim_{h\to 0}\frac{e^h - 1}{h} = 1 \qquad \lim_{x\to 0}(1 + x)^{\frac{1}{x}} = e$$

10. Rules of Differentiation:

a. **Definition of the Derivative of a Function:**

$$f'(x) = \lim_{h\to 0}\frac{f(x + h) - f(x)}{h}$$

b. **Power Rule:** $\dfrac{d}{dx}(x^n) = nx^{n-1}$

c. **Sum & Difference Rules:**

$$\frac{d}{dx}(u \pm v) = \frac{du}{dx} \pm \frac{dv}{dx}$$

d. **Product Rule:**

$$\frac{d}{dx}(uv) = v\frac{du}{dx} + u\frac{dv}{dx}$$

e. **Quotient Rule:**

$$\frac{d}{dx}\left(\frac{u}{v}\right) = \frac{v\dfrac{du}{dx} - u\dfrac{dv}{dx}}{v^2},\; v \neq 0$$

Summary of Sum, Difference, Product, and Quotient Rules:

$$(u \pm v)' = u' \pm v' \qquad (uv)' = u'v + v'u$$

$$\left(\frac{u}{v}\right)' = \frac{u'v - v'u}{v^2}$$

f. **Chain Rule:**

$$\frac{d}{dx}[f(g(x))] = f'(g(x)) \cdot g'(x)$$

or $\dfrac{dy}{dx} = \dfrac{dy}{du} \cdot \dfrac{du}{dx}$

11. Inverse Function and Derivatives:

$$\left(f^{-1}\right)'(x) = \frac{1}{f'(f^{-1}(x))} \;\text{ or }\; \frac{dy}{dx} = \frac{1}{dx/dy}$$

12. Differentiation and Integration Formulas:
Integration Rules

a. $\displaystyle\int f(x)dx = F(x) + C \Rightarrow F'(x) = f(x)$

b. $\displaystyle\int af(x)dx = a\int f(x)dx$

c. $\displaystyle\int -f(x)dx = -\int f(x)dx$

d. $\displaystyle\int [f(x) \pm g(x)]\,dx = \int f(x)dx \pm \int g(x)dx$

Differentiation Formulas:

a. $\dfrac{d}{dx}(x) = 1$

b. $\dfrac{d}{dx}(ax) = a$

c. $\dfrac{d}{dx}(x^n) = nx^{n-1}$

d. $\dfrac{d}{dx}(\cos x) = -\sin x$

e. $\dfrac{d}{dx}(\sin x) = \cos x$

f. $\dfrac{d}{dx}(\tan x) = \sec^2 x$

g. $\dfrac{d}{dx}(\cot x) = -\csc^2 x$

h. $\dfrac{d}{dx}(\sec x) = \sec x \tan x$

i. $\dfrac{d}{dx}(\csc x) = -\csc x\,(\cot x)$

j. $\dfrac{d}{dx}(\ln x) = \dfrac{1}{x}$

k. $\dfrac{d}{dx}(e^x) = e^x$

l. $\dfrac{d}{dx}(a^x) = (\ln a)\,a^x$

m. $\dfrac{d}{dx}\left(\sin^{-1} x\right) = \dfrac{1}{\sqrt{1-x^2}}$

n. $\dfrac{d}{dx}\left(\tan^{-1} x\right) = \dfrac{1}{1+x^2}$

o. $\dfrac{d}{dx}\left(\sec^{-1} x\right) = \dfrac{1}{|x|\sqrt{x^2-1}}$

Integration Formulas:

a. $\displaystyle\int 1\,dx = x + c$

b. $\displaystyle\int a\,dx = ax + c$

c. $\displaystyle\int x^n dx = \dfrac{x^{n+1}}{n+1} + c,\ n \neq -1$

d. $\displaystyle\int \sin x\,dx = -\cos x + c$

e. $\displaystyle\int \cos x\,dx = \sin x + c$

f. $\displaystyle\int \sec^2 x\,dx = \tan x + c$

g. $\displaystyle\int \csc^2 x\,dx = -\cot x + c$

h. $\displaystyle\int \sec x\,(\tan x)\,dx = \sec x + c$

i. $\displaystyle\int \csc x\,(\cot x)\,dx = -\csc x + c$

j. $\displaystyle\int \dfrac{1}{x}\,dx = \ln|x| + c$

k. $\displaystyle\int e^x dx = e^x + c$

l. $\displaystyle\int a^x dx = \dfrac{a^x}{\ln a} + c\ \ a > 0,\ a \neq 1$

m. $\displaystyle\int \dfrac{1}{\sqrt{1-x^2}}\,dx = \sin^{-1} x + c$

n. $\displaystyle\int \dfrac{1}{1+x^2}\,dx = \tan^{-1} x + c$

o. $\displaystyle\int \dfrac{1}{|x|\sqrt{x^2-1}}\,dx = \sec^{-1} x + c$

More Integration Formulas:

a. $\displaystyle\int \tan x\,dx = \ln|\sec x| + c\ \text{or} -\ln|\cos x| + c$

b. $\displaystyle\int \cot x\,dx = \ln|\sin x| + c\ \text{or}\ -\ln|\csc x| + c$

c. $\displaystyle\int \sec x\,dx = \ln|\sec x + \tan x| + c$

d. $\displaystyle\int \csc x\,dx = \ln|\csc x - \cot x| + c$

e. $\displaystyle\int \ln x\,dx = x\ln|x| - x + c$

f. $\displaystyle\int \dfrac{1}{\sqrt{a^2-x^2}}\,dx = \sin^{-1}\left(\dfrac{x}{a}\right) + c$

g. $\displaystyle\int \dfrac{1}{a^2+x^2}\,dx = \dfrac{1}{a}\tan^{-1}\left(\dfrac{x}{a}\right) + c$

h. $\displaystyle\int \dfrac{1}{x\sqrt{x^2-a^2}}\,dx = \dfrac{1}{a}\sec^{-1}\left|\dfrac{x}{a}\right| + c$ or $\dfrac{1}{a}\cos^{-1}\left|\dfrac{a}{x}\right| + c$

i. $\displaystyle\int \sin^2 x\,dx = \dfrac{x}{2} - \dfrac{\sin(2x)}{4} + c.$

Note: $\sin^2 x = \dfrac{1 - \cos 2x}{2}$

Note: After evaluating an integral, always check the result by taking the derivative of the answer (i.e., taking the derivative of the antiderivative).

13. Intergration By parts $\int u dv = uv - \int v du$ (and follow CIPET Rule)

14. The Fundamental Theorems of Calculus

$$\int_a^b f(x)dx = F(b) - F(a),$$

where $F'(x) = f(x)$.

If $F(x) = \int_a^x f(t)dt$, then $F'(x) = f(x)$.

15. Trapezoidal Approximation:

$$\int_a^b f(x)dx$$

$$= \frac{b-a}{2n}\left[\begin{array}{c} f(x_0) + 2f(x_1) + 2f(x_2)\cdots \\ +2f(x_{n-1}) + f(x_n) \end{array}\right]$$

16. Average Value of a Function:

$$f(c) = \frac{1}{b-a}\int_a^b f(x)dx$$

17. Mean Value Theorem:

$$f'(c) = \frac{f(b) - ft(a)}{b-a} \text{ For some } c \text{ in } (a, b)$$

Mean Value Theorem for Integrals:

$$\int_a^b f(x)dx = f(c)(b-a) \text{ For some } c \text{ in } (a, b)$$

18. Area Bounded by 2 Curves:

$$\text{Area} = \int_{x_1}^{x_2} (f(x) - g(x))dx, \text{ where } f(x) \geq g(x)$$

19. Volume of a Solid with Known Cross Section:

$$V = \int_a^b A(x)dx,$$

where $A(x)$ is the cross section.

20. Disc Method:

$$V = \pi \int_a^b (f(x))^2\, dx, \text{ where } f(x) = \text{radius}$$

21. Washer Method:

$$V = \pi \int_a^b \left((f(x))^2 - (g(x))^2\right)dx$$

where $f(x) = $ outer radius and $g(x) = $ inner radius

22. Distance Traveled Formulas:

- Position Function: $s(t); s(t) = \int v(t)dt$
- Velocity: $v(t) = \dfrac{ds}{dt}; v(t) = \int a(t)dt$
- Acceleration: $a(t) = \dfrac{dv}{dt}$
- Speed: $|v(t)|$
- Displacement from t_1 to $t_2 = \displaystyle\int_{t_1}^{t_2} v(t)$
  $= s(t_2) - s(t_1)$
- Total Distance Traveled from t_1 to

$$t_2 = \int_{t_1}^{t_2} |v(t)|dt$$

23. Business Formulas:

$P(x) = R(x) - C(x)$	Profit = Revenue − Cost
$R(x) = px$	Revenue = (price) (items sold)
$P'(x)$	Marginal Profit
$R'(x)$	Marginal Revenue
$C'(x)$	Marginal Cost

$P'(x), R'(x), C'(x)$ are the instantaneous rates of change of profit, revenue, and cost respectively.

24. Exponential Growth/Decay Formulas:

$$\frac{dy}{dt} = ky, y > 0 \text{ and } y(t) = y_0e^{kt}$$

25. Logistic Growth Models

$$\frac{dP}{dt} = kP\left(1 - \frac{P}{M}\right) \quad \text{or} \quad \frac{dP}{dt} = \left(\frac{k}{M}\right)(P)(M - P)$$

$$P = \frac{M}{1 + Ae^{-kt}}$$

Appendix II

Bibliography

Advanced Placement Program Course Description. New York: The College Board, 2000.

Anton, Howard. *Calculus*. New York: John Wiley & Sons, 1984.

Apostol, Tom M. *Calculus*. Waltham, MA: Blaisdell Publishing Company, 1967.

Berlinski, David. *A Tour of the Calculus*. New York: Pantheon Books, 1995.

Boyer, Carl B. *The History of the Calculus and Its Conceptual Development*. New York: Dover, 1959.

Finney, R., Demana, F. D., Waits, B. K., Kennedy, D. *Calculus Graphical, Numerical, Algebraic*. New York: Scott Foresman Addison Wesley, 1999.

Kennedy, Dan. *Teacher's Guide–AP Calculus*. New York: The College Board, 1997.

Larson, R. E., Hostetler, R. P., Edwards, B. H. *Calculus*. New York: Houghton Mifflin Company, 1998.

Leithold, Louis. *The Calculus with Analytic Geometry*. New York: Harper & Row, 1976.

Sawyer, W. W. *What Is Calculus About?* Washington, DC: Mathematical Association of America, 1961.

Spivak, Michael. *Calculus*. New York: W. A. Benjamin, Inc., 1967.

Stewart, James. *Calculus*. New York: Brooks/Cole Publishing Company, 1995.

Appendix III

Websites

http://apcentral.collegeboard.com

http://www.maa.org/features/mathed_disc.html

http://www.askdrmath.com/

http://www.askdrmath.com/calculus/calculus.html

http://www.askdrmath.com/library/topics/svcalc/

www.calculus-help.com/funstuff/phobe.html

http://mathforum.com/epigone/ap-calc/

http://www.sparknotes.com/math/calcab/.dir/